1613554480

AF598235

# NONLINEAR BIOMEDICAL SIGNAL PROCESSING

Volume II

## IEEE Press Series on Biomedical Engineering

The focus of our series is to introduce current and emerging technologies to biomedical and electrical engineering practitioners, researchers, and students. This series seeks to foster interdisciplinary biomedical engineering education to satisfy the needs of the industrial and academic areas. This requires an innovative approach that overcomes the difficulties associated with the traditional textbook and edited collections.



### Books in the IEEE Press Series on Biomedical Engineering

Akay, M., *Time Frequency and Wavelets in Biomedical Signal Processing*
Hudson, D. L. and M. E. Cohen, *Neural Networks and Artificial Intelligence for Biomedical Engineering*
Khoo, M. C. K., *Physiological Control Systems: Analysis, Simulation, and Estimation*
Liang, Z-P. and P. C. Lauterbur, *Principles of Magnetic Resonance Imaging: A Signal Processing Perspective*
Akay, M. *Nonlinear Biomedical Signal Processing: Volume I, Fuzzy Logic, Neural Networks, and New Algorithms*
Akay, M. *Nonlinear Biomedical Signal Processing: Volume II, Dynamic Analysis and Modeling*
Ying, H. *Fuzzy Control and Modeling: Analytical Foundations and Applications*

# NONLINEAR BIOMEDICAL SIGNAL PROCESSING

## *Dynamic Analysis and Modeling*

Volume II

Edited by

**Metin Akay**
*Dartmouth College*
*Hanover, NH*

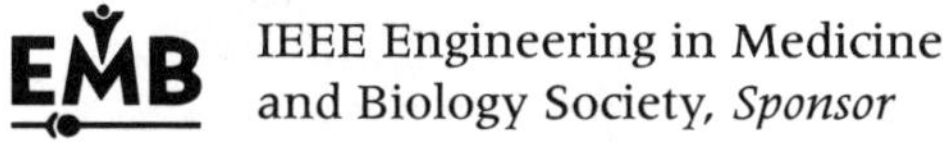

The Institute of Electrical and Electronics Engineers, Inc., New York

This book and other books may be purchased at a discount
from the publisher when ordered in bulk quantities. Contact:

IEEE Press Marketing
Attn: Special Sales
445 Hoes Lane
P.O. Box 1331
Piscataway, NJ 08855-1331
Fax: +1 732 981 9334

For more information about IEEE Press products, visit the
IEEE Online Catalog & Store: http://www.ieee.org/ieeestore.



Printed in the United States of America.

10 9 8 7 6 5 4 3 2 1

**ISBN 0-7803-6012-5**
**IEEE Order No. PC5862**

**Library of Congress Cataloging-in-Publication Data**
Nonlinear biomedical signal processing/edited by Metin Akay.
v. < > cm. — (IEEE Press series on biomedical engineering)
Includes bibliographical references and index.
Contents: v. 1. Fuzzy logic, neural networks, and new algorithms — v. 2. Dynamic analysis and modeling.
ISBN 0–7803–6012–5
1. Signal processing. 2. Biomedical engineering. 3. Fuzzy logic. 4. Neural networks. I. Akay, Metin. II. Series.

R857.S47 N66 2000
610′.285′632—dc21 00-027777

*This book is dedicated to the memory of my father, Ibrahim Akay, for his unbounded support, encouragement, compassion, love, and never-ending generosity in teaching me the importance of education.*

*May his soul rest in peace.*

"No way you can sleep nights in Varna,
no way you can sleep:
for the wealth of stars
so close and brilliant,
for the rustle of dead waves on the sand,
of salty weeds
with their pearly shells
and pebbles,
for the sound of motorboats throbbing like a heart at sea,
for the memories filling my room,
coming from ISTANBUL,
      passing through the Bosphorus,
            and filling my room,
some with green eyes,
some in handcuffs
or holding a handkerchief—
the handkerchief smells of lavender.
No way you can sleep in Varna, my love,
in Varna at the Bor Hotel."

Nazim Hikmet, June 2, 1957

IEEE Press
445 Hoes Lane, P.O. Box 1331
Piscataway, NJ 08855-1331



IEEE Engineering in Medicine and Biology Society, *Sponsor*
EMB-S Liaison to IEEE Press, Metin Akay

Cover design: William T. Donnelly, *WT Design*

**Technical Reviewers**

Eric W. Abel, *University of Dundee, United Kingdom*
Tom Brotherton, *Orincon Corporation, San Diego, CA*
Richard D. Jones, *Christchurch Hospital, Christchurch, New Zealand*
Suzanne Keilson, *Loyola College, MD*
Kristina M. Ropella, *Marquette University, Milwaukee, WI*
Alvin Wald, *Columbia University, New York, NY*

**Books of Related Interest from IEEE Press**

*RANDOM PROCESSES FOR IMAGE AND SIGNAL PROCESSING*
Edward R. Dougherty
An SPIE Press book published in cooperation with IEEE Press
A volume in the SPIE/IEEE Series on Imaging Science & Engineering
1999 Hardcover 616 pp IEEE Order No. PC5747 ISBN 0-7803-3495-7

*THE IMAGE PROCESSING HANDBOOK,* Third Edition
John C. Russ
A CRC Press handbook published in cooperation with IEEE Press
1998 Hardcover 800 pp IEEE Order No. PC5775 ISBN 0-7803-4729-3

# CONTENTS

## CHAPTER 2 SEARCHING FOR THE ORIGIN OF CHAOS 40

*Tomoyuki Yambe, Makoto Yoshizawa, Kou-ichi Tabayashi, and Shin-ichi Nitta*



## CHAPTER 3 APPROXIMATE ENTROPY AND ITS APPLICATION TO BIOSIGNAL ANALYSIS 72

*Yang Fusheng, Hong Bo, and Tang Qingyu*

# PREFACE

*Nonlinear Biomedical Signal Processing, Volume II* discusses the concepts, recent advances, and implementations of nonlinear dynamical analysis methods. This volume covers the phase plane, fractional Brownian motion analysis, critical points, limit cycles, domains of attraction, bifurcations, Poincaré section, linear and nonlinear stability, and asymptotic analysis using multiple time-scale and averaging method. In addition, the development of an original mathematical simulation and modeling of physiological systems are presented. Finally, several biomedical examples and related MATLAB® and Pascal programs are included.

Chapter 1 by Henry et al. offers an extensive overview of widely used nonlinear dynamical analysis techniques and the dynamical systems theory and their implementations.

Chapter 2 by Yambe et al. investigates the origin of chaos in the cardiovascular system and nonlinear dynamical analysis of the baroreflex system and other clinical applications.

Chapter 3 by Fusheng et al. reviews the approximate entropy and its application to the analysis of heart rate variability signals.

Chapter 4 by Celka et al. discusses the parsimonious modeling of biomedical signals and systems and its applications to the cardiovascular system.

Chapter 5 by Maier et al. reviews the nonlinear behavior of heart rate variability as registered after heart transplantation as an alternative approach to assess and predict the restoration of the normal characteristics of heart rate variability in heart transplantation subjects.

Chapters 6 by Teich et al. focuses on several measures and models of the heart rate variability signals for assessing the presence and likelihood of cardiovascular disease.

Chapter 7 by Schulz et al. discusses the ventricular-arterial interaction after acute increase of the aortic input impedance using recurrence plot analysis.

Chapter 8 by Sörnmo et al. presents the analysis of electrocardiography and vector cardiographic signals influenced by respiratory movements using the maximum likelihood method for eliminating the respiratory-induced variations, to obtain accurate morphologic measurements.

Chapter 9 by Porta et al. is devoted to the detection of nonlinear dynamical analysis of sympathetic activity using several methods, including superposition plot, recurrence map and space-time separation plot.

Chapter 10 by Signorini et al. presents the use of nonlinear dynamical analysis of heart rate variability signals for the detection and prediction of cardiovascular diseases.

Chapter 11 by Lovell et al. discusses the applications of nonlinear analysis methods for the analysis of arterial blood pressure for investigating the existence of nonlinear deterministic behavior in the control mechanisms associated with arterial blood pressure.

Chapter 12 by Iasemidis et al. presents an extensive overview of measurement and quantification of spatiotemporal dynamical analysis of epileptic seizures using the short-term largest Lyapunov exponent to predict human epileptic seizures.

Chapter 13 by Wang et al. discusses nonlinear dynamical analysis of gastric myoelectrical activity for controlling impaired gastric motility disorders.

I am grateful to the contributors for their support and help. I also thank Linda Matarazzo, IEEE Press Associate Acquisitions Editor, Savoula Amanatidis, Production and Manufacturing Manager, Surendra Bhimani, Production Editor, and Cathy Faduska, Senior Acquisitions Editor for their help and understanding throughout the preparation of this volume.

Finally, many thanks to my wife and friend, Dr. Yasemin M. Akay of Dartmouth Medical School, and my son, Altug R. Akay, for their support, care and sacrifices.

Metin Akay
*Dartmouth College*
*Hanover, NH*

# LIST OF CONTRIBUTORS

M. Abo
University of Texas Medical Branch at Galveston
Division of Gastroenterology
Galverston, TX 77555-0632 USA

Magnus Astrom
Lund University
Signal Processing Group
Dept. of Applied Electronics
Box 118
S-221 00 Lund, SWEDEN
Tel: +46 46 222 90 23, +46 70 660 50 39
Fax: +46 46 222 47 18

Giuseppe Baselli
Polytechnic University Milano
Department of Biomedical Engineering
Piazza Leonardo da Vinci 32
20133 Milano, ITALY
E-mail: baselli@biomed.polimi.it

Robert Bauernschmitt
University of Heidelberg
Dept. of Cardiac Surgery
Experimental Laboratory
INF 326
69120 Heidelberg, GERMANY
Tel: +49-(0)6221/56-6260
Fax: +49-(0)6221/439987

Hong Bo
Tsinghua University
Dept. of Electrical Engineering
Beijing, 100084, CHINA

Fernando Camacho
University of New South Wales
Graduate School of Biomedical Engineering
Sydney NSW 2052, AUSTRALIA
Tel: +61-2-93853922
Fax: +61-2-96632108

Drew Carlson
University of Maryland
Dept. of Surgery
Baltimore, MD 21201 USA

Elena Carro
Lund University
Signal Processing Group
Dept. of Applied Electronics
Box 118
S-221 00 Lund, SWEDEN
Tel: +46 46 222 90 23, +46 70 660 50 39
Fax: +46 46 222 47 18

Patrick Celka
Queensland University of Technology
School of Electrical & Electronic Syst. Eng.
Signal Processing Research Centre
GPO 2434, Brisbane 4001
AUSTRALIA
Tel: +61 7 3864 5351
Fax: +61 7 3864 1748
http://www.sprc.qut.edu.au/
http://www.eese.qut.edu.au/~celka/

Branko Celler
University of New South Wales
Biomedical Systems Laboratory
School of Electrical Engineering
Sydney 2052, AUSTRALIA

Sergio Cerutti
Polytechnic University Milano
Department of Biomemdical Engineering
Piazza Leonardo da Vinvi 32
20133 Milano, ITALY
Tel: +39-2-2399-3339

Fax: +39-2-2399-3360
E-mail: cerutti@icil64.cilea.it

Jiande D.Z. Chen, Ph.D.
GI Research Laboratory
108 The Strand
221 Microbiology Building
Galveston, TX 77555-0632 USA
Tel: +1 409 747 3071
Fax: +1 409 747 3084
E-mail: jianchen@utmb.edu

Martha Connolly
University of Maryland
Dept. of Surgery
Baltimore, MD 21201 USA

Prof. Dr. Ing. Hartmut Dickhaus
University of Heidelberg
Med. Informatics
Fachhochschule Heilbrönn
Max-Planck-Str. 39
D74081 Heilbronn, GERMANY
Tel: 00497131/504394
Fax: 00497131/252470
E-mail: dickhaus@fh-heilbronn.de

H. Duplain
University Hospital of Lausanne
Dept. of Internal Medicine
Lausanne, SWITZERLAND

Ysang Fusheng
Tsinghua University
Dept. of Electrical Engineering
Beijing, 100084, CHINA

R. Grueter
Swiss Federal Institute of Technology
Signal processing Laboratory
SWITZERLAND

Prof. Conor Heneghan
University College Dublin
Dept. of Electronic and Electrical Engineering
Belfield, Dublin 4, IRELAND
Tel: 353-1-706-1925
Fax: 353-1-283-0921
E-mail: Conor.Heneghan@ucd.ie

Bruce Henry
University of New South Wales
School of Mathematics
Sydney NSW 2052, AUSTRALIA

L.D. Iasemidis
University of Florida
Depts. of Electrical Engineering and Neuroscience
Gainesvilee, FL 32610 USA
Also:
VA Medical Center
Research Service
Gainesville, FL 32608 USA

Bradley M. Jost
Chapter written while affiliated with:
Department of Electrical & Computer Engineering
Boston University
Current address and affiliation:
Air force Research Laboratory
DELA
3550 Aberdeen Avenue SE
Kirtland AFB, NM 87117 USA
E-mail: bjost@ieee.org

L.M. Khadra
JUST
Dept. of Electrical Engineering
P.O. Box 3030
Irbid 221 10, JORDAN

Uwe Kienecke
University of Karlsruhe
Institute of Industrial Information Technique
P.O. Box 69 80, 76128 Karlsruhe, GERMANY

Pablo Laguna
Universidad de Zaragoza
Grupo de Tecnologias de las Comunicaciones
Departmento de Ingenieria, Electronica y Comunicaciones
Centro Politecnico Superior
Zaragoza, SPAIN

Dr. Nigel Lovell
Graduate School of Biomedical Engineering
University of New South Wales
Sydney NSW 2052, AUSTRALIA
Tel: +61-2-93853922
Mobile: 0419 627583

Fax: +61-2-96632108
E-mail: N.Lovell@unsw.edu.au

Dr. Steven B. Lowen
Chapter written while affiliated with:
Department of Electrical & Computer Engineering
Boston University
Current address and affiliation:
Department of Psychiatry, Harvard Medical School
Development Biopsychiatry Research Program
McLean Hospital
115 Mill St
Belmont, MA 02478 USA
E-mail: lowen@mclean.org

T. Maayah
JUST
Dept. of Electrical Engineering
P.O. Box 3030
Irbid 221 10 JORDAN

C. Maier
University of Heidelberg
Dept. of Medical Informatics
Heidelberg, GERMANY
E-mail: christoph.maier@fh-heilbronn.de

Alberto Malliani
Dipartmento di Scienze Precliniche
Universita' degli Studi di Milano
LITA di Vialba
via G.B. Grassi 74
20157 Milano, ITALY
Tel: +39 02 38210945
Fax: +39 02 38210533

Nicola Montano
Dipartimento di Scienze Precliniche
Universita' degli Studi di Milano
LITA di Vialba
via G.B. Grassi 74
20157 Milano, ITALY
Tel: +39 02 38210945
Fax: +39 02 38210533

Shin-ichi Nitta
Tohoku University
Dept. of Medical Engineering and Cardiology
Institute of Development, Aging and Cancer
4-1 Seiryo-machi, Aoba-ku, Sendai 980-8575
JAPAN
Tel: +81-22-717-8517
Fax: +81-22-717-8518

Alberto Porta, MS, PhD
Dipartimento di Scienze Precliniche
Universita' degli Studi di Milano
LITA di Vialba
via G.B. Grassi 74
20157 Milano, ITALY
Tel: +39 02 38210945
Fax: +39 02 38210533

Jose C. Principe, Ph.D.
BellSouth Professor and Director
Computational NeuroEngineering Laboratory
EB 451, Bldg #33
University of Florida
Gainesville, FL 32611 USA
Tel: 352-392-2662
Fax: 352-392-0044
E-mail: principe@cnel.ufl.edu

E. Pruvot
University Hospital of Lausanne
Division of Cardiology
Lausanne, SWITZERLAND

Tang Qingyu
Tsinghua University
Dept. of Electrical Engineering
Beijing, 100084, CHINA

J.C. Sackellares
University of Florida
Depts. of Neuroscience and Neurology
Gainesville, FL 32610 USA
Also:
VA Medical Center
Neurology Service
Gainesville, FL 32608 USA

U. Scherer
University Hospital of Lausanne
Dept. of Internal Medicine
Lausanne, SWITZERLAND

Roberto Sassi
Dipartimento di Bioingegneria
Politecnico di Milano

p.za Leonardo da Vinci 32
220133 Milano, ITALY
Tel: +39-2-23993342
Fax: +39-2-23993360

Stephan Schulz
University of Heidelberg
Dept. of Cardiac Surgery
Experimental Laboratory
INF 326
69120 Heidelberg, GERMANY
Tel: +49-(0)6221/56-6260
Fax: +49-(0)6221/439987
E-mail: Stephan.Schulz@urz.uni-heidelberg.de

Andreas Schwarzhaupt
University of Karlsruhe
Institute of Industrial Information Technique
P.O. Box 69 80
76128 Karlsruhe, GERMANY

Maria G. Signorini, Ph.D.
Dipartmento di Bioingegneria
Politecnico di Milano
p.za Leonardo da Vinci 32
220133 Milano, ITALY
Tel: +39-2-23993342
Fax: +39-2-23993360
E-mail: signorini@biomed.polimi.it

Leif Sörumo, Ph.D.
Signal Processing Group
Dept. of Applied Electronics
Lund University
Box 118
S-221 00 Lund, SWEDEN
Tel: +46 46 222 9023, +46 70 660 50 39
Fax: +46 46 222 47 18
E-mail: leif.sornmo@tde.lth.se

Martin Stridh
Signal Processing Group
Dept. of Applied Electronics
Lund University
Box 118
S-221 00 Lund, SWEDEN
Tel: +46 46 222 90 23, +46 70 660 50 39
Fax: +46 46 222 47 18

Koupichi Tabayashi
Tohoku University School of Medicine
Dept. of Thoracic and Cardiovascular Surgery
1-1 Seiryo-machi, Aoba-ku, Sendai 980-8574
JAPAN
Tel: +81-22–717-7220

Malvin C. Teich
Boston University
ECE Dept.
8 Saint Marys St.
Boston, MA 02215-2421 USA
Tel: +1 617-353-1236
Fax: +1 617-353-1459
E-mail: teich@bu.edu

Karin Vibe-Theymer
Department of Electrical and Computer Engineering
Boston University
8 Saint Marys St.
Boston, MA 02215-2421 USA
E-mail: kvibe@ieee.org

G. Thonet
Swiss Federal Institute of Technology
Signal Processing Laboratory
SWITZERLAND

C.F. Vahl
University of Heidelberg
Dept. of Cardiac Surgery
Experimental Laboratory
INF 326
69120 Heidelberg
GERMANY
Tel: +49-(0)6221/56-6260
Fax: +49-(0)6221/439987

J.M. Vesin
Swiss Federal Institute of Technology
Signal Processing Laboratory
SWITZERLAND

R. Vetter
Swiss Federal Institute of Technology
Signal Processing Laboratory
SWITZERLAND

Zhishun Wang, Ph.D.
GI Research Laboratory
221 Microbiology Building
1108 The Strand
Galveston, TX 77555-0632 USA
Tel: +1 409-747-3098
Fax: +1 409-747-3084
http://members.xoom.com/ZSWang

Tomoyuki Yambe
Tohoku University
Dept. of Medical Engineering and Cardiology
Institute of Development, Aging and Cancer
4-1 Seiryo-machi, Aoba-ku, Sendai 980-8575
JAPAN
Tel: +81-22-717-8517
Fax: +81-22-717-8518
E-mail: yambe@idac.tohoku.ac.ip

Makoto Yoshizawa
Tohoku University
Graduate School of Engineering
4-1 Seiryo-machi, Aoba-ku, Sendai 980-8575
JAPAN
Tel: +81-22-217-7073
PHS: +81-50-042005633
Fax: +81-22-263-9289
E-mail: yoshi@abe.ecci.tohoku.ac.ip

Chapter 1

# NONLINEAR DYNAMICS TIME SERIES ANALYSIS

Bruce Henry, Nigel Lovell, and Fernando Camacho

## 1. INTRODUCTION

Much of what is known about physiological systems has been learned using linear system theory. However, many biomedical signals are apparently random or aperiodic in time. Traditionally, the randomness in biological signals has been ascribed to noise or interactions between very large numbers of constituent components.

One of the most important mathematical discoveries of the past few decades is that random behavior can arise in deterministic nonlinear systems with just a few degrees of freedom. This discovery gives new hope to providing simple mathematical models for analyzing, and ultimately controlling, physiological systems.

The purpose of this chapter is to provide a brief pedagogic survey of the main techniques used in nonlinear time series analysis and to provide a MATLAB tool box for their implementation. Mathematical reviews of techniques in nonlinear modeling and forecasting can be found in Refs. 1–5. Biomedical signals that have been analyzed using these techniques include heart rate [6–8], nerve activity [9], renal flow [10], arterial pressure [11], electroencephalogram [12], and respiratory waveforms [13].

Section 2 provides a brief overview of dynamical systems theory including phase space portraits, Poincaré surfaces of section, attractors, chaos, Lyapunov exponents, and fractal dimensions. The forced Duffing–Van der Pol oscillator (a ubiquitous model in engineering problems) is investigated as an illustrative example. Section 3 outlines the theoretical tools for time series analysis using dynamical systems theory. Reliability checks based on forecasting and surrogate data are also described. The time series methods are illustrated using data from the time evolution of one of the dynamical variables of the forced Duffing–Van der Pol oscillator. Section 4 concludes with a discussion of possible future directions for applications of nonlinear time series analysis in biomedical processes.

## 2. DYNAMICAL SYSTEMS THEORY

### 2.1. Deterministic Chaos

A *dynamical system* is any system that evolves in time. Dynamical systems whose behavior changes continuously in time are mathematically described by a coupled set of first-order autonomous ordinary differential equations

$$\frac{d\vec{x}(t)}{dt} = \vec{F}(\vec{x}(t), \vec{\mu}) \tag{1}$$

The components of the vector $\vec{x}(t)$ are the dynamical variables of the system, the components of the vector $\vec{\mu}$ are parameters, and the components of the vector field $\vec{F}$ are the dynamical rules governing the behavior of the dynamical variables. There is no loss of generality in the restriction to *autonomous systems*, where $\vec{F}$ is not an explicit function of $t$, since a nonautonomous system in $\mathbb{R}^n$ can be transformed into an autonomous system in $\mathbb{R}^{n+1}$. If the vector field is *affine*, i.e.,

$$\vec{F}(\vec{x}, \vec{\mu}) = \underline{\underline{A}}(\vec{\mu})\vec{x} + \vec{b}(\vec{\mu}) \tag{2}$$

for some constant matrix $\underline{\underline{A}}$ and vector $\vec{b}$, then the dynamical system is said to be *linear*. Otherwise it is *nonlinear*. In linear dynamical systems any linear combination of solutions is also a solution.

An example of a nonlinear dynamical system with numerous applications in engineering is the single well-forced Duffing–Van der Pol oscillator [14]

$$\frac{d^2y}{dt^2} \quad \mu(1 - y^2)\frac{dy}{dt} + y^3 - f\cos\omega t \tag{3}$$

where $\mu$, $f$, $\omega$ are parameters. This second-order nonautonomous equation can be written as the first-order system

$$\frac{dx_1}{dt} = x_2 \tag{4}$$

$$\frac{dx_2}{dt} = \mu(1 - x_1^2)x_2 - x_1^3 + f\cos x_3 \tag{5}$$

$$\frac{dx_3}{dt} = \omega \tag{6}$$

by defining dynamical variables $x_1 = y$, $x_2 = dy/dt$, $x_3 = \omega t$.

Dynamical systems whose behavior changes at discrete time intervals are described by a coupled set of first-order autonomous difference equations

$$\vec{x}(n+1) = \vec{G}(\vec{x}(n), \vec{\mu}) \tag{7}$$

In this equation $\vec{G}$ describes the dynamical rules and time is represented by the integer $n$. A discrete dynamical system may be obtained from a continuous dynamical system (1) by sampling the solution of the continuous dynamical system at regular time intervals $T$—the dynamical rule relating successive sampled values of the dynamical variables is called a *time T map*, and (2) by sampling the solution of the continuous dynamical system in $\mathbb{R}^n$ at successive transverse intersections with a surface of section of dimension $\mathbb{R}^{n-1}$—the dynamical rule relating successive sampled values of the dynamical variables is called a *Poincaré map* or a *first return map*. For example, in the forced Duffing–Van der Pol oscillator a surface of section could be defined by $x_3 = \theta_0$ where

$\theta_0 \in (0, 2\pi)$ is a constant. In this case the Poincaré map is equivalent to a time $T$ map with $T = 2\pi/\omega$.

Under modest smoothness assumptions about the dynamical rules, the solutions of dynamical systems are unique and the dynamical system is *deterministic*; that is, the state of the dynamical system for all times is uniquely determined by the state at any one time. The *existence* of unique solutions does not necessarily mean that *explicit algebraic representations* exist. However, if explicit algebraic solutions do exist and they can be used to predict the future behavior of the system for all initial conditions, then the system is said to be *integrable*. All *linear* dynamical systems are integrable. Explicit solutions can be constructed for linear systems by first transforming to a new set of dynamical variables in which the governing equations become decoupled.

One of the surprising and far-reaching mathematical discoveries of the past few decades has been that the solutions of deterministic *nonlinear* dynamical systems may be as random (in a statistical sense) as the sequence of heads and tails in the toss of a fair coin [15]. This behavior is called *deterministic chaos*. The discovery of deterministic chaos is surprising because randomness has been traditionally associated with unknown external disturbances (noise). What makes the discovery far reaching is that most dynamical systems are nonlinear and most nonlinear systems have random solutions. Deterministic chaos has immediate ramifications for constructing mathematical models for systems characterized by random signals. A fundamental question in this regard is: Are all random signals equally random? It turns out that they are not. Random signals generated by noise are fundamentally different from random signals generated by deterministic dynamics with small numbers of dynamical variables. The difference is not revealed by statistical analysis but is instead revealed by dynamical analysis based on phase space reconstruction.

## 2.2. Phase Space—Attractors

Phase space is an abstract mathematical space spanned by the dynamical variables of the system. The state of the dynamical system at a given instant in time can be represented by a point in this phase space. If there are $n$ dynamical variables, then the state at a given time can be represented by a point in the Euclidean space $\mathbb{R}^n$. As the dynamical variables change their values in time, the representative point traces out a path in the phase space—a continuous curve in the case of a continuous dynamical system and a sequence of points in the case of a discrete dynamical system.

For an idealized simple pendulum there are two physical dynamical variables, the angular position $\theta$ and velocity $\dot{\theta}$, so the phase space can be taken to be $\mathbb{R}^2$. If the energy is conserved in this system and the angular oscillations are small, then $E = \frac{1}{2}\dot{\theta}^2 + \frac{1}{2}\theta^2$ constrains the phase space variables to lie on a circle. The radius of the circle is determined by the system energy. A realistic pendulum dissipates energy to the surroundings via friction and air resistance. The phase space path in this case is a spiral in toward a final resting point called a fixed point attractor. The starting radius for the spiral again depends on the initial energy in the system, but the location of the fixed point attractor is independent of this starting energy. Most physical systems are dissipative and their long-time dynamical behavior can be described by an attractor in phase space. In Figure 1 phase space portraits are shown for the Duffing–Van der Pol oscillator, Eqs. 4–6 for initial conditions $x_1 = 1$, $x_2 = 0$, $x_3 = 0$, and four sets of parameters: (a) $\mu = 0.0, f = 0.0$; (b) $\mu = 0.2, f = 0.0$; (c) $\mu = 0.2, f = 1.0, \omega = 0.9$; and (d)

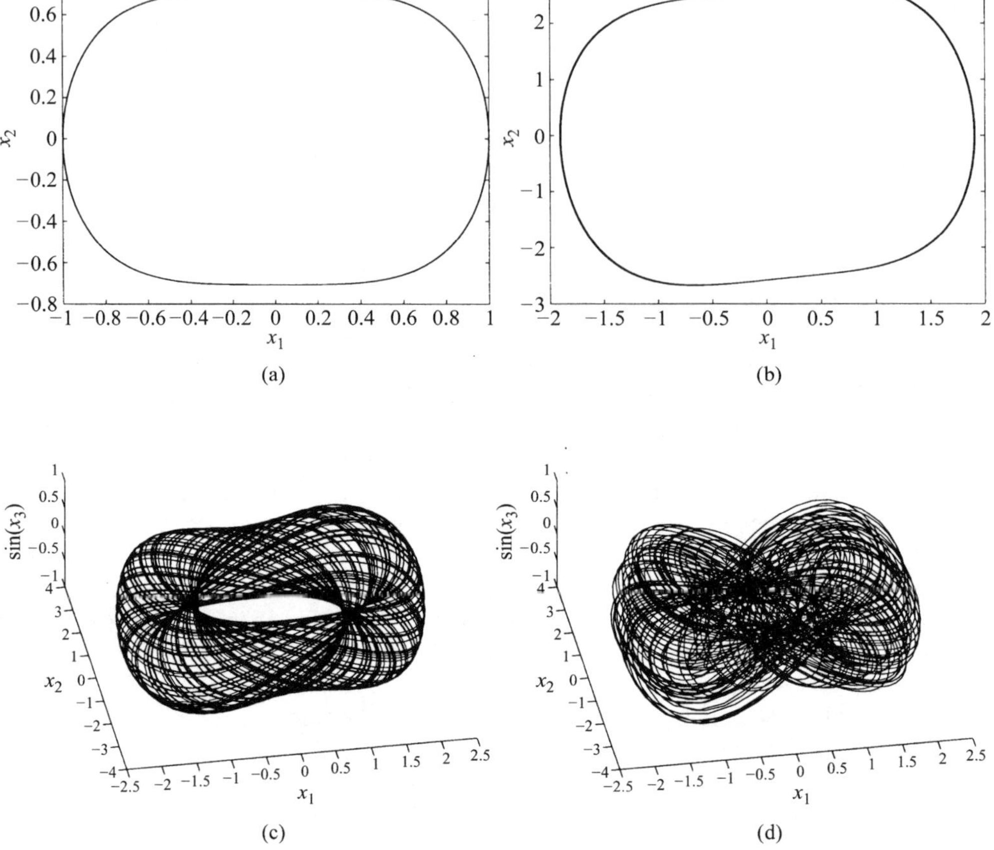


**Figure 1** Phase space portraits for the Duffing–Van der Pol oscillator for parameters (a) $\mu = 0.0, f = 0.0$; (b) $\mu = 0.2, f = 0.0$; (c) $\mu = 0.2, f = 1.0, \omega = 0.90$; (d) $\mu = 0.2, f = 1.0, \omega = 0.94$.

$\mu = 0.2, f = 1.0, \omega = 0.94$. In cases (c) and (d) it is convenient to choose as coordinate axes the dynamical variables $x_1$, $x_2$, $\sin(x_3)$. The initial transient behavior is not shown in these figures.

For the parameters in case (a), there is no dissipation and the path is topologically similar to that of the idealized pendulum. The distortion away from a circle is due to nonlinear restoring forces. In case (b), dissipation is included ($\mu \neq 0$) and the path is a limit cycle attractor. In case (c), the dissipation ($\mu \neq 0$) is balanced by external forcing ($f \neq 0$) and another periodic orbit results [14]. The conditions in case (d) are very similar to the conditions in case (c) but the slightly different forcing frequency in case (d) results in a very different orbit—a chaotic strange attractor [14]. The Poincaré surface of section equivalent to the time $2\pi/\omega$ map for the two sets of parameters, case (c) and case (d), is shown in Figure 2a and b, respectively. Again, the initial transient behavior is not shown.

Appendix I contains MATLAB programs for numerically integrating three coupled differential equations (Appendix I.A), generating three-dimensional phase

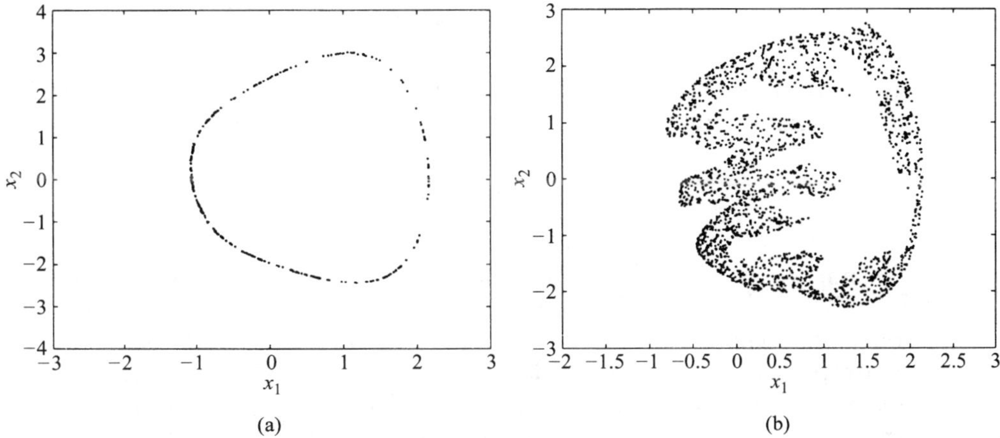


**Figure 2** Poincaré surfaces of section for the Duffing–Van der Pol oscillator for parameters (a) $\mu = 0.2, f = 1.0, \omega = 0.9$; (b) $\mu = 0.2, f = 1.0, \omega = 0.94$.

space portraits (Appendix I.B), and constructing two-dimensional Poincaré surfaces of section (Appendix I.C).

The concepts of dissipation and attractors are defined more precisely for a *continuous* dynamical system as follows. A dynamical system is *conservative* if

$$\vec{\nabla}.\vec{F} = 0 \tag{8}$$

A dynamical system is *dissipative* if on average (where the average is over all initial conditions)

$$\vec{\nabla}.\vec{F} < 0 \tag{9}$$

In a dissipative dynamical system, phase space volume elements contract as the system evolves. In the forced Duffing–Van der Pol oscillator,

$$\vec{\nabla}.\vec{F} = -\mu(x_1^2 - 1) \tag{10}$$

so that the system is dissipative for $x_1^2 > 1$; $\mu > 0$.

A point $\vec{x}$ is a *fixed point* (or equilibrium point) if

$$\vec{F}(\vec{x}) = 0 \tag{11}$$

A phase space trajectory $\vec{x}(t)$ is *periodic* if

$$\vec{x}(t) = \vec{x}(t + T) \tag{12}$$

for some nonzero $T$. The period is the minimum nonzero $T$ for which Eq. 12 holds. It is convenient to represent a phase space path by

$$\vec{x}(t) = \phi^{t,t_0}(\vec{x_0}) \tag{13}$$

where $\phi$ is the flow that maps an initial point $\vec{x}_0 \in \mathbb{R}^n$ and time $t \in \mathbb{R}$ onto the solution $\vec{x}(t)$. An $\omega$-limit point of $\vec{x}_0$ is a phase space point $\vec{x}$ such that

$$\lim_{t\to\infty} \phi^{t,t_0}(\vec{x}_0) \to \vec{x} \tag{14}$$

An $\omega$-limit set is a set of $\omega$-limit points corresponding to a set of initial points $\vec{X}_0 = \{\vec{x}_0^{(1)}, \vec{x}_0^{(2)}, \ldots, \vec{x}_0^{(k)}\}$. An *attracting set* is an $\omega$-limit set to which all orbits starting in the neighborhood of the set of initial points $\vec{X}_0$ tend as $t \to \infty$. An *attractor* is an attracting set that contains a dense orbit. A system must be dissipative in order to have an attractor. Since phase space volume elements contract in time in dissipative systems, it follows that attractors must occupy zero volume in phase space. A *limit cycle* attractor is a periodic attractor.

A *strange attractor* is an aperiodic attractor with the additional properties that (1) phase space paths through all points on the attractor diverge on average at an exponential rate and (2) the dimension of the set of points comprised by the attractor is not an integer.

## 2.3. Lyapunov Exponents

Lyapunov exponents quantify the average exponential separation between nearby phase space trajectories. An exponential divergence of initially nearby trajectories in phase space coupled with folding of trajectories (to ensure that solutions remain finite) is the generic mechanism for generating deterministic randomness and unpredictability. Indeed, the existence of a positive Lyapunov exponent for almost all initial conditions in a bounded dynamical system is a widely used definition of deterministic chaos.

Let $\vec{x}_0(t)$ denote a reference trajectory passing through $\vec{x}_0(0)$ at time $t = 0$ and let $\vec{x}_1(t)$ denote a trajectory passing through $\vec{x}_1(0)$ at time $t = 0$. The (maximum) Lyapunov exponent $\lambda(\vec{x}_0)$ is defined with respect to the reference orbit $\vec{x}_0$ by [16,17]

$$\lambda(\vec{x}_0) = \lim_{t\to\infty} \lim_{\|\Delta\vec{x}(0)\|\to 0} \frac{1}{t} \log \frac{\|\Delta\vec{x}(t)\|}{\|\Delta\vec{x}(0)\|} \tag{15}$$

where $\|\Delta\vec{x}(0)\|$ is the Euclidean distance between the trajectories $\vec{x}_0(t)$ and $\vec{x}_1(t)$ at an initial time $t = 0$ and $\|\Delta\vec{x}(t)\|$ is the Euclidean distance between the trajectories $\vec{x}_0(t)$ and $\vec{x}_1(t)$ at a later time $t$. In this definition $\vec{x}_1(t)$ can be any trajectory that is initially infinitesimally close to $\vec{x}_0(0)$ at time $t = 0$. The correspondence between sensitivity to initial conditions and a positive Lyapunov exponent is obvious in the rearrangement

$$\|\Delta\vec{x}(t)\| \sim \|\Delta\vec{x}(0)\| e^{\lambda t} \tag{16}$$

A dynamical system in $\mathbb{R}^m$ has associated with it $m$ Lyapunov exponents

$$\lambda_1 \geq \lambda_2 \geq \cdots \geq \lambda_m \tag{17}$$

To define the full set of exponents, consider an infinitesimal $m$-dimensional sphere of initial conditions that is anchored to a reference trajectory. As the sphere evolves, it becomes deformed into an ellipsoid. Let $p_i(t)$ denote the length of the $i$th principal axis, ordered from most rapidly growing to least rapidly growing. Then [16,18]

$$\lambda_i = \lim_{t\to\infty} \frac{1}{t} \log\left(\frac{p_i(t)}{p_i(0)}\right) \qquad i = 1, 2, \ldots, m \tag{18}$$

defines the set of Lyapunov exponents ordered from largest to smallest. Volume elements in phase space evolve in time as [19]

$$V(t) \sim V(0) \exp\left(\sum_i^m \lambda_i t\right) \tag{19}$$

The sum of the Lyapunov exponents is equal to the divergence of the vector field; i.e.,

$$\sum_i^m \lambda_i = \vec{\nabla}.\vec{F} \tag{20}$$

Thus, in dissipative systems the average of the sum of the Lyapunov exponents is negative. Furthermore, for bounded trajectories that do not approach a fixed point at least one of the Lyapunov exponents is zero.

In numerical computations of Lyapunov exponents the limit $\|\Delta\vec{x}(0)\| \to 0$ can be effectively realized by evolving $\Delta\vec{x}(t)$ with the linear variational equation [20]

$$\frac{d\Delta\vec{x}(t)}{dt} = D_x\vec{F}(\vec{x}_0(t))\Delta\vec{x}(t) \tag{21}$$

In this equation $\vec{x}_0(t)$ is the reference orbit obtained by integrating the original dynamical system and $D_x\vec{F}$ is a matrix with components $\partial F_i/\partial x_j$ evaluated along the reference orbit. If the system has a positive Lyapunov exponent, then direct numerical integration of the linear variational equation will eventually lead to numerical overflow. This problem can be avoided by renormalizing the solution of the linear variational equation at a periodic intervals $\tau$. The maximum Lyapunov exponent is then equivalently given by [20]

$$\lambda_{\max} = \lim_{n\to\infty} \frac{1}{(n+1)\tau} \sum_{j=0}^{n} \log \|\Delta\vec{x}_j(\tau)\| \tag{22}$$

where $\Delta\vec{x}_j(\tau)$ are solutions of the variational equations for renormalized initial vectors

$$\Delta\vec{x}_j(0) = \frac{\Delta\vec{x}_{j-1}(\tau)}{|\Delta\vec{x}_{j-1}(\tau)|} \tag{23}$$

A similar renormalization using the Gram-Schmidt reorthonormalization scheme can be employed to measure the full set of Lyapunov exponents [21,22]. In Table 1 the full

**TABLE 1** Lyapunov Exponents for the Duffing–Van der Pol Oscillator for Parameters

| Parameters[a] | $\lambda_1$ | $\lambda_2$ | $\lambda_3$ |
|---|---|---|---|
| (a) | 0.000 | 0.000 | |
| (b) | 0.000 | −0.133 | |
| (c) | 0.000 | 0.000 | −0.049 |
| (d) | 0.0254 | 0.000 | −0.0285 |

[a] (a), $\mu = 0.0, f = 0.0$; (b), $\mu = 0.2, f = 0.0$; (c), $\mu = 0.2, f = 1.0, \omega = 0.90$; (d), $\mu = 0.2, f = 1.0, \omega = 0.94$.

set of Lyapunov exponents is listed for each of the four sets of parameters used in Figure 1. Note that case (d) has a positive Lyapunov exponent and the sum of the exponents is negative. This is consistent with chaotic dynamics on a strange attractor.

A MATLAB code for measuring the Lyapunov exponents for three coupled differential equations is listed in Appendix I.D.

## 2.4. Fractal Dimensions

A geometrical object can be fully represented by a set of points in a Euclidean space $\mathbb{R}^m$ provided that $m$ is sufficiently large to be able to uniquely locate the position of each point in the object. Each set in $\mathbb{R}^m$ has assigned to it a topological dimension $d$ that is an integer in the range $[0, m]$. If the set is all of $\mathbb{R}^m$, then $d = m$. In Euclidean geometry, points have dimension $d = 0$, lines have dimension $d = 1$, plane surfaces have dimension $d = 2$, solids have dimension $d = 3$, etc.

A *fractal dimension* $D$ is any dimension measurement that allows noninteger values [23]. A *fractal* is a set with a noninteger fractal dimension [23]. Standard objects in Euclidean geometry are not fractals but have integer fractal dimensions $D = d$. The primary importance of fractals in dynamics is that strange attractors are fractals and their fractal dimension $D$ is simply related to the minimum number of dynamical variables needed to model the dynamics of the strange attractor.

The simplest way (conceptually) to measure the dimension of a set is to measure the *Kolmogorov capacity* (or box-counting dimension). In this measurement a set is covered with small cells (e.g., squares for sets embedded in two dimensions, cubes for sets embedded in three dimensions) of size $\epsilon$. Let $M(\epsilon)$ denote the number of such cells that contain part of the set. The dimension is then defined as

$$D = \lim_{\epsilon \to 0} \frac{\log(M(\epsilon))}{\log(\frac{1}{\epsilon})} \tag{24}$$

For $n$ isolated points, $M(\epsilon) = n$ and $D = 0$; for a straight line of length $L$, $M(\epsilon) = L/\epsilon$ and $D = 1$; for a plane region of area $A$, $M(\epsilon) = A/\epsilon^2$ and $D = 2$. In practical applications the limit $\epsilon$ is not attainable. Instead, the number $M(\epsilon)$ is measured for a range of small values of $\epsilon$ and the dimension $D$ is estimated as the slope of the straight line portion of the plot of $\log(M(\epsilon))$ versus $\log(1/\epsilon)$.

A mathematical example of a set with a noninteger fractal dimension is the Cantor set, which is defined as the limiting set in a sequence of sets. Consider the set in $\mathbb{R}^2$ defined by the following sequence of sets. At stage $k = 0$ (Figure 3a) let $S_0$ denote a square with sides of length $l$. At stage $k = 1$ (Figure 3b) divide the set $S_0$ into nine squares of uniform size and remove the middle square. The remaining set is labeled $S_1$.

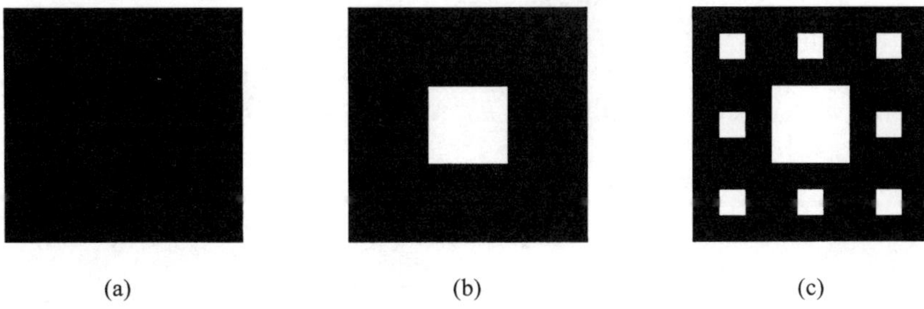

(a) (b) (c)

**Figure 3** First three stages in the construction of a Cantor set in $\mathbb{R}^2$; (a) stage $k = 0$; (b) stage $k = 1$; (c) stage $k = 2$.

At stage $k = 2$ (Figure 3c) divide each remaining square in $S_1$ into nine squares of uniform size and remove the middle squares. This new set is labeled $S_2$. The process of subdividing and removing is continued iteratively to obtain a sequence of sets $S_0, S_1, S_2, \ldots$ and the Cantor set is defined as the limiting set

$$S = \lim_{n \to \infty} S_n \tag{25}$$

It is straightforward to measure the Kolmogorov capacity for this Cantor set. At stage $k = 0$ the set $S_0$ is covered with one square of size $a$. Thus for $k = 0$; $\epsilon = a$ and $M(\epsilon) = 1$. At stage $k = 1$ the set $S_1$ is covered with eight squares of size $a/3$. Thus for $k = 1$; $\epsilon = a/3$ and $M(\epsilon) = 8$. At stage $k = 2$ the set $S_2$ is covered with 64 squares of size $a/9$. Thus for $k = 2$; $\epsilon = a/9$ and $M(\epsilon) = 64$. For general $k$ it follows by induction that the set $S_k$ is covered with $\epsilon = a/3^k$ and $M(\epsilon) = 8^k$. The fractal dimension of the limiting Cantor set is thus

$$D = \lim_{k \to \infty} \frac{\log(8^k)}{\log\left(\frac{a}{3^k}\right)} \tag{26}$$

$$= 1.892 \tag{27}$$

The fractal dimension less than two means that this Cantor set does not fill an area in $\mathbb{R}^2$.

When computing the box-counting dimension, a box is counted whenever it contains part of the set. This counting does not differentiate between whether a box contains many points of the set or few points of the set. More elaborate dimension measurements are available that take into account inhomogeneities or correlations in the set. The *dimension spectrum* defined by Hentschel and Procaccia [24],

$$D_q = \lim_{r \to 0} \frac{1}{q-1} \frac{\log \sum_{i=1}^{M(r)} p_i^q}{\log r}, \qquad q = 0, 1, 2, \ldots \tag{28}$$

provides a set of fractal dimension measurements that take into account higher order correlations as $q$ is increased. In the dimension spectrum, $M(r)$ is the number of $m$-dimensional cells of size $r$ (e.g., hypercubes of side $r$) needed to cover the set and $p_i = N_i/N$ is the probability of finding a point of the set in hypercube $i$; $N$ is the total number of points in the set and $N_i$ is the number of points of the set in hypercube $i$. It can be readily seen that the box-counting dimension is equivalent to $D_0$.

The dimension $D_1$ is called the *information dimension*. This is defined by taking the limit $q \to 1$, i.e.,

$$D_1 = \lim_{q \to 1} D_2 \tag{29}$$

$$= \lim_{r \to 0} \frac{\sum_{i=1}^{M(r)} p_i \log p_i}{\log r} \tag{30}$$

The information dimension has also been related to the Lyapunov exponents through a conjecture of Kaplan and Yorke [25]:

$$D_1 = j + \frac{\sum_{i=1}^{j} \lambda_i}{|\lambda_{j+1}|} \tag{31}$$

In Eq. 31, $\lambda_i$ are the Lyapunov exponents of the attractor ordered from largest to smallest and $\sum_{i=1}^{j} \lambda_i \geq 0$; $\sum_{i=1}^{j+1} \lambda_i < 0$. As an example consider the strange attractor in the Duffing–Van der Pol oscillator with parameters $\mu = 0.2$, $f = 1.0$, $\omega = 0.94$, Figure 1d. The Lyapunov information dimension for this case is, from the values in Table 1 (d), $D_1 = 2.84$. The noninteger value confirms that the attractor is a strange attractor.

The dimension $D_2$ is called the *correlation dimension*. This can be written as

$$D_2 = \lim_{r \to 0} \frac{\log C(r)}{\log r} \tag{32}$$

where

$$C(r) = \sum_{i=1}^{M(r)} p_i^2 \tag{33}$$

is the correlation sum, which is essentially (exact in the limit $N \to \infty$) the probability that two points of the set are in the same cell.

For a given set, the dimensions are ordered $D_0 \geq D_1 \geq D_2 \geq \cdots$. If the set is a homogeneous attractor then

$$p_i = \frac{1}{M} \tag{34}$$

and all dimensions, Eq. 28, are equal; otherwise the set is called a *multifractal*. The major difficulty in calculating $D_q$ is the practical difficulty of covering the set with cells of very small size. In general, this requires too much computer storage and time to obtain convergence to a limit $r \to 0$.

When $q = 2$ the dimension estimate can be made computationally tractable by using an algorithm proposed by Grassberger and Procaccia [26]. This algorithm has become the most widely used method for estimating fractal dimensions of experimental data sets.

### 2.5. Grassberger–Procaccia Algorithm

The Grassberger–Procaccia algorithm [26] is based on the following approximation: The probability that two points of the set are in the same cell of size $r$ is approximately equal to the probability that two points of the set are separated by a distance $\rho$ less than or equal to $r$. Thus $C(r)$ is approximately given by

$$C(r) \approx \frac{\sum_{i=1,j>i}^{N} \Theta\left(r - \rho(\vec{x}_i, \vec{x}_j)\right)}{\frac{1}{2}N(N-1)} \tag{35}$$

where the Heaviside function is defined as

$$\Theta(s) = \begin{cases} 1 & \text{if } s \geq 0 \\ 0 & \text{if } s < 0 \end{cases} \tag{36}$$

The approximation in Eq. 35 is exact in the limit $N \to \infty$; however, this limit cannot be realized in practical applications. The limit $r \to 0$ used in the definition of $D_2$ is also not possible in practice. Instead, Procaccia and Grassberger propose the (approximate) evaluation of $C(r)$ over a range of values of $r$ and then deduce $D_2$ from the slope of the straight line of best fit in the linear scaling region of a plot of $\log C(r)$ versus $\log r$.

The most common metric employed to measure the distance $\rho$ in Eq. 35 is the Euclidean metric,

$$\rho(\vec{x}_i, \vec{x}_j) = \sqrt{\sum_{k=1}^{m} \left(x_i(k) - x_j(k)\right)^2} \tag{37}$$

However, other metrics have also been considered. In any case, the choice of metric should not affect the scaling of the correlation sum with $r$.

The reliability of estimating the slope in the linear scaling region is the most serious possible source of error in the Grassberger–Procaccia algorithm. Clearly, the linear scaling regime will be bounded above by the maximum separation distance between points in the set and bounded below by the minimum separation distance between points in the set. Essentially, the idea is to measure the slope for the smallest $r$ values possible while avoiding the problems of sparse numbers when $r$ is close to the minimum separation. One ad hoc scheme that has received some popularity is to plot $\log C(r)$ versus $\log r$ for a number of equally spaced values of $\log r$ between $\log r_{\min}$ and $\log r_{\max}$. Then deduce the slope of the straight line of best fit over the middle third of the vertical range of the plot. This method should be used with caution as it is possible that the middle third straddles two different straight line behaviors—noise and deterministic chaos. In particular, if the system contains noise on a scale $r^*$ then for an $m$-dimensional embedding the correlation sum will scale as

$$C(r) \sim \begin{cases} r^m & \text{for } r < r^* \\ r^D & \text{for } r > r^* \end{cases} \tag{38}$$

Thus a plot of $\log C(r)$ versus $\log r$ will reveal a change of slope from $m$ to $D$ with the crossover at $r^*$ providing an estimate of the level of noise in the system.

Figure 4 shows a plot of $\log C(r)$ versus $\log r$ for each of the phase space paths of the Duffing–Van der Pol oscillator shown in Figure 1. The straight lines of best fit using data across the domain from $10^{-2}$ to $10^{-1}$ are also shown. Estimates of the correlation dimension based on the slopes of these straight line portions are (a) $D_2 = 1.01$, (b) $D_2 = 1.01$, (c) $D_2 = 2.17$, (d) $D_2 = 2.67$.

A MATLAB code for implementing the Grassberger–Procaccia algorithm to measure the correlation dimension of a phase space trajectory is listed in Appendix I.E.

There have been several estimates of the minimum number of data points $N_{\min}$ required for estimates of $D$ to be reliable using the Grassberger–Procaccia algorithm. A "rule of thumb" estimate due to Ruelle [27] is that

$$N_{\min} = 10^{(D/2)} \tag{39}$$

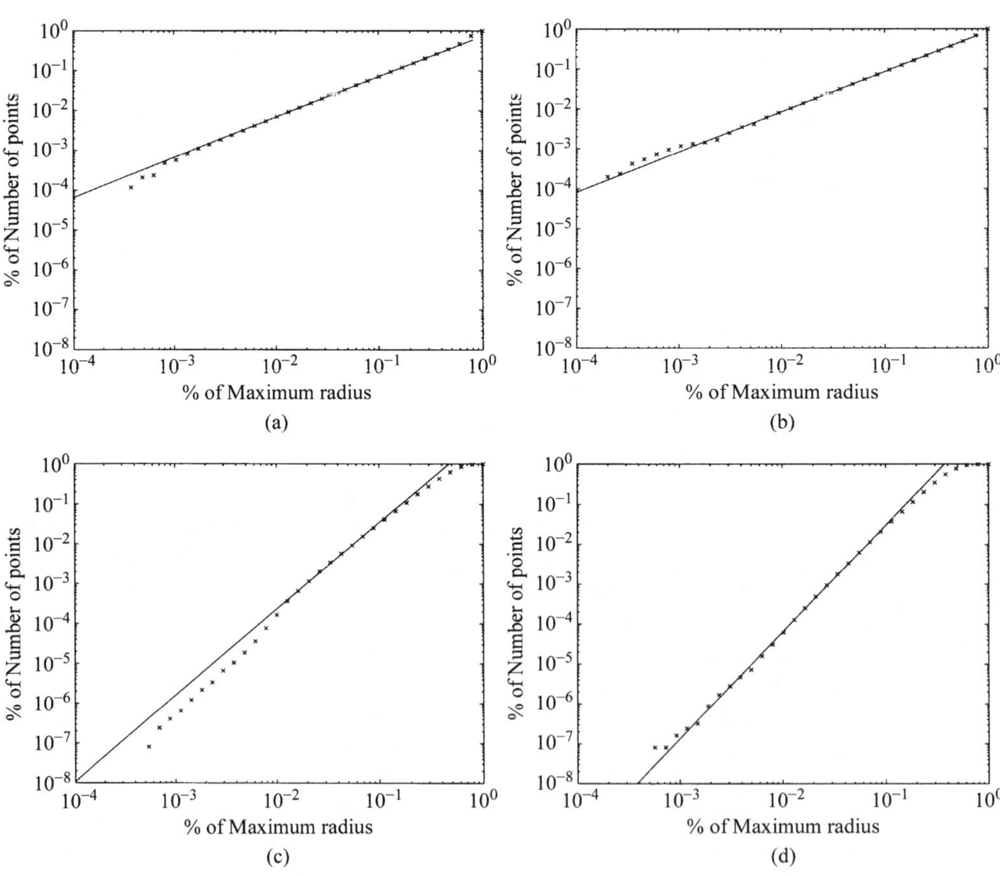


**Figure 4** Plots of the logarithm of the correlation sum versus the logarithm of the separation distance for phase space trajectories of the Duffing–Van der Pol oscillator for parameters (a) $\mu = 0.0$, $f = 0.0$; (b) $\mu = 0.2$, $f = 0.0$; (c) $\mu = 0.2, f = 1.0$, $\omega = 0.90$; (d) $\mu = 0.2, f = 1.0$, $\omega = 0.94$.

Thus, estimates of $D \geq 2\log_{10} N$ obtained from time series analysis should be regarded as unreliable. A simple derivation of this estimate is as follows: The correlation sum scales as

$$C(r) \sim Ar^D \tag{40}$$

Assume that this scaling applies up to the maximum separation distance $r_{\max}$. Then

$$A \sim \frac{C(r_{\max})}{r_{\max}^D} \tag{41}$$

but at the limiting separation

$$C(r_{\max}) \sim \tfrac{1}{2} N^2 \tag{42}$$

Combining the preceding three equations now yields

$$C(r) \sim \frac{N^2}{2} \left( \frac{r}{r_{\max}} \right)^D \tag{43}$$

Clearly, the correlation sum is bounded below by $C(r_{\min}) = 1$, hence $C(r) > 1$ and

$$2 \log N > D(\log r_{\max} - \log r) \tag{44}$$

The *Ruelle conjecture*, Eq. 39, now follows immediately from the reasonable expectation that the linear scaling regime in the log-log plot (if it exists) should persist over at least one decade in the range of $r$. The theoretical basis of the Ruelle conjecture has been questioned and other theoretical requirements on the sample size have been proposed [28,29]. However the rule of thumb, Eq. 39, has been found to be relevant in many experimental studies [27]. Moreover, all theoretical results for requirements on sample size reveal an exponential growth with dimension.

## 3. TIME SERIES ANALYSIS

The techniques in this section are illustrated using data from numerical integrations of the Duffing–Van der Pol oscillator Eqs. 4–6 with initial conditions $x_1 = 1, x_2 = 0, x_3 = 0$ and the four sets of parameters (a) $\mu = 0.0, f = 0.0$; (b) $\mu = 0.2, f = 0.0$; (c) $\mu = 0.2$, $f = 1.0$, $\omega = 0.90$; (d) $\mu = 0.2, f = 1.0$, $\omega = 0.94$. A time series $y_n$ is constructed from the numerical solution for the single dynamical variable $x_1(t)$ sampled at intervals $\Delta t = 0.1$, i.e., $y_n = x_1(n\Delta t)$. The results in this section based on the time series analysis of a single variable can be compared with the results in Section 2 based on the direct analysis of the evolution of all dynamical variables. Time series for the Duffing–Van der Pol oscillator with this set of parameters are shown in Figure 5.

### 3.1. Power Spectrum and Autocorrelation

The power spectrum reveals periodic components of a signal and it should always be employed in time series analysis whether the primary analysis is statistical or dyna-

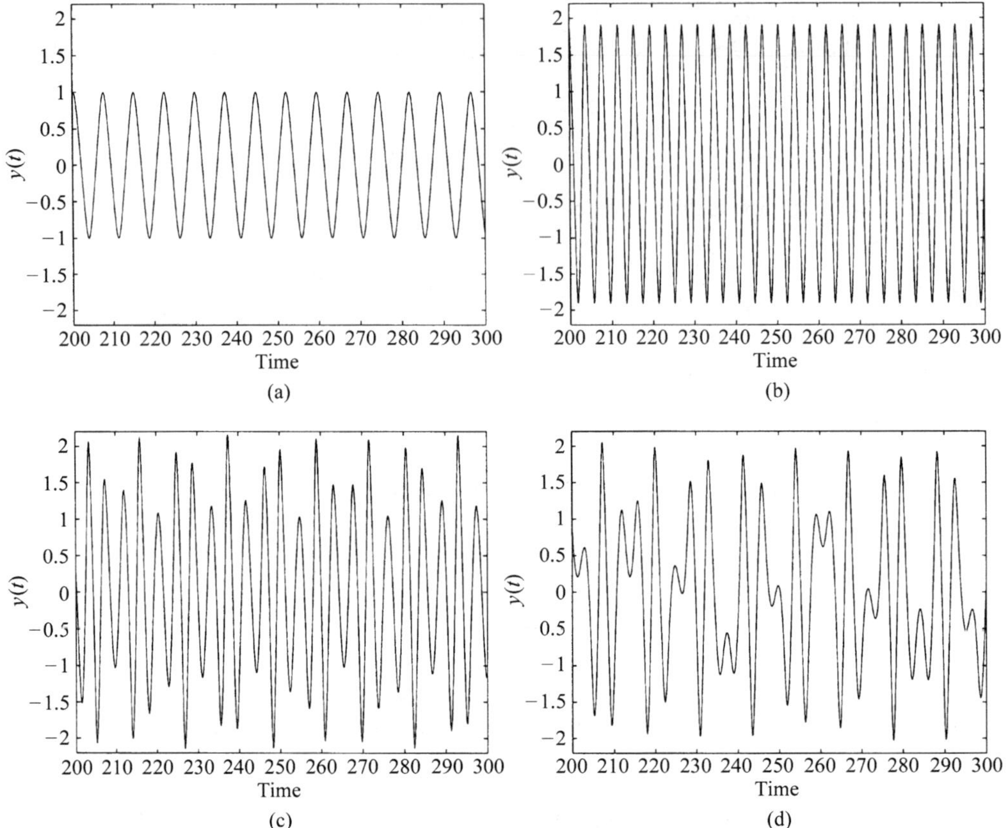


**Figure 5** Time series data for the Duffing–Van der Pol oscillator for parameters (a) $\mu = 0.0, f = 0.0$; (b) $\mu = 0.2, f = 0.0$; (c) $\mu = 0.2, f = 1.0, \omega = 0.90$; (d) $\mu = 0.2, f = 1.0, \omega = 0.94$.

mical. If the signal is periodic then the power spectrum will consist of discrete lines, whereas in a stochastic signal the power will be spread over a continuous range of frequencies. Consider a time series

$$y_n = y(n\Delta t), \qquad n = 1, 2, \ldots, N \tag{45}$$

The discrete Fourier transform is defined by

$$Z_m = \frac{1}{N} \sum_{n=0}^{N-1} y_n \exp\left(-i2\pi(m-1)(n-1)/N\right) \tag{46}$$

and the power spectrum is defined as

$$P_m = |Z_m|^2 = X_m^2 + Y_m^2 \tag{47}$$

where $X$ and $Y$ are the real and imaginary parts of $Z$. Each value of $m$ for which there is a peak in the power spectrum corresponds to a frequency component

$$f_m = \frac{m}{N\Delta t} \tag{48}$$

in the original time series.

The autocorrelation function also provides a diagnostic tool for discriminating between periodic and stochastic behavior. In a periodic signal the autocorrelation is periodic, whereas in a stochastic signal the autocorrelation will be irregular. The autocorrelation function is defined by

$$c_j = \frac{1}{N} \sum_{i=1}^{N} y_i y_{i+j} \tag{49}$$

where periodic boundary conditions

$$y_{N+k} = y_k \tag{50}$$

are imposed to extend the times series beyond $y_N$. This function provides a simple quantitative measure of the linear correlation between data points.

Fourier transforms of the time series data $y_n = x_1(n\Delta t)$ in the Duffing–Van der Pol oscillator are shown for four sets of parameters in Figure 6. The autocorrelation functions for the same sets of data are shown in Figure 7. The "grassy appearance" of the Fourier transform in Figure 6d and the aperiodicity of the autocorrelation function in Figure 7d are characteristic of a chaotic signal, whereas the sharp peaks in Figure 6a–c and the periodicities in Figure 7a–c are characteristic of periodic signals. The horizontal lines in Figure 7 show the value $1/e$.

### 3.2. Phase Space Reconstruction

The essential problem in nonlinear time series analysis is to determine whether or not a given time series is a deterministic signal from a low-dimensional dynamical system. If it is, then further questions of interest are: What is the dimension of the phase space supporting the data set? Is the data set chaotic?

The key to answering these questions is embodied in the method of phase space reconstruction [30], which has been rigorously justified by the embedding theorems of Takens [31] and Sauer et al. [32,33]. Takens' embedding theorem asserts that if a time series is one component of an attractor that can be represented by a smooth $d$-dimensional manifold (with $d$ an integer) then the topological properties of the attractor (such as dimension and Lyapunov exponents) are equivalent to the topological properties of the embedding formed by the $m$-dimensional phase space vectors

$$\vec{X}_i = (y(i\Delta t),\ y(i\Delta t + \tau),\ y(i\Delta t + 2\tau), \ldots,\ y(i\Delta t + (m-1)\tau)) \tag{51}$$

whenever $m \geq 2d + 1$. In Eq. 51 $\tau$ is called the delay time and $m$ is the embedding dimension. Different choices of $\tau$ and $m$ yield different reconstructed trajectories. Takens' theorem has been generalized by Sauer et al. to the case where the attractor

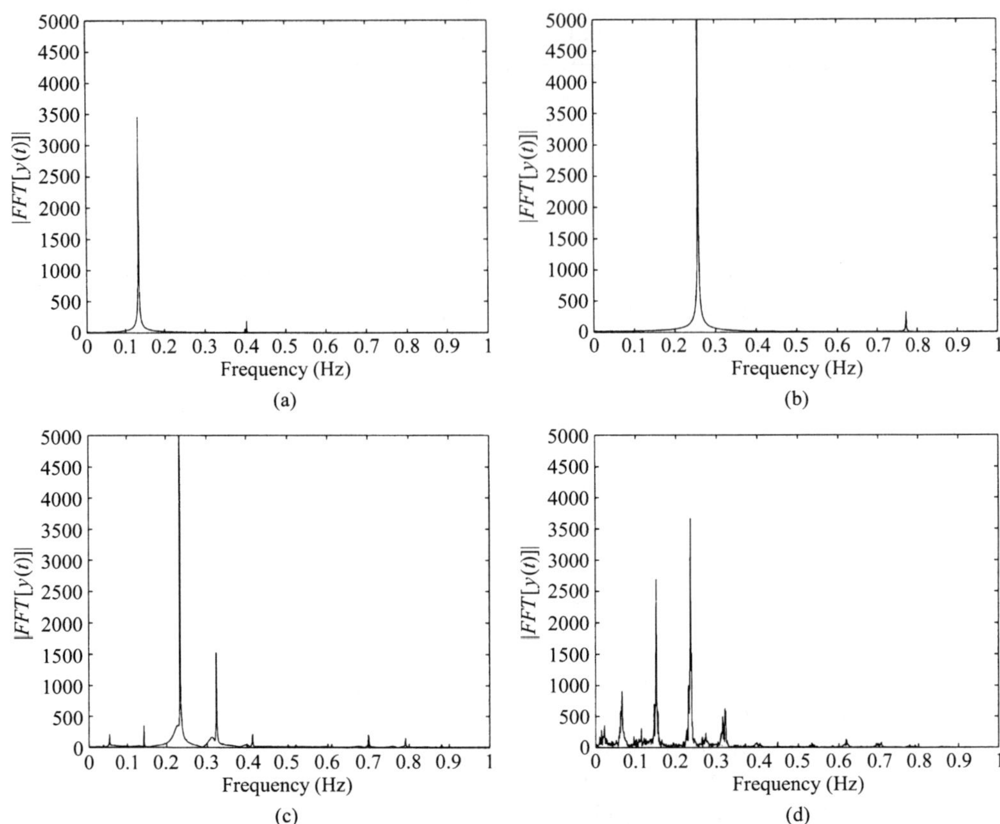


**Figure 6** Absolute value Fourier transforms for time series data from the Duffing–Van der Pol oscillator for parameters (a) $\mu = 0.0$, $f = 0.0$; (b) $\mu = 0.2$, $f = 0.0$; (c) $\mu = 0.2, f = 1.0$; $\omega = 0.90$; (d) $\mu = 0.2, f = 1.0$, $\omega = 0.94$.

is a strange attractor with a fractal dimension $D$. The embedding of a strange attractor using time delay coordinates is one to one if $m \geq 2D + 1$.

There are several technical issues to bear in mind in nonlinear time series analysis. Foremost is the quality of the time series itself. If there is a simple deterministic rule governing the evolution of the time series, then the time interval between data points should be sufficiently small, the length of the data should be sufficiently long, and the level of noise should be sufficiently low to allow detection of the deterministic dynamics and subsequent forecasting. In biomedical signals the sampling rate, the signal length, and the noise level are typically limited by technological considerations.

A MATLAB code for phase space reconstruction is listed in Appendix II.A.

The next two sections describe the problem of finding optimal values for $m$ (Section 3.2.1) and $\tau$ (Section 3.2.2). Two possible definitions of an optimal embedding are that (1) an embedding is optimal if it delivers the best possible estimates for the topological properties of the attractor and (2) an embedding is optimal if it provides the most accurate forecast of the time series.

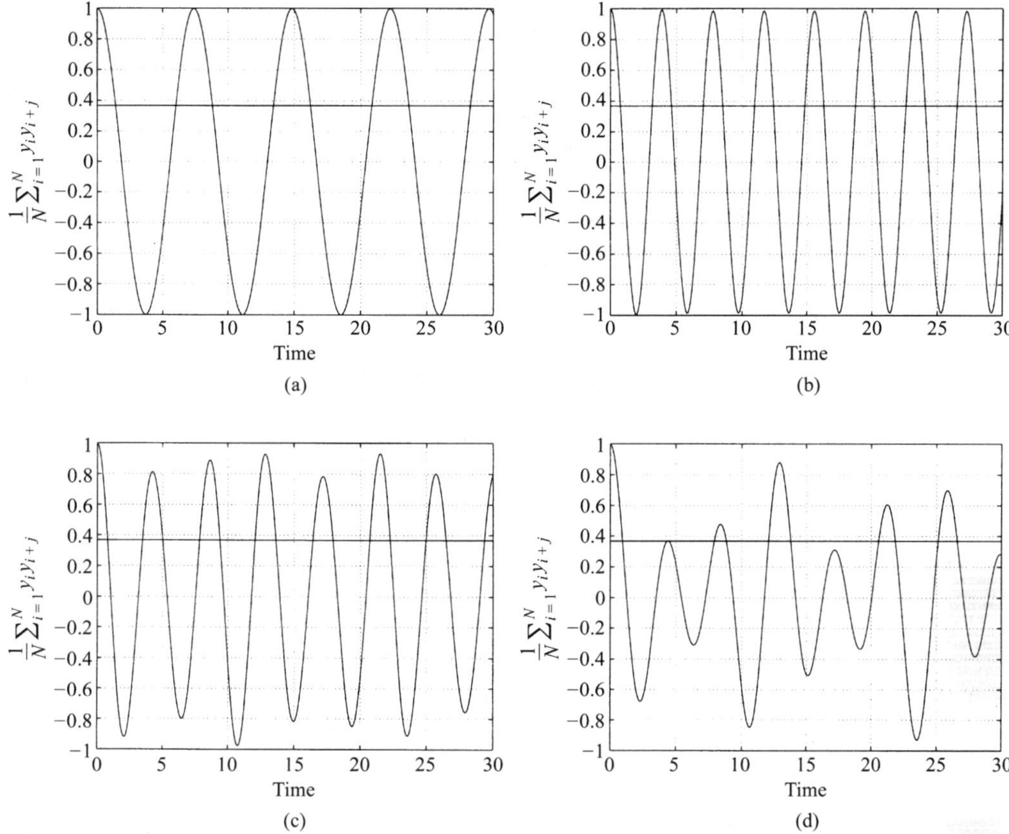


**Figure 7** Autocorrelation functions for time series data from the Duffing–Van der Pol oscillator for parameters (a) $\mu = 0.0$, $f = 0.0$; (b) $\mu = 0.2$, $f = 0.0$; (c) $\mu = 0.2, f = 1.0, \omega = 0.90$; (d) $\mu = 0.2, f = 1.0$; $\omega = 0.94$. The horizontal line is the value $1/e$.

### 3.2.1. Optimal Embedding Dimension

Partial answers to the problem of finding an optimal embedding dimension have been provided by Sauer et al. [33]. If the attractor has box-counting dimension $D_0$, then an embedding dimension of $m \geq 2D_0 + 1$ is sufficient to ensure that the reconstruction is a one-to-one embedding. The one-to-one property is in turn a necessary requirement for forecasting. If the attractor has correlation dimension $D_2$, then an embedding dimension of $m \geq D_2$ is sufficient to measure the correlation dimension from the embedding. The condition $m \geq D_2$ is also a necessary but not sufficient condition for forecasting (Section 3.3.1). A clear difficulty with these formal bounds is that the fractal dimensions $D_0$ and $D_2$ are generally unknown a priori.

In practical applications the Grassberger–Procaccia algorithm can be employed to measure the correlation dimension of reconstructions for different embedding dimensions. The minimum embedding dimension of the attractor is $m + 1$, where $m$ is the embedding dimension above which the measured value of the correlation dimension $D_2$ remains constant.

An optimal embedding dimension for forecasting can be found in the following utilitarian fashion [34,35]. Forecasts based on the first half of a time series can be constructed for a range of different embedding dimensions. These forecasts can then be compared with the actual time series data from the second half of the time series to find the best forecast (Section 3.3.1) for a given embedding dimension.

### 3.2.2. Optimal Delay Time

A one-to-one embedding can be obtained for any value of the delay time $\tau > 0$. However, very small delay times will result in near-linear reconstructions with high correlations between consecutive phase space points and very large delays might obscure the deterministic structure linking points along a single degree of freedom. If the delay time is commensurate with a characteristic time in the underlying dynamics, then this too may result in a distorted reconstruction. The optimal delay time for forecasting can be determined using the approach of Holton and May described in the previous section for finding an optimal embedding dimension.

There have been various proposals for choosing an optimal delay time for topological properties based on the behavior of the autocorrelation function. These include the earliest time $\tau$ at which the autocorrelation drops to a fraction of its initial value [36] or has a point of inflcction [37]. These definitions seek to find times where linear correlations between different points in the time series are negligible, but they do not rule out the possibility of more general correlations.

Fraser and Swinney [38] argue that a better value for $\tau$ is the value that corresponds to the first local minimum of the mutual information where the mutual information is a measure of how much information can be predicted about one time series point given full information about the other. Liebert and Schuster [39] have shown that the values of $\tau$ at which the mutual information has a local minimum are equivalent to the values of $\tau$ at which the logarithm of the correlation sum (Eq. 33) has a local minimum. In seeking a local minimum of $C(r, \tau)$ as a function of $\tau$ it is necessary to fix $r$. Liebert and Schuster suggest employing the smallest value of $r$ where $C(r, \tau)$ scales as $r^{-D}$.

Some authors have suggested that it is more appropriate to define an optimal embedding window $\tau(m-1)$ rather than optimal values for $m$ and $\tau$ separately [40–42].

It is not clear which method if any is superior for all topological properties. However, optimal values based on the behavior of the autocorrelation function are the easiest to compute.

Figure 8a–d show phase space reconstructions using a time delay of $\tau = 1$ for the Duffing–Van der Pol time series with parameters as in Figure 5a–d, respectively. These phase space reconstructions compare favorably with the original phase space portraits in Figure 1a–d. The time delay $\tau = 1$ is optimal in the sense that these are the times at which the autocorrelation function has first decreased to $1/e$ of its original value (see Figure 7). Also shown in Figure 8 are reconstructions of the chaotic time series (Figure 5d) with nonoptimal choices of the time delay. In Figure 8e, $\tau = 0.1$ and the reconstruction is stretched along the diagonal of the embedding, whereas in Figure 8f, $\tau = 20$, and the reconstruction appears to fill a region of the embedding space.

Figure 9 shows a plot of the fractal dimension, measured using the Grassberger–Procaccia algorithm, versus the embedding dimension for the reconstructed phase portrait shown in Figure 8d. The fractal dimension saturates at about $D \approx 2.5$. This

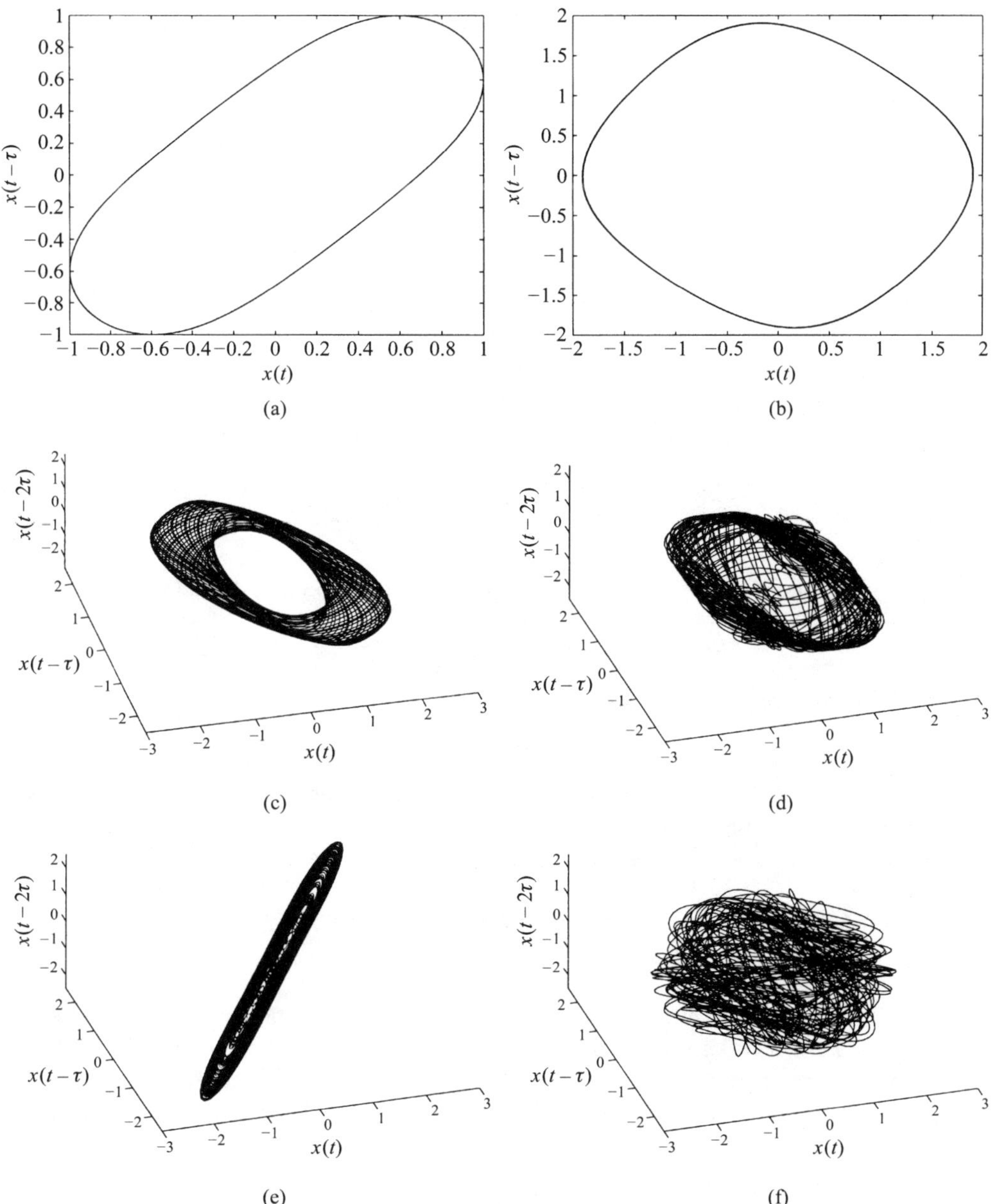


**Figure 8** Phase space reconstructions (a)–(d) with embedding dimension three and $\tau = 1$ for time series data from the Duffing–Van der Pol oscillator shown in Figure 5a–d, respectively. Reconstructions are also shown for the time series in Figure 5d with (e) $\tau = .1$ and (f) $\tau = 20$.

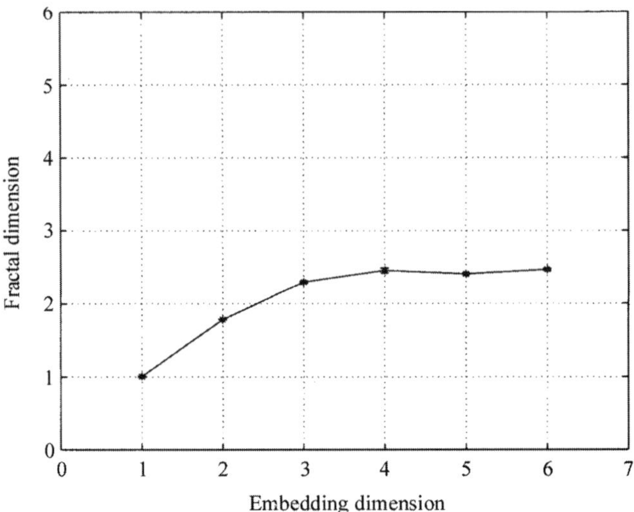


**Figure 9** Plot of the fractal dimension versus the embedding dimension for phase space reconstructions of the time series data from the Duffing–Van der Pol oscillator with parameters $\mu = 0.2$, $f = 1.0$, $\omega = 0.94$. The time delay was set to $\tau = 1$ in the reconstructions.

suggests that the minimum embedding for the time series in Figure 5d is three. The value of the fractal dimension based on the time series for just one of the system's dynamical variables is also in good agreement with the measured value of the fractal dimension using all three dynamical variables (Section 1.5).

#### 3.2.3. Measuring Lyapunov Exponents from Time Series

The first algorithms for calculating Lyapunov exponents from a time series were proposed independently by Wolfe et al. [18] and Sano and Sawada [43] in 1985. The first step in these methods is to construct an appropriate embedding of the experimental time series using the method of time delays described in Section 3.2. The maximum Lyapunov exponent can now be calculated as follows. Choose a reference point labeled $\vec{X}(0)$ and the "closest" (see further comments below) neighboring point labeled $\vec{X}^{(1)}(0)$ from the set of reconstructed phase space vectors and calculate

$$\|\Delta\vec{X}_0(0)\| = \|\vec{X}(0) - \vec{X}^{(1)}(0)\| \tag{52}$$

Evolve the two points $\vec{X}(0)$ and $\vec{X}^{(1)}(0)$ forward in the reconstructed phase space for a time $T_1$ and calculate the new separation distance

$$\|\Delta\vec{X}(T_1)\| = \|\vec{X}(T_1) - \vec{X}^{(1)}(T_1)\| \tag{53}$$

An approximate renormalization is now performed by finding a point $\vec{X}^{(2)}(0)$ that satisfies the dual requirement that (1) $\vec{X}^{(2)}(0)$ is a neighboring point to $\vec{X}(T_1)$ and (2)

$$\Delta\vec{X}_0(T_1) = \vec{X}(T_1) - \vec{X}^{(2)}(0) \tag{54}$$

and $\Delta\vec{X}(T_1)$ are in approximately the same direction. The two points $\vec{X}(T_1)$ and $\vec{X}^{(2)}(0)$ are now evolved for a time $T_2$ in the reconstructed phase space to calculate

$$\|\Delta\vec{X}(T_1 + T_2)\| = \|\vec{X}(T_1 + T_2) - \vec{X}^{(2)}(T_2)\| \tag{55}$$

The renormalization process of finding a neighboring point to the current point that has a similar orientation to the previous replacement point is repeated $N$ times and then the maximum Lyapunov exponent is calculated as [18]

$$\lambda = \frac{1}{\sum_{k=1}^{N} T_k} \sum_{k=1}^{N} \log \frac{\|\Delta\vec{X}\left(\sum_{j=1}^{k} T_j\right)\|}{\|\Delta\vec{X}_0(T_k)\|} \tag{56}$$

This calculation should then be averaged over several different initial starting points. In implementing the method it is important not to accept as the "closest" neighboring point a point that is temporally separated by a distance less than the delay time $\tau$. This is to avoid choosing adjacent points on the same trajectory. Thus the times $T_1, \ldots, T_N$ should be greater than $\tau$. On the other hand, these times should be small enough to obtain exponential separations.

If the system is ergodic so that the reconstruction phase space attractor is sampled uniformly over time, then the numerical renormalization process can be avoided by following the short-time evolutions of neighboring pairs of points on the attractor and estimating the maximum Lyapunov exponent from [43,44]

$$\lambda(i) = \frac{1}{i\Delta t}\frac{1}{M-i} \sum_{j=1}^{M-i} \log \frac{d_j(i)}{d_j(0)} \tag{57}$$

where $d_j(i)$ is the separation distance between the $j$th pair of "nearest" neighbors after $i$ discrete time steps.

### 3.3. Reliability Checks

#### 3.3.1. Forecasting

Holton and May [35] argue that prediction is the sine qua non of determinism and hence the reliability of forecasts should be a fundamental tool for discriminating between deterministic chaos and noise-induced randomness. The following algorithm for predicting time series essentially follows Farmer and Sidorowich [45]. Consider a time series

$$y(\Delta t), y(2\Delta t), \ldots, y(n\Delta t) \tag{58}$$

where $\Delta t$ is the sampling interval. The aim is to predict $y(n\Delta t + T)$ for some small time $T$. The first step is to embed the time series to obtain the reconstruction

$$\overrightarrow{X}(i\Delta t) = (y(i\Delta t), y(i\Delta t - \tau), \ldots, y(i\Delta t - (m-1)\tau)) \qquad i = n, n-1, \ldots, n-n^* \quad (59)$$

where $n^* = n - (m-1)\tau$ is the number of reconstructed vectors for a time series of length $n$. Note that the reconstruction slides backward through the data set here. The next step is to measure the separation distance between the vector $\overrightarrow{X}(n\Delta t)$ and the other reconstructed vectors and thus to order the reconstructed vectors $\overrightarrow{X}^{(1)}, \overrightarrow{X}^{(2)}, \ldots \overrightarrow{X}^{(n^*-1)}$ so that the separation distance is from smallest to largest i.e.,

$$\|\overrightarrow{X} - \overrightarrow{X}^{(1)}\| \leq \|\overrightarrow{X} - \overrightarrow{X}^{(2)}\| \leq \cdots \leq \|\overrightarrow{X} - \overrightarrow{X}^{(n^*-1)}\| \quad (60)$$

The metric $\|.\|$ is the usual Euclidean metric. Since the $\overrightarrow{X}^{(i)}$ are ordered with respect to $\overrightarrow{X}(n\Delta t)$, they may be written as $\overrightarrow{X}^{(i)}(n\Delta t)$. The next step is to map the $k \geq m+1$ nearest neighbors of $\overrightarrow{X}(n\Delta t)$ forward in the reconstructed phase space for a time $T$. These evolved points are $\overrightarrow{X}^{(i)}(n\Delta t + T)$. Suppose that the components of these vectors are as follows:

$$\overrightarrow{X}^{(i)}(n\Delta t + T) = \left(x_1^{(i)}(n\Delta t + T), x_2^{(i)}(n\Delta t + T), \ldots, x_m^{(i)}(n\Delta t + T)\right) \quad (61)$$

Now assume a local linear approximation and fit an affine model of the form

$$x_1^{(1)}(n + \Delta t + T) = a_0 + a_1 x_1^{(1)}(n\Delta t) + \cdots + a_m x_m^{(1)}(n\Delta t) \quad (62)$$

$$\vdots$$

$$x_1^{(k)}(n + \Delta t + T) = a_0 + a_1 x_1^{(k)}(n\Delta t) + \cdots + a_m x_m^{(k)}(n\Delta t) \quad (63)$$

The unknown coefficients $a_j$ can be solved using a least-squares method. Finally, the coefficients $a_j$ can be used to construct the prediction

$$y(n\Delta t + T) = x_1(n\Delta t + T) = a_0 + a_1 x_1(n\Delta t) + \cdots + a_m x_m(n\Delta t) \quad (64)$$

The system of equations, Eqs. 62, 63, may be inconsistent or incomplete, in which case the forecast is ignored or $k$ is increased.

Holton and May [35] use the reliability of forecasts to determine optimal values for the delay time and the embedding dimension. The first step in this procedure is to reconstruct the attractor with a delay time $\tau$ and an embedding dimension $m$. Forecasts are then made over times $t$ for $N$ different starting points using the first half of the time series data. The correlation coefficient

$$\rho(\tau, m; t) = \frac{< (x(k,t) - < x(k,t) >)(y(k,t) - < y(k,t) >) >}{\sqrt{< (x(k,t) - < x(k,t) >)^2 >}\sqrt{< (y(k,t) - < y(k,t) >)^2 >}} \quad (65)$$

is then computed as a function of the forecast time $t$. In Eq. 65, $x(k, t)$ denotes the first component of the evolved vector $x$ in the embedding for the $k$th forecast after time $t$, $y(k, t)$ denotes the actual data value corresponding to the $k$th forecast after time $t$, and the angular brackets $< >$ denote an average over the $k$ forecasts.

A MATLAB code for forecasting a time series is listed in Appendix II.B.

### 3.3.2. Surrogate Data

The method of using surrogate data in nonlinear time series analysis was introduced by Theiler et al. in 1992 [46]. This section contains a brief sketch of some of the ideas and methods in that reference. The starting point is to create an ensemble of random nondeterministic surrogate data sets that have the same mean, variance, and power spectrum as the experimental time series data of interest. The measured topological properties of the experimental time series are then compared with the measured topological properties of the surrogate data sets. If both the experimental time series data and the surrogate data sets yield the same values for the topological properties (within the standard deviation measured from the surrogate data sets), then the null hypothesis that the experimental data set is random noise cannot be ruled out.

The method of calculating surrogate data sets with the same mean, variance, and power spectrum but otherwise random is as follows: First construct the Fourier transform of the experimental time series data, then randomize the phases, then take the inverse Fourier transform. An explicit algorithm for achieving this is as follows [47]:

1. Input the experimental time series data $x(t_j)$, $j = 1, \dots N$ into a complex array

$$z(n) = x(n) + iy(n), \qquad n = 1, \dots N \tag{66}$$

   where $x(n) = x(t_n)$ and $y(n) = 0$.
2. Construct the discrete Fourier transform

$$Z(m) = X(m) + iY(m) = \frac{1}{N} \sum_{n=1}^{N} z_n e^{-2\pi i(m-1)(n-1)/N} \tag{67}$$

3. Construct a set of random phases

$$\phi_m \in [0, \pi], \qquad m = 2, 3, \dots \frac{N}{2} \tag{68}$$

4. Apply the randomized phases to the Fourier transformed data

$$Z(m)' = \begin{cases} Z(m) & \text{for } m = 1 \text{ and } m = \frac{N}{2} + 1 \\ |Z(m)|e^{i\phi_m} & \text{for } m = 2, 3, \dots, \frac{N}{2} \\ |Z(N-m+2)|e^{-i\phi_{N-m+2}} & \text{for } m = \frac{N}{2} + 2, \ \frac{N}{2} + 3, \dots, N \end{cases} \tag{69}$$

5. Construct the inverse Fourier transform of $Z(m)'$

$$z(n)' = x(n)' + iy(n)' = \frac{1}{N} \sum_{m=1}^{N} Z'_m e^{2\pi i(m-1)(n-1)/N} \tag{70}$$

A MATLAB code for creating surrogate data using the preceding algorithm is listed in Appendix II.C. Figure 10a shows a surrogate data time series sharing the same spectral properties as the time series data for the Duffing–Van der Pol oscillator with parameters as in Figure 5d. A phase space reconstruction for the surrogate data time series using an embedding dimension $m = 3$ and a time delay $\tau = 1$ is shown in Figure 10b.

The phase space portrait for this surrogate data set appears to be more space filling (noisy) than the phase space portrait for the original time series (Figure 8d). Forecasts based on the original time series data and the surrogate time series data using phase space reconstructions with $m = 3$ and $\tau = 1$ in each case are compared in Figure 11a and b, respectively. In Figure 11c the correlation coefficient is computed for 100 forecasts for both the original time series data (solid line only) and the surrogate data (line with crosses). From Figure 11 it is clear that the forecast for the original time series is clearly superior. This is consistent with the apparent randomness in the original time series being due to nonlinear dynamics rather than noise.

## 4. DISCUSSION

This chapter provided a tutorial-style introduction to the problem of detecting, analyzing and forecasting low-dimensional deterministic chaos in experimental time series. The chapter also contains a set of MATLAB programs for this type of analysis. The detection of deterministic chaos may be considered as a first but very important step

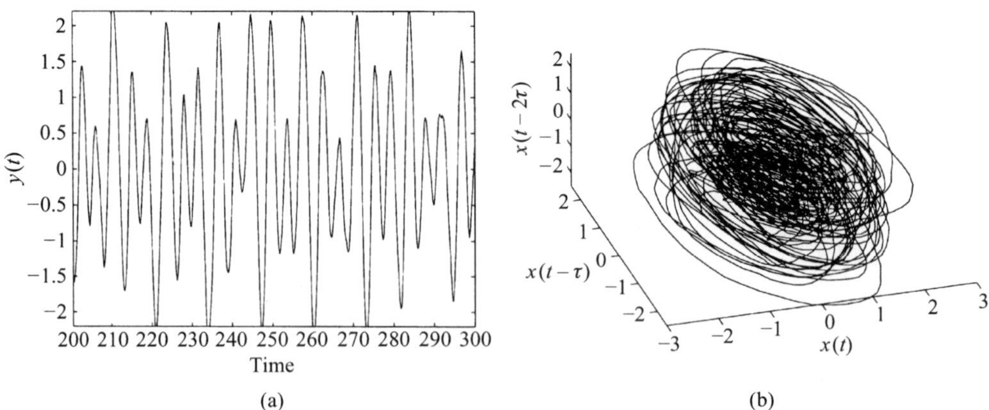


**Figure 10** Surrogate data time series (a) and phase space reconstruction (b) with $m = 3$ and $\tau = 1$. The surrogate data have the same spectral properties as the time series data from the Duffing–Van der Pol oscillator for parameters $\mu = 0.2, f = 1.0, \omega = 0.94$.

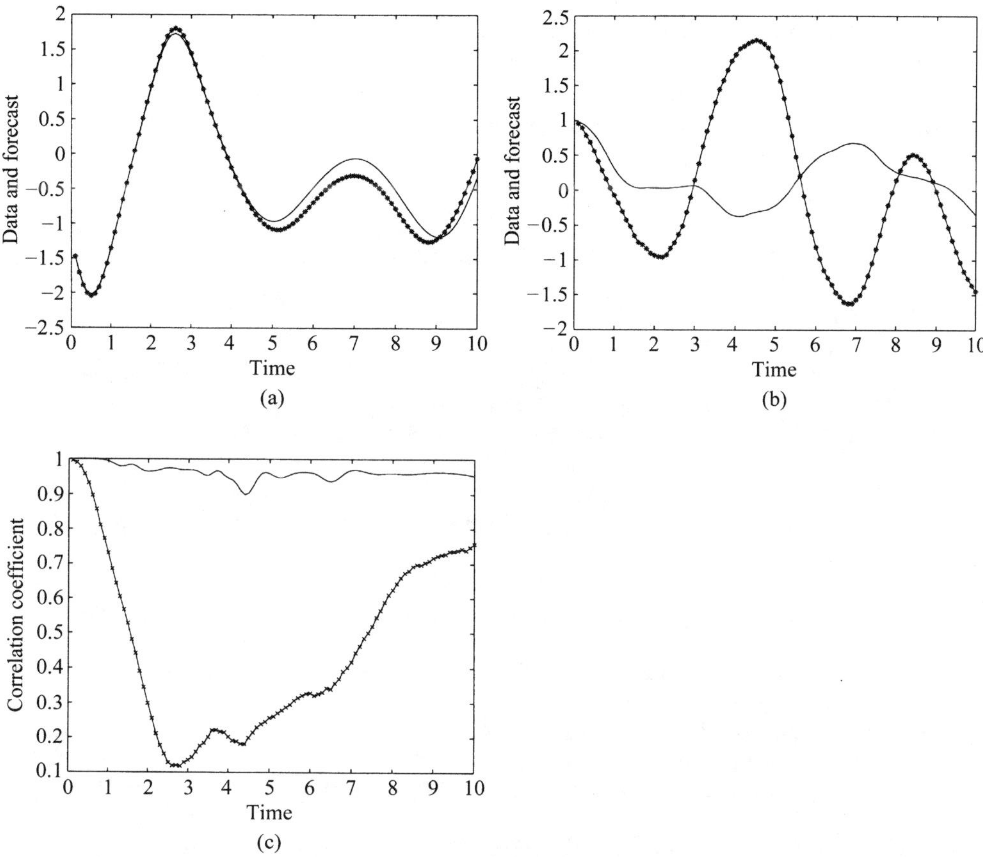


**Figure 11** Comparison between forecasts (*) and actual data (solid lines) for (a) time series data from the Duffing–Van der Pol oscillator for parameters $\mu = 0.2, f = 1.0, \omega = 0.94$ and (b) surrogate data time series sharing the same spectral properties. The correlation coefficient based on 100 such forecasts is shown in (c) for the original time series data (solid line only) and the surrogate data (line with crosses).

[48] in a more ambitious program aimed at modeling [49] and ultimately controlling [50] time series data. The methods, algorithms, and computer programs in this chapter constitute a tool box for nonlinear time series analysis in much the same way that standard statistical packages are part of the trade of social scientists. On the other hand, the tool box should not be treated as a black box that can be applied indiscriminately. Some of the questions to address in this process include: Is the time interval between data points in the experimental data set sufficiently small and is the data set sufficiently free from noise to retain deterministic structure if it exists? Has the experimental signal been filtered in any way? Is the time series sufficiently long to permit a reliable reconstruction of the full phase space trajectory? Is the scaling regime in fractal dimension measurements unambiguous? Is the convergence of the fractal dimension of the reconstructed trajectory unambiguous? Is the measured maximum Lyapunov exponent homogeneous over the reconstructed trajectory? If it has a positive value, is this

value significantly different from zero? Can the irregularities in the experimental data be accounted for equally well using linear statistical analysis?

With careful attention to these and other questions, nonlinear time series analysis will provide a valuable adjunct to linear statistical analysis of apparently random-looking biomedical data.

## REFERENCES

[1] H. Tong, A personal overview of non-linear time series analysis from a chaos perspective. *Scand. J. Stat.* 22:399–445, 1995.

[2] T. Mullin, A dynamical systems approach to time series analysis. In *The Nature of Chaos*, T. Mullin, (ed.), Chap. 2. Oxford: Oxford University Press, 1993.

[3] V. Isham, Statistical aspects of chaos: a review. In *Networks and Chaos—Statistical and Probabilistic Aspects*, O. E. Barndorff-Nielsen, J. L. Jensen, and W. S. Kendall, (eds.), Chap. 3. London: Chapman & Hall, 1993.

[4] M. Casdagli, Chaos and deterministic versus stochastic non-linear modelling. *J. R. Stat. Soc. B.* 54:303–328, 1992.

[5] N. Gershenfeld, An experimentalist's introduction to the observation of dynamical systems. In *Directions in Chaos*, Vol. 2, Hao Bai-Lin, (ed.), pp. 310–383. Singapore: World Scientific, 1988.

[6] M. R. Guevara, L. Glass, and A. Shrier, Phase locking, period doubling bifurcations, and irregular dynamics in periodically stimulated cardiac cells. *Science* 214:1350–1352, 1981.

[7] D. R. Chialvo, D. C. Michaels, and J. Jalife, Supernormal excitability as a mechanism of chaotic dynamics of activation in cardiac Purkinje fibers. *Circ. Res.* 66:525–545, 1990.

[8] K. M. Stein, N. Lippman, and P. Kligfield, Fractal rhythms of the heart. *J. Electrocardiol.* 24(Suppl.):72–76, 1992.

[9] Z. S. Huang, G. L. Gebber, S. Zhong, and S. Barman, Forced oscillations in sympathetic nerve discharge. *Am. J. Physiol.* 263:R564–R571, 1992.

[10] K. P. Yip, N. H. Holstein-Rathlou, and D. J. Marsh, Chaos in blood flow control in genetic and renovascular hypertensive rats. *Am. J. Physiol.* 261:F400–F408, 1991.

[11] C. D. Wagner, R. Mrowka, B. Nafz, and P. B. Persson, Complexity and "chaos" in blood pressure after baroreceptor denervation of conscious dogs. *Am. J. Physiol.* 269:H1760–H1766, 1996.

[12] C. A. Skarda and W. J. Freeman, How brains make chaos in order to make sense of the world. *Behav. Brain Sci.* 10:161–195, 1987.

[13] D. Hoyer, K. Schmidt, R. Bauer, U. Zwiener, M. Kohler, B. Luthke, and M. Eiselt, Nonlinear analysis of heart rate and respiratory dynamics. *IEEE Eng. Med. Biol. Mag.* 16(1): 31–39, 1997.

[14] W. Szemplinska-Stupnicka and J. Rudowski, Neimark bifurcation, almost periodicity and chaos in the forced van der Pol–Duffing system in the neighbourhood of the principal resonance. *Phys. Lett. A* 192:201–206, 1994.

[15] J. Ford, How random is a coin toss? *Phys. Today* 36(4):40, 1983.

[16] V. I. Oseledec, A multiplicative ergodic theorem. *Trans. Moscow. Math. Soc.* 19:197–231, 1968.

[17] G. Benettin, L. Galgani, and J. M. Strelcyn, Kolmogorov entropy and numerical experiments. *Phys. Rev. A* 14:2338–2345, 1976.

[18] A. Wolfe, J. B. Swift, H. L. Swinney, and J. A. Vastano, Determining Lyapunov exponents from a time series. *Physica D* 16:285–317, 1985.

[19] H. Fujisaka, Multiperiodic flows, chaos and Lyapunov exponents. *Prog. Theor. Phys.* 68:1105–1119, 1982.

[20] M. Casartelli, E. Diana, L. Galgani, and A. Scotti, Numerical computations on a stochastic parameter related to Kolmogorov entropy. *Phys. Rev. A* 13:1921–1925, 1976.

[21] I. Shimada and T. Nagashima, A numerical approach to ergodic problem of dissipative systems. *Prog. Theor. Phys.* 61:1605–1613, 1979.

[22] G. Benettin, L. Galgani, A. Giorgilli, and J.-M. Strelcyn, Lyapunov characteristic exponents for smooth dynamical systems: A method for computing all of them. *Meccanica* 15:9–19, 1980.

[23] B. Mandelbrot, *The Fractal Geometry of Nature*. New York: Freeman, 1982.

[24] H. G. E. Hentschel and I. Procaccia, The infinite number of generalized dimensions of fractals and strange attractors. *Physica D* 8:435–444, 1983.

[25] J. Kaplan and J. Yorke, Functional differential equations and the approximation of fixed points. In *Lecture Notes in Mathematics*, No. 730. Berlin: Springer-Verlag, 1979.

[26] P. Grassberger and I. Procaccia, Characterization of strange attractors. *Phys. Rev. Lett.* 50:346–349, 1983.

[27] D. Ruelle, Deterministic chaos: The science and the fiction. *Proc. R. Soc. Lond. A* 427:241–248, 1990.

[28] C. Essex and M. A. H. Nerenberg, Comments on "Deterministic chaos: the science and the fiction" by D. Ruelle. *Proc. R. Soc. Lond. A* 435:287–292, 1991.

[29] S.-Z. Hong and S.-M. Hong, An amendment to the fundamental limit on dimension calculations. *Fractals* 2:123–125, 1994.

[30] N. H. Packard, J. P. Crutchfield, J. D. Farmer, and R. S. Shaw, Geometry from a time series. *Phys. Rev. Lett.* 45:712–716, 1980.

[31] F. Takens, Detecting strange attractors in turbulence. In *Dynamical Systems and Turbulence*, D. A. Rand and L. S. Young, eds. Berlin: Springer, 1981.

[32] T. Sauer and J. A. Yorke, Rigorous verification of trajectories for computer simulation of dynamical systems. *Nonlinearity* 4:961–979, 1991.

[33] T. Sauer, J. Yorke, and M. Casdagli, Embedology. *J. Stat. Phys.* 65:579–616, 1994.

[34] G. Sugihara and R. M. May, Nonlinear forecasting as a way of distinguishing chaos from measurement error in time series. *Nature* 344:734–741, 1990.

[35] D. Holton and R. M. May, Distinguishing chaos from noise. In *The Nature of Chaos*, Chap. 7. Oxford: Oxford University Press, 1993.

[36] A. M. Albano, J. Muench, C. Schwartz, A. I. Mees, and P. E. Rapp, Singular-value decomposition and the Grassberger–Procaccia algorithm. *Phys. Rev. A* 38:3017, 1988.

[37] G. P. King, R. Jones, and D. S. Broomhead, Phase portraits from a time series: A singular system approach. *Nucl. Phys. B* 2:379, 1987.

[38] A. M. Fraser and H. Swinney, Independent coordinates for strange attractors from mutual information. *Phys. Rev. A* 33:1134–1139, 1986.

[39] W. Liebert and H. G. Schuster, Proper choice of the time delay for the analysis of chaotic time series. *Phys. Lett. A* 142:107, 1989.

[40] A. M. Albano, A. Passamante, and M. E. Farrell, Using higher-order correlations to define an embedding window. *Physica D* 54:85, 1991.

[41] J. M. Martinerie, A. M. Albano, A. I. Mees, and P. E. Rapp, Mutual information, strange attractors, and the optimal estimation of dimension. *Phys. Rev. A.* 45:7058, 1992.

[42] M. T. Rosenstein, J. J. Collins, and C. J. de Luca, Reconstruction expansion as a geometry-based framework for choosing proper delay times. *Physica D* 73:82–98, 1994.

[43] M. Sano and Y. Sawada, Measurement of the Lyapunov spectrum from a chaotic time series. *Phys. Rev. Lett.* 55:1082–1085, 1985.

[44] M. T. Rosenstein, J. J. Collins, and C. J. de Luca, A practical method for calculating largest Lyapunov exponents from small data sets. *Physica D* 65:117–134, 1994.

[45] J. D. Farmer and J. J. Sidorowich, Predicting chaotic times series, *Phys. Rev. Lett.* 59: 845–848, 1987.

[46] J. Theiler, B. Galdrikian, A. Longtin, S. Eubank, and J. D. Farmer, Using surrogate data to detect nonlinearity in time series. In *Nonlinear Modeling and Forecasting*, M. Casdagli and S. Eubank, eds. New York: Addison-Wesley, 1992.

[47] J. Mazaraki, Dynamical methods for analysing and forecasting chaotic data. Honours thesis, Applied Mathematics, University of New South Wales, 1997.

[48] D. Ruelle, Where can one hope to profitably apply the ideas of chaos? *Phys. Today* July: 24–30, 1994.

[49] B. R. Noack, F. Ohle, and H. Eckelmann, Construction and analysis of differential equations from experimental time series of oscillatory systems. *Physica D* 56:389–405, 1992.

[50] E. Ott and M. Spano, Controlling chaos. *Phys. Today* May: 34–40, 1995.

## APPENDIX

# I. Dynamical Systems Analysis—MATLAB Programs

### A. Numerical Integration of Three Coupled ODEs

```
% File: VanDerPolSolv.m
%
% Numerical integration of Duffing-Van Der Pol Oscillator.
% Ordinary differential equation solver using in-built
% Matlab function (ode45).
%
% Uses: VanDerPol.m
% Set initial conditions for the three differential equations
x0=[1;0;0];
% Set integration time
timePoints=5000:0.3:7500;
% Solve equations
options=odeset('RelTol',1e-10,'AbsTol',1e-12);
[t,x]=ode45('VanDerPol',timePoints,x0);

function xp-VanDerPol(t,x)
%
% Duffing-Van Der Pol equations expressed as a first order system
% (see equations 3-5). Parameter values (mu, f and omega)
% can be chosen to demonstrate various trajectories (see text).
%
% IN:
% t: time (not used, but necessary for ODE solver)
% x: Input vector
%
% OUT:
% xp: dx/dt
% Current parameters demonstrate chaotic behavior of the oscillator.
mu=0.2; f=1.0; omega=0.94;
% define equations
xp(1)=x(2);
xp(2)=mu*(1-x(1)^2)*x(2)-x(1)^3+f*cos(x(3));
xp(3)=omega;
% transpose into column vector for ODE solver
xp=xp';
```

### B. Three-Dimensional Phase Space Plots

```
% File: PSplot.m
%
% Three-dimensional phase space plot of three coupled ODEs
% (data assumed to be stored in matrix 'x' from VanDerPolSolve.m).
%
% Transform x3 to sin(x3)
xnew=x;
xnew(:,3)=sin(x(:,3));
% Generate 3D plot
plot3(xnew(:,1),xnew(:,2),xnew(:,3));
xlabel('x1');
ylabel('x2');
zlabel('sin(x3)')
title('Duffing-Van der Pol oscillator.');
rotate3d on
view([-5,58]);
```

### C. Two-Dimensional Surface of Section

```
% File: SurfaceOfSection.m
%
% Two dimension Poincare surface of section on plane x3=0
% (data assumed to be stored in matrix 'xnew' generated from PSplot.m).
%
% Start with empty surface
clear Surface
OldDistance=0;
k=1;
for i=1:size(xnew)*[1;0]
     NewDistance=xnew(i,3);
     if (NewDistance>=0 * OldDistance<0)
     % Add new point to the surface
     TotalDistance=NewDistance-OldDistance;
     Surface(k,:)=xnew(i-1,:)-(OldDistance/Total Distance)*...
        (xnew(i,:)-xnew(i-1,:));
     k=k+1;
   end
   OldDistance=NewDistance;
end
% Generate 2D plot
plot(Surface(:,1),Surface(:,2),'*');
xlabel('x1');
ylabel('x2');
title('Poincare Surface of Section.');
```

### D. Lyapunov Exponents for Three Coupled ODEs

```
% File: LyapunovSolver.m
%
% Calculate all the Lyapunov exponents for the Duffing-Van der Pol oscillator.
%
% Uses: IntegrateVDPSystem.m, GramSchmidt.m
% Step 1: Construct a unit hypersphere (Dimension=Dim; axis at time 0:
% a(1)=[1;0;0...;0], a(2)=[0;1;0...;0],...,
a(n)=[0;0;0...;1]
% centered on an initial point (x(1,:))
% n=number of iterations to average
n=5000;
% tau
tau=0.1;
Sigma=zeros(n,3);
% Initial conditions
x0=[1;0;0];
a0=eye(3);
x=zeros(3,n);
% j=1 (Iteration variable)
for j=1:n
   % Step2: Integrate the nonlinear equation of motion over a
   % characteristic time scale tau to obtain x(j*tau).
   % Integrate the variational equations to obtain the evolved
   % axis vectors (a1,a2,...,an) for time tau.
   [xpoint,a]=IntegrateVDPSystem(x0,a0,tau);
   % x: numerical solution of the system {NumberOfPoints.Dim}
   x(:,j)=xpoint;
   % Step3: Apply Gram-Schmidt re-orthonormalization to the axis vectors.
   [aUnit]=GramSchmidt(a);
   %Step4: Compute Sigma where
   % Sigma(j,1)=log(norm(a(1)),
   Sigma(j,2)=log(norm(a(2)),...,
```

```
   % Sigma(j,n)=log(norm(a(n))
   for i=1:3
      aOrth(:,i)=dot(a(:,i),aUnit(:,i))*aUnit(:,i);
   end
   for i=1:3
      Sigma(j,i)=log(norm(aOrth(:,i),2));
   end
   x0=xpoint;
   a0=aUnit;
end
% Step 5: Compute the Lyapunov exponents
Lyapunov=(1/(n*tau))*sum(Sigma,1);


function [x,a]=IntegrateVDPSystem(x0,a0,tau)
%
% This function integrates the Duffing-Van der Pol equations
% and the variational equations returning the final result at time tau
% for initial conditions x0 and a0 (3 vectors associated with variational
% equations).
%
% IN:
% x0: Vector {3} of initial conditions
% a0: Matrix {3,3} with initial vectors associated with variational equations.
% tau: Final time.
%
% OUT:
% x: Vector {3} with solution at time tau
% a: Matrix {3,3} with solution of 3 vectors at time tau.
%
% Uses: VanDerPolVarEqu.m
s0=zeros(12,1);
s0=x0;
for i=1:3
   s0(i*3+1:i*3+3)=a0(:,i);
end
t0=0
tfinal=t0+tau;
options=odeset('RelTol', 1e-10,'AbsTol',1e-12);
[t,s]=ode45('VanDerPolVarEqu',[t0 tfinal], s0);
DataLength=size(s)*[1;0];
x=s(DataLength,1:3)';
for i=1:3
   a(:,i)=s(DataLength,i*3+1:i*3+3)';
end

function sp=VanDerPolVarEqu(t,s)
%
% Duffing-Van Der Pol equations expressed as a first order system
% (see equations 3-5). Parameter values (mu, f and omega)
% can be chosen to demonstrate various trajectories (see text).
% Includes Variational Equations necessary to solve Jacobian sp(4:12).
%
% Current parameters demonstrate chaotic behavior of the oscillator.
%
% IN:
% t: time (not used, but necessary for ODE solver)
% s: Input Vector
%
% OUT:
% sp: ds/dt
%
% Dimension of the system
n=3;
% Initial parameters describing chaotic behavior
mu=0.2; f=0.0; omega=1.0;
```

```
% define equations
sp(1)=s(2);
sp(2)=mu*(1-s(1)^2)*s(2)-s(1)^3+f*cos(s(3));
sp(3)=omega;
for i=1:n
  sp(n*i+1)=s(n*i+2);
  sp(n*i+2)=(-2*mu*s(1)*s(2)-3*s(1)^2)*s(n*i+1)+...
    mu*(1-s(1)^2)*s(n*i+2)+f*s(n*i+3);
  sp(n*i+3)=0;
end
% transpose into column vector for ODE solver
sp=sp';

function [ReNormMatrix]=GramSchmidt(Matrix)
%
% Compute a set of normal-orthogonal vectors using the Gram-Schmidt algorithm.
% Matrix contains vector Matrix(:,1),Matrix(:,2),..., Matrix(:,3)
%
% IN:
% Matrix: Matrix containing original vectors.
%
% OUT:
% ReNormMatrix: Matrix with the renormalized set of vectors
Dim=size(Matrix)*[1;0];
ReNormMatrix=zeros(size(Matrix));
ReNormMatrix(:,1)=Matrix(:,1)/norm(Matrix(:,1),2);
for j=2:Dim
  z=Matrix(:,j);
  for i=j-1:-1:1
    z=z-dot(Matrix(:,j),ReNormMatrix(:,i))*ReNormMatrix(:,i);
  end
  ReNormMatrix(:,j)=z/norm(z,2);
end
```

### E. Grassberger–Procaccia Algorithm

```
% File: CorrDim.m
%
% Implements the Grassberger-Procaccia algorithm to measure the correlation
% dimension (data assumed to be stored in matrix 'x' from VanDerPolSolve.m).
%
% Uses: Distance.m, BinFilling.m, Slope.m
% Transform x3 to sin(x3)
xnew=x;
xnew(:,3)=sin(x(:,3));
% Number of valid points in xnew matrix.
NoPoints=[size(xnew)*[1;0] size(xnew)*[1;0] size(xnew)*[1;0]];
% Calculates 32 equi-spaced bins on a logarithmic scale
[Radius]=Distance(xnew, NoPoints);
% Fills bins with the number of pairs of points with separation given by Radius.
[BinCount]=BinFilling(xnew,NoPoints,Radius);
MaxEmDim=3;
% Normalizes the matrix Radius, by its maximum value
% (for each dimension independently).
for i=1:MaxEmDim
  max=Radius(32,i);
  RadiusNormal(:,i)=Radius(:,i)/max;
end
% Plots the BinCount for specific Radius for all Embedding Dimensions.
figure(1);
for i-1:MaxEmDim
  if i==1
      hold off
  end
  loglog(RadiusNormal(:,i),BinCount(:,i),'+-');
```

```
   if i==1
      hold on
   end
end
% Plot correlation Integral
ok=1;
title(strcat('Correlation Integral'));
xlabel(strcat('% of Maximum Radius, MaxEmDim=',num2str (MaxEmDim),')'));
ylabel('% of Number of Points');
% Find the slope for all the dimensions.
for i=1:MaxEmDim
   [Slopes(i), SlopeErrors(i)]=Slope(Radius(:,i),
   BinCount(:,i), .6 .125);
end
% Plot Fractal Dimensions
figure(2)
Dim=1:MaxEmDim;
errorbar(Dim,Slopes,SlopeErrors, 'b*-');
axis([0 MaxEmDim+1 0 MaxEmDim]);
grid on
zoom on;
title(strcat('Data Set: ','Duffing VanDerPol Oscillator'));
xlabel('Dimension');
ylabel('Fractal Dimension');

function [Radius]=Distance(Portrait, NoPoints);
%
% IN:
% Portrait: Is the matrix {Number of data points, MaxEmDim} in which
%    the trajectories are contained.
% NoPoints: Is the vector {1,MaxEmDim} in which the number of valid
%    points for each dimension is contained.
%
% OUT:
% Radius: Is the matrix {32,MaxEmDim} in which the difference between
%    the maximum and minimum distances from one point to any other, is divided
%    into 32 logarithmically equal intervals (for all dimensions).
%
% Uses: DistVectPoint.m, ElimZero.m
MaxEmDim=size(Portrait)*[0;1];
NoPointsEnd=[NoPoints 1];
MinDist=ones(1,MaxEmDim)*1e20;
MaxDist=zeros(1,MaxEmDim);
Radius=zeros(32,MaxEmDim);
for EmDim=1:MaxEmDim
   minval=zeros(1,EmDim);
   minloc=zeros(1,EmDim);
   maxval=zeros(1,EmDim);
   maxloc=zeros(1,EmDim);
   for i=NoPointsEnd(EmDim):-1:NoPointsEnd(EmDim+1)+1
      % Calculates the distances from point Portrait(i,1: EmDim) to all the
      % points in matrix Portrait (1:i-1,1:EmDim)
      Distances=DistVectPoint(Portrait(1:1-1,1:EmDim), Portrait(i,1:EmDim));
      % Eliminates all points with distance less than Tolerance=1e-10
      DistanceNoZero=ElimZero(Distances, 1e-10);
      [minval,minloc]=min(DistanceNoZero,[],1);
      [maxval,maxloc]=max(Distances,[],1);
      for j=1:EmDim;
         if MinDist(j)>minval(j)
              MinDist(j)=minval(j);
         end
         if MaxDist(j)<maxval(j)
              MaxDist(j)=maxval(j);
         end
      end
```

```
  end
end
% Fill 32 bins (equally spaced logarithmically)
for k-1:32
  Radius(k,:)exp(log(MinDist)+k*(log(MaxDist)-log(MinDist))/32);
end

function [Distance]=DistVectPoint(data,point);
%
% This function calculates the distance from all elements of the matrix
% {n,MaxDim} to the point {1,MaxDim}, for all the dimensions.
%
% IN:
% data: Is a matrix {n,MaxDim} of n points to which 'point' is going
    to be compared.
% point: Is a vector {1,MaxDim} that represents the 'point' in MaxDim
%    dimensions.
%
% OUT:
% Distance: Is a matrix {n,MaxDim} that contains the distances from
%    'point' to all other points in 'data' (for dimensions 1 to MaxDim).
%
% Example:  data=[      0           0           0
%                       0           0           1
%                       0           1           1
%                       1           0           1
%                       1           1           1 ];
%              point=[  0           0           0 ];
%         Distance=[    0           0           0
%                       0           0           1.0000
%                       0           1.0000      1.4142
%                       1.0000      1.0000      1.4142
%                       1.0000      1.4142      1.7321 ]
Diffe=zeros(size(data));
for i=1:size(data)*[0;1]
  Diffe(:,i)=data(:,i)-point(i);
end
% Calculate Euclidean distance
Diffe=Diffe.^2;
Distance=cumsum(Diffe,2);
Distance=sqrt(Distance);

function DistanceNoZero=ElimZero(Distance, Tolerance);
%
% Replaces all points with distance less than Tolerance with 1e20.
% This is necessary in the bin-filling algorithm.
%
% IN:
% Distance: Is a matrix {n,MaxDim} that contains the distances from
%    'point' to all other points in 'data' (for dimensions 1 to MaxDim).
% Tolerance: Is a scalar that determines the minimum distance to be
considered.
%
% OUT:
% DistanceNoZero: Is a matrix {n,MaxDim} equal to Distance, but with all
%    elements smaller than Tolerance replaced with 1e20
SigDist=Distance-Tolerance;
SigDist=((sign(sign(SigDist.*-1)-0.5))+1)*1e20;
DistanceNoZero=Distance+SigDist;

function [BinCount]=BinFilling(Portrait,NoPoints,Radius)
%
% IN:
% Portrait: Is the matrix {Number of data points, MaxEmDim} in which the
%    trajectories are contained.
```

```
% NoPoints: Is the vector {1,MaxEmDim} in which the number of points for each
%    dimension is contained.
% Radius: Is the matrix {32,MaxEmDim} in which the difference between the
%    maximum and minimum distances from one point to any other, is divided
into
%    32 logarithmically equal intervals (for all dimensions)
%
% OUT:
% BinCount: Is a matrix {32,MaxEmDim} with the total count of pair of points
%    with a distance smaller than that specified by Radius for the 32
intervals.
%
% Uses: DistVectPoint.m, CountPoints.m
MaxEmDim=size(Portrait)*[0;1];
BinCount=zeros(32,MaxEmDim);
NoPointsEnd=[NoPoints 1];
for EmDim=1:MaxEmDim
     for i=NoPointsEnd(EmDim):-1:NoPointsEnd(EmDim+1)+1
        Distances=zeros(i-1,EmDim);
        Distances=DistVectPoint(Portrait(1:i-1,1:EmDim),
        Portrait(i,1:EmDim));
        for j=1:32
           BinCount(j,1:EmDim)=BinCount(j,1:EmDim)+...
              CountPoints(Distances,Radius(j,1:EmDim));
        end
     end
end
BinCount=BinCount./(((ones(32,1)*NoPoints).*(ones(32,1)*NoPoints-1))/2);

function [CountVect]=CountPoints(Distances,Threshold);
%
% IN:
% Distance: Is a matrix {n,MaxDim} that contains the distances from 'point'
%    to all other points in 'data' for dimensions 1 to MaxDim.
% Threshold: Is the upper bound on Distance.
%
% OUT:
% CountVect: Is a vector {1,MaxDim] with the count of distances smaller
%    than Threshold
VectLength=length(Threshold);
NumOfPoints=size(Distances)*[1;0];
CountVect=zeros(1,VectLength);
ThresholdMatr=ones(NumOfPoints,1)*Threshold;
CountVect=sum((Distances<ThresholdMatr),1);

function [Slope, SlopeError]=Slope(RadiusV, BinCountV, center, high)
%
% This function gives the slope and error for a line (in a logarithmic scale)
% given by the points RadiusV, BinCountV. The only relevant points are the ones
% that are contained in the band center-high/2, center+high/2.
%
% The values for center define the position of the center of the band and can
% range from 0 to 1 with 1 at the top.
%
% IN:
% RadiusV: Vector with the radii limits of a specific Dimension.
% BinCountV: Vector containing the counts of pairs of elements with distance
%    smaller than radius.
% Center: Center position where the slope is to be evaluated.
% High: Band size around center.
%
% OUT:
% Slope: Slope evaluated in the region center-high/2, center+high/2.
% SlopeError: Error of the evaluated fitted line to the original data.
lnRadiusV=log(RadiusV);
lnBinCountV=log(BinCountV);
```

```
Max=0;
Min=lnBinCountV(1);
IntervalHigh=(Max-Min)*high;
Top=-((Max-Min)*(1-center))+(IntervalHigh/2);
Base=-((Max-Min)*(1-center))-(IntervalHigh/2);
k=1;
for i=1:32
     if((lnBinCountV(i)>=Base & lnBinCountV(i)<=Top))
          RelDataX(k)=lnRadiusV(i);
          RelDataY(k)=lnBinCountV(i);
          k=k+1;
     end
end
[P,S]=polyfit(RelDataX,RelDataY,1);
Slope=P(1);
SlopeError=S.normr;
```

## II. Time Series Analysis—MATLAB Programs

### A. Phase Space Reconstruction

```
function [Portrait, NoPoints, MaxPosEmDim]=...
    Trajectory(SigData, MaxEmDim, TimeDelay)
%
% This function creates a matrix containing the MaxEmDim trajectories generated
% for a specified time delay and a given set of data (SigData).
%
% IN:
% SigData: Vector (of the form n,1) to be analyzed.
% MaxEmDim: The maximum embedding dimension for which the trajectory
(portrait)
%    is to be constructed.
% TimeDelay: Time delay in number of points.
%
% OUT:
% Portrait: Matrix in which each row is a point in the reconstructed trajectory.
%    Each point in the row is the coordinate of that point.
% NoPoints: Number of points for each dimension. For any dimension EmDim,
%    NoPoints=length(SigData)-(EmDim-1)*TimeDelay.
% MaxPosEmDim: Maximum possible embedding dimension for the number of
points in
%    SigData.
DataLength=length(SigData);
MaxPosEmDim=floor(2*log10(DataLength));
clear NoPoints
for i=1:MaxEmDim
    NoPoints(i)=DataLength-((i-1)*TimeDelay);
end
clear Portrait;
Portrait=zeros(NoPoints(1),MaxEmDim);
for i=1:MaxEmDim
    Portrait(1:DataLength-((i-1)*TimeDelay),i)=...
      SigData(((i-1)*TimeDelay)+1:DataLength);
end
```

### B. Forecasting

```
function [yFuture]=Forecast(y,NoPointsForcast,TimeDelay,EmDim,Redund)
%
% IN:
% y: Vector with original data.
% NoPointsForcast: Number of points to forecast.
% TimeDelay: Time delay for the trajectory reconstruction.
% EmDim: Dimension for the trajectory reconstruction.
% Redund: Number of redundant points to evaluate forecasting parameters.
%
% OUT:
% yFuture: Vector {NoPointsForcast} with forecast data.
%
% Uses: BackEmbedding.m, ClosestPoints.m, EvalFuture.m,
yFuture=zeros(NoPointsForcast,1);
% Backward embedding
[x,NoPoints]=BackEmbedding(y,TimeDelay,EmDim);
% Find closest points
x0=x(NoPoints,:);
% EmDim+1 is the minimum to solve the problem.
% Can be increased to evaluate more points.
```

```
k=(EmDim+1)+Redund;
[xClosest,PosClosest]=ClosestPoints(x(1:NoPoints-NoPointsForcast,:),x0,k);
for i=1:NoPointsForcast
[xClosestFuture]=EvalFuture (x,PosClosest,i,k);
     % Calculate the set of parameters 'a' that best generate the forecast of
     % xClosestFuture from xClosest.
     k=size(xClosest)*[1;0];
     a=regress(xClosestFuture(:,1),[ones(k,1),xClosest]);
     % Forecast y
     yFuture(i,:)=[1, x0]*a;
end

function [xClosest, PosClosest]=ClosestPoints(x,x0,N)
%
% Searches for the N closest points to x0 in matrix x.
%
% IN:
% x: Matrix {of the form n,EmDim} with n points of dimension EmDim.
% x0: Vector {of the form 1, EmDim} to where the closest points
%    in x are search.
% N: Number of points to look for.
%
% OUT:
% xClosest: Matrix {of the form N, EmDim} with the N closest points
%    of x to x0.
% PosClosest: Vector {of the form N,1} with the position of the
%    closest points.
[NoPoints, EmDim]=size(x);
Distance=sum(((x-ones(NoPoints,1)*x0).^2),2);
[Distance, I]=sort(Distance);
xClosest=zeros(N,EmDim);
PosClosest=zeros(N,1);
for i=1:N
     xClosest(i,:)=x(I(i),:);
end
PosClosest=I(1:N);

function [xClosestFuture]=EvalFuture(x,PosClosest,NoPoints Forcast,N)
%
% Evaluate the trajectory (x) for each point indicated by the vector
% PosClosest N steps in the future.
%
% IN:
% x: Matrix {of the form n,EmDim} with n points of dimension
%    EmDim.
% PosClosest: Vector {of the form N,1} with the position of the
%    closest points.
% NoPointsForcast:
% N: Number of points.
%
% OUT:
% xClosestFuture: Matrix {of the form N, EmDim} with the evolved
%    points in matrix PosClosest, NoPoints ahead in the future.
[NoPoints,EmDim]=size(x);
xClosestFuture=zeros(length(PosClosest),EmDim);
for i=1:N
     xClosestFuture(i,:)=x(PosClosest(i)+NoPointsForcast,:);
end
```

### *C. Surrogate Data*

```
function [VectorOut]=SurrDataFFT(VectorIn)
%
% This function assigns random phases for all frequencies of the input
% vector (VectorIn{n,1}) in its Fourier representation.
%
% IN:
% VectorIn: Vector {of the form n,1} with original data.
%
% OUT:
% VectorOut: Vector {of the form n,1} with surrogate data.
VectorLength=length(VectorIn);
% FFT of original Signal
Vectorfft=fft(VectorIn);
% Randomize Phase
NRel=ceil((VectorLength-1)/2)+1;
NChange=VectorLength-NRel;
RelVector=zeros(NRel,1);
RelVector=Vectorfft(1:NRel);
RandAngles=rand(NChange,1)*2*pi;
RelVector(2:NChange+1)=(cos(RandAngles)+sin(RandAngles)*i).*...
     abs(Vectorfft(2:Nchange+1));
VectorRandom=zeros(VectorLength,1);
NRel=ceil((VectorLength-1)/2)+1;
VectorRandom(1:NRel)=RelVector;
for i=VectorLength:-1:NRel+1
     j=VectorLength-i+2;
     VectorRandom(i)=conj(RelVector(j));
end
% IFFT to generate new signal
VectorOut=real(ifft(VectorRandom));
% VectorOut: Vector {of the form n,1} with surrogate data.
```

Chapter 2

# SEARCHING FOR THE ORIGIN OF CHAOS

Tomoyuki Yambe, Makoto Yoshizawa, Kou-ichi Tabayashi, and Shin-ichi Nitta

The clinical significance of chaotic dynamics in the cardiovascular system has attracted attention. The circulatory system is a kind of complex system having many feedback circuits, so it has been difficult to investigate the origin of chaos in the circulatory system. In this study, we investigated the origin of chaos by using the methodology of open-loop analysis with an artificial heart, which did not have any fluctuation in its own pumping rate and contraction power, in chronic animal experiments using healthy adult goats. In the circulatory time series data of the artificial heart, which did not have any fluctuation, low-dimensional deterministic chaos was discovered by nonlinear mathematical analysis, suggesting the importance of the blood vessel system in the chaotic dynamics of the cardiovascular system. To investigate the origin of chaos further, sympathetic nerve activity was directly measured in the animal experiments with artificial heart circulation. Chaotic dynamics was also recognized in the sympathetic nerve action potentials, even during artificial heart circulation, suggesting the importance of a central nervous system. In the next step, in order to generate chaotic dynamics in the circulation, an electrical simulation model of left heart circulation with a feedback loop with some delay was tested, and we found that we can make chaos in the circulation by a feedback loop. Thus, we constituted an artificial baroreflex system with an artificial heart system and found that we could introduce chaos in an animal model. Finally, we proposed clinical application of these methodologies for the diagnosis of an autonomic nervous system and found that it may be useful. These findings may be useful for analyzing the nonlinear dynamics in the circulation.

## 1. OPEN-LOOP ANALYSIS BY USE OF AN ARTIFICIAL HEART

### 1.1. Artificial Heart for Investigation of the Cardiovascular Control System

In order to develop an optimal automatic control algorithm for an artificial heart system, investigation of the basic characteristics of the cardiovascular system may be important. The clinical significance of the chaotic dynamics attracted attention, especially in the cardiovascular system, which had nonlinear dynamic behaviors [1, 2]. The circulatory system is a complex system having many feedback circuits, so it was difficult to investigate the origin of chaos in the circulatory system.

In this study, we investigated the origin of chaos by using the methodology of open-loop analysis with an artificial heart, which did not have any fluctuation in its pumping rhythm and contraction power, in chronic animal experiments using healthy adult goats [3, 4]. First, we investigated the hemodynamic time series data for a natural heart and an artificial heart circulation. For a comparison of the circulatory dynamics with a natural heart and an artificial heart we adopted a biventricular bypass type of total artificial circulation model with which observation and comparison are possible in the same animals.

If deterministic chaos existed in the circulation with an artificial heart, it might originate from the dynamics of the peripheral vessels because there were no fluctuations in the pumping rhythm and contractility of the artificial heart. To search for the origin of chaos in the circulation, we used nonlinear mathematical methodologies including chaos and fractal theories [1, 2]. Our use of these methodologies and results are reported here.

## 1.2. Experimental Preparation for Open-Loop Analysis

The experimental goats used in this study weighed from 60 to 70 kg with a mean of 65 kg. They were kept fasting for 2 days before the experiments. Three goats were anesthetized by halothane inhalation. After tracheal tube intubation by tracheotomy, they were placed on a respirator. Electrodes for electrocardiography (ECG) were attached to the legs and later implanted in the pericardium. The left pleural cavity was opened by a left fourth rib resection. Arterial blood pressure was monitored continuously with catheters inserted into the aorta through the left internal mammalian artery. The fluid-filled catheter through the internal mammalian vein measured central venous pressure (CVP). For left artificial heart implantation, the intercostal arteries were separated to free the descending aorta. A polyvinyl chloride (PVC) outflow cannula was sutured to the descending aorta. A PVC inflow cannula was inserted into the left atrium through the left atrial appendage. Both cannulae were connected to our TH-7B pneumatically driven sac-type blood pump by built-in valve connectors. The PVC outflow and inflow cannulae were inserted into the pulmonary artery and the right atrium, respectively. Then both cannulae were connected to the right pump. Pump output was measured with an electromagnetic flowmeter attached to the outflow cannula.

The TH-7B pneumatically driven sac-type blood pumps were used in this experiment to constitute a biventricular bypass (BVB) type of artificial heart model (Figure 1). The blood-contacting surface of the pump was coated with Cardiothane, and the outer casing of the pump is made of polycarbonate. Silicone ball valves were affixed to the inflow and outflow connectors. After the chest was closed, these pumps were placed paracorporeally on the chest wall, and the goats were then placed in a cage and extubated after waking. After the influence of the anesthesia was thought to be terminated (2–3 days after the operation), the goats received intravenous heparin (100 U/kg) and control time series data were recorded without the biventricular assist device. Data were recorded under awake conditions with the goats standing and in a preprandial condition. Time series data for the hemodynamic variabilities were recorded with an ink-jet recorder and magnetic tape after stabilization of all hemodynamic derivatives without the artificial heart (20–30 minutes after the biventricular

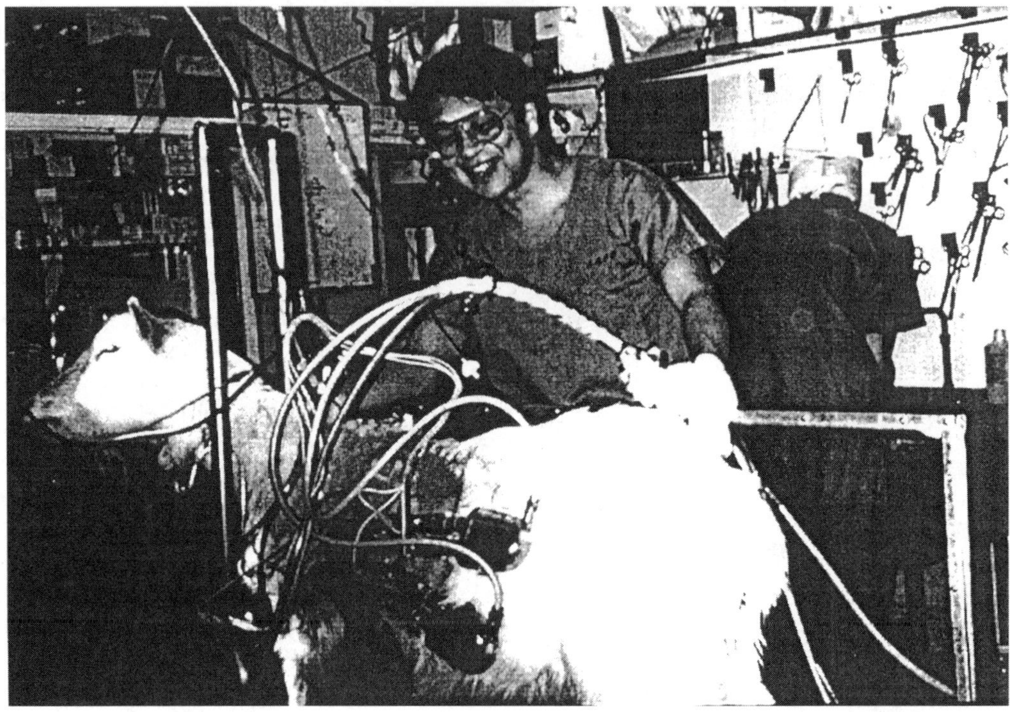

**Figure 1** Photograph of a goat with a biventricular bypass type of total artificial heart. Two pneumatically actuated sac-type blood pumps are shown in this photograph. The upper pump was a left heart bypass pump. Blood was received from the left atrium and pumped to the descending aorta. The lower pump was the right heart bypass pump. Blood was received from the right atrium and pumped into the pulmonary artery.

assist devices were stopped). After the control data were recorded, bilateral ventricular assistance was started and ventricular fibrillation was induced electrically. Confirming the stabilization of the hemodynamics with stable TAH drive conditions (20–30 minutes after stabilization of the hemodynamics), time series data for the hemodynamic variabilities were recorded. Driving conditions of the pump were manually operated both to maintain satisfactory pump output (80–100 ml/min/kg) and to maintain the hemodynamic parameters within normal limits. Driving conditions of both pumps were fixed when the time series data for the hemodynamic parameters were recorded.

After the recording, all time series data were analyzed in the personal computer system (PC9801BA) by off-line analysis. By use of an analog-to-digital (AD) converter, time series of the hemodynamic parameters were input into the computer system. Then, quantification and statistical handling were performed. By use of nonlinear mathematical methodology, chaotic dynamics and fractal theory analysis were performed. Time series data were embedded in the phase space by the method of Takens et al. For the quantitative analysis, Lyapunov exponents of the reconstructed attractors were calculated using the algorithm of Wolf et al.

## 1.3. Time Series Data Obtained from Open-Loop Analysis

By use of a biventricular bypass type of total artificial circulation model, comparison of the natural heart circulation and artificial heart circulation was realized in the same animals. To evaluate the nonlinear dynamics in this animal model, we used reconstruction of the strange attractor in the phase space, Lyapunov exponents analysis, correlation dimension analysis, and fractal dimension analysis of the return map.

During artificial heart circulation, the natural heart was fibrillated by electrical stimulation and systemic and pulmonary circulation was maintained with a biventricular assist pump. Ventricular assist devices used in this experiment were designed for human clinical use and pneumatically driven by a console outside the body of the experimental animals. During the recording of the time series data of the artificial heart circulation, the driving condition of the artificial heart was fixed; thus, there were no fluctuations in driving rhythm and contraction power and we could observe the circulation without fluctuations of the natural heartbeat. Of course, the artificial heart was stopped during the observation of the natural heart circulation.

Figure 1 shows a photograph of an experimental goat with the biventricular assist type of total artificial circulation model. Two pneumatically actuated sac-type blood pumps are shown. The upper pump was a left heart bypass pump. Blood was received from the left atrium and pumped to the descending aorta. The lower pump was a right heart bypass pump. Blood was received from right atrium and pumped into the pulmonary artery. All hemodynamic parameters were recorded in the magnetic tape data recorder and off-line analysis was performed through the AD converter. Reconstructed attractors of the arterial blood pressure during the natural heartbeat embedded in the four-dimensional phase space and projected into the three-dimensional phase space are shown in Figure 2. Before reconstructing this attractor, two-dimensional reconstruction and three-dimensional reconstruction were also checked and it was found that more than three dimensions may be desirable.

A strange attractor may be shown in the figure. The basic cycle in this attractor of arterial blood pressure was, of course, coincident with the cardiac cycle. Rhythmic fluctuations such as respiratory fluctuations and Mayer waves were seen in the time series data and may be shown as the bandwidth of this attractor. Chaotic systems characteristically exhibit sensitive dependence upon initial conditions [1–4]. To evaluate

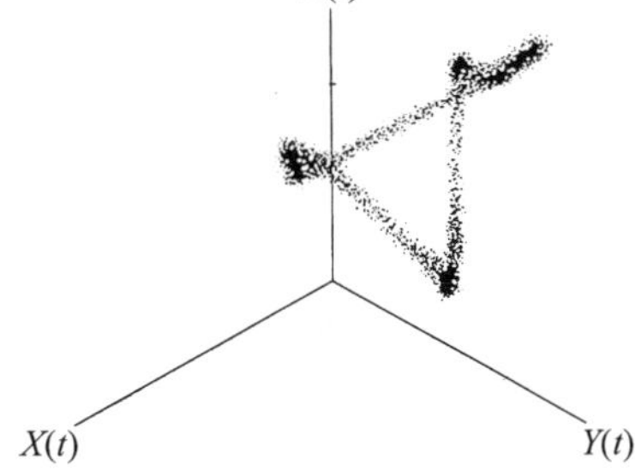


**Figure 2** Example of a reconstructed attractor of the arterial blood pressure during the natural heartbeat embedded in four-dimensional phase space and projected into three-dimensional phase space.

this, we calculated the Lyapunov exponents from the reconstructed attractor. Lyapunov exponents were a quantitative measure of the rate of separation of the attractors in the phase space. In this study, we calculated the Lyapunov exponents by the method of Wolf et al. [5, 6]. In the example in Figure 3, it converges in a positive value, suggesting sensitive dependence on initial conditions and the existence of deterministic chaos.

Various investigators have shown chaotic dynamics in hemodynamic parameters in animals. Our result in this study supports these reports. Of course, it is the next issue that is interesting here.

## 1.4. Chaos and Artificial Organs

Can an artificial heart produce chaos? The results are illustrated in Figures 4 and 5. Figure 4 shows an example of a reconstructed attractor of the arterial blood pressure during artificial heart circulation embedded in four-dimensional phase spaces and projected into three-dimensional phase spaces. Banding may suggest fluctuations in hemodynamic parameters and fractal structures, suggesting the existence of deterministic chaos. Lyapunov exponents of the reconstructed attractor of the arterial blood pressure during artificial heart circulation embedded in four-dimensional phase spaces are shown in Figure 5. The results suggest lower dimensional chaotic dynamics compared with that during the natural heartbeat.

The existence of chaos in the circulation without a natural heart is an interesting result. The chaos must affect the baroreceptors in the arterial wall, and this information is communicated to the central nervous system. The brain regulating sinus nodes must respond to this input and produce chaos, too. If other information is added in this stage, higher dimensional chaos may be generated as in Figures 2 and 3. Of course,

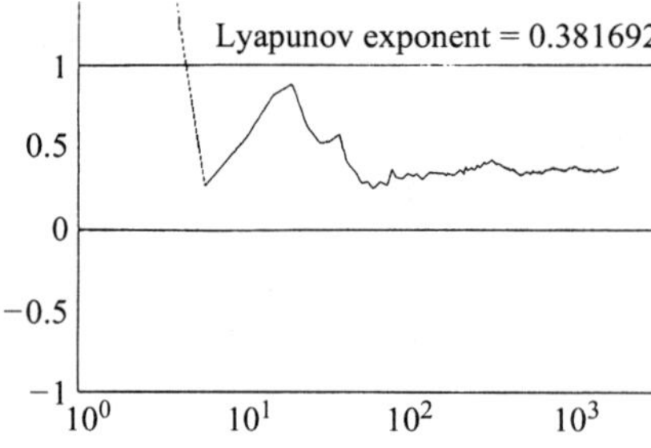


**Figure 3** Lyapunov exponents of the reconstructed attractor of the arterial blood pressure during the natural heartbeat embedded in four-dimensional phase space.

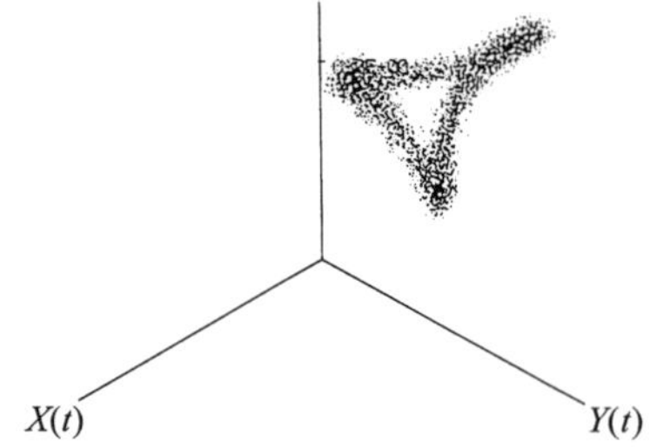


**Figure 4** Example of a reconstructed attractor of the arterial blood pressure during artificial heart circulation embedded in four-dimensional phase space and projected into three-dimensional phase space.

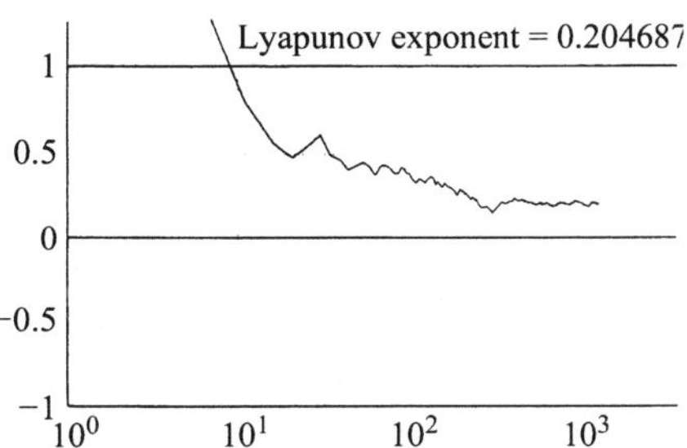


**Figure 5** Lyapunov exponents of the reconstructed attractor of the arterial blood pressure during artificial heart circulation embedded in four-dimensional phase space.

other possibilities are also considered. The natural heartbeat is regulated not only by the central nervous system but also by other regulatory systems such as hormonal factors, preload, and afterload. If other information were added to the chaotic dynamics in the artificial heart circulation, it might produce higher dimensional chaotic dynamics in the circulation. This is very interesting and is a result understood with this experiment for the first time. In the next step, we wish to consider the central nervous system, which mediates the circulation.

## 2. DIRECT RECORDING OF SYMPATHETIC NERVE DISCHARGES TO SEARCH FOR THE ORIGIN OF CHAOS

### 2.1. Direct Recording of Autonomic Nerve Discharges

Peripheral vascular resistance has been reported to be mediated by the autonomic nervous system. Thus, in the next step, we directly recorded autonomic nerve discharges during the artificial heart circulation. To investigate further the origin of chaos, sympathetic nerve activity was measured in the animal experiments with the artificial heart circulation by use of the renal sympathetic nerve.

The left pleural cavity was opened through the fifth intercostal space in four adult mongrel dogs weighing 15–35 kg after intravenous administration of thiopental sodium (2.5 mg/kg) and ketamine sodium (5.0 mg/kg). Electrodes for ECG were implanted, and aortic pressure (AoP) and left atrial pressure (LAP) were monitored continuously with catheters inserted into the aorta and left atrium through the left femoral artery and the left appendage, respectively.

For left artificial heart (LAH) implantation, a PVC outflow cannula with a T-shaped pipe was inserted into the descending aorta and secured by ligature. A PVC inflow cannula was inserted into the left atrium through the left atrial appendage. Both cannulae were connected to our TH-7B pneumatically driven sac-type blood pump via built-in valve connectors. A PVC outflow cannula was inserted into the pulmonary artery, and a PVC inflow cannula was inserted into the right atrium through the right atrial appendage. Both cannulae were then connected to the right pump.

The TH-7B pneumatically driven sac-type blood pump was used in these experiments to constitute a BVB-type complete artificial circulation model. Driving pressure and systolic duration of both blood pumps were selected to maintain hemodynamic derivatives within the normal range, and pump output was controlled to maintain the cardiac output before electrical ventricular fibrillation. After BVB pump implantation, the left flank was opened and the left renal artery was then exposed [7, 8].

After the left renal sympathetic nerves were separated, the nerve sheath was removed and the nerve was placed on bipolar stainless steel electrodes for recording of renal sympathetic nerve activity (RSNA) [7, 8]. The renal sympathetic nerve discharges were amplified by means of a preamplifier and main amplifier (DPA-21, DPA-100E; DIA Medical System Co.) with a bandwidth of 30 Hz to 3 kHz and were displayed on an oscilloscope. Time series data of the hemodynamic parameters and sympathetic never activity were recorded with an ink-jet recorder and magnetic tape after the stabilization of all hemodynamic derivatives (almost 20–30 minutes after the preparation). After the control data recording, ventricular fibrillation was induced electrically. After confirming the stabilization of the hemodynamics by the stable BVAD drive condition, time series data were recorded.

## 2.2. Recording Data of the Sympathetic Nerve Discharges

During artificial circulation with the biventricular assist type of total artificial heart (TAH) model under electrically induced ventricular fibrillation, hemodynamic parameters were relatively easily maintained within the normal range by the manually operated pneumatic drive console. Figure 6 shows an example of time series data of the hemodynamic parameters and sympathetic nerve recording.

Figure 6 shows the ECG, arterial blood pressure, integrated nerve activity, and renal sympathetic nerve discharges. During the artificial heart circulation, sympathetic nerve discharges were synchronized with the artificial heart driving rhythm, although the sympathetic nerve activity was synchronized with the heart rate and respiration during the natural heart circulation. The driving rate of the artificial heart was fixed during the measurement, thus the sympathetic discharges kept the fixed rhythm. This

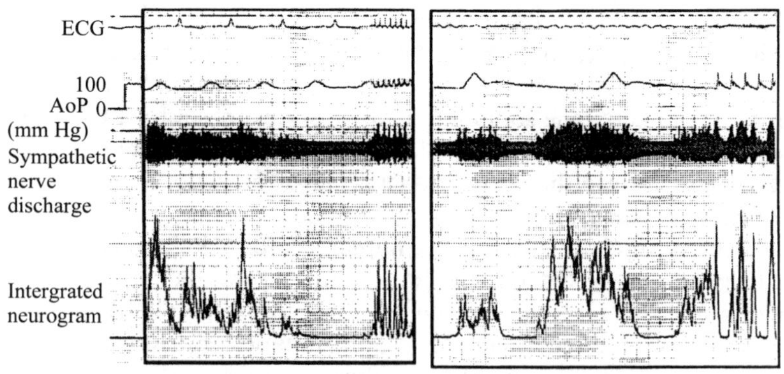


**Figure 6** Example of time series data of hemodynamic parameters and sympathetic nerve recording. The upper tracing shows the electrocardiogram, arterial blood pressure, integrated nerve activity, and renal sympathetic nerve discharges. During artificial heart circulation, sympathetic nerve discharges were synchronized with an artificial heart driving rhythm, whereas the sympathetic nerve activity was synchronized with heart rate and respiration during natural heart circulation.

phenomenon allowed us to reconstruct the attractor of the sympathetic nerve activity in phase space.

Figure 7 shows an example of the reconstructed attractor of the integrated sympathetic nerve discharges. The time constant of integration was 0.1 second. It was thought that it might be a strange attractor, which is characteristic of deterministic chaos, because it had fractal characteristics in the orbit formation. Lyapunov numerical analysis suggests its sensitive dependence upon initial conditions. These results suggest that sympathetic nerve time series signals show characteristics of deterministic chaos.

The central nervous system is thought to play an important role in the chaotic dynamics of the circulation. The chaotic dynamics in sympathetic nerve activity are thought to mediate the properties of peripheral vessels and produce chaos in the circulation.

### 2.3. Nonlinear Relationship in Biological Systems

To analyze this relationship, mutual information analysis was performed in this study [9].

According to an algorithm proposed by Fraser and Swinney [9], we measured the dependence of the values of $y(t+T)$ on the values of $x(t)$, where $T$ is a time delay. The assignment $[s, q] = [x(t), y(t+T)]$ was made to consider a general coupled system $(S, Q)$. The mutual information was made to answer the question, Given a measurement of $s$, how many bits on the average can be predicted about $q$?

$$I(S, Q) = \int P_{sq}(s, q) \log\left[P_{sq}(s, q)/(P_s(s)P_q(q))\right] ds\, dq \tag{1}$$

where $S$ and $Q$ denote systems; $P_s(s)$ and $P_q(q)$ are the probability densities at $s$ and $q$, respectively; and $P_{sq}(s, q)$ is the joint probability density at $s$ and $q$. If the value of mutual information for $(S, Q)$ is larger, it means that the mutual dependence between $S$ and $Q$ is stronger.

The results suggest that in some driving conditions, the coupling of sympathetic nerve and peripheral vessel properties became strong compared with that during the natural heartbeat. Thus, sympathetic nerves were suggested to contribute to the chaotic dynamics in the circulation.

If chaotic dynamics were generated in the peripheral vessels, it must influence the arterial baroreflex system on the arterial wall. This information was communicated to

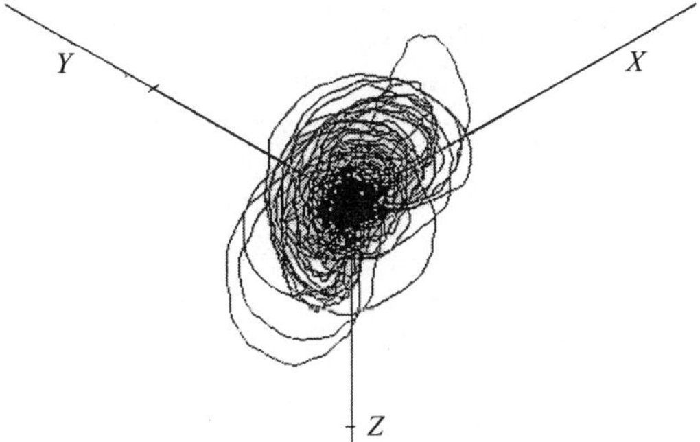


**Figure 7** Example of the reconstructed attractor of the integrated sympathetic nerve discharges. The time constant of integration was 0.1 second. It was thought that this might be a strange attractor, which is characteristic of deterministic chaos, because it had fractal characteristics in this orbit formation.

the brain, and the response to this information was sent to the sinus nodes, finally resulting in chaos in heart rate variability.

These results suggest the importance of the brain in generating chaos in the circulation. In the next step, we considered control of an artificial heart by use of this information from the brain from the viewpoint of nonlinear dynamics.

## 3. PREDICTIVE CONTROL OF AN ARTIFICIAL HEART SYSTEM

Tohoku University has been involved in artificial heart (AH) research for over 20 years and has developed an automatic AH control system using various types of microsensors. However, the control algorithm for the AH using hemodynamic parameters and hormonal factors could not avoid the possibility of a time delay compared with the natural heart.

### 3.1. Prediction with Information about Sympathetic Nerve Discharges

We therefore proceeded to evaluate the autonomic nervous system using the sympathetic neurogram during AH pumping and developed a new control algorithm for the ventricular assist device (VAD) using sympathetic tone. The present study was designed to establish a real-time control system for the TAH, which works at a speed like that of the natural heart without any time delay, using sympathetic tone and hemodynamic parameters. Not all cardiac nerves could be used for measuring the sympathetic neurogram because TAH implantation removed the cardiac nervous system and retrograde generation extinguishes them [11]. Therefore, we used the renal sympathetic nerve, already employed for autonomic nervous system evaluation, as the parameter of sympathetic tone. The renal sympathetic nerve is almost completely composed of efferent sympathetic nerve fibers [12]. This may be useful for assessment of efferent sympathetic tone in long-term animal experiments. Moving averages of LAP and AoP were calculated and taken as the parameters of the preload and afterload of the natural heart. Multiple linear regression analysis was used to set up the equations, which indicate following cardiac output.

In this study, the functional formula for an accurate estimate of the following cardiac output was designed, and the possibility of a real-time automatic TAH control system was analyzed [13]. Seven adult mongrel dogs of both sexes weighing 15–35 kg were anesthetized with thiopental sodium (2.5 mg/kg) and intravenous ketamine sodium (5.0 mg/kg). After tracheal tube intubation, they were placed on a volume-limited respirator (Acoma ARF-850). The anesthesia was maintained with nitrous oxide inhalation under mechanical ventilation. ECG electrodes were implanted in the left foreleg and both hind legs. AoP and LAP were monitored continuously from a catheter inserted into the aorta and left atrium through the left femoral artery and the left appendage, respectively.

An ultrasonic flowmeter was used in this experiment to avoid electromagnetic noise contamination of the nerve waveform. The ultrasonic flowmeter (Advance; Transonic T101) was placed on the pulmonary artery to measure cardiac output because the right flow rate must be controlled in animals with a TAH to maintain lung oxygenation. The left flank was opened between the iliac crest and the costovertebral angle, and the left renal artery was then exposed. The left renal sympathetic nerve was dissected free from the left renal artery and surrounding connective tissues. After removal of the nerve sheath, the nerve was attached to a bipolar stainless steel electrode to record renal sympathetic nerve activity (RSNA). Discharges were recorded after amplifying the original signal with a differential preamplifier and the main amplifier (DIA Medical System Co.: DPA-21, DPA-100E) of bandwidth 30 Hz to 3 kHz and displayed on an oscilloscope. The output from the amplifier was passed through a gate circuit to remove baseline noise and rectified by an absolute value circuit. Then an R-C integrator circuit (time constant, 0.1 second) integrated the rectified output. The output of the integrated nerve discharges was calibrated in l/V, and their areas were measured for a given period and expressed as RSNA per time unit.

ECG; systolic, diastolic, and mean AoP; mean LAP; mean pulmonary artery flow; and RSNA were measured, RSNA was quantified from the area of the integrated nerve discharge waveform per time unit. All data were analyzed in the computer system (NEC PC9801VM21). Moving averages of the mean LAP and AoP were calculated as the parameters of the preload and afterload of the natural heart, respectively. Multiple regression analysis was done using the preload, afterload, RSNA, and following cardiac output. Differences were analyzed by the paired $t$-test and the $F$-test and were considered significant when $p < 0.05$.

## 3.2. Future Prediction by an Autonomic Nerve and Hemodynamics

An example of the computer-analyzed time series data obtained for an adult mongrel dog is shown in Figure 8. The ECG, AoP, left ventricular pressure (LVP), integrated RSNA, marker RSNA, pulmonary artery flow (PAF), and LAP were quantified in the computer system. Grouped discharges, which appeared synchronously with heartbeat and respiration, were observed in the spontaneous activity of the renal sympathetic nerve. Figure 9 illustrates the responses of the RSNA and heart rate to clamp-

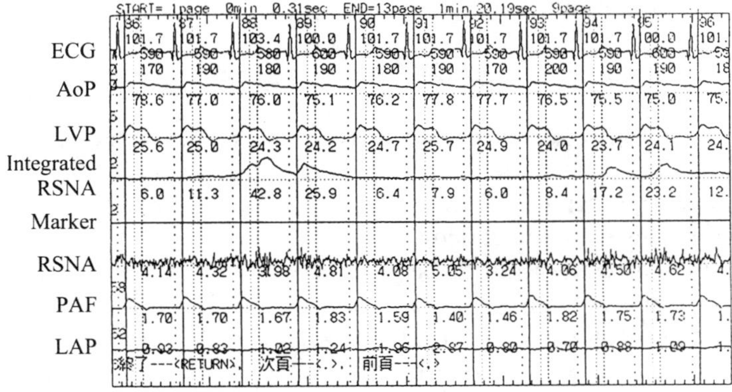


**Figure 8** Example of computer-analyzed time series data for an adult mongrel dog. From the upper tracing the ECG, AoP, left ventricular pressure (LVP), integrated RSNA, marker, RSNA, pulmonary artery flow (PAF), and LAP were quantified in the computer system. Grouped discharges, which appeared synchronously with heartbeat and respiration, were observed in the spontaneous activity of the renal sympathetic nerve.

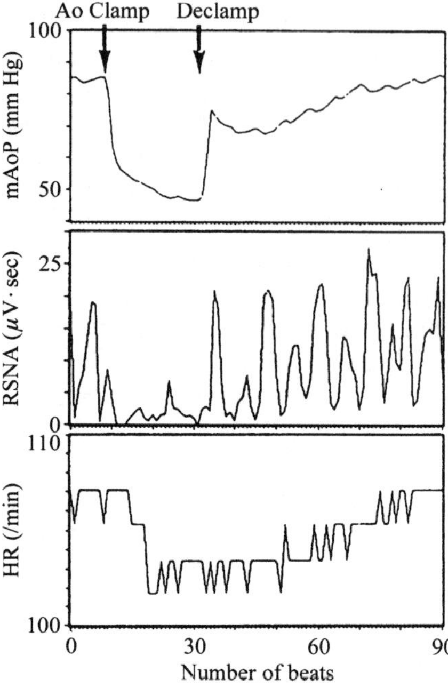


**Figure 9** Responses of the RSNA and heart rate to clamping the descending aorta. The AoP dropped immediately after Ao clamping and rose again soon after declamping. But the RSNA responses had some time delay, which probably indicates the response time of the baroreflex. The heart rate needed longer to respond to sympathetic tone, probably indicating the response time of the sinus node to the automatic nervous system.

ing the descending aorta. AoP dropped immediately after Ao clamping and rose again soon after declamping. But there was some time delay in the RSNA responses, which probably indicates the response time of the baroreflex. The heart rate response took longer to respond to sympathetic tone, probably indicating the response time of the sinus node to the autonomic nervous system.

Assuming the moving averages of AoP, LAP, and integrated RSNA as the explanatory variables and the following cardiac output as the criterion variable, multiple linear regression analysis was performed using the time series data of this experiment. An example of the multiple regression equation for the following cardiac output is

$$Y = a1 \times mRSNA + a2 \times mLAP - a3 \times mAoP + b \tag{2}$$

where $Y$ is the following cardiac output (1/min), mRSNA the moving average of mean RSNA ($\mu$V sec), mLAP the moving average of mean LAP (mm Hg), and mAoP the moving average of mean AoP (mm Hg). For example, from the time series data of these experiments, they were calculated as $a1 = 0.024$, $a2 = 0.413$, $a3 = 0.004$, and $b = 0.967$. The multiple correlation coefficient was $R = 0.700$, and this equation was significant ($p < 0.01$); in the explanatory variables, mRSNA and mLAP were significant but not mAoP. A multi-colinearity problem between mRSNA and mAOP eliminates mAOP among the explanatory variables by the backward elimination method.

Thus, an example of a new multiple regression equation is

$$Y = a1 \times mRSNA + a2 \times mLAP + b \tag{3}$$

This was worked out as $a1 = 0.022$, $a2 = 0.430$, and $b = 0.700$. The multiple correlation coefficient was $R = 0.682$ and the contribution rate was $R = 0.465$, so this equation was significant ($p < 0.01$). Standardized partial regression coefficients in this dog were $a1 = 0.432$ and $a2 = 0.454$, so the contributions of these parameters to this equation were not significantly different.

Figure 10 illustrates the correlation between estimated cardiac output calculated in this equation and the following cardiac output detected about 2.9 seconds (five beats) after the moving averages of LAP and AoP. The average and standard deviation of the standardized partial regression coefficients in the seven dogs were calculated as $a1 = 0.363 \pm 0.062$ and $a2 = 0.471 \pm 0.067$. Both parts of this equation were significant. Therefore mean LAP and RSNA, recorded five beats before, were useful for estimating the following cardiac output.

A variety of TAH control methods have been evaluated in various laboratories [13, 14]. Many investigators have used a control algorithm based on Starling's law and others have proposed a control method in which AoP is maintained as a constant value [15]. However, these control strategies could not avoid the possibility of a time delay compared with the natural heart because they were based on hemodynamic data, which are by definition the result of the behavior of the vascular system in the immediate past. In the natural heart, cardiac output can be changed by various mechanisms: sympathetic stimulation, with an increase in venous return and heart rate or stroke volume or both, and Starling's law.

The sympathetic nervous system detects physiological stress, transmits this information to the peripheral organs, and adapts the circulation and metabolism to meet demand [11, 12]. However, in most TAH systems, there is no feedback loop to transmit this information to the control system. We believe that it is necessary to detect the sympathetic tone for a real-time TAH control system. In this study, we used the sympathetic tone and Starling's law as the information for establishing the target cardiac output for the TAH real-time control system. Estimated cardiac output, calculated by multiple linear regression analysis of the RSNA and preload, was correlated significantly with the following cardiac output, actually measured later. This suggests the possibility of a real-time control method for the TAH system using sympathetic tone and hemodynamic parameters.

Although RSNA is useful for detecting efferent sympathetic tone, it is not the cardiac nerve, so it is always open to discussion whether RSNA tone acts like that of the cardiac nerve. But as the cardiac nerves do not exist after TAH implantation, RSNA is useful for determining the sympathetic efferent tone because it is almost entirely composed of efferent sympathetic nerve fibers [12].

Another important problem concerns hormonal factors such as the catecholamines. If there is to be a useful sensor of hormonal factors, we must add explanatory

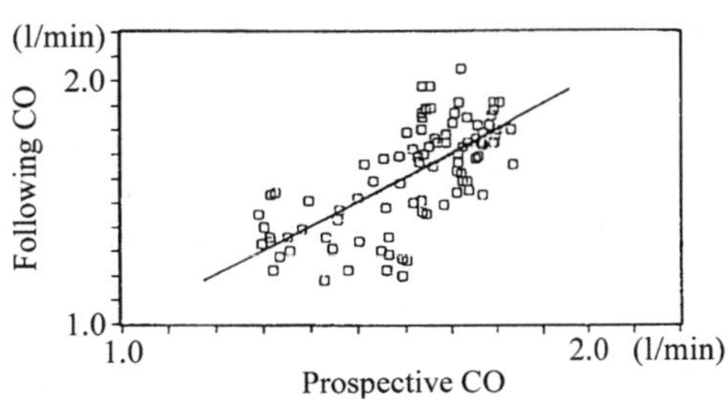


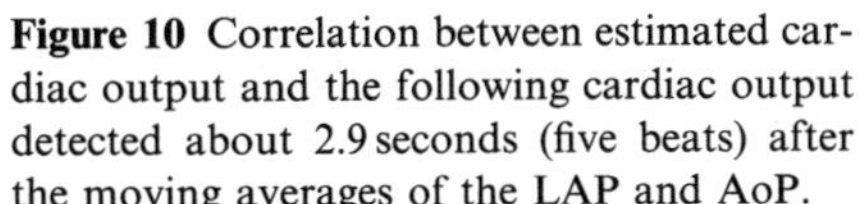
**Figure 10** Correlation between estimated cardiac output and the following cardiac output detected about 2.9 seconds (five beats) after the moving averages of the LAP and AoP.

variables of the equation that estimate the following cardiac output. Important problems in the development of a permanent electrode for the neurogram still remain unsolved. For this reason, we proposed a new TAH control methodology based on the neural information and shown in the following.

### 3.3. Mayer Wave Control for an Artificial Heart

A control algorithm for the TAH has been discussed more and more from various viewpoints [13–15]. Several investigators have proposed an automatic TAH control algorithm to prevent thrombus formation and suggested the importance of full stroke driving [14, 15]. Some researchers reported the importance of left and right heart balance, and some investigators have controlled their artificial hearts according to Starling's law [14]. Other researchers have controlled their devices to maintain arterial blood pressures [15]. Many developers of a TAH showed the importance of the optimal drive of their devices; however, few researchers have pointed out the importance of optimization from the physiological and pathophysiological viewpoints [16, 17].

In view of the widespread clinical application of TAHs such as their bridge use for transplantation, the importance of quality of life (QOL) for a patient with an artificial heart attracts attention as a next step in development [13, 15]. Various cardiac outputs are needed during various activities [11]. If the cardiac output of patients with an artificial heart cannot follow their physical activity, congestive heart failure may occur and patients cannot enjoy QOL. For this reason, we must develop a new control algorithm that can match physical activities.

It was very important to consider the optimal drive of a TAH, not only an optimal drive for mechanical devices but also an optimal drive from the physiological viewpoint. For that purpose, information on the autonomic nervous system may be needed; however, it is very difficult to monitor autonomic nerve discharges continuously [7, 8].

### 3.4. Information of the Hemodynamic Fluctuation

In this study, we propose a new basic concept for an automatic control algorithm for the TAH using fluctuations in the circulatory system [18, 19]. It was reported that hemodynamic fluctuations reflect ongoing information of the autonomic nervous system [18, 19]. Several investigators suggested that the Mayer wave around 0.1 Hz reflects sympathetic nerve information and the respiratory wave reflects parasympathetic nerve information [18]. To detect information from the autonomic nervous system, we considered these components of rhythmical fluctuation in this study. However, of course, we could not measure heart rate variabilities in animals with an artificial heart because there was no heart. Accordingly, the only information that could easily be measured for the hemodynamics reflecting the central nervous system might be the properties of peripheral vessels. Therefore, we paid attention to the fluctuations of the vascular resistance, which could be measured even during artificial heart circulation.

By recording time series data for hemodynamic parameters of healthy adult goats in the awake condition, fluctuations in the peripheral vascular resistance were calculated in this study. The probability of a predictive automatic TAH control system using autonomic nerve information was considered.

The goats used weighed from 60 to 70 kg with a mean of 65 kg. These goats were kept fasting for 2 days before the experiments. Three goats were anesthetized by halothane inhalation. After tracheal tube incubation by tracheotomy, they were placed on a respirator. ECG electrodes were attached to the legs and later implanted in the pericardium. The left pleural cavity was opened by a left fourth rib resection. Arterial blood pressure was monitored continuously with catheters inserted into the aorta through the left internal thoracic artery. Central venous pressure was measured by a fluid-filled catheter inserted through the internal thoracic vein. Cardiac output was measured with an electromagnetic flowmeter attached to the ascending aorta. After the chest was closed, the goats were placed in a cage and extubated after waking. All hemodynamic time series data were monitored continuously during the experiments (Fukuda Denshi; MCS-5000) and recorded in the awake condition.

All time series data were recorded in a magnetic tape data recorder (TEAC; RD-130TE). Quantitative analysis, statistics processing, and spectral analysis were carried out in the personal computer system (PC9801BA) through the AD converter. The Mayer wave peak was clearly recognized in all goats in the spectrum of vascular resistance. A band-pass filter was used to convert this information for automatic control. The band-pass frequencies were 0.04–0.15 Hz, which was reported to be the frequency band of the Mayer wave.

Of course, an important problem for predictive control of the TAH using neural information was the time lag of the prediction. In this study, we calculated the cross-correlation function of the band-pass value of the peripheral vascular resistance and cardiac output. Time delay was considered for an automatic TAH control system using information on the fluctuations in the peripheral vascular resistance. Finally, we proposed a predictive TAH control algorithm using autonomic nerve information.

## 3.5. Time Series Data Analysis

First, we checked the spectral peaks in the power spectrum of the peripheral vascular resistance, which could be obtained even during artificial heart circulation. If we could not obtain spectral peaks by use of fast Fourier transform (FFT) methodologies, we could not use this information for artificial heart control. In all experimental animals, Mayer wave peaks, reflecting autonomic nerve regulation, were observed, suggesting that we might be able to use this information. Figure 11 shows an example of the time series data for the hemodynamic parameters: heart rate, arterial blood pressure, cardiac output, peripheral vascular resistances, and resistances after band-pass processing. The band-pass filter was established as 0.04–0.15 Hz, reported to be the Mayer wave fluctuations in hemodynamic time series data. The behavior of the time series data suggested that band-passed data gave us information different from peripheral vascular resistances. After intense changes of the filtered peripheral vascular resistances, cardiac output was shown to be increasing.

Second, we check the time lag between the changes of the Mayer wave and alteration of the cardiac output. Figure 12 shows an example of the cross-correlation of the band-passed peripheral vascular resistances and cardiac output. A clear peak was observed after almost 5 seconds, suggesting the possibility of predicting the cardiac output form the information concerning the Mayer wave in vascular resistances.

Finally, we predicted the cardiac output in the future by using the Mayer wave in the peripheral vascular resistances, which could easily be obtained from the animals with an artificial heart. Figure 13 shows an example of the prediction after 5 seconds.

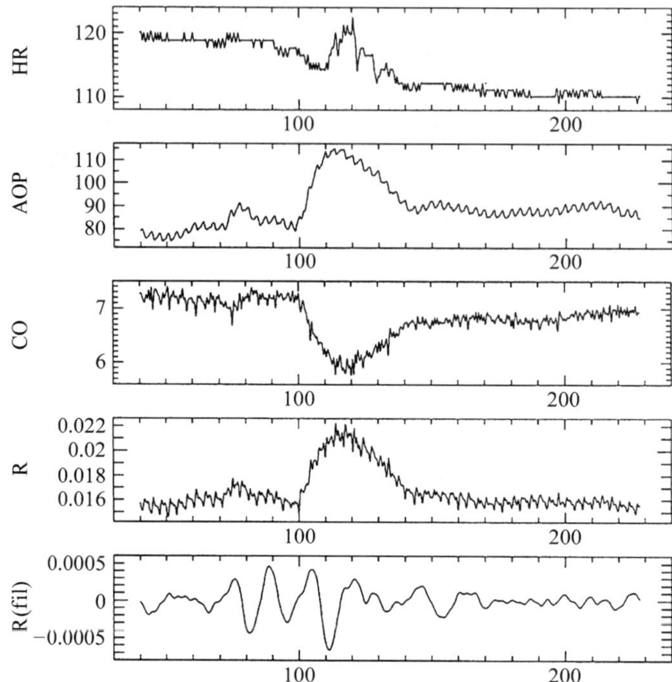


**Figure 11** Example of time series data for the hemodynamic parameters. The upper tracing shows heart rate, arterial blood pressure, cardiac output, peripheral vascular resistances, and resistances after band-pass processing.

The $X$ axis shows band-passed peripheral vascular resistances and the $Y$ axis measured cardiac output recorded 5 seconds later. A significant correlation is observed in the figure, suggesting the probability of the realization of predictive automatic control for an artificial heart system. In all four goats, a significant correlation was observed between cardiac output after 5 seconds and filter-treated peripheral vascular resistance, suggesting reproducibility of this prediction algorithm.

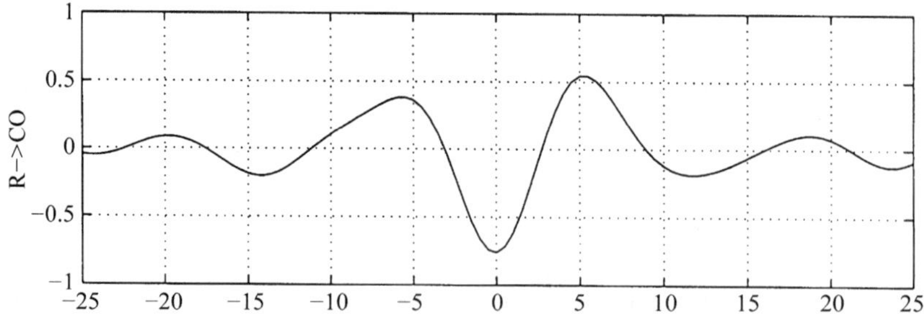


**Figure 12** Example of cross-correlation of the band-passed peripheral vascular resistances and cardiac output. A clear peak was observed after almost 5 seconds, suggesting that the cardiac output might be predicted from the information in the Mayer wave in vascular resistances.

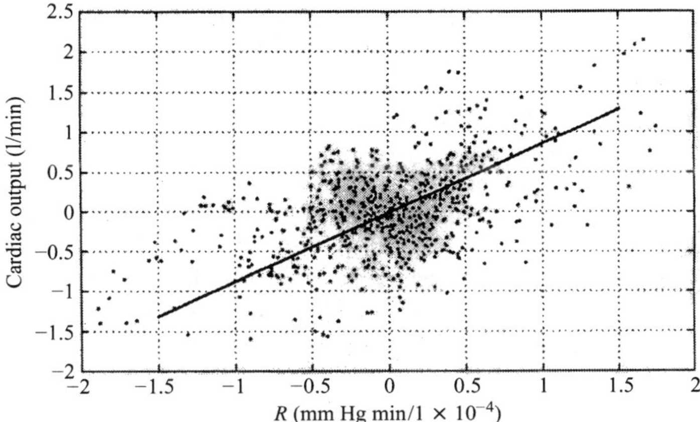


**Figure 13** Example of the prediction after 5 seconds. The $x$ axis shows band-passed peripheral vascular resistances and the $y$ axis measured cardiac output recorded 5 seconds later. A significant correlation is observed in the figure, suggesting the probability of the realization of predictive automatic control for an artificial heart system.

## 3.6. Consideration of Predictive Control Algorithm

One of the major findings of this study is the probability of actualization of the predictive control algorithm for an artificial heart by the use of information from the autonomic nervous system. For the QOL of the patient with an artificial heart, information from the biological system may be important because the artificial heart must respond to physical activity. The pump output of an artificial heart must increase when the individual runs and must decrease when he sleeps. To control an artificial heart, biological information may be necessary. However, it is very difficult to detect autonomic nerve activity directly in the chronic stage. In this study, stable measurement of autonomic nerve information was realized by the use of hemodynamic fluctuations. It may become insensitive compared with direct measurement as we showed before, and we selected the stability.

In this study, a time series curve of the Mayer wave of vascular resistance was provided. This index was reported to be very useful for the parameter of the sympathetic nervous system. It was compared with the time series curve of cardiac output. After a change of the Mayer wave, an increase in the cardiac output was observed after 5 seconds. This phenomenon may be interpreted in terms of sympathetic nerve control of the changes in cardiac output. These results suggest that an artificial heart may be controlled by the measurement of the Mayer wave of the vascular resistance.

Of course, the development of another control algorithm for an artificial heart by use of direct measurement is ongoing in Tohoku University. For example, we recorded the sympathetic nerve discharges in animals with artificial hearts for the first time [7, 8]. Some investigators carried out a supplementary examination. Recently, we have measured vagal nerve activity in chronic animal experiments in the awake condition. It may give useful information for controlling an artificial heart in the future. Research on direct measurement of an autonomic nerve is ongoing; however, at this stage, indirect measurement may be desirable for stability.

Mayer wave fluctuations in HRV were reported to originate from the fluctuations in the arterial blood pressure [18]. When the blood pressure changed, this information was detected in the baroreceptors in the arterial wall and carotid sinus and sent to the central nervous system (CNS). After that, orders were sent to the sinus nodes and peripheral vessels. When the sinus nodes and vascular resistances were altered, blood pressure was, of course, changed, and this information was sent to the CNS. Thus, it makes a negative feedback loop. In this study, we made a kind of feedback circuit by use of the Mayer wave in resistances.

If we consider the nonlinear dynamics in the circulation, a feedback loop plays an important role. Thus, we tried to make chaos in an electrical simulation circuit with a feedback loop as discussed next.

## 4. MAKING OF CHAOS

### 4.1. Feedback Loop Making Chaos

We could constitute the feedback loop for an artificial heart automatic control algorithm by use of autonomic nerve discharges or the Mayer wave in peripheral vascular resistances. It may make complicated structures in time series data as a "complex system" such as chaos or fractals. However, is it really making a complex system such as the circulatory regulatory system in animals?

### 4.2. Chaos and Time Delay

A few investigators have challenged this problem. For example, Cavalcanti et al. published very interesting articles in 1996 [20, 21] in which they tried to explain the profile. A model of the baroreceptor reflex series was constituted with a simple mathematical model. Simulation with various delays in the feedback loop was carried out. The behavior of the hemodynamics repeats a bifurcation from the simple limit cycle vibration. Before long, it changes to a condition that is chaotic. With alteration only of delay, the time series showed a period doubling bifurcation and reached chaos. This kind of experiment was not carried out in the past.

If it is considered from the viewpoint of artificial heart control, a baroreflex delay means a control delay of an artificial heart. Of course, hemodynamics is a time series signal such as blood pressure or heart rate. We tried to generate deterministic chaos in the feedback loop model of artificial heart control.

A section equivalent to the cardiovascular system has been simulated with the three-element Windkessel represented with a simple electric circuit. A three-element Windkessel is expressed as the systemic vascular resistance ($R$), arterial compliance ($C$), and characteristic impedance ($r$). In this model, the input is aortic flow and the output is aortic pressure. In other words, it is similar to a cardiovascular system with an electric circuit model. The formulas formed with this model are Eqs. 4 and 5. Each value of these three elements was adopted from data in the physiology literature.

$$dP(t) = \omega_t[RQ(t) - P_s(t)], \qquad \omega_t = 1/RC \tag{4}$$

$$P(t) = P_s + Q(t) \tag{5}$$

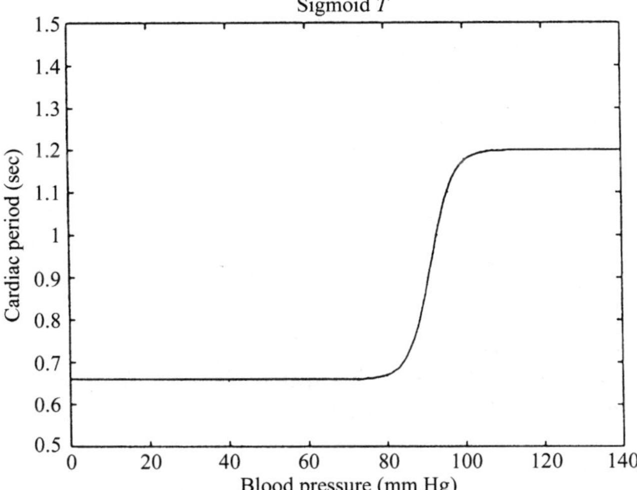


**Figure 14** Nonlinear relationship between blood pressure and cardiac cycle in an electrical simulation circuit for left heart circulation.

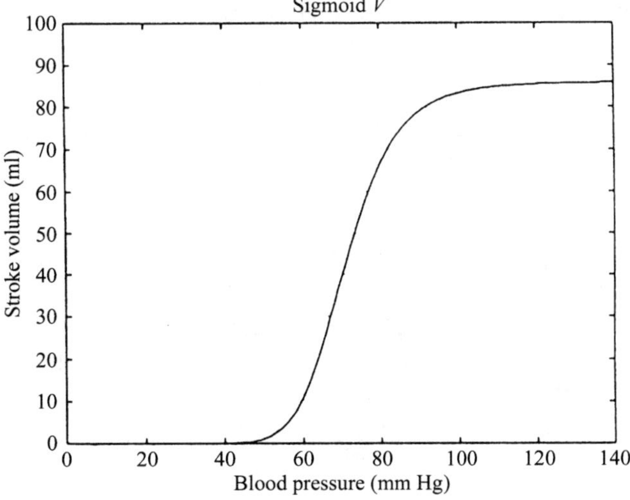


**Figure 15** Nonlinear relationship between blood pressure and stroke volume (SV) in an electrical simulation circuit for left heart circulation.

In this simulation, a nonlinear curve was used for determination. Cardiac cycle ($T$) and stroke volume ($SV$) are determined from arterial blood pressure ($P$), which is baroreflex data. In this study, a nonlinear curve as a sigmoid function was used for that purpose as we show in Eqs. 6 and 7. For each constant, we sought the value from the physiology literature. Figures 14 and 15 show the relationships.

$$T(P) = T_s + (Tm - Ts)/\left(1 + \gamma e^{-\alpha P/Pn}\right) \quad (6)$$

$$SV(P) = Sv_{\max}/(1 + \beta(P/P_v - 1)^{-k}) \quad (7)$$

Delay $t$ was introduced in the feedback loop in this study. After the delay time $r$, it was input into the next element. This was included in the feedback loop branch. This time delay $r$ was altered as a parameter when we carried out the simulation.

## 4.3. Simulation Model

In this study, it was attempted to cause deterministic chaos only by alteration of delay by use of this very simple model. A block diagram is shown in Figure 16. Simulation was carried out about this model. Time series data of the simulated hemodynamic parameters, three-dimensional attractors, and frequency spectrum were used here. We used MATLAB for simulation and analysis.

The behavior of the hemodynamic parameters was simulated in the simple electrical circuit model with changes of the delay time $r$. Various interesting phenomena were shown in the simulated data. Some typical data are shown in the figures.

Figure 17 shows simulated time series data for blood pressure (BP), RR interval, and stroke volume (SV) with time delay 0.6 second. At the right side is the reconstructed attractor of the simulated time series with BP, HR, and SV. In the time series data of these parameters, all signals showed a damping oscillation. The whorl that converges in a point was shown in the reconstructed three-dimensional attractor.

Figure 18 shows simulated time series data with time delay 1.5 second. In the time series data of these parameters, all signals showed a constant periodic oscillation. In the

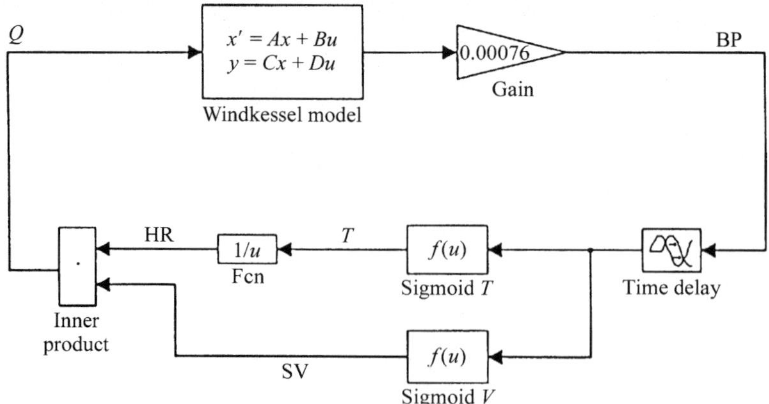


**Figure 16** Block diagram of the electrical circuit simulation for systemic circulation with a feedback loop.

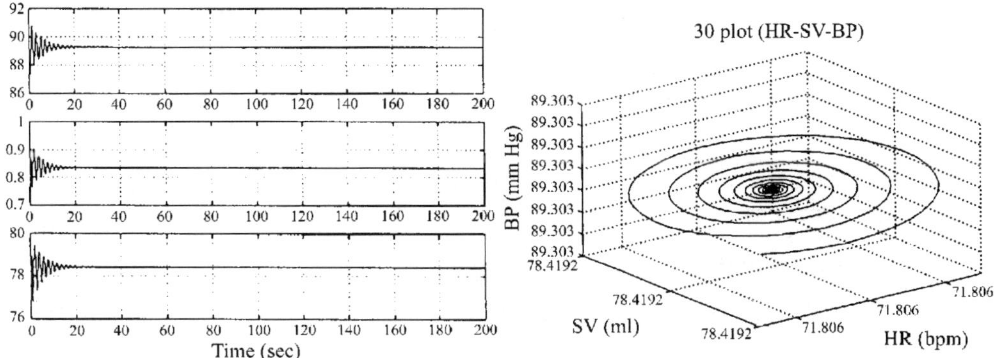


**Figure 17** Simulated results: $r = 0.6$ second.

reconstructed three-dimensional attractor, the orbit converged and formed a circle. This shows a so-called the limit cycle attractor.

A complex result with time delay 1.8 second is presented in Figure 19. The simulated cardiovascular signals showed an interesting time series suggesting a periodic oscillation with two mountains. When we reconstructed these data into a three-dimensional map, the orbit converged and became a track with two rounds.

A complex time series with time delay 2.5 seconds is presented in Figure 20. The simulated cardiovascular signal showed a complex oscillation without any rule. We tried to embed these time series into three dimensions. At the right side of the figure, the orbit, which does not converge is drawn.

We considered the behavior with time delay 2.5 seconds. The time series data and power spectrum analysis of the simulated heart rate variabilities were shown in Figure 9. The time series waveform of HR becomes a complex oscillation like that of other cardiovascular signals. In the frequency spectrum the biggest peak is approximately 0.12 Hz with many mountains.

Note that we used the variable initial value change with a 2.5 second delay. Figure 10 shows correlation estimated cardiac output and the following cardiac output detected at 2.9 seconds (five beats) after rolling averages of LAP and AoP.

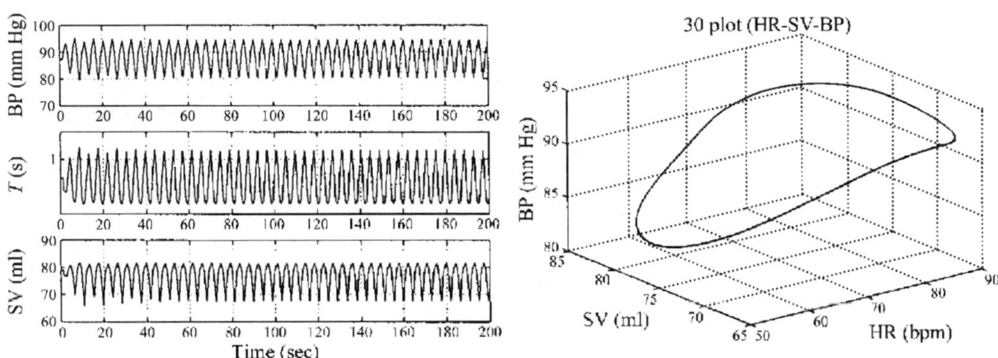


**Figure 18** Simulated results: $r = 1.5$ seconds.

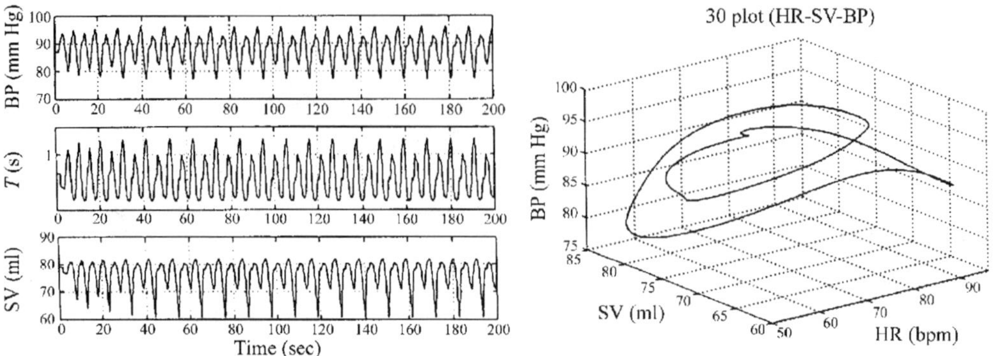


**Figure 19** Simulated results: $r = 1.8$ seconds.

## 4.4. Model and Time Lag

By use of information concerning the autonomic nervous system such as sympathetic nerve activity of the Mayer wave in the peripheral vascular resistances, we can make a feedback loop for an artificial heart automatic control system. In this study, we made a simulation model by use of an electrical circuit model. Behavior changes in the simulation model altering the delay time could be shown in figures. When the delay in simulation was comparatively small, all signals showed damping oscillation. The whorl that converges in a point was shown in the reconstructed three-dimensional attractor. If we set up a larger delay in the simulation circuit, periodic oscillation and then a period doubling phenomenon were shown in the reconstructed attractor. After repetition of the period doubling according to the increase in the delay time, we could get a strange attractor suggesting the existence of deterministic chaos when we set the delay time of 2.5 seconds in this simulation model electrical circuit. This complex oscillation did not disappear even when the delay was set to be more than 2.5 seconds.

There is a point showing the characteristics that indicated deterministic chaos when one considers this condition. At first, the frequency spectrum of the simulated heart rate variability showed it. As we have illustrated, the power spectrum has a peak

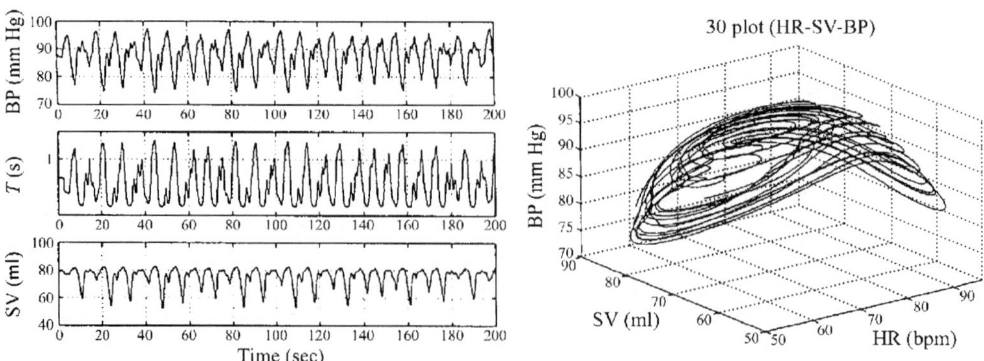


**Figure 20** Simulated results: $r = 2.5$ seconds.

on the section that was not of harmonics composition, and many other small mountains were found. These many spectra were considered to have formed a complex signal. This spectral pattern was, of course, one of the characteristics of deterministic chaos. Only the harmonics composition shows a peak in the case of a limit cycle.

A second consideration was the behavior of the simulated data as the initial value changes. One of the fundamental characteristics of chaos is a sensitive initial value dependence. We altered the initial value a little and performed simulation. The time series waveform showed oscillation that was different. This phenomenon backed up the existence of deterministic chaos.

When a spectrum was considered, a peak having the largest power in 0.12 Hz was found. This was, of course, a peak equivalent to a Mayer wave. Fluctuation of the period for approximately 10 seconds occurred, which was very special in the animals. Use of a model with this simple electrical circuit caused this interesting fluctuation. In this simulation, we might consider that the delay would cause this phenomenon.

### 4.5. Nonlinearity and Simulation

A nonlinear curve, which is, of course, common in animals, was considered at this time. From blood pressure, determination of the cardiac cycle and SV was performed with nonlinear curves of sigmoid style. Of these two curves, the curve representing flow determined from blood pressure was drawn. In this simulation, the operating range of blood pressure was from 70 to 100 mm Hg, similar to that in humans. The section of graph had to be considered. In this range, the curve was convex, resembling a logistic function. This indicated that "stretching" and "holding" might have occurred. It is considered to have caused the condition, which was chaotic.

Of course, this electrical simulation model still has some problems:

1. The complex cardiovascular system was simulated with a simple electric circuit model.
2. Stroke volume was determined from arterial blood pressure.
3. Simulation is only of the systemic circulation by the left heart.

It was thought that the first problems might be important. We need to consider whether this model was able to simulate the original living body and to what degree. The model of the baroreflex series is made using real data from animals, and further simulation needs to be tried in the future. The SV depends on the pressure of the atrium, which is preload. In the electrical simulation model of this study, it is determined only by the arterial blood pressure, which is afterload. And, of course, the behavior of the pulmonary circulation is important. Accordingly, this baroreceptor reflex model is not yet adequate. The baroreceptor reflex system of the living body will not be expressed in this simple electrical simulation model.

However, chaotic dynamics may be caused with a simple three-element circuit with some delay. These results suggest that we could achieve chaotic dynamics such as that in the human body by use of a feedback automatic control algorithm of an artificial heart system. These results may be very important in the future when artificial organs become very common and the QOL of people with artificial organs becomes important. Therefore, studies should be continued with more refined procedures.

## 5. MAKING THE FEEDBACK LOOP IN CIRCULATION BY AN ARTIFICIAL HEART SYSTEM

### 5.1. Biological Information for Artificial Heart Control

Is information about the biological system necessary for an artificial heart? It may not be if one does not think about QOL. A patient with an implanted artificial heart will eventually return to society. Such a patient will experience various kinds of external stimulation and the artificial heart must respond to these conditions. If homeostasis is disturbed by stimulation, it must be restored. If these requirements are not filled, the circulation cannot be maintained and, of course, homeostasis cannot be maintained. Accordingly, external stimulation must be responded to. Thus, some sensor is necessary to respond to turbulence of the circulation. A manifestation of turbulence of the circulation is alteration of the blood pressure. If blood pressure cannot be maintained, circulation cannot be maintained. There is an arterial blood pressure reflex mechanism to maintain blood pressure in biological systems. This system will become necessary to maintain blood pressure in an artificial heart circulation.

### 5.2. Automatic Control for an Artificial Heart

In 1994, Abe et al. designed an 1/R control algorithm for an artificial heart [3, 22]. With this system, vascular resistance is calculated and fed back to the artificial heart control algorithm. Accordingly, this algorithm can become a control mechanism like the arterial baroreflex system of biological systems.

However,, we must maintain the optimal drive point of the artificial heart to prevent thrombus formation, and left and right heart flow balances are important. This concept has been thought to have priority.

In this research, optimal control of the artificial heart, right and left flow balance, and automatic control, which maintained hemodynamics within the normal range, was produced experimentally. In addition, we fed back vascular resistance at every one heart rate. In other words, we created the control algorithm to have the arterial blood pressure reflex while maintaining circulation. We examined this algorithm in Moc circulation and through acute and chronic animal experiments.

The next important issue of this study is discussed below. Several investigators suggested that hemodynamic fluctuations such as heart rate variability (HRV) originated from some biological feedback mechanisms. In particular, the Mayer wave around 0.04–0.15 Hz in HRV was suggested to originate from the arterial baroreflex system. It was an interesting problem to confirm this thesis and investigate the spectral analysis of this artificial baroreflex system.

### 5.3. Automatic Control Experiments

Before starting the animal experiments, we tested this automatic control algorithm in a Moc circulation model during changes of the peripheral vascular resistance. After confirming the control algorithm, we started acute animal experiments first.

Three goats were kept fasting 2 days before the experiments. After inhalation anesthesia with halothane, the left chest cavity was opened through the fourth intercostal space. Inflow and outflow PVC cannulae were inserted into the left atrium and descending aorta, respectively. Then, right inflow and outflow PVC cannulae were inserted into the right atrium and pulmonary artery, respectively. Both sets of cannulae were connected to our TH-7 pneumatically driven sac-type blood pump by built-in silicone ball valves. ECG, central venous pressure, left atrial pressure, arterial pressure, and both sides of pump flow were recorded during the experiments. Both pump flows were measured with an electromagnetic flowmeter on the cannulae. After the animal preparation, the ascending aorta and pulmonary artery were completely ligated to constitute total biventricular bypass with the artificial heart.

The TAH automatic control algorithm was based on this concept. First, the automatic control algorithm maintained the optimal drive point of the inner sac for the pneumatic driver. It was needed to prevent thrombus formation in the inner sac. Second, by altering the stroke volume, left and right flow balances were maintained. Stroke volume changes were maintained within an optimal operating point. Finally, maintaining the hemodynamics within a normal range and an automatic TAH control algorithm based on the artificial baroreflex concept were added to this basic control algorithm. By drug administration, arterial blood pressure was altered to observe the behavior of the automatic TAH control algorithm.

Chronic animal experiments using three healthy adult goats weighing $46 \pm 5$ kg were performed to confirm the usefulness of this automatic control algorithm. After anesthesia was induced by halothane inhalation, the left pleural cavity was opened through the fourth intercostal space under mechanical ventilation. ECG electrodes were put on the pericardium. Biventricular bypass was implanted as in the acute animal experiments. Before we closed the chest cavity, the pulmonary artery was completely ligated to constitute the total right heart bypass and the ascending aorta was almost ligated to prevent myocardial ischemia. Only coronary flow was maintained with the natural heart. Left and right atrial pressures and arterial blood pressures were measured from the side hole of the cannulae, and pump flow was measured with an electromagnetic flowmeter on the cannulae.

After the chest was closed, the goats were placed in a cage and extubated after waking. The hemodynamic derivatives were recorded with a data recorder. The data were input to the personal computer system through an AD converter.

### 5.4. Producing Chaos in Animal Experiments

Hemodynamics was maintained within the normal range by this automatic control algorithm. Right and left balance, too, was maintained. The reaction to an administered agent in the acute animal experiment is shown in Figure 21. When a vasoconstrictive agent is administered, blood pressure rises. Methotrexate was administered here. The elevation of vascular resistance is shown in Figure 21. Vascular resistance is calculated every one beat for the input in this automatic control algorithm. The reaction of the artificial heart after the elevation of vascular resistance was that the heart rate (beats/min) fell immediately. Thus blood pressure falls and homeostasis is maintained.

The most important results of the study were that the artificial baroreflex system was embodied by the basic TAH automatic control algorithm with optimal TAH drive and left-right balances, hemodynamic parameters were kept within the normal range and 1/R control was added to these basic concepts. Of course, it was important to maintain the optimal TAH drive to prevent thrombus formation and endurance of the

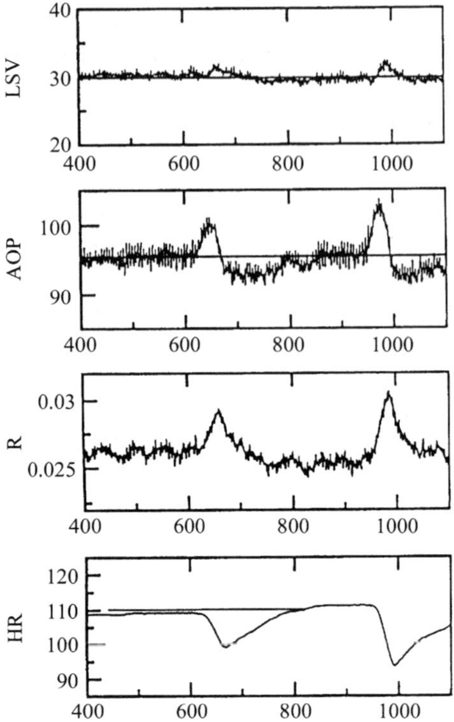


**Figure 21** Reaction to an administered agent in the acute animal experiment. When a vasoconstrictive medicine agent is administered, blood pressure rises. Vascular resistance is calculated every one beat for the input in this automatic control algorithm. The reaction of the artificial heart was immediate. The artificial heart rate (beats/min) fell immediately after elevation of the vascular resistance. Hence, blood pressure falls, and homeostasis is maintained.

inner sac. Left and right heart balances were important to maintain peripheral and pulmonary circulation and prevent pulmonary congestion. And basic hemodynamic parameters were needed to maintain the peripheral circulation and blood supply for each organ. Finally, the baroreflex system was important to maintain homeostasis in the patient with an artificial heart. In this study, this stepwise automatic control was embodied.

Various investigators have studied the circulatory regulatory system during artificial circulation. However, the automatic control algorithms may alter all previous reports. Thus, for a physiologically optimal TAH drive, further study may be needed in the future.

In the next step, we performed chronic animal experiments and nonlinear dynamic analysis.

## 6. NONLINEAR DYNAMICS OF AN ARTIFICIAL BAROREFLEX SYSTEM

### 6.1. Approaches to Nonlinear Dynamics

Nose [23] concluded from many experiments that no man-made computer servo system should be incorporated for TAH control. To evaluate the TAH automatic control algorithm, the total system must be analyzed as an entity, not as single parts such as the autonomic nervous system, hormonal factors, and so on. Progress in nonlinear mathematics enables us to estimate the whole system [1, 2].

During the past two decades, many researchers have investigated nonlinear dynamics in the cardiovascular system [1–3]. Guevara et al. [1] described the cellular and subcellular mechanisms that produce the cardiac action potential, known to be characterized by a chaotic attractor. Other investigators reported that nonoscillatory cardiac tissues also manifest nonlinear dynamics [2–4]. Of course, cardiovascular function is characterized by a complex interaction of many control mechanisms that permit adaptation to the changing environment. These complexities from the field of nonlinear dynamics made it difficult to analyze the circulatory regulatory system quantitatively.

Nonlinear dynamics coinciding with deterministic chaos may generate random-like time series, which on closer analysis appear to be highly ordered and critically dependent on the initial conditions. Mathematically, all nonlinear dynamic systems with more than two degrees of freedom can generate deterministic chaos, becoming unpredictable [1–3]. Using nonlinear mathematical methodology, we can obtain attractors of nonlinear dynamic behavior in the phase space. Attractors displaying chaotic behavior are termed strange attractors and have a sensitive dependence on initial conditions.

## 6.2. Baroreflex System for an Artificial Heart

In linear systems, attractors are, of course, not strange. Attractors of ordinary differential equations with a degree of freedom below 2 are limited to either a fixed point or a limited cycle and have proved not to be strange. In a strange attractor, the stretching and folding operations smear out the initial volume, thereby destroying the initial information as the system evolves and the dynamics create new information. In this study, to evaluate the circulation with the TAH with an experimentally constructed baroreflex system from the viewpoint of the whole system, the hemodynamic parameters with TAH were analyzed by nonlinear mathematical techniques useful in the study of deterministic chaos. Hemodynamic parameters were recorded and analyzed in chronic animal experiments with healthy adult female goats weighing 48 and 60 kg.

After resection of the natural heart under extracorporeal circulation, pneumatically driven sac-type blood pumps were connected and set outside the body on the chest wall. The only medication used in these experiments was antibiotics to avoid infection. Pump output was measured at the outlet port of the left pump with an electromagnetic flowmeter. Aortic pressure and left and right atrial pressure were measured with pressure transducers through the side tube of the cannulae. These data were fed through an AD converter into a personal computer in which the control logic was run.

In these experiments, the TAH automatic control algorithm maintained the optimal drive point of the inner sac for the pneumatic driver in order to prevent thrombus formation in the inner sac. Second, by the alteration of the stroke volume, left and right flow balances were maintained. Stroke volume changes were maintained within an optimal operating point. Finally, to maintain the hemodynamics within the normal range, an automatic TAH control algorithm based on the artificial baroreflex concept was added to the basic control algorithm. The TAH drive parameters used in the experiments were pulse rates, drive air pressure, and systolic-diastolic rate. Dynamic balancing of blood volumes between the pulmonary and systolic circulations controlled the left-right balance.

In this study, we used the embedding technique proposed by Takens [24] to evaluate the hemodynamics of 1/R control. For a variable $x(t)$ denoting a time series datum of the hemodynamics, we considered new variables $y(t) = x(t + T)$, $z(t) = x(t + 2T)$,

$w(t) = x(t + 3T), \ldots$, where $T$ is the order of the correlation time. Using this embedding technique, we reconstructed the attractor in the phase space. It is well known that time series data of a chaotic system are similar to time series data of a random system, but the phase portrait of the chaotic system shows a pattern completely different from that of the random system. The reconstructed attractor of the chaotic dynamics is found to be a strange attractor.

The Lyapunov exponent calculated from the reconstructed attractor in the phase space was widely used in the application of nonlinear mathematics for a pragmatic measure of chaos as shown before.

### 6.3. Nonlinear Dynamical Behavior of an Artificial Baroreflex System

In this study, all the hemodynamic parameters were embedded in the phase space and projected into the three-dimensional phase space. Time series data of the pressure pattern were recorded through the side branch of the cannulae, and continuous monitoring of the hemodynamics was easily obtained with this side branch system, while the mean pressures were useful for the automatic control of the TAH. However, the water hammer effect of the prosthetic valves in the TAH cannot be avoided. Thus, the phenomenon most evident in the pressure pattern attractor is thought to be the effect of the water hammer, so we cannot detect the hemodynamic behavior exactly using this method.

Therefore, the phase portrait of the pump output flow pattern during the TAH fixed driving condition was embedded in the four-dimensional phase space and projected into the three-dimensional phase space as shown in Figure 22. This attractor was projected from a higher dimensional phase space, so we cannot describe the dimensions of the axis. A simple attractor suggesting a limited cycle attractor coinciding with a periodic system is shown. Figure 23 shows the reconstructed attractor of the pump output with automatic TAH control. To evaluate the patterns of the phase portrait, we calculated the Lyapunov exponents from the reconstructed attractor of

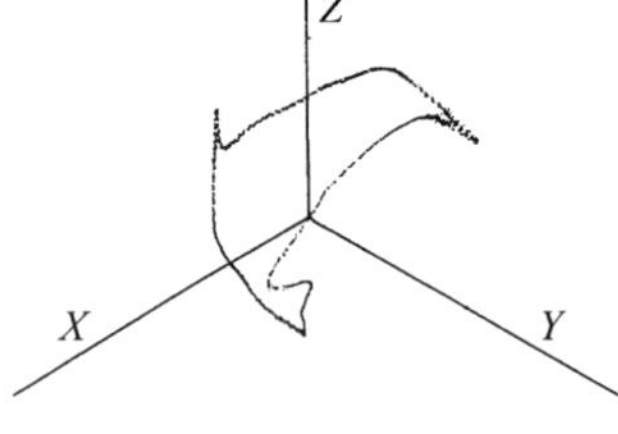


**Figure 22** Phase portrait of the artificial pump outflow embedded in the four-dimensional phase space and projected into three-dimensional phase space during a TAH fixed drive condition.

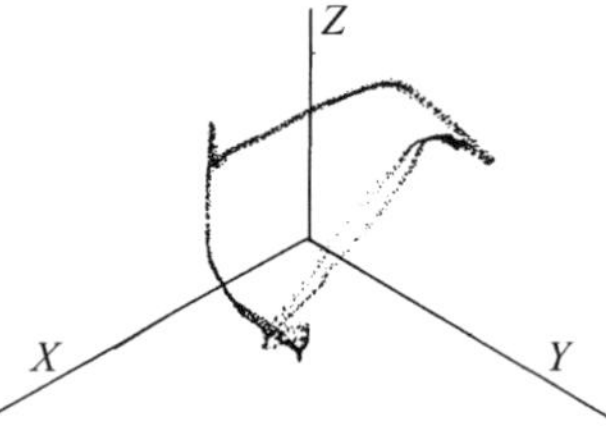


**Figure 23** Phase portrait of the artificial pump outflow embedded in the four-dimensional phase space and projected into three-dimensional phase space during automatic TAH control.

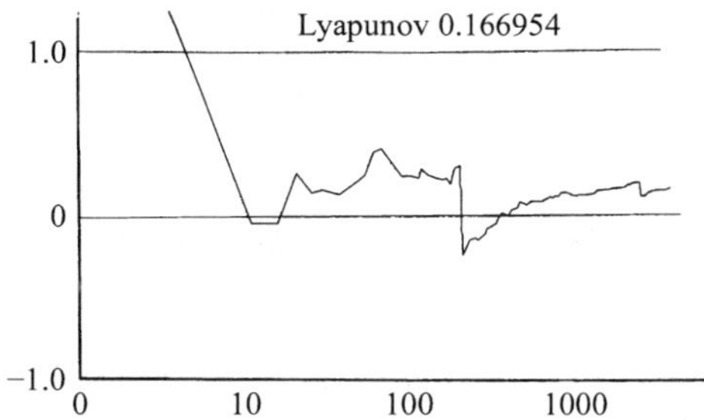


**Figure 24** Lyapunov exponents calculated from the time series data for the pump output during 1/R TAH automatic control.

the time series data. At least one positive large Lyapunov exponent suggests a strange attractor, which is a feature of nonlinear dynamics with deterministic chaos. Figure 24 shows the calculation of the largest Lyapunov exponents during TAH automatic control. Positive Lyapunov exponents suggest a sensitive dependence on the reconstructed attractor.

One of the main findings in this study is that the reconstructed attractor of the artificial heart pump output suggests the formation of a strange attractor in the phase space during the automatic TAH control algorithm. The strangeness of the reconstructed attractor was measured by reconstructing it and using Lyapunov exponents. A wider band in the reconstructed attractor and positive Lyapunov exponents during the segmented time series were seen. Our results suggest that the strangeness of the reconstructed attractor appears to have the largest value, indicating higher dimensional strong chaos with the TAH automatic control algorithm.

In this study, the chaos was determined by reconstructing the attractor in the four-dimensional phase space and calculating the largest Lyapunov exponent. A simple reconstructed attractor in the phase space during fixed TAH driving and the positive largest Lyapunov exponents suggest a lower dimensional chaotic system. Compared with the fixed driving condition, the TAH automatic control algorithm showed more complicated nonlinear dynamics suggesting higher dimensional deterministic chaos. Although several investigators have been studying nonlinear dynamics including deterministic chaos, there is no universally accepted definition. The widely accepted characteristics of deterministic chaos are

1. A deterministic dynamic system
2. A sensitive dependence on initial conditions
3. An attractor

Many investigators have noted the functional advantages of the chaotic system [2–6]. In some clinical cases, pathologies exhibit increasing periodic behavior and loss of variability [2–6]. In nonlinear dynamics, a chaotic system operates under a wide range of conditions and is therefore adaptable and flexible compared with a periodic system [6]. When unexpected stimuli are applied from outside, this plasticity allows whole systems to cope with the requirements of an unpredictable and changing environment [2–6]. For example, Tsuda et al. [6] found deterministic chaos in the pressure waveform of the finger capillary vessels in normal and psychiatric patients and proposed the notion of a "homeochaotic" state. From this standpoint, a cardiovascular regulatory system with TAH fixed driving shows a lower dimensional limit cycle attractor, suggesting a lower dimensional dynamic system, so fixed TAH driving may be in the lower dimensional homeochaotic state. Thus, hemodynamic parameters can be thought of as irritated by external turbulence compared with automatic TAH driving. These results suggest that TAH automatic control algorithm may be suitable when unexpected stimuli are applied from outside.

## 7. CLINICAL APPLICATION OF THE ORIGIN OF CHAOS

### 7.1. Approaches for Clinical Application

The foregoing results showed that chaotic dynamics in the circulation may be generated from the peripheral vascular properties mediated by sympathetic discharges. If we consider the clinical application of these results, autonomic function mediating cardiovascular function may be a suitable target. Patients with diabetes mellitus and patients with myocardial ischemia showed cardiac denervation, which is an important problem for these patients. In the clinic, we used heart rate variability to analyze the autonomic function; however, there are some problems.

In clinical medicine the trend is toward low invasiveness and short time frames. For example, the therapy for serious coronary artery disease was a bypass operation a short while ago. Now, percutaneous transluminal coronary angioplasty (PTCA) by catheter intervention has become the mainstay. Operation with low invasiveness by abdominal scope or laparoscope attracts attention in abdominal surgery. In diagnostics, more correct information obtained by more indirect procedures is demanded. Accordingly, nonlinear dynamics must now be pursued using an indirect index.

An example is heart rate variability. In ECG, a biological electrical phenomenon can be measured with a circuit, which is simple and easy. Information related to the pathophysiology, such as myocardial ischemia from a waveform, can be detected. This methodology is, of course, in broad clinical use. For the purpose of detecting correct biological system information, this methodology does not always fill the requirements completely. The RR interval used in research on heart rate variability is an example. It is assumed that the R wave of the ECG corresponds to systole.

Unfortunately, accumulation of data in clinical medicine suggests that there can be patho-physiologic situations in which this prerequisite is not right. The state of a disease of electromechanical dissociation may be a typical example.

### 7.2. Nonlinear Dynamics in the Time Series Data of the Left Ventricular Stroke Volume

Correct instrumentation is assumed in research on the nonlinear dynamics of hemodynamics. Because systole may not always be associated with the ECG in clinical medicine, the RR interval method was excluded. First, movement of the heart must be detected to examine the alteration of hemodynamics. However, it is not always simple to observe the behavior of the heart directly. There is image analysis by conductance catheter, nuclear medicine, magnetic resonance imaging, or computed tomography. But there are some problems related to the noninvasiveness or observation for a long time. These methods are not suitable for long-time measurements with stability.

In this research, instrumentation of the stroke volume measurement by use of cardiac ultrasonography was attempted. This methodology was used for the analysis of nonlinear hemodynamics. The M-mode image of the left ventricle provided a time series curve of the systolic dynamics of the left ventricular volume. A tracing image of the M-mode inside diameter was input into a computer with an image reader. Based on the theory of Takens, the time series curve provided was embedded in a phase space of high dimension, and nonlinear analysis was attempted. The value at time $t$, $X(t)$, is represented on the $x$-axis. $X(t+T)$, which is the value after time $T$, is represented in the $y$-axis, and $X(t+2T)$, after $2T$, on the $z$-axis. In this way, plotting was performed.

A change of trajectory with time in the phase space was observed. Of course, it is the determination of time lag that becomes important here. Many procedures have been proposed in the past, involving fundamental periods, attractors, and autocorrelation functions.

An attractor related to systole sought for with such a procedure is shown in Figure 25. The upper section is for a person who is not physically handicapped and the lower for a myocardial infarct patient with a past history of left anterior descending artery occlusion.

At first glance, we understand that the attractor becomes moderated. Even in the two-dimensional phase plane it seems possible to describe the findings. Alteration to a lower dimension of the dynamic system controlling attractor is suggested.

Accordingly, application to quantitative diagnosis is prospective. Abnormality of innervation in a patient with a myocardial infarct has been observed in the past by nuclear medicine. It is thought that a more interesting result would be obtained by comparison with the current procedure.

By further application of these nonlinear mathematical analysis methodologies, we can anticipate development of the diagnostics for patients with cardiovascular regulatory system disturbances, especially from the viewpoint of the whole system.

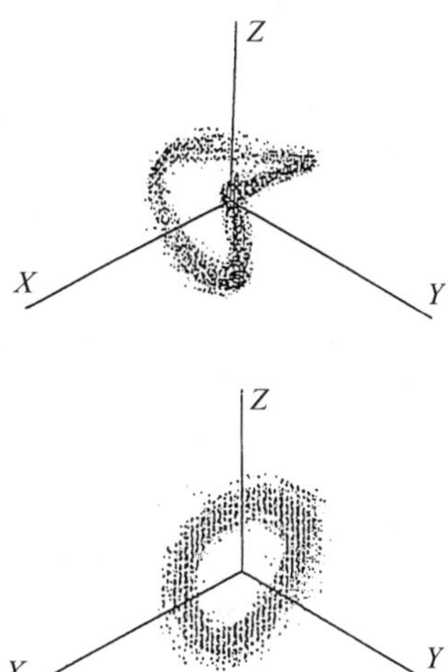


**Figure 25** Examples of the attractors of the stroke volume time series data. The upper section is for a person who is not physically handicapped and the lower for a myocardial infarct patient with anamnesis with the left anterior descending artery occluded.

## ACKNOWLEDGMENTS

The mathematical methodology of this chapter was discussed in the workshop "Various Approaches to Complex Systems" held at the International Institute for Advanced Studies in Kyoto, Japan. The authors thank Mr. Kimio Kikuchi for experimental cooperation and Miss Rie Sakurai, Mrs. Takako Iijima, and Mrs. Hisako Iijima for their excellent assistance and cooperation. This work was partly supported by a Grand-in-Aid for Encouragement of Young Scientists (70241578), a Research Grant for Cardiovascular Diseases from the Ministry of Health and Welfare, and the Program for promotion of Fundamental Studies in Health Sciences of the Organization for Drug ADR relief, R&D promotion, and product review of Japan.

## REFERENCES

[1] M. R. Guevara, L. Glass, and A. Shirer, Phase locking period doubling bifurcations, and irregular dynamics in periodically stimulated cardiac cells. *Sciences* 214: 1981.

[2] J. P. Crutchfield, J. D. Farmer, N. H. Packard, and R. S. Shaw, Chaos. *Sci. Am.* 255: 1986.

[3] T. Yambe, Y. Abe, M. Yoshizawa, K. Imachi, K. Imachi, K, Tabayashi, H. Takayasu, H. Takeda, K. Gouhara, and S. Nitta, Strange hemodynamic attractor parameter with 1/R total artificial heart automatic control algorithm. *Int. J. Artif. Organs* 19: 1996.

[4] T. Yambe, S. Nitta, T. Sonobe, S. Naganuma, Y. Kakinuma, K. Izutsu, H. Akiho, S. Kobayashi, N. Ohsawa, S. Nanka, M. Tanaka, T. Fukuju, M. Miura, N. Uchida, N. Sato, K. Tabayashi, S. Koide, K. Abe, H. Takeda, and M. Yoshizawa, Deterministic chaos in the hemodynamics of an artificial heart. *ASAIO J.* 41: 1995.

[5] A. Wolf, J. B. Swift, H. L. Swinney, and J. A. Vastano, Determining Lyapunov exponents from a time series. *Physica D* 16: 1985.

[6] I. Tsuda, T. Tahara, and H. Iwanaga, Chaotic pulsation in human capillary vessels and its dependence on mental and physical conditions. *Int. J. Bifurc. Chaos* 2: 1992.

[7] T. Yambe, S. Nitta, Y. Katahira, M. Tanaka, N. Satoh, M. Miura, H. Mohri, M. Yoshizawa, and H. Takeda, Effect of left ventricular assistance on sympathetic tone. *Int. J. Artif. Organs* 13: 1990.

[8] T. Yambe, S. Nitta, Y. Katahira, T. Sonobe, S. Naganuma, H. Akiho, H. Hayashi, M. Tanaka, M. Miura, N. Satoh, H. Mohri, M. Yoshizawa, and H. Takeda, Postganglionic sympathetic activity with correlation to heart rhythm during left ventricular assistance. *Artif. Organs* 15: 1991.

[9] A. M. Fraser and H. L. Swinney, Independent coordinate for strange attractors from mutual information. *Phys. Rev. A* 33: 1986.

[10] D. T. Kaplan and R. J. Cohen, Is fibrillation chaos? *Circ. Res.* 67: 1990.

[11] W. C. Randall, *Nervous Control of Cardiovascular Function*. Oxford: Oxford University Press, 1984.

[12] L. Barajas and P. Wang, Localization of tritiated norepinephrine in the renal arteriolar nerves. *Anat. Rec.* 195: 1979.

[13] T. Yambe, S. Nitta, S. Chiba, Y. Saijoh, S. Naganuma, H. Akiho, H. Hayashi, M. Tanaka, M. Miura, N. Satoh, H. Mohri, M. Yoshizawa, and H. Takeda, Estimation of the following cardiac output using sympathetic tone and hemodynamics for the control of total artificial heart. *Int. J. Artif. Organs* 15: 1992.

[14] K. W. Hiller, W. Siegel, and W. J. Kolff, A servomechanism to drive an artificial heart inside the chest. *ASAIO Trans.* 8: 1962.

[15] A. J. Snyder, G. Rosenberg, J. Reibson, J. H. Donachy, G. A. Prophet, J. Areas, B. Daily, S. McGary, O. Kawaguchi, R. Quinn, and W. S. Pierce, An electrically powered total artificial heart: Over 1 year survival in the calf. *ASAIO J.* 38: 1992.

[16] T. Yambe, S. Nitta, T. Sonobe, S. Naganuma, Y. Kakinuma, S. Kobayashi, S. Nanka, N. Ohsawa, H. Akiho, M. Tanaka, T. Fukuju, M. Miura, N. Uchida, N. Sata, H. Mohri, M. Yoshizawa, S. Koide, K. Abe, and H. Takeda, Chaotic behavior of the hemodynamics with ventricular assist device. *Int. J. Artif. Organs* 18: 1995.

[17] T. Yambe, S. Nitta, T. Sonobe, S. Naganuma, Y. Kakinuma, S. Kobayashi, S. Nanka, N. Ohsawa, H. Akiho, M. Tanaka, T. Fukuju, M. Miura, N. Uchida, N. Sato, H. Mohri, M. Yoshizawa, S. Koide, K. Abe, and H. Takeda, Fractal dimension analysis of the oscillated blood flow with vibrating flow pump. *Artif. Organs* 19: 1995.

[18] S. Akselrod, D. Gordon, J. B. Madwell, N. C. Snidman, D. C. Shannon, and R. J. Cohen, Hemodynamic regulation: Investigation by special analysis. *Am. J. Physiol.* 249: 1985.

[19] T. Yambe, S. Nitta, T. Sonobe, S. Naganuma, Y. Kakinuma, S. Kobayashi, S. Nanka, N. Ohsawa, H. Akiho, M. Tanaka, T. Fukuju, M. Miura, N. Uchida, N. Sato, H. Mohri, M. Yoshizawa, S. Koide, K. Abe, and H. Takeda, Origin of the rhythmical fluctuations in hemodynamic parameters in animal without natural heartbeat. *Artif. Organs* 17: 1993.

[20] S. Cavalcanti and E. Belardinelli, Modeling of cardiovascular variability using a different delay equation. *IEEE Trans. Biomed. Eng.* 43: October 1996.

[21] S. Cavalcanti, S. Severi, and C. Boarini, Mathematical analysis of the autonomic influence on the heart rate variability. *Proceedings 18th Annual Conference IEEE Engineering in Medical Biological Soc* (CD-ROM), 1996.

[22] Y. Abe, T. Chinzei, K. Imachi, et al., Can artificial heart animals control their TAH by themselves? One year survival of a TAH goat using a new automatic control method (1/R control). *ASAIO J.* 40: 1994.

[23] Y. Nose, Computer control of cardiac prosthesis: Is it really necessary? *Artif. Organs* 18: 1994.

[24] F. Takens, Detecting strange attractors in turbulences. In *Lecture Notes in Mathematics*, D. A. Rand and Y. S. Young, eds, Berlin: Springer-Verlag, 1981.

Chapter 3

# APPROXIMATE ENTROPY AND ITS APPLICATION IN BIOSIGNAL ANALYSIS

Yang Fusheng, Hong Bo, and Tang Qingyu

## 1. INTRODUCTION

Nonlinear dynamic analysis may be a powerful tool to reveal the characteristics and mechanism of biosignals because physiological systems are basically nonlinear in nature. That is why it has been used widely in recent years to analyze various biosignals. But a very long data sequence is needed to estimate accurately many of the nonlinear parameters currently used (according to Wolf et al. [1], the quantity of experimental data required for accurate calculation is $10^m$–$30^m$ points at least, where $m$ is the dimension of the attractor), which hinders practical use. In engineering practice, the data obtained experimentally are often short in length and our main interests are not in a precise overall description of the attractor. The goal is usually to classify the data for pattern recognition. Moreover, the dynamic evolution of the ongoing process must be described. For these purposes, what we actually need are some nonlinear parameters that can be estimated robustly using shorter data. In this respect, approximate entropy may be one of the parameters that meet the need.

Approximate entropy (abbreviated ApEn hereafter) is a statistic that can be used as a measure to quantify the complexity (or irregularity) of a signal. It was first proposed by Pincus [2] in 1991 and was then used mainly in the analysis of heart rate variability [3–5] and endocrine hormone release pulsatility [6, 7]. It is also used to analyze various other physiological signals such as the electrocardiogram (ECG), the electroencephalogram (EEG), and respiration. Its application is spreading rapidly as shown in Table 1, which lists the number of papers related to ApEn entered in the Science Index (ScI) from 1994 to 1997.

The following salient features of ApEn make it attractive for use in signal processing:

1. A robust estimate of ApEn can be obtained by using shorter data—in the range 100–5000 points, with 1000 used most often.
2. It is highly resistant to short strong transient interference (i.e., outliers or wild points).
3. The influence of noises can be suppressed by properly choosing the relevant parameter in the algorithm.

**TABLE 1** Number of Papers Related to ApEn That Were Entered in the Science Index

| Year | 1994 | 1995 | 1996 | 1997 |
|---|---|---|---|---|
| Number | 8 | 10 | 14 | 22 |

4. It can be applied to both deterministic (chaotic) and stochastic signals and to their combinations.

The first three properties make it suitable to the analysis of experimentally obtained, noise-contaminated short data. The last property is beneficial for biosignal analysis, because the biological system is so complex that the output usually consists of both deterministic and random components.

Definition, properties, a fast algorithm, and examples of application will be introduced in this chapter.

## 2. DEFINITION AND INTERPRETATION

The definition of ApEn will be given first, followed by some interpretations.

### 2.1. Definition

The definition of ApEn is closely related to its mathematical formulation. Therefore, we will introduce its definition step by step along with its computing procedures.

Let the original data be $\langle x(n) \rangle = x(1), x(2), \ldots, x(N)$, where $N$ is the total number of data points. Two parameters must be specified before ApEn can be computed: $m$, the embedding dimension of the vector to be formed, and $r$, a threshold that is, in effect, a noise filter.

1. Form $m$-vectors $X(1)$, $X(N-m+1)$ defined by:

$$X(i) = [x(i), x(i+1), \ldots, x(i+m-1)] \qquad i = 1, N-m+1$$

2. Define the distance between $X(i)$ and $X(j)$, $d[X(i), X(j)]$, as the maximum absolute difference between their corresponding scalar elements, i.e.,

$$d[X(i), X(j)] = \max_{k=0,m-1} \left[ |x(i+k) - x(j+k)| \right]$$

(Note that all the other differences between the corresponding elements will then be less than $d$.)

3. For a given $X(i)$, find the number of $d[X(i), X(j)]$ $(j = 1, N-m+1)$ that is $\leq r$ and the ratio of this number to the total number of $m$-vectors $(N-m+1)$, i.e.,

$$\text{Let } N^m(i) = \text{no. of } d[X(i), X(j)] \leq r, \quad \text{then } C_r^m(i) = N^m(i)/(N-m+1)$$

This step is performed over all $i$. That is, find $C_r^m(i)$ for $i = 1, N-m+1$.

4. Take the natural logarithm of each $C_r^m(i)$, and average it over i:

$$\phi^m m(r) = \frac{1}{N-m+1} \sum_{i=1}^{N-m+1} \ln C_r^m(i)$$

5. Increase the dimension to $m+1$. Repeat steps 1–4 and find $C_r^{m+1}(i)$, $\phi^{m+1}(r)$.
6. Theoretically, the approximate entropy is defined as

$$ApEn(m,r) = \lim_{N\to\infty} \left[\phi^m(r) - \phi^{m+1}(r)\right]$$

In practice, the number of data points $N$ is finite and the result obtained through the preceding steps is an estimate of ApEn when the data length is $N$. This is denoted by

$$ApEn(m,r,N) = \phi^m(r) - \phi^{m+1}(r)$$

Obviously, the value of the estimate depends on $m$ and $r$. As suggested by Pincus, $m$ can be taken as 2 and $r$ be taken as $(0.1, 0.25)SD_x$, where $SD_x$ is the standard deviation of the original data $\langle x(n) \rangle$, i.e.,

$$SD_x = \sqrt{\frac{1}{N-1} \sum_{n=1}^{N} \left[ x(n) - \frac{1}{N} \sum_{n=1}^{N} x(n) \right]^2}$$

## 2.2. Interpretation

Let us explain intuitively the meaning of ApEn with the aid of Figure 1, which shows a time series with 30 points.

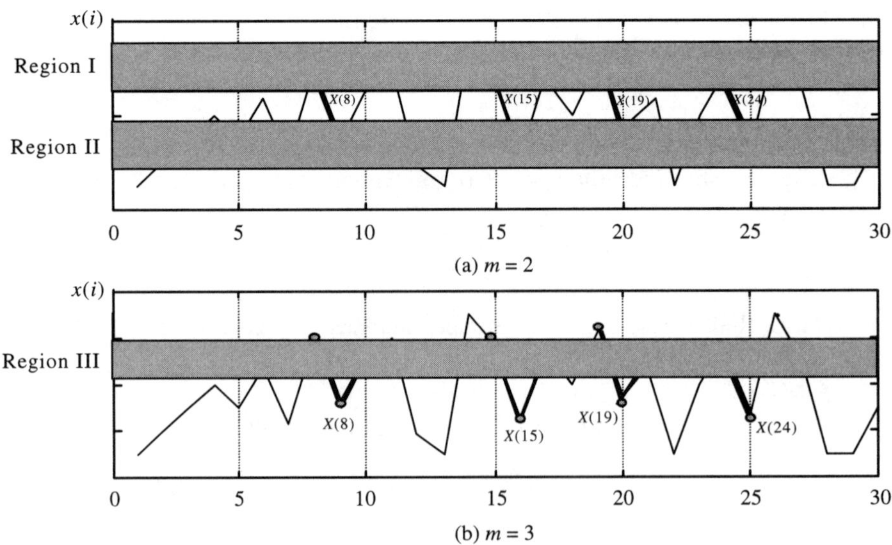


**Figure 1** Graphical interpretation of ApEn.

For $m = 2$, each $X(i) = [x(i)\ x(i+1)]$ is a line segment (i.e., two-point pattern) joining every two neighboring data points [e.g., when $i = 8$, $X(8) = [x(8)\ x(9)]$ is shown as a thick line segment in Figure 1a]. Two horizontal bands I and II, each of width $2r$, can be drawn around $x(i)$ and $x(i+1)$. They are the tolerance regions that satisfy the requirement $d[X(i), X(j)] \leq r$. In other words, when the first element $x(j)$ of a vector $X(j) = [x(j)\ x(j+1)]$ falls inside band I and the second element $x(j+1)$ falls inside band II, then $X(j)$ is a vector satisfying the requirement $d[X(i), X(j)] \leq r$. As shown in Figure 1a, in addition to $X(i = 8)$ itself, there are three other vectors ($X(15) = [x(15)\ x(16)]$, $X(19) = [x(19)\ x(20)]$, and $X(24) = [x(24)\ x(25)]$) satisfying the requirement. Thus $N^{m=2}(i = 8) = 4$ and

$$C_r^{m=2}(i = 8) = 4/(N - m + 1) = 4/29 = 0.1379$$

Intuitively, $N^{m=2}(i)$ is the total number of line segments (i.e., two-point patterns) formed by all the neighboring points in the sequence that are "close" to $X(i)$ within the tolerance $\pm r$ and $C_r^{m=2}(i)$ is the frequency of its occurrence.

Similarly, when $m = 3$, $X(i) = [x(i)\ x(i+1)\ x(i+2)]$ is a three-point pattern formed by joining the three neighboring points (an example $X(8) = [x(8)\ x(9)\ x(10)]$ is shown in Figure 1b) and $N^{m=3}(i)$ is the total number of such three-point patterns $X(j) = [x(j)\ x(j+1)\ x(j+2)]$ in the data sequence that are close to $X(i)$ within the tolerance $\pm r$. In fact, such a pattern could be found by restricting the search to only the $X(j)$'s that satisfy $d \leq r$ when $m = 2$ and checking only the distance between their third element $x(i+2)$ [e.g., $x(10)$ for $i = 8$ in the example] and $x(j+2)$. As shown in Figure 1b for the given example, only $X(15)$ and $X(19)$ satisfy the requirement, but $X(24)$ fails because its third element, $x(26)$, falls outside the tolerance band III of $x(10)$. When $N$ is large, $C_r^{m=3}(i)$ is approximately the probability of occurrence of three-point patterns in the sequence that are close (within the tolerance $\pm r$) to the three-point pattern $X(i) = [x(i)\ x(i+1)\ x(i+2)]$.

Thus,

$$\phi^m(r) = \frac{1}{N - m + 1} \sum_{i=1}^{N-m+1} \ln C_r^m(i)$$

portrays the average frequency that all the $m$-point patterns in the sequence remain close to each other. Why the logarithm is taken before averaging will be explained later.

Finally, the meaning of $ApEn(m, r) = \lim_{N\to\infty} [\phi^m(r) - \phi^{m+1}(r)]$ can be interpreted from two different but related points of view:

First, statistically, ApEn($m$, $r$) can be proved to be the logarithm of the following conditional probability averaged over $i$:

$$\mathrm{Prob}[|x(i+m) - x(j+m)|] \leq r \tag{1}$$

subject to the condition

$$\left|x(i+k) - x(j+k)\right| \leq r, k = 1, m - 1$$

*Proof*

$$
\begin{aligned}
ApEn(m, r) &= \lim_{N\to\infty} [\phi^{m+1} - \phi^m(r)] \\
&= \lim_{N\to\infty} \left[ \frac{1}{N-m} \sum_{i=1}^{N-m} \ln C_r^{m+1}(i) - \frac{1}{N-m+1} \sum_{i=1}^{N-m+1} \ln C_r^m(i) \right] \\
&\cong \lim_{N\to\infty} \left\{ \frac{1}{N-m} \sum_{i=1}^{N-m} [\ln C_r^{m+1}(i) - \ln C_r^m(i)] \right\} \\
&= \lim_{N\to\infty} \left\{ \frac{1}{N-m} \sum_{i=1}^{N-m} \left[ \ln \frac{C_r^{m+1}(i)}{C_r^m(i)} \right] \right\}
\end{aligned}
$$

As mentioned before, when $N$ is large, $C_r^m(i)$ is the probability of the $m$-point patterns in the sequence that are close to the $m$-point pattern formed by $X(i) = [x(i)\ x(i+1)\ \ldots\ x(i+m-1)]$. Thus, $C_r^{m+1}(i)/C_r^m(i)$ is the conditional probability defined by Eq. 1 and ApEn is its logarithm averaged over $i$.

Second, stated in another way $ApEn(m, r) = \phi^m(r) - \phi^{m+1}(r)$ $[m = 2]$ is the difference between the *frequency* that all the two-point patterns in the sequence are close to each other and the *frequency* that all the three-point patterns in the sequence are close to each other. Thus, ApEn($m = 2, r$) expresses the degree of new pattern generation when the dimension $m$ of the pattern decreases from 3 to 2. A larger value of ApEn means the chance of new pattern generation is greater, so the sequence is more irregular (more complex) and vice versa. For instance, ApEn of a periodic signal is low because the possibility of two-point patterns being similar is nearly the same as the probability of three-point patterns being similar. And for white noise, $\phi^{m=2}(r)$, and $\phi^{m=3}(r)$ may differ greatly, so its ApEn is large.

In summary, there are two ways to look at ApEn. From one point of view, it is a statistical characteristic (average of the logarithm of a conditional probability), which makes it applicable to both deterministic and stochastic processes. From the other point of view, it reflects the rate of new pattern generation and is thus related to the concept of entropy.

## 3. FAST ALGORITHM

It is not difficult to write a computer program for ApEn estimation following the definition given in the last section. In fact, a Fortran program is given in Ref. 8 for use. Although there are no complicated calculations involved in the program, considerable time is needed for its execution. So a fast algorithm is needed to speed up the calculation, especially when real-time dynamic estimation of ApEn is required.

The most time-consuming step for ApEn estimation is the calculation of $C_r^m(i)$. Thus, the following discussion will be focused on a fast method for $C_r^m(i)$ calculation.

Let $\langle x(n) \rangle = x(1), x(2), \ldots, x(N)$ be the given data sequence for which ApEn is to be estimated. One may find $C_r^m(i)$ through the following steps:

1. The absolute difference between any two data points $|x(i) - x(j)|$ $(i, j = 1, N)$ is calculated and entered into the corresponding position $d(i, j)$ of an $N * N$ distance

matrix $D$ (Figure 2a). Clearly, $D$ is a symmetrical matrix, $d(j, i) = d(i, j)$, and the elements on its main diagonal $d_{ii}$ ($i = 1, N$) are equal to zero.

2. Compare each $d(i, j)$ with the threshold $r$. The results are entered into the corresponding position of a binary matrix $S$ according to the following rule (Figure 2b):

$$s(i,j) = \begin{cases} 0 & \text{if d(i, j)} > \text{r} \\ 1 & \text{if d(i, j)} \leq \text{r} \end{cases} \qquad i = 1, N, j \leq i$$

$$s(j, i) = s(i, j)$$

$$s(i, i) = 1 \qquad i = 1, N$$

(In actual practice, it is not necessary to set up matrix $D$. Matrix $S$ can be constructed directly from the given data).

3. Find $C_r^m(i)$ for $m = 2$. When $i$ is specified, the $i$th and $(i + 1)$st rows of matrix $S$ can be used to find $C_r^2(i)$ (Figure 3) through the following steps:
   - Find $C_r^2(i, j)$ by $C_r^m(i, j) = s(i, j) \cap s(i + 1, j+1)$ (see the two circled matrix elements in Figure 3). Obviously

$C_r^2(i, j) = 1$, if both $s(i, j)$ and $s(i + 1, j + 1)$ equal 1, which means $d[X(i), X(j)] \leq r$. So $X(j)$ is a vector satisfying the distance tolerance.

$C_r^2(i, j) = 0$, if either $s(i, j)$ or $s(i + 1, j + 1)$ does not satisfy the requirement.

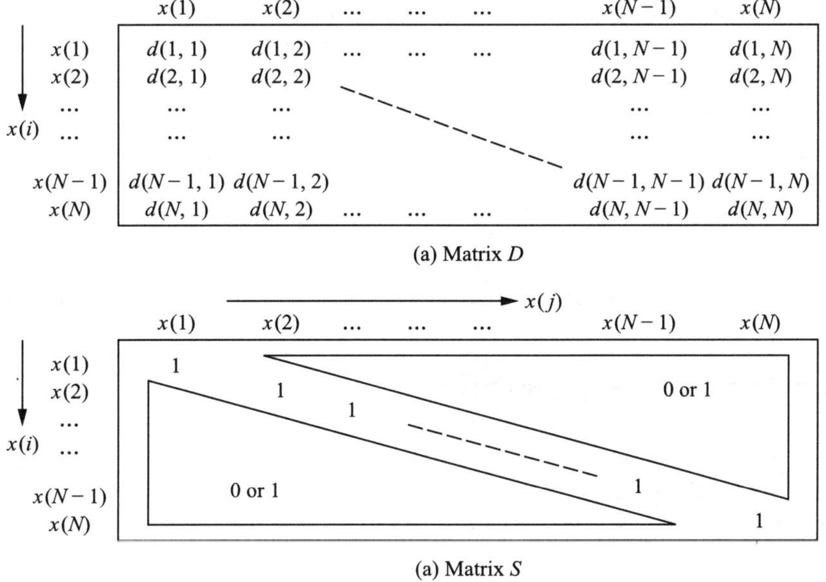


**Figure 2** $D$ matrix and $S$ matrix.

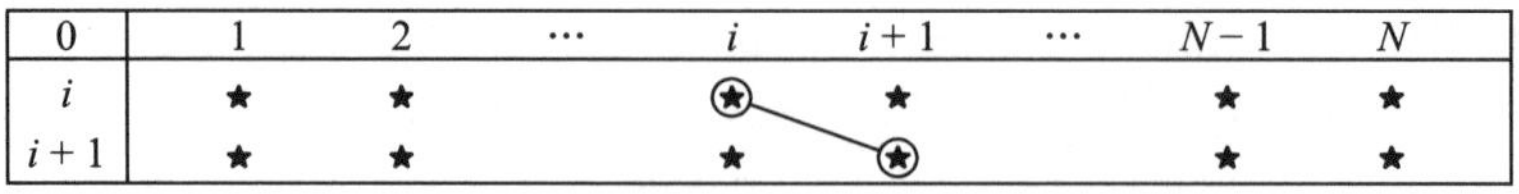

| 0 | 1 | 2 | ··· | $i$ | $i+1$ | ··· | $N-1$ | $N$ |
|---|---|---|---|---|---|---|---|---|
| $i$ | ★ | ★ | | ⊛ | ★ | | ★ | ★ |
| $i+1$ | ★ | ★ | | ★ | ⊛ | | ★ | ★ |

**Figure 3** $i$th and (i + 1)st rows of Matrix $S$.

- Find $N^{m=2}(i) = \sum_{j=1}^{N-m+1} C_r^{m=2}(i,j)$.

$$C_r^{m=2}(i) = N^{m=2}(i)/(N-m+1)$$

4. Find $C_r^m(i)$ for $m = 3$. Only the two-point patterns $X(j)$ that are selected out in step 2 with $C_r^2(i,j) = 1$ need further checking. And, since the first two points in the three-point patterns have been checked in step 3, only the newly introduced third point needs to be checked. that is, for the $X(j)$'s whose $C_r^2(i,j) = 1$, check if $C_r^3(i,j) = C_r^2(i,j) \cap s(j+2) = 1$ and

$$N^3(i) = \sum_j C_r^3(i,j) \qquad j \in \text{ the } X(j) \text{ for which } C_r^2(i,j) = 1$$

$$C_r^3(i) = N^3(i)/(N-m)$$

When $C_r^2(i)$ and $C_r^3(i)$, $i = 1, N-m$ are found, $\phi^2(r)$, $\phi^3(r)$, and ApEn(2, $r$) can then be found.

Figure 4 is a flow graph of the algorithm.

When dynamic estimation is needed, a moving data window is applied to the data sequence and ApEn of the data within the window is calculated repeatedly. As the window moves point by point along the time axis, the change of ApEn with the time is obtained. Figure 5a shows the data points involved in the window when the window moves by one point. Figure 5b shows the corresponding $D$ matrix before and after the sliding. They are highly overlapped, and the new elements to be included in the $D$ matrix after the displacement are

$$d(i,j): \qquad i = 2, N,\ j = N+1 \quad \text{and} \quad i = N+1,\ j = 2, N+1$$

So recursive processing can be used to simplify the ApEn calculation further. It should be emphasized that the threshold $r$ will vary when the window slides because the $SD$ of the data within the window does not keep constant. But the $SD$ can also be calculated recursively.

## 4. CROSS APPROXIMATE ENTROPY

The definition of approximate entropy of a single time series can be easily generalized to cross approximate entropy (abbreviated Co-ApEn) describing the pattern similarity between two time series.

**Definition.** Let $\langle x(n) \rangle = x(1), x(2), \ldots, x(N)$ and $\langle y(n) \rangle = y(1), y(2), \ldots, y(N)$ be two $N$-point time series.

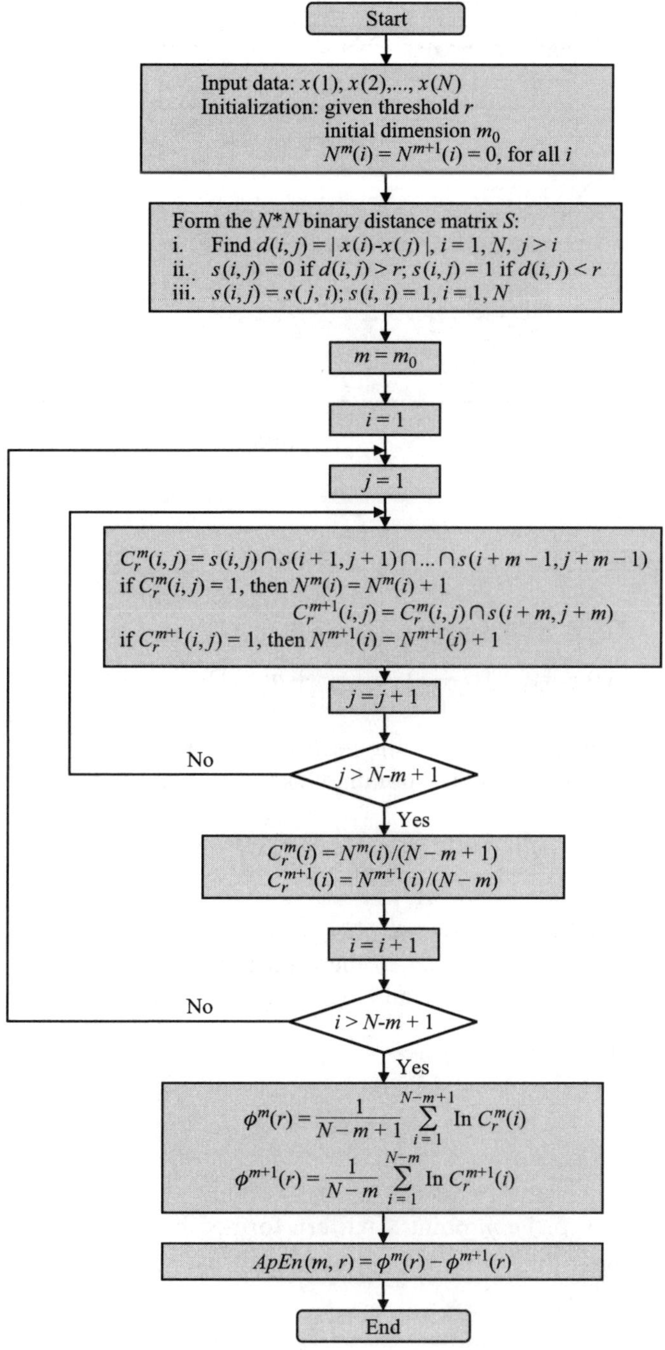


**Figure 4** Flowchart of ApEn computation.

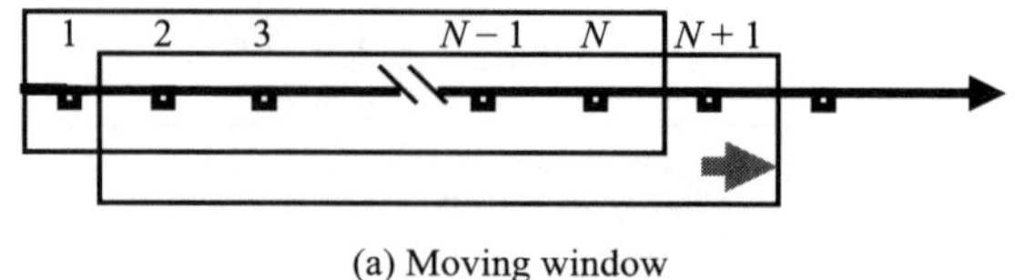


(a) Moving window

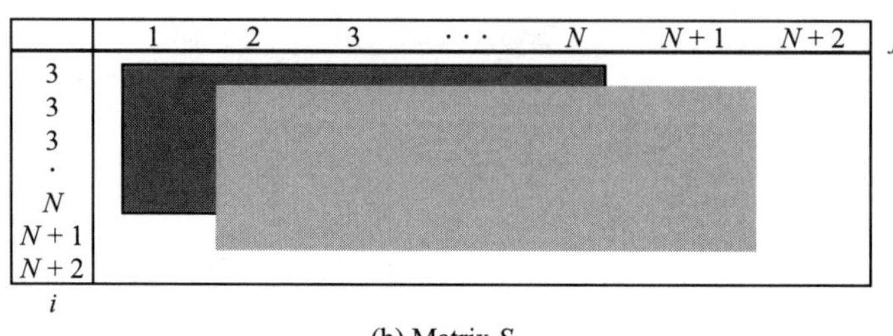


(b) Matrix S

**Figure 5** Dynamic estimation.

1. For a given $m$, form two sets of $m$-vectors;

$$X(i) = [x(i)\ x(i+1)\ \ldots\ x(i+m-1)] \qquad i = 1, N-m+1$$

$$Y(j) = [y(j)\ y(j+1)\ \ldots\ y(j+m-1)] \qquad j = 1, N-m+1$$

2. Define the distance between the vectors $X(i)$, $Y(j)$ as the maximum absolute difference between their corresponding elements, i.e.,

$$d[X(i), Y(j)] = \max_{k=0,m-1}\left[\left|x(i+k) - y(j+k)\right|\right]$$

3. With a given $X(i)$, find the number of $d[X(i), Y(j)]$ ($j = 1$–$N - m + 1$) that is $\leq r$ and the ratio of this number to the total number of $m$-vectors ($N - m + 1$).

   Let $N_{xy}^m(i) =$ the number of $Y(j)$ satisfying the requirement $d[X(i), Y(j)] \leq r$, then

$$C_{xy}^m(i) = \frac{N_{xy}^m(i)}{N - m + 1}$$

   $C_{xy}^m(i)$ measures the frequency of the $m$-point $y$ pattern being similar (within a tolerance $\pm r$) to the $m$-point $x$ pattern formed by $X(i)$.

4. Take the logarithm of $C_{xy}^m(i)$ and average it over $i$ to get $\phi_{xy}^m(r)$, i.e.,

$$\phi_{xy}^m(r) = \frac{1}{N - m + 1}\sum_{i=1}^{N-m+1} \ln C_{xy}^m(i)$$

5. Increase $m$ by 1 and repeat steps 1–4 to find $C_{xy}^{m+1}(i)$, $\phi_{xy}^{m+1}(r)$.
6. Finally, $Co\text{-}ApEn_{xy}(m, r) = \lim_{N\to\infty}[\phi_{xy}^m(r) - \phi_{xy}^{m+1}(r)]$

   and for $N$-point data, its estimate is

$$Co\text{-}ApEn_{xy}(m, r, N) = \phi_{xy}^{m}(r) - \phi_{xy}^{m+1}(r)$$

Cross ApEn is potentially an important measure, not simply in assessing the independence of realizations but also in comparing sequences from two distinct yet intertwined processes in a system.

Threshold $r$ can be taken as (0.1, 0.25) times the cross-covariance between $x(n)$ and $y(n)$. If the amplitudes of $x(n)$ and $y(n)$ differ greatly, it may be better to standardize them to:

$$x'(n) = \frac{x(n) - m_x}{SD_x}$$

$$y'(n) = \frac{y(n) - m_y}{SD_y}$$

(where $m$ and $SD$ stand for *mean* and *standard deviation*, respectively) and then find the Co-ApEn between $x'(n)$ and $y'(n)$. In this case, $r$ can be taken as 0.2.

## 5. PROPERTIES OF APPROXIMATE ENTROPY

### 5.1. Basic Properties

1. *Approximate entropy and complexity (or nonregularity).* As stated in Section 2, ApEn can be used as a measure of the complexity of a signal. The more regular the signal is, the smaller its ApEn will be. Figure 6a shows the waveform of a periodic signal, $x(t) = \sin 100\pi t + \sin 240\pi t$, and Figure 6b is the same signal with white noise $N(0, 1)$ added. The waveform of the latter is clearly more irregular than that of Figure 6a. The estimated ApEn are:

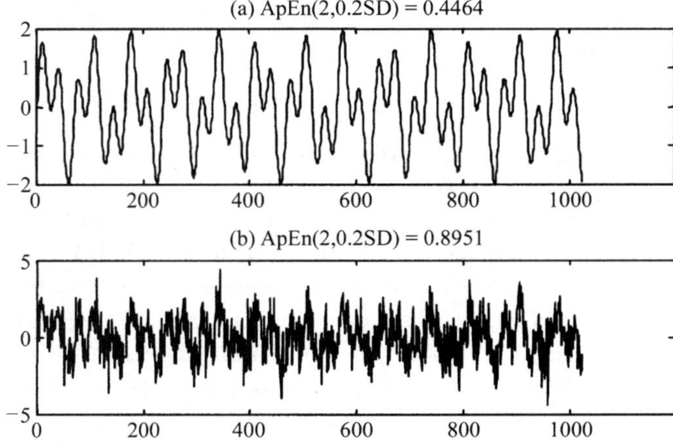


**Figure 6** (a) A periodic signal. (b) The signal in (a) with added noise.

$$\text{For Figure 6a:}\quad \text{ApEn}[2, 0.2SD, 1000] = 0.4469$$
$$\text{For Figure 6b:}\quad \text{ApEn}[2, 0.2SD, 1000] = 0.895$$

The latter is much larger than the former.

2. *Ability to obtain a robust estimate using shorter data.* The record of a stationary EEG is used to show this feature. Segments of different length are used for ApEn estimation. The results are shown in Table 2. From the table, the standard deviation of the estimated ApEn is only 0.0261 when the data length $N$ varies from 200 to 1000. The reason why a more robust estimate can be obtained by using a shorter data length is that it is essentially an estimate related to conditional probability. It is well known that less data is required to get a good estimate of the conditional probability density function than is required for the joint probability density function, because the former is a projection of the latter onto a subspace. One may argue that a difference in joint probability function may not be reflected in its conditional probability function. But our main purpose in using ApEn is to get a generally applicable, statistically operable algorithm to differentiate between various data series obtained experimentally, and ApEn is sufficient for that purpose.

**TABLE 2** ApEn Estimate of a Stationary Signal Using Different Data Length

| Length of data $N$ | 1000 | 800 | 600 | 400 | 200 | SD of ApEn |
|---|---|---|---|---|---|---|
| ApEn(2, 0.25SD, N) | 0.4762 | 0.4802 | 0.5199 | 0.5126 | 0.4777 | 0.0261 |

3. *Better ability to reject outliers (wild points).* This is because the pattern formed by wild points will rarely be repeated in the waveform and thus can be rejected during the calculation of $N^m(i)$ and $C_r^m(r)$.
4. *Denoising ability.* As said before, this is caused by a suitably chosen threshold $r$. In other words, threshold detection is basically a noise rejection operation. When the amplitude of the noise is lower than $r$, its influence will be eliminated effectively. This is shown by the following simulation:

   An EEG signal $s(k)$ is contaminated by noise $n(k)$ of variable intensity. The observed data $x(k)$ are

   $$x(k) = s(k) + c \cdot n(k)$$

   where $n(k)$ is uniformly distributed in $-1$ to $+1$ and $c$ is the factor controlling its intensity. Figure 7a and b show respectively the waveforms of $s(k)$, $n(k)$ and Figure 7c shows the ApEn of $x(k)$ as a function of $c$. At the beginning of the curve when $c$ is small (within the rectangle shown in the figure) ApEn of $x(k)$ does not increase with $c$, which manifests the denoising ability of ApEn with respect to a weak noise.
5. *Applicability to either deterministic (chaotic) signals or stochastic signals.* Moreover, it has the ability to differentiate between different mixed processes composed of deterministic and random components occurring with different probability. To show this, define a mixed process $Mix_k(P)$ by

   $$Mix_k(P) = (1 - z_k)x_k + z_k y_k$$

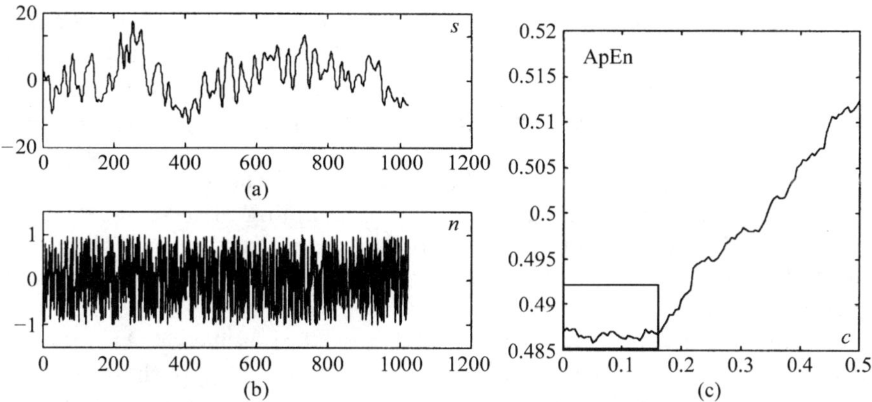


**Figure 7** Denoising ability. (a) Signal $s$; (b) noise $n$; (c) ApEn of $x = s + cn$.

where $x_k$ is a periodic process defined by $x_k = \sqrt{2}\sin(2\pi/12k)$

$y_k$ is an independently identically distributed random variable uniformly distributed over $(-\sqrt{3}, +\sqrt{3})$

$z_k$ is a binary random variable, $z_k = 1$ with probability $P$ and $z_k = 0$ with probability $1 - P$

$Mix_k(P)$ is thus a composite of deterministic and stochastic components, and $P$ is the parameter defining the *concentration* of its components. The mean value and *SD* of $Mix_k(P)$ are always 0 and 1 independent of $P$. As $P$ increases, the process apparently becomes more irregular. ApEn quantifies the increase in irregularity as $P$ increases. Figure 8 shows the increase of ApEn with $P$.

The ability of ApEn to distinguish different mixed processes is attractive for biosignal analysis because many biosignals are composed of deterministic and stochastic components. Kolomorov-Sanai (K-S) entropy and correlation dimension are unable to distinguish such processes. The K-S entropy of $Mix(P)$ equals $\infty$ for $P > 0$ and equals 0 for $P = 0$. The correlation dimension of $Mix(P)$ equals 0 for $P < 1$ and equals $\infty$ for $P = 1$.

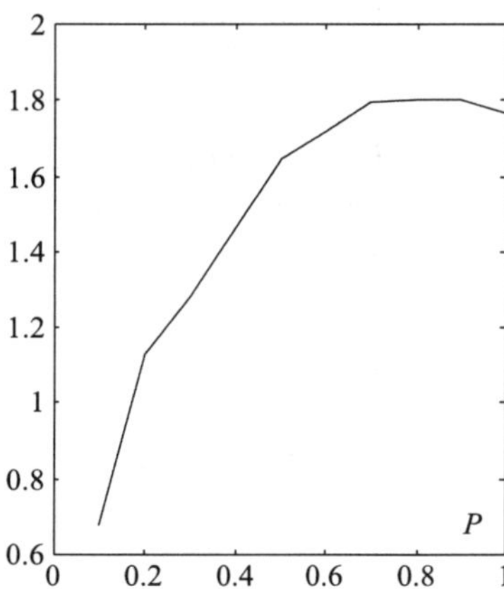


**Figure 8** Relation between ApEn of $Mix(P)$ and $P$.

Besides these properties, Pincus [9] claimed that ApEn of a system reflects the degree of its isolation from the surroundings. Greater signal regularity may indicate increased system isolation. He invoked the hypothesis that in a variety of systems, decreased complexity and greater regularity correspond to greater component autonomy and isolation. A healthy system has good lines of communication (marked by numbers of external interacting influences and by the extent of this interaction). Disease and pathology would cause system decoupling or lessening of external input and, in effect, isolating a central system from its ambient universe. If this hypothesis is true, ApEn could be a measure of the coupling between the system and the external world. Pincus [9] made a series of simulations to support his hypothesis (including an autonomous oscillation system such as a Van der Pol oscillator, AR and MA systems, and two systems with different degrees of coupling). The hypothesis is also supported by the facts that ApEn decreases with age and that a decrease in the ApEn of heart rate variability is an indication of a pathological disorder.

## 5.2. Relation with Other Characteristic Parameters

### 5.2.1. Relation with Correlation Function

Sometimes it seems that cross-correlation and cross-ApEn are similar in that both can be used to measure the similarity between two signals. For example, signals $s_1$, $s_2$ in Figure 9a are two EEGs measured from two electrodes near each other on the scalp. They are similar in waveform. Signal $s_3$ in Figure 9b is the same as $s_2$ for the first 400 points, but the rest of it is taken from another EEG. This is evidenced by their appearances: $s_1$ and $s_3$ differ substantially after the first 400 points.

The cross–ApEn between $s_1$, $s_2$ and that between $s_1$, $s_3$ are estimated respectively. The results are: for $s_1$, $s_2$ Co-ApEn = 0.0313; for $s_1$. $s_3$, Co-ApEn = 0.1094 ($r = 0.25$SD, $N = 1000$). The latter is much larger than the former, which means Co-ApEn can be used to express the similarity between signals. In addition, a 200-point moving window is used to estimate the dynamic evolution of Co-ApEn between $s_1$ and $s_3$. The result is shown in Figure 9c: Co-ApEn between $s_1$ and $s_3$ increases starting from the point

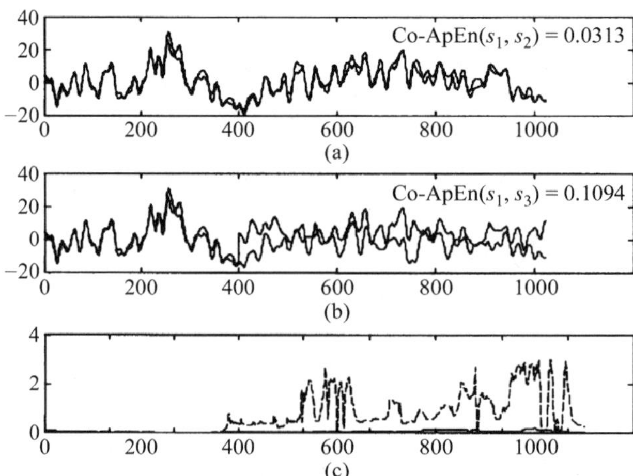


**Figure 9** Cross-ApEn between signals. (a) $s_1$ and $s_2$; (b) $s_1$ and $s_3$; (c) dynamic estimation.

$n \cong 250$. This means Co-ApEn can be used to show the dynamic variation of the pattern similarity between two time sequences.

But we must be aware of the fact that there are still significant differences between the properties of the two parameters. Let us illustrate this by an example. The signal P1 in Figure 10a is 20 half-wave sinusoidal pulses distributed uniformly along the time axis. Figure 10b shows the same pulses randomly distributed and called P2. Figure 11 shows the autocorrelation function (ACF) of P1, P2 and the cross-correlation function (CCF) between them. The ACF of P1 is periodic but the ACF of P2 has only a single peak in the vicinity of $\tau = 0$ and fluctuates randomly with a low amplitude at larger $\tau$. Although their ACFs differ greatly, their ApEns are small and close to each other. This is due to the fact that ApEn can characterize the pattern similarity in a signal regardless of how the patterns are distributed. For the same reason, Co-ApEn between P1 and P2 is rather small (Co-ApEn $= 0.0548$) and the CCF fluctuates with small periodic peaks (Figure 11c).

### 5.2.2. ApEn and Standard Deviation

Standard deviation is a statistical parameter used to quantify the degree of scattering of the data around their mean value. The time order of the data is immaterial. On the other hand, the time order of the data (i.e., pattern) is a crucial factor affecting ApEn. When the time order of the data is shuffled, the SD remains unchanged but the ApEn changes appreciably. Discerning changes in order from apparently random to regular is the primary focus of this parameter.

### 5.2.3. Relation with Correlation Dimension and K-S Entropy

The initial steps in the estimation of the three parameters are similar. The original data are embedded in an $m$-dimensional space, forming $m$-vectors $X(i)$. Find the distances between these vectors:

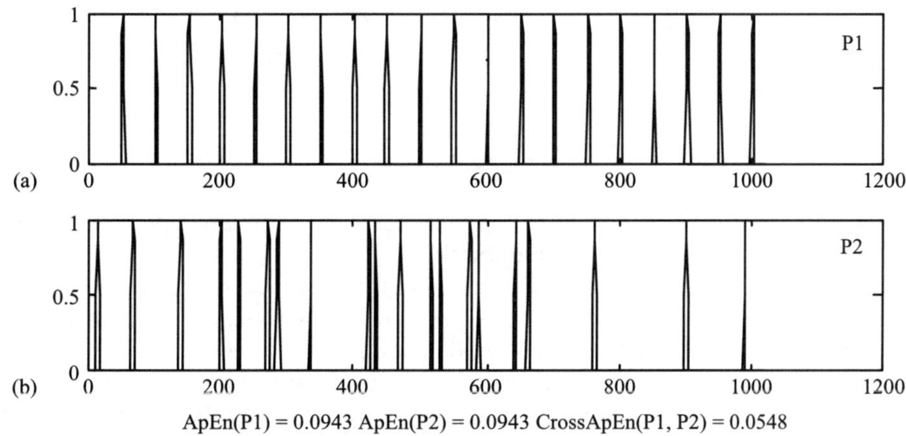


**Figure 10** (a) P1: pulses uniformly distributed. (b) P2 pulses randomly distributed.

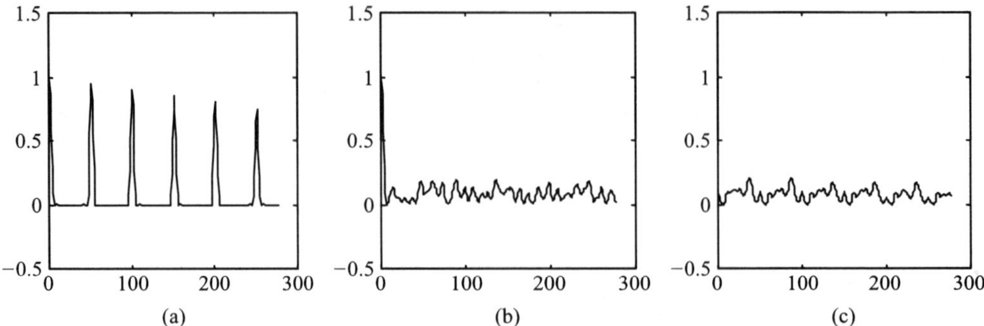

**Figure 11** Correlation functions of P1 and P2. (a) ACF of P1; (b) ACF of P2; (c) CCF between P1 and P2.

$$d[X(i), X(j)] = \max_{k=1,N}\left[\left|x(i+k) - x(j+k)\right|\right]$$

and the number $N^m(i)$ of the distances $\leq$ threshold $r$. Then

$$C_r^m(i) = N^m(i)/(N - m + 1)$$

The difference between the three comes from the remaining steps in the estimation:

For correlation dimension:

$$\phi^m(r) = \frac{1}{N-m+1}\sum_{i=1}^{N-m+1} C_r^m(i)$$

$$CD = \lim_{N\to\infty}\lim_{r\to 0}[\ln \phi^m(r)/\ln(r)]$$

For ApEn:

$$\phi^m(r) = \frac{1}{N-m+1}\sum_{i=1}^{N-m+1} \ln C_r^m(i)$$

$$ApEn(m, r) = \lim_{N\to\infty}\left[\phi^m(r) - \phi^{m+1}(r)\right]$$

For E-R entropy:

$$ER = \lim_{N\to\infty}\lim_{m\to\infty}\lim_{r\to 0}\left[\phi^m(r) - \phi^{m+1}(r)\right]$$

(E-R entropy refers to an algorithm proposed by Eckmann and Ruelle to calculate directly the K-S entropy of a time series.)

E-R entropy and CD are invariant measures defined by some limiting procedure, but ApEn can be estimated with fixed $m$ and $r$. For signals generated from the same model [e.g., *Mix*(*P*), AR, or MA model, logistic functions with given parameter], a robust estimate of ApEn can usually be obtained by using a different segment of the signal. When $N = 1000$, the SD of the estimate will be about 0.05. But the robustness of the estimate of CD and ER is much worse; a much longer data length is required. In addition, the estimates of ER and CD are sensitive to contaminating noises. Besides,

CD and ER cannot be applied to stochastic process, but ApEn can be applied to mixed processes consisting of chaotic and stochastic components. Thus, ApEn is not an approximation of the K-S entropy. It is to be considered as a family of statistics and system comparisons are intended with fixed $m$ and $r$.

#### 5.2.4. ApEn and Ziv's Complexity Measure

There are various complexity measures. Lempel and Ziv proposed [10] an operable algorithm for its estimation as follows: First, convert the original signal into a binary sequence by the following steps: Find the average value $m_x$ of the sequence and then compare each point of the data with the average value: If the magnitude of the data $\geq m_x$, the output will be 1. If it is smaller than $m_x$, the output will be 0. The binary sequence is thus formed. Ziv's algorithm then goes further to find the complexity measure by counting the rate of new pattern generation from the binary sequence [10].

Compared with the calculation of ApEn, both algorithms find the complexity measure through a binary conversion of the original data. Ziv's algorithm converts the original sequence directly into binary, so a great deal of information embedded in the waveform is lost. ApEn forms the binary conversion from the distance of the waveform pattern, and more information embedded in the original sequence can then be preserved.

## 6. EXAMPLES OF APPLICATION

Some examples are given here to illustrate the application of approximate entropy.

### 6.1. Analysis of Heart Rate Variability (HRV) Signals [4]

Figure 12 shows two heart rate tracings, both obtained on 4-month-old infants during quiet sleep. Figure 12a is from an infant who had an aborted sudden infant death syndrome (SIDS) episode 1 week before the recording, and Figure 12b is from a healthy infant. The overall variabilities (SDs) of the two tracings are approximately the

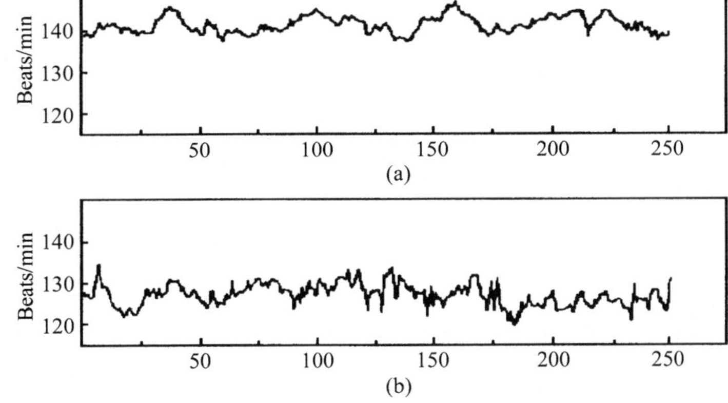


**Figure 12** HRV signals of two infants: (a) SIDS and (b) healthy.

same (2.49 beats/min for Figure 12a, 2.61 beats/min for Figure 12b), and whereas the aborted-SIDS infant has a somewhat higher mean heart rate, both are well within the normal range. Yet the tracing in Figure 12a appears to be more regular than that in Figure 12b. The ApEns estimated with $N = 1000$, $m = 2$, $r = 0.15\text{SD}$ differ greatly: 0.826 for Figure 12a and 1.426 for Figure 12b. It is thus illustrated that ApEn is an effective characterizing parameter for describing complexity.

## 6.2. Dynamic Analysis of HRV Signal

An HRV signal is used to detect its dynamic information. The original data are taken from Ref. 11. The HRV signal is from a subject suffering from an ischemic heart attack, as shown in Figure 13a. *Start* on the figure marks the start of the ischemic attack and *End* stands for its end. The data length of the signal is 1000 points.

A sliding window of 256 points is applied to the data and the ApEn of data within the window is calculated. As shown in Figure 13b, ApEn decreases when the attack begins and recovers gradually to the normal value after the attack, which reflects the regulation of the autonomic nervous system during the ischemic attack.

## 6.3. Analysis of Evoked Potential

A stimulating sound (a click) is sent to a testee through earphones mounted over the ears. It is sent to either the left ear or the right ear with the same probability. Thirty EEG channels are recorded from the scalp. The positions of the electrodes are shown in Figure 14. Each time a new click is sent out, 2.56 seconds of data are recorded. With a sampling frequency of 400 Hz, the number of data point in each record is 1024. The original signal is preprocessed through the following two steps:

Step 1, Laplacian processing. In order to localize the scalp area that affects the signal measured by each channel, the contributions of the potentials from all other electrodes are subtracted from the potential measured from the $i$th electrode according to the following formula:

$$x_i' = x_i - \frac{\sum_{\substack{j=1 \\ j \neq i}}^{N} x_j / d_{ij}}{\sum_{\substack{j=1 \\ j \neq i}}^{N} 1 / d_{ij}}, \qquad i = 1, 30, N = 30$$

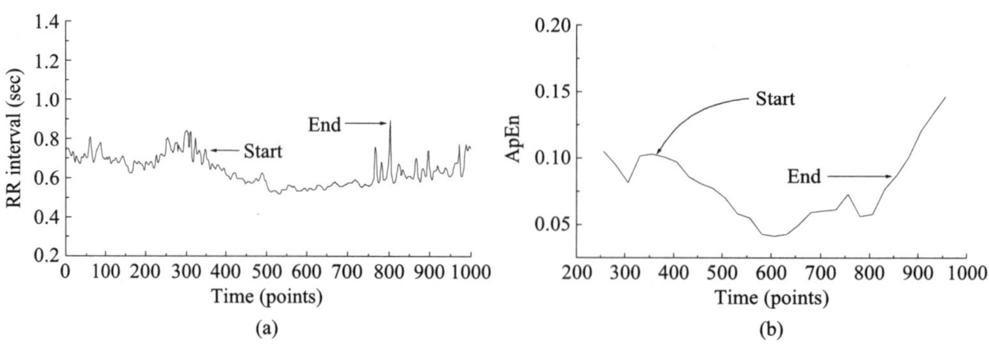


**Figure 13** Dynamic ApEn during an ischemic attack: (a) HRV signal and (b) running ApEn.

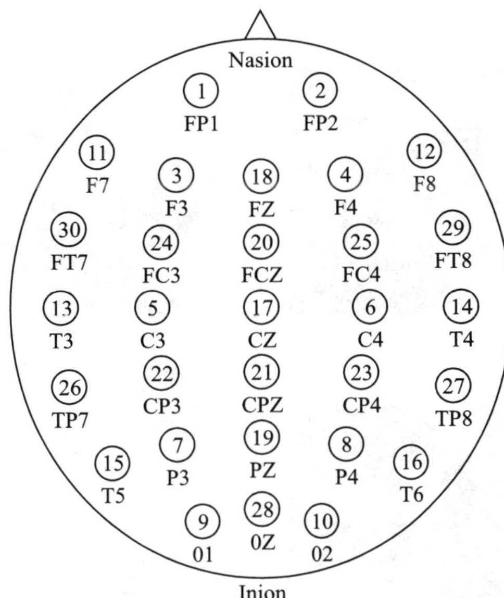


**Figure 14** Positions of the electrodes.

where $d_{ij}$ is the scalp distance between electrodes $i$ and $j$.

Step 2, singular value decomposition (SVD). To suppress contaminating noise, SVD is applied to the data matrix $X$ formed by the 30 channels of data with 1024 points each, i.e.,

$$\underset{30\times 1024}{X} = \underset{30\times 30}{U} \underset{30\times 1024}{\sum} \underset{1024\times 1024}{V^T} \tag{2}$$

where $U$ and $V$ are the orthonormal matrices formed by the eigenvectors and $\Sigma$ is the quasidiagonal matrix of singular values. The magnitudes of the singular values are arranged in descending order. Set the smaller singular values (which are considered to be contributed by noise) to zero and substitute back into Eq. 2, replacing $\Sigma$ (denote it by $\Sigma'$). Then find the data matrix $X' = U\Sigma'V^T$. The result is the noise-reduced data.

Find the ApEn for each channel and use different brightnesses to represent its magnitudes: brighter for larger values and darker for smaller values. Entering the brightness into the space grid formed by the electrodes (Figure 15d), we get the spatial distribution of ApEn over the scalp. Figure 15a–c show the results of three independent tests. The results are all similar. Note that the region with the largest ApEn is always located at the position that is consistent with the physiological finding [12].

## 7. CONCLUSIONS

The definition, properties, and fast algorithm for the approximate entropy have been discussed in detail. It is a parameter that can give a robust estimate from short data and is attractive for dynamic analysis. There are, of course, other nonlinear parameters that can be applied to short data. The coarse-grained entropy rate proposed by Palus [13] is

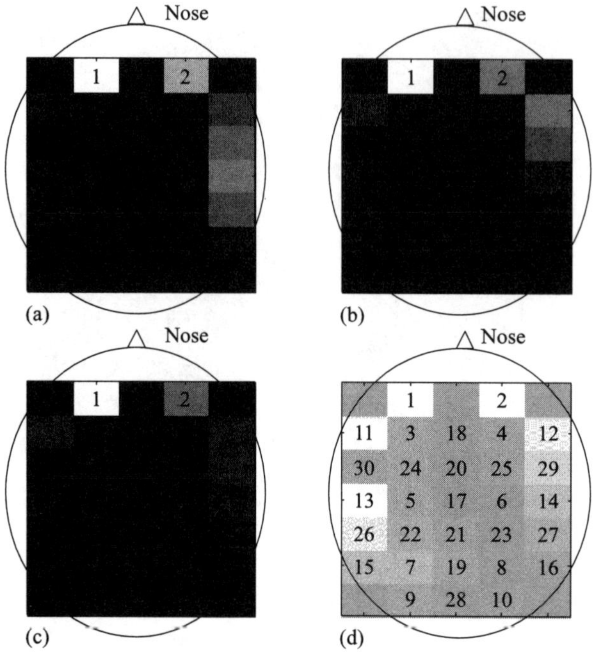


**Figure 15** Spatial distribution of ApEn. (a–c) ApEn; (d) space grid.

one of them. But it appears that ApEn has potential widespread utility to practical data analysis and clinical application because of its salient features.

## REFERENCES

[1] A. Wolf, et al., Determining Lyapunov exponents from a time series. *Physica* 16D:285–317, 1985.

[2] S. M. Pincus, Approximate entropy as a measure of system complexity. *Proc. Natl. Acad. Sci. USA* 88:2297–2301, 1991.

[3] D. Sapoznikov, et al., Detection of regularities in heart rate variations by linear and nonlinear analysis: Power spectrum versus approximate entropy. *Comput. Methods Programs Biomed.* 48:201–209, 1995.

[4] S. M. Pincus, Heart rate control in normal and aborted-SIDS infants. *Am. J. Physiol.* 33:R638–R646, 1991.

[5] S. M. Pincus, Physiological time series analysis: What does regularity quantify? *Am. J. Physiol.* 266:H1643–H1656, 1994.

[6] S. M. Pincus, Older males secrete luteinizing hormone and testosterone more irregularly and joint more asynchronously, than younger males. *Proc. Natl. Acad. Sci. USA* 93:14100–14105, 1996.

[7] J. D. Veldhuis, et al., Differential impact of age, sex steroid-hormones, and obesity on basal versus pulsatile growth-hormone secretion in men as assessed in an ultrasensitive chemiluminescence assay. *J. Clin. Endocrinol. Metab.* 80:3209–3222, 1995.

[8] S. M. Pincus, et al., A regularity statistic for medical data analysis. *J. Clin. Monitor.* 7:335–345, 1991.

[9] S. M. Pincus, Greater signal regularity may indicate increased system isolation. *Math. Biosci.* 122:161–181, 1994.

[10] A. Lempel, et al., On the complexity of finite sequences. *IEEE Trans.* IT-22:75–81, 1976.
[11] A. M. Bianchi, et al., Time-invariant power spectrum analysis for the detection of transient episodes in HRV signal. *IEEE Trans.* BME-40:136–144, 1993.
[12] M. Terunobu, *Physiology*. People's Heath Care Press, 1987 [in Chinese].
[13] M. Palus, Coarse-grained entropy rates for characterization of complex time series. *Physica D* 93:64–77, 1996.

Chapter 4

# PARSIMONIOUS MODELING OF BIOMEDICAL SIGNALS AND SYSTEMS: APPLICATIONS TO THE CARDIOVASCULAR SYSTEM

P. Celka, J. M. Vesin, R. Vetter, R. Grueter, G. Thonet, E. Pruvot, H. Duplain, and U. Scherrer

## 1. INTRODUCTION

Science is the expression of the human desire to discover the underlying laws that govern the mechanisms of nature. Time series modeling is in keeping with this desire. A signal processing practitioner tries to reveal the relationships between successive samples of a signal or between various signals expressing the time evolution of a system.

In the early days of modeling [1], there was a general tendency to consider the time series as an abstract sequence of numbers and to develop models that were able, for instance, to make correct predictions but without necessarily resorting to the physical phenomenon responsible for this time series. Let us say it was the epoch of modeling per se, when new types of models were tested on a few benchmark series such as the sunspot or the Canadian lynx series [2].

Later, there was a progressive shift toward a more utilitarian view of time series modeling, when the techniques developed started to be widely applied in areas such as economics [3]. The incentive was sometimes to be able to predict future events but also more often than not to get insights into the causes of the specific time evolution of the system under study.

We may say that a further step has been taken with the modeling of biomedical time series. A model in this context helps in analyzing the physiological mechanisms at work, but it can also be used for classification purposes. If the model is accurate enough and the physiological mechanisms are modified because of some perturbation, then this modification with respect to a reference state can be detected because the model itself is modified.

Another major evolution in time series modeling has been the transition from linear to nonlinear modeling. Linear modeling presents some very valuable advantages: the underlying mathematical machinery is well established, it furnishes at no extra cost an interpretation in the frequency domain, and it works fairly well in a broad range of applications. However, it cannot properly deal with some aspects of real-world signals. Saturation for instance, which simply expresses the fact that most physical quantities cannot grow without bound, cannot be taken into account by a linear model. In some cases also, a nonlinear model can grasp the main features of a time series with a few parameters, whereas a linear model would become cumbersome. Recent years have witnessed a huge number of theoretical developments in nonlinear modeling as well as many applications to various fields in engineering (e.g., [2, 4]). But the price to pay when switching to nonlinear modeling is the following: nonlinearity is a nonproperty, which means that one knows only what the model is not. The structure of a linear model is imposed, but there is an infinite number of possible nonlinear ones in a particular application. At this point, prior knowledge of the problem at hand is very helpful in performing model class preselection. Once the class of nonlinear models is chosen, another crucial problem arises: nonlinear models may easily have a complex structure, that is, be characterized by a large number of terms and parameters. The temptation is great (and many researchers indeed have not resisted this temptation) to use all the possibilities of the nonlinear model in order to obtain a good fit of the data set, i.e., to *overfit* it, at the risk of extracting information in the time series where it is not present.

We present in this chapter a nonlinaer modeling framework (model class, model selection and estimation) that we hope is well suited to the analysis and classification of biomedical time series. We use polynomial nonlinear autoregressive moving average (NARMA) models [5] and a NARMA model selection procedure we have developed, based on Rissanen's minimum description length (MDL) criterion [6]. This procedure allows one, in a reasonable computing time, to determine a *parsimonious* NARMA model, i.e., a model that contains only information-bearing terms. The NARMA model class has the advantage of including the linear ARMA models so that, through model selection, it is possible to determine the opportunity to use a nonlinear model instead of a linear one and even to quantify the amount of nonlinearity in the time series or system under study. In addition, we have incorporated in our modeling framework the concept of subband modeling. Many natural phenomena, especially in biomedicine, contain components at various time scales (or equivalently at various frequency bands). it is thus advantageous, in terms of both modeling accuracy and a posteriori analysis, to model these components separately. We describe how this separation can be performed efficiently using an appropriate wavelet-based scheme. Finally, we apply our technique to the cardiovascular system, more precisely to the prediction of heartbeat intervals from past values or from other signals such as blood pressure, respiration, and muscle sympathetic nerve activity. These experiments are carried out for several drug-induced physiological conditions.

This chapter is organized as follows: Section 2 describes the NARMA model class and efficient methods for NARMA model estimation. In Section 3 the derivation of the MDL criterion and its variants is summarized. Then we introduce our model selection method. Its performance is assessed on synthetic and benchmark time series. In Section 4 the principle of subband signal decomposition is described. In Section 5 the results of the application of our selection method to cardiovascular signals are presented and analyzed. Section 6 concludes the chapter.

## 2. POLYNOMIAL EXPANSION MODELS

### 2.1. Problem Statement

Facing the problem of modeling nonlinear systems is not a trivial task. By opposition to linear modeling, there is no unified theory behind nonlinear modeling. Depending on the number and type of available signals, the a priori knowledge about the system, and the goals to achieve, several approaches can be chosen. The two main statistical classes of signals are stationary and nonstationary ones. Nonstationary signals requires the use of time-dependent models, and recursive algorithms are the most suitable in this case [7]. Although adaptive algorithms are mostly appropriate for nonstationary signals, they also apply to stationary signals. A majority of the work on nonlinear adaptive models for nonstationary signals has employed multilayer perceptrons or radial basis function networks [4]. These models are usually nonlinear in the parameters and thus require nonlinear search approaches for their estimation. Stationarity can be assessed or hypothesized prior to any choice of the nonlinear modeling class. We will assume here that the recorded signals reflect some steady state of the corresponding system and thus focus on stationary conditions. In this framework, three model classes have been identified by Billings [8] and Korenberg et al. [10]:

1. Block structured approaches based on Wiener- or Hammerstein-type models [8,9]
2. Orthogonal or nonorthogonal functional expansions such as the polynomial Wiener or Volterra expansions [8, 10]
3. Parametric approaches such as the NARMA model [5]

The first class was introduced by Wiener [9] and has been studied for many years [8]. Models in this class are represented by cascaded blocks (dynamic linear–static nonlinear function). Although parsimonious models can be obtained because of the small number of parameters in the nonlinear static blocks, the type of nonlinearity has still to be chosen a priori. If one assumes that the nonlinear functions are smooth, they can be approximated with limited Taylor series. The overall model can thus be represented as a polynomial expansion.

The second class includes orthogonal Wiener-type and nonorthogonal Volterra-type polynomial expansions. Wiener-type expansions are based on Laguerre functions. They have the nice property of being orthogonal on a given interval, so that parameters in the expansions can be computed independently. Volterra-type series are nonorthogonal expansions. The orthogonality may be a restriction in some cases and an advantage in others. Anyway, within this class, the models are linear in the parameters and classical linear tools can be used to find a solution.

The third class, by opposition to the first one, does not impose a fixed structure. On the other hand, it has the disadvantage of being expensive in the number of parameters, implying a high computational load. Very interesting and efficient approaches have been developed by Korenberg and Billings in using fast orthogonal search algorithms to overcome the drawback [11]. They improve upon classical NARMA and Volterra series estimation methods by providing both parameter estimation and model selection. Their scheme applies only to linear-in-the-parameter models and is based on a modified Gram-Schmidt (MGS) procedure or on a singular value decomposition (SVD). It allows selection of the best model without using an exhaustive search in the full parameter space. Model selection is one of the principal subjects of this chapter, and Section 3 is devoted entirely to this topic.

These fast algorithms have been employed successfully in identifying NARMA, Volterra, or other polynomial expansion systems in biomedicine [12, 13]. Among the several models introduced above, Wiener, Volterra, and NARMA polynomial expansions have the advantage of including the linear model class. This allows one to quantify directly the nonlinear content of the model (see Section 5). For all these reasons, we have chosen to use NARMA models for our study and now present briefly their formulation.

## 2.2. NARMA Models

Under some restrictions [5], any nonlinear discrete-time stochastic system with $M$ inputs $x_m$, $m = 1, \ldots, M$, one output $y$, an output dynamical noise source $e_{\rm d}$, and an output measurement noise source $e_{\rm m}$, can be represented by a NARMAX model (NARMA with exogenous inputs) [5]. Note that in the control terminology, the noise signal $e_{\rm d}$ is considered as an input excitation signal, and the $x_m$ are the exogenous input signals. The noises $e_{\rm d}$ and $e_{\rm m}$ are supposed to be independent of each other and of the signals $y$ and $x_m$. The output signal $y(n)$ at time $n$ of the system characterized by the function $f(.)$ can be expressed by

$$\begin{aligned} y_s(n) &= f\big(\mathbf{y}_s(n-1), \mathbf{x}(n), \mathbf{e}_{\rm d}(n-1)\big) + e_{\rm d}(n) \\ y(n) &= y_s(n) + e_{\rm m}(n) \end{aligned} \tag{1}$$

where

$$\mathbf{y}_s^T(n-1) = \big[y_s(n-1), \ldots, y_s(n-l_y)\big] \tag{2}$$

$$\mathbf{y}^T(n-1) = \big[y(n-1), \ldots, y(n-l_y)\big] \tag{3}$$

$$\mathbf{x}^T(n) = \big[\mathrm{x}_1^T(n), \ldots, \mathrm{x}_M^T(n)\big] \tag{4}$$

$$\mathbf{e}_{\rm d}^T(n-1) = [e_{\rm d}(n-1), \ldots, e_{\rm d}(n-l_e)] \tag{5}$$

and

$$\mathbf{x}_m^T(n) = \Big[x_m(n), \ldots, x_m(n-l_x^{(m)})\Big] \; \forall m \in \{1, 2, \ldots, M\} \tag{6}$$

The $m$th input maximum lag corresponds to $l_x^{(m)}$ and the output maximum lag is $l_y$. Observe that we permit different lags for the various input signals. In Eq. 1, we observe that the noise signal $e_{\rm d}$ acts at the same level as the input signal $x_m$. This may introduce some confusion in the formalism. Whereas the dynamical noise $e_{\rm d}$ is part of the dynamical system (1) and therefore modifies its dynamics, the measurement noise $e_{\rm m}$ does not interfere with the system itself.

As the nonlinear function $f(.)$ is unknown, and under the assumption that it is "weakly" nonlinear, it is usual to approximate it using a polynomial model [11]. In the following, the polynomial expansion is characterized by its *maximal polynomial degree* $D$, maximum lags $l_x^{(m)}$ and $l_y$, and related parameters $\theta^{(d)}(u_1, \ldots, u_d)$ and $\phi_m^{(d)}(u_1, \ldots, u_d)$. With respect to the control terminology, the modeling part that takes into account the $e_{\rm d}$ signal is called the moving average (MA) part. In the following ,we will not consider this contribution to the modeling. The MA part will be the input-related parameter set. The output signal $y_s(n)$ at time $n$ is given by

$$y_s(n) = \theta^{(0)} + \sum_{d=1}^{D} \sum_{u_1,\ldots,u_d=1}^{l_y} \theta^{(d)}(u_1,\ldots,u_d) y_s(n-u_1)\cdots y_s(n-u_d)$$
$$+ \sum_{m=1}^{M} \sum_{d=1}^{D} \sum_{u_1,\ldots,u_d=0}^{l_x^{(m)}} \phi_m^{(d)}(u_1,\ldots,u_d) x_m(n-u_1)\cdots x_m(n-u_d) + e_{\mathrm{d}}(n) \tag{7}$$

where we have separated the autoregressive (AR) part characterized by the $\theta^{(d)}(u_1, \ldots, u_d)$ and the moving average (MA) part by $\phi_m^{(d)}(u_1, \ldots, u_d)$. The parameter $\theta^{(0)}$ is a constant bias. What is actually measured and used as the output signal is $y = y_s + e_{\mathrm{m}}$. Equation 7 is called a NARX structure in the control formalism (there is no MA part). The noise signal $e_{\mathrm{d}}$ is also called the modeling error. Modeling and measurement errors are actually mixed in the observed data. We will drop the measurement error $e_{\mathrm{m}}$ and call $e = e_{\mathrm{d}}$ the overall noise contribution. As a result, $y = y_s$.

In order to demonstrate the complexity of this model, let us compute the total number of its coefficients. This number $N_{\mathrm{Par}}$ is the sum of the numbers of parameters for each degree $d$. As we have $M$ input signals, we have $C^d_{L_x^{(M)}+d-1}$ coefficients for the MA part ($C_n^r = n!/(r!(n-r)!)$), where $L_x^{(M)} = \sum_{m=1}^{M}(l_x^{(m)} + 1)$, whereas we have $C^d_{l_y+d-1}$ coefficients for the AR part of each degree $d$ in Eq. 7. Thus $N_{\mathrm{Par}}$ is given by

$$N_{\mathrm{Par}} = \sum_{d=1}^{D} \left( C^d_{L_x^{(M)}+d-1} + C^d_{l_y+d-1} \right) + 1 \tag{8}$$

The constant 1 is due to the zero-degree term. To give an example, let us take $M = 3$, $D = 3$, $l_x^{(m)} = l_y = 10$ for $m = 1, 2, \ldots, M$. Equation 8 gives $N_{\mathrm{Par}} = 7920$, and if we take into account only the linear terms ($D = 1$) we get $N_{\mathrm{Par}} = 44$. This undoubtedly justifies the use of a selection criterion (see Section 3) to reduce the complexity of the nonlinear model.

Let us introduce the following compact notations highlighting the linear-in-the-parameter structure of the NARMA model described in Eq. 7

$$\begin{aligned} y(n) &= \mathbf{p}_y(n)\boldsymbol{\Theta} + \mathbf{p}_x(n)\boldsymbol{\Phi} + e(n) \\ &= \left[\mathbf{p}_y(n), \mathbf{p}_\mathbf{x}(\mathbf{n})\right] \cdot \begin{bmatrix} \boldsymbol{\Theta} \\ \boldsymbol{\Phi} \end{bmatrix} + e(n) \\ &= \mathbf{p}(n)\boldsymbol{\Xi} + e(n) \end{aligned} \tag{9}$$

where $\mathbf{p}_x(n)$ is the vector of the input signal terms, $\mathbf{p}_y\,(n)$ is the vector of output terms, $\boldsymbol{\Theta}$ is the parameter vector of the AR part, and $\boldsymbol{\Phi}$ is the parameter vector of the MA part. The input polynomial vector $\mathbf{p}_x(n)$ is

$$\mathbf{p}_x(n) = \left[\mathbf{p}_{x_1}(n), \ldots, \mathbf{p}_{x_M}(n)\right] \tag{10}$$

with

$$\mathbf{p}_{x_m}(n) = \left[v_{(m;0)}(n), \ldots, v_{(m;l_x^{(m)})}(n), v_{(m;00)}(n), \ldots, v_{m;l_x^{(m)}\ldots l_x^{(m)})}(n)\right] \tag{11}$$

and $v_{(m;u_1\ldots u_d)}(n) = x_m(n-u_1)\cdots x_m(n-u_d)$ for $m = 1, 2, \ldots, M$. The output polynomial vector $\mathbf{p}_y(n)$ is

$$\mathbf{p}_y(n) = \left[1, w_{(1)}(n), \ldots, w_{(1_y)}(n), w_{(11)}(n), \ldots, w_{(l_y\cdots l_y)}(n)\right] \tag{12}$$

with $w_{(u_1\ldots u_d)}(n) = y(n-u_1)\cdots y(n-u_d)$. The parameter vectors are expressed by

$$\mathbf{\Theta}^T = \left[\theta^{(d=0)}, \theta^{(d=1)}(1), \cdots, \theta^{(d)}(l_y, \ldots, l_y)\right] \tag{13}$$

$$\mathbf{\Phi}^T = \left[\phi_1^{(d=1)}(0), \cdots, \phi_M^{(d)}\left(l_x^{(M)}, \ldots, l_x^{(M)}\right)\right] \tag{14}$$

Equation 9 is linear in $\mathbf{\Theta}$ and $\mathbf{\Phi}$ and stochastic because of the presence of the output noise source $e$. Assuming the noise source $e$ to be independent of the input signals and zero mean, the least-mean-squares solution $\mathbf{\Xi}^T = [\hat{\mathbf{\Theta}}^T, \hat{\mathbf{\Phi}}^T]$ to Eq. 9 is given by

$$\begin{aligned}\begin{bmatrix} \mathbf{\Theta} \\ \hat{\mathbf{\Phi}} \end{bmatrix} &= \begin{bmatrix} E[\mathbf{p}_y^T \mathbf{p}_y] & E[\mathbf{p}_y^T \mathbf{p}_x] \\ E[\mathbf{p}_x^T \mathbf{p}_y] & E[\mathbf{p}_x^T \mathbf{p}_x] \end{bmatrix}^{-1} \cdot \begin{bmatrix} E[y\mathbf{p}_y^T] \\ E[y\mathbf{p}_x^T] \end{bmatrix} \\ &= \mathbf{R}_{xy}^{-1} \mathbf{p}_{xy}^T \end{aligned} \tag{15}$$

where $E[x]$ stands for the expectation of $x$. The solution 15 supposes that one can estimate the various auto- and cross-correlation matrices and that the matrix $\mathbf{R}_{xy}$ is full rank. Having computed $\hat{\mathbf{\Xi}}$ the estimated output signal at time $n$ is given by

$$\hat{y}(n) = \mathbf{p}(n)\hat{\mathbf{\Xi}} \tag{16}$$

and the residual error $\epsilon$ is

$$\epsilon(n) = y(n) - \hat{y}(n) = \mathbf{p}(n)(\mathbf{\Xi} - \hat{\mathbf{\Xi}}) + e(n) \tag{17}$$

From Eq. 17, we observe that even in the case of unbiased estimation, i.e., $E[\hat{\mathbf{\Xi}}] = \mathbf{\Xi}$, the residual error variance is bounded by the noise floor, i.e.,

$$\mathrm{Var}(\epsilon) = \sigma_\epsilon^2 = \mathrm{Var}(e) = \sigma_e^2 \quad \text{if } E[\hat{\mathbf{\Xi}}] = \mathbf{\Xi} \tag{18}$$

The value Var($\epsilon$) is usually called the mean-square error (MSE), or residual error variance, corresponding to the solution of Eq. 9. The MSE depends on the inputs and output signal powers. In order to avoid this, one can normalize the MSE with respect to the output signal power $E[y^2]$. Indeed, as we allow even terms in the polynomial expansion, the output signal may be nonzero mean even if the inputs are zero mean. One obtains the relative residual variance (RRV) described by the following formula [14]:

$$\mathrm{RRV} = \frac{\sigma_\epsilon^2}{\mathrm{E}[\mathrm{y}^2]} \tag{19}$$

As $\mathrm{RRV} \in [\sigma_e^2/\mathrm{E}[\mathrm{y}^2], 1]$, it is usually expressed as a percentage. Several other indices related to RRV will be defined in Section 5. A small number of RRV indicates a good fit of the model to the data, but this does not mean that one has correctly identified the system. More often than not, when dealing with nonlinear models, one can achieve a very small RRV value but with a tremendous number of parameters. This problem is called *overfitting* and reflects the fact that the number of parameters may be almost equivalent to the number of samples available for the modeling (see Section 3).

In practice, $N$ signal samples, $n = 1, \ldots, N$, are available. Equation 9 can be written for $n = l_x + 1, \ldots, N$ with $l_x = \mathrm{Max}\,\{l_x^{(1)}, \ldots, l_x^{(M)}, l_y\}$. All these equations can be grouped into

$$\mathbf{y} = \mathbf{P} \cdot \mathbf{\Xi} + \mathbf{e} \tag{20}$$

using the following notations

$$\mathbf{y} = \begin{bmatrix} y(l_x+1) \\ \vdots \\ y(N) \end{bmatrix}; \qquad \mathbf{e} = \begin{bmatrix} e(l_x+1) \\ \vdots \\ e(N) \end{bmatrix}; \qquad \mathbf{P} = \begin{bmatrix} \mathrm{p}(l_x+1) \\ \vdots \\ \mathrm{p}(N) \end{bmatrix} \tag{21}$$

The parameter vector $\mathbf{\Xi}$ is a straightforward extension of the previously defined vector in Eq. 9.

As soon as the significant terms inside the model of the system are found, the regression matrix $\mathbf{P}$ can be formed. In order to find the parameter vector estimate $\hat{\mathbf{\Xi}}$, a linear least-squares problem can be defined

$$\hat{\mathbf{\Xi}} = \min_{\Xi} \|\mathbf{y} - \mathbf{P} \cdot \mathbf{\Xi}\|^2 \tag{22}$$

where $\|\cdot\|$ is the Euclidean norm. The solution $\hat{\mathbf{\Xi}}$ satisfies the normal equation

$$\mathbf{P}^T\mathbf{P} \cdot \hat{\mathbf{\Xi}} = \mathbf{P}^T \cdot \mathbf{y} \tag{23}$$

where $\mathbf{P}^T\mathbf{P}$ is called the information matrix and is positive definite. Similarly to the stochastic approach described by 2.15, one can estimate the RRV using $\hat{\mathbf{y}} = \mathbf{P}\hat{\mathbf{\Xi}}$

$$\begin{aligned} \mathrm{RRV} &= \frac{(\mathbf{y} - \hat{\mathbf{y}})^T(\mathbf{y} - \hat{\mathbf{y}})}{\mathbf{y}^T\mathbf{y}} \\ &= 1 - \frac{2\hat{\mathbf{\Xi}}^T\mathbf{P}^T \cdot \mathbf{y} - \hat{\mathbf{\Xi}}^T\mathbf{P}^T\mathbf{P}\hat{\mathbf{\Xi}}}{\mathbf{y}^T\mathbf{y}} \end{aligned} \tag{24}$$

In this case $\hat{\mathbf{\Xi}}$ is a solution of Eq. 23. Several methods have been proposed in the literature to solve Eq. 23, including Cholesky decomposition, Householder projection, MGS orthogonalization, and SVD decomposition (see Ref. 11 for a review of these methods). We present briefly the orthogonalization and SVD methods in Sections 2.3 and 2.4.

### 2.3. Orthogonal Least-Squares Method

Orthogonal parameter estimation methods are well known to provide simpler computational analysis to find appropriate model terms [11]. The advantage of orthogonal compared with nonorthogonal expansions is the successive approximation property. In this case, when adding one term to the model, the former parameters do not need to be recalculated. The error reduction provided by the new term can be derived separately from the other terms.

Let us assume that the matrix **P** is of full rank and therefore can be decomposed using an orthogonal matrix **A**. Equation 23 becomes

$$\mathbf{A}\mathbf{D}\mathbf{A}^{\mathrm{T}} \cdot \hat{\boldsymbol{\Xi}} = \mathbf{A}\mathbf{D} \cdot \mathbf{g} = \mathbf{P}^{\mathrm{T}} \cdot \mathbf{y} \tag{25}$$

where **D** is a positive diagonal matrix, and define $\mathbf{g} = \mathbf{A}^{\mathrm{T}} \cdot \hat{\boldsymbol{\Xi}}$. A Gram-Schmidt or modified Gram-Schmidt procedure can be used to determine **g** when orthogonalizing **P** [11], and finally the parameter vector is estimated with

$$\hat{\boldsymbol{\Xi}} = \mathbf{A} \cdot \mathbf{g} \tag{26}$$

Suppose that our model already contains $K$ polynomial terms; i.e., the matrix **P** has $K$ columns. If one wants to add a $(K+1)$st term in the expansion, this amounts to adding a new column in **P**. The resulting MSE reduction $Q_{K+1}$ is directly obtained from the quantity [11]

$$Q_{K+1} = \mathbf{g}(K+1)^2 \cdot \frac{\mathbf{D}(K+1, K+1)}{N - l_x} \tag{27}$$

The Gram-Schmidt–based approach has the advantage of yielding directly the MSE reduction brought by this additional term and therefore provides low computational complexity. However, it is very sensitive to the possible ill conditioning of **P**. A robust numerical scheme is presented in the next subsection.

### 2.4. Singular Value Decomposition

Singular value decomposition is a well-known scheme for solving the normal equation 23 and is known to provide more stable parameter estimates even when **P** is not fullrank. It can be shown that the $(N - l_x) \times K$ matrix **P** can be factorized as

$$\mathbf{P} = \mathbf{U}\mathbf{S}\mathbf{V}^{T} \tag{28}$$

where **U** is an $(N - l_x) \times K$ orthogonal matrix, **V** is a $K \times K$ orthogonal matrix, and $\mathbf{S} = \mathrm{diag}\{s_1, \ldots, s_K\}$ with $s_i > 0$ called the $i$th singular value of **P**. The pseudoinverse of **P** is defined as

$$\mathbf{P}^{+} = \mathbf{V}\mathbf{S}^{+}\mathbf{U}^{T} \tag{29}$$

where the elements of the diagonal matrix $\mathbf{S}^+$ are

$$s_i^+ = \begin{cases} 1/s_i & \text{for } s_i > 0 \\ 0 & \text{for } s_i = 0 \end{cases} \tag{30}$$

Finally, the solution to 23 is

$$\hat{\Xi} = \mathbf{P}^+\mathbf{y} \tag{31}$$

Prior to the final parameter estimation, spurious parameters must be excluded in order to extract an optimal model and conserve the essentials of the signal dynamics. This problem is often called "model selection," and the results of this selection will be called the *minimal model.*

Selection requires the introduction of appropriate criteria. Section 3 describes in detail Rissanen's minimum description length (MDL) criterion and compares it with other well-known information-based criteria.

## 3. MODEL SELECTION

### 3.1. Minimum Description Length

Constructing a model for the prediction of time series or system identification involves both the selection of a model class and the selection of a model within the selected model class. Successful selection of a model class appears to be a very difficult task without prior information about the time series (see Section 2). Selection of a model inside a class appears to be a more manageable problem. A parametric model may be constructed under the assumption that there is a class of conditional probability function $P(\mathbf{y}|\Xi)$, each assigning a probability to any possible observed time series or sequence $\mathbf{y} = [y(1), y(2), \ldots, y(N)]^T$ of $N$ sample points. The parameters $\Xi = [\xi_1, \xi_2, \ldots, \xi_k]^T$ are to be estimated to optimize the model. This can be done by maximizing $P(\mathbf{y}|\Xi)$ or its logarithm with respect to $\Xi$ only if we ignore prior information about $P(\Xi)$, which is known as the maximum likelihood (ML) approach

$$\hat{\Xi} = \max_{\Xi}\{\ln[P(\mathbf{y}|\Xi)]\} \tag{32}$$

It can be shown that the ML estimation criterion can also be expressed in coding theoretic terms [15]. For any parameter vector $\hat{\Xi}$, one can assign a binary code sequence to $\mathbf{y}$ that is uniquely decodable. The corresponding mean code length of this sequence, which is equal to $L(\mathbf{y}|\Xi) = -\ln[P(\mathbf{y}|\Xi)]$, is called its *entropy*. The minimal value of the entropy is attained for $\Xi = \hat{\Xi}$. Hence, determining the ML estimate and finding the most efficient encoding of y in a binary code sequence are equivalent tasks. Up to now, the number of parameters $k$ has been supposed to be known. Things get more complicated if it is not so, which is by far the most frequent situation. When applying the ML estimator consecutively for all increasing $k$ values, one may end up with as many parameters as sample points and a very short code sequence for $\mathbf{y}$. But if a binary code sequence for $\mathbf{y}$ is constructed for some $\Xi$, this parameter vector has to be known at the decoder side for successful decoding. A more realistic encoding demands that the parameters be encoded themselves and added to the code sequence. In this case, the total code string length is

$$L(\mathbf{y}, \Xi) = L(\mathbf{y}|\Xi) + L(\Xi) \tag{33}$$

The crucial point in this representation of the total code length is the balance between the code length for the data $L(\mathbf{y}|\Xi)$ and the code length for the parameter vector $L(\Xi)$. For a rough description of the data with a parsimonious number of parameters, the latter are encoded with a moderate code length, whereas the data need a relatively long code length. On the other hand, describing data with a high number of parameters may demand a short code length for the data, but the price must be paid with a longer code for the parameters. From that point of view, it is reasonable to look for the parameters that minimize the total code length. This gives rise to the *minimum description length* (MDL) of data [6]. When minimizing Eq. 33, the general derivation of the MDL leads to

$$\hat{\Xi} = \min_{\Xi}\left\{-\ln[P(\mathbf{y}|\Xi)] - \ln[P(\Xi)] - \sum_{j=1}^{k}\ln[\delta_j]\right\} \tag{34}$$

where[1] $P(\Xi)$ is the probability distribution of the parameters and $\delta_j$ is the precision on the $j$th parameter. The first term comes from $L(\mathbf{y}|\Xi)$ (the data) and the two other terms come from $L(\Xi)$ (the parameters). The last term in Eq. 34 decreases if coarser precision is used (larger $\delta_j$), while the first term generally increases. An estimation of the precision coefficients $\delta_j$ can be found by solving the following equation [16]:

$$\left(\frac{\partial^2 Q}{\partial \Xi^2}\cdot\boldsymbol{\delta}\right)_j = 1/\delta_j \quad \text{with } Q = L(\mathbf{y}|\hat{\Xi}) \tag{35}$$

where $\boldsymbol{\delta}^T = [\delta_1, \ldots, \delta_k]$. Equation 35 comes from the minimization of the log-likelihood function $L(\mathbf{y}, \Xi)$ using a second-order approximation of $L(\mathbf{y}|\Xi)$, i.e.,

$$L(\mathbf{y}|\Xi) \leq L(\mathbf{y}|\hat{\Xi}) + \boldsymbol{\delta}^T\mathbf{Q}\boldsymbol{\delta}/2 \tag{36}$$

In our case, we have a linear-in-the-parameter model, and the matrix $\mathbf{Q}$ can be written as

$$\mathbf{Q} = \mathbf{P}^T\mathbf{P}/\sigma_\epsilon^2 \tag{37}$$

From the second term on the right-hand side of Eq. 36, we can define what is usually called the confidence ellipsoid. As the $\mathbf{P}$ matrix contains all the polynomial up to degree $D$, the confidence ellipsoid may be highly eccentric for $D > 1$ (very thin in some directions). In order to avoid ill conditioning of $\mathbf{Q}$ and therefore inaccuracies in estimating $\delta_j$, scaling of the ellipsoid must be used [17].

[1] The requirement that the code length should be an integer is ignored.

Different MDL expressions may arise when computing the minimum of the description length given in Eq. 34. When considering a computer floating-point representation of the model parameters and under the assumption that the modeling error $\epsilon$ is normally distributed, simplifications occur and Eq. 34 is given by [16]

$$\mathrm{MDL(k)} = \left(\frac{1}{2}\mathrm{N} - 1\right) \cdot \ln[\sigma_\epsilon^2] + (\mathrm{k}+1) \cdot \left(\frac{1}{2} + \ln[\gamma]\right) - \sum_{\mathrm{j}=1}^{\mathrm{k}} \ln[\delta_\mathrm{j}] \tag{38}$$

where $\sigma_\epsilon^2$ is the estimated model error variance and $\gamma$ is the upper bound of the code length for the parameters. A value $\gamma = 32$, which corresponds to the usual floating-point binary representation, has been advised [16]. With respect to model complexity, the first term (the data) acts in the opposite direction to both the second (the number of parameters) and the third one (parameter precision). Note that in the second term, one finds $k+1$ and not $k$ because the error variance must also be considered as a model parameter. The best model is selected according to the model complexity that yields a minimum value for MDL(k).

It can be shown that for *large N* and after omitting all the constants, a simplified MDL criterion is obtained

$$\mathrm{MDL_s(k)} = \mathrm{N} \cdot \ln[\sigma_\epsilon^2] + \mathrm{k} \cdot \ln[\mathrm{N}] \tag{39}$$

This simplified MDL criterion no longer includes the precisions $\delta_j$ because their influence becomes very small. However, in practice, it is hard to determine the conditions under which Eq. 39 becomes valid. We will come back to this problem in Subsection 3.4 when we compare the two MDL criteria given by Eqs. 38 and 39.

## 3.2. Other Selection Criteria

A valid model must be able to describe the dynamics of a time series correctly. Most often, real-world time series are embedded in noise, which makes it difficult to achieve this goal. A good model selection criterion should be able to distinguish between signal and noise.

Statistical hypothesis testing is a classical framework for the selection of one model out of two candidate ones. Well-known statistical tests are the likelihood ratio and the log determinant ratio tests [18]. These methods are restricted to the comparison of two particular models and are not appropriate for the selection of the best model inside a model class. In Ref. 19, a so-called ERR test is introduced. An additional term is accepted in the model if the corresponding error reduction lies above a given threshold. In addition, to verify that the selected model provides the correct fit to the data, correlation-based validity tests can be applied [20]. These tests detect potential correlations in the modeling error $\epsilon$, also on the basis of a predetermined threshold.

The model selection methods that have been mentioned use significance levels and subjective thresholds that do not depend on the amount of data. Criteria that are independent of these subjective choices are called *objective criteria* [18]. With an objective criterion, model selection can be applied directly while minimizing the criterion.

Besides MDL, a well-known objective criterion is the Akaike information criterion (AIC) [21]. For data with a Gaussian distribution the AIC is given by

$$\mathrm{AIC(k)} = \mathrm{N} \cdot \ln[\sigma_\epsilon^2] + \mathrm{k} \cdot \mathrm{c} \tag{40}$$

with $c = 2$. The expression for the AIC looks very similar to the simplified MDL criterion. The change is in the penalizing term $k \cdot c$, which differs by a factor from $c/\ln[N]$ of the corresponding MDL term in Eq. 39. However, the origin of the AIC is different. It is based on the maximization of the Kullback information [22], which expresses the difference between the true probability density function and the one given by the model in the light of the observed data.

The model with the largest possible number of parameters makes the term $N \cdot \ln[\sigma_\epsilon^2]$ minimum, whereas the term $k \cdot c$ that penalizes the complexity of the model is maximum. As with the MDL criterion, model selection can be performed automatically by selecting the model that minimizes the AIC. However, the AIC has been criticized because it can be shown in rather general contexts that the estimates of the number of parameters it provides are not consistent, unlike those of the MDL criterion. Moreover, AIC tends to overestimate the true number of parameters [23]. Note that in some applications, such as parametric spectral estimation, some overparameterization is necessary [24]. In such cases, AIC seems more suitable.

More parsimonious models are selected by increasing the constant $c$ in the penalizing term $k \cdot c$. It can be shown that $c$ is related to the probability of selecting more parameters than the true model possesses. This probability is 0.0456 when $c = 4$ [18]. Finally, the objectivity of the AIC is only as good as the significance level it corresponds to.

In Subsection 3.3, we compare the model selection behavior of the AIC given by Eq. 40 with the MDL criteria given by Eqs. 38 and 39, respectively.

### 3.3. Polynomial Search Method

In Section 2, parameter estimation methods were presented. In addition, a search method that scans through all the possible model candidates is needed.

Obviously, one can perform an exhaustive search, that is, examine all the possible combinations of candidate terms. However, for polynomial models, the number of terms may be so large (see Eq. 8) that this exhaustive search becomes infeasible.

Therefore, we suggest applying *an exhaustive method per term* starting with small models that will be expanded term by term. We set a priori the maximal degree $D_{\max}$ and lag $lag_{\max}$ (for all degrees $d$). This defines the set of possible polynomial terms in the model on which the search will be performed.

In the first step, the signal is modeled only by its average value. Iteratively, one adds in turn each polynomial of $\mathbf{p}$ not already selected. The polynomial giving the minimum MDL value is retained. This iterative process is pursued until no further reduction of MDL is observed. Suppose that we already have $K$ columns in $\mathbf{P}$; the polynomial selection scheme is continued as follows:

1. Add one column $K + 1$ in $\mathbf{P}$.
2. Solve the normal equation 23.
3. Go to step 2 until all polynomials are tested and store the polynomial corresponding to the minimum MDL.
4. If this MDL is smaller then the previous one, retain the selected polynomial, and discard otherwise.
5. Go to step 1 until no further MDL reduction is obtained.

## 3.4. Validation of the Selection Criteria

In this subsection, we present some illustrative examples with synthetic systems and experimental data. Using the polynomial model selection algorithm presented above, we extract the most significant polynomial terms according to the previously described selection criteria. In pure synthetic experiments, we take normally distributed random data and use the Gram-Schmidt procedure.

### 3.4.1. Numerical Experiments

The first example consists of a fifth-order linear MA filter with a single ($M = 1$) independent, identically distributed Gaussian input. According to Eqs. 7 and 9, $D = 1$, $l_x^{(1)} = 4$, $\Theta = 0$ for the true model

$$y(n) = \sum_{i=0}^{4} \phi_1^{(1)}(i)x_1(n-i) + e(n) \tag{41}$$

where $e(n)$ represents the output noise signal. The signal-to-noise ratio has been set to $-3$ dB. The performance of any model selection algorithm clearly depends on the noise level.

Figure 1a depicts the average number of parameters selected by the different model selection criteria as a function of the signal length in a Monte Carlo simulation with 30 runs. The algorithm is initiated with $D_{\max} = 3$ and $l_{\max} = 10$, such that the algorithm does not know a priori whether the system is linear. Figure 1a shows that the AIC with $c = 2$ overfits the true model for all signal lengths, and with $c = 4$ the results are slightly better but still present overfitting. $\text{MDL}_\text{S}$ converges to the real synthetic model with increasing data lengths, and MDL ($\gamma = 4$ or $\gamma = 32$) is quite robust even for short data lengths (almost no overfitting is observed). We now proceed to a synthetic nonlinear polynomial system with a signal-to-noise ratio of $-3$ dB described by

$$\begin{aligned} y(n) = 0.5 * x_1(n-1) * x_1(n-2) + 0.4 * x_1(n-1) - 0.4 * x_1(n-2) * x_1(n-5) \\ + 0.3 * x_1(n-5) - 0.3 * x_1(n-1) * x_1(n-3) * x_1(n-4) + e(n) \end{aligned} \tag{42}$$

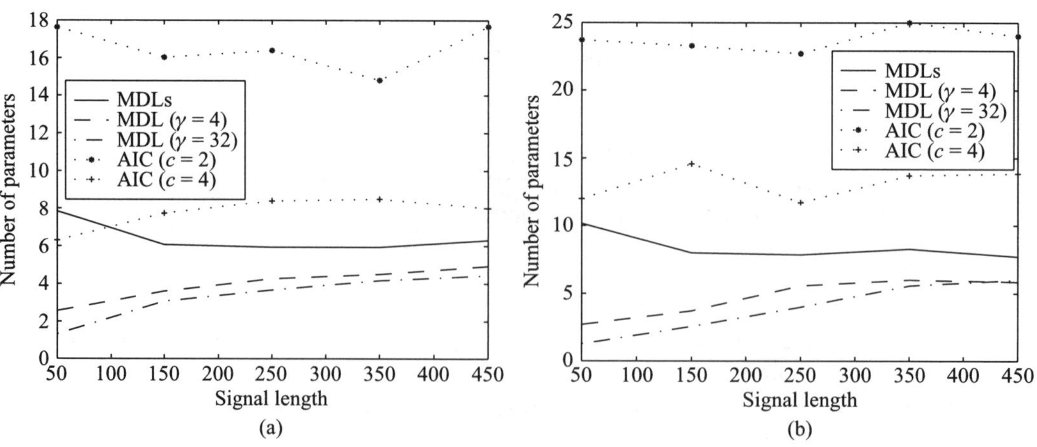


**Figure 1** Number of parameters as a function of the signal length.

Figure 1b shows again the number of parameters selected by applying the different selection criteria. The same observation can be made as in the previous linear example. MDL selects the correct number of parameters, and the behavior of $MDL_S$ is asymptotically correct. AIC overestimates the true model order by a wide margin. A typical example of a model selected by $MDL_S$ for a signal length of $N = 250$ is given next, with the amount of RRV reduction ($\Delta$RRV) indicated below each term

$$\begin{aligned} y(n) = 0.006 &+ \underbrace{0.510 * x_1(n-1) * x_1(n-2)}_{11.63\%} + \underbrace{0.400 * x_1(n-1)}_{7.45\%} \\ &\underbrace{-0.392 * x_1(n-2) * x_1(n-5)}_{6.70\%} + \underbrace{0.298 * x_1(n-5)}_{3.71\%} \\ &\underbrace{-0.268 * x_1(n-1) * x_1(n-3) * x_1(n-4)}_{3.01\%} + \underbrace{\epsilon(n)}_{67.50\%} \end{aligned} \tag{43}$$

Besides the fact that the correct terms are selected, one may notice the capacity of the method not to overfit the data. After the algorithm stops, the residual error variance is correctly attributed to random variations that do not belong to the system structure. In the preceding example, the residual error variance accounted for 67.5% of the output signal power, a signal-to-noise ratio of about −3 dB.

### 3.4.2. Sunspot Time Series

Several benchmark time series are widely used in the literature to study the linear and nonlinear modeling of time series. Therefore, we tested our NARMA models together with our search method (see Subsection 3.3) on one famous benchmark series, namely the sunspot series. The annual sunspot time series represents the variations in sunspot activity over the last 300 years. The sunspot activity is quantified by the Wolfer number, which measures both the number and size of the sunspots. The sunspot series used here consists of 250 observations covering the years 1700–1949. It displays a cycle of approximately 11.3 years. We want to compare results presented in Ref. 25 with those provided by our model selection method. In Ref. 25, several AIC selected models are presented for the sunspot series. A linearity test based on the null hypothesis that the estimated bispectrum of the series is zero at all frequencies is presented. Table 1 summarizes the number of parameters and the MSE provided by the selected models. With respect to AIC with $c = 4$, the best model was found to be a ninth-order linear AR, AR(9). However, in Ref. 25, the best linear model including only terms with lags 1, 2, and 9 is described. Results using these three parameters for the prediction are quite similar to those of the complete AR(9) model (Table 1). In the same reference, a bilinear model (second-order extension of the ARMA model) is applied to the prediction of the

**TABLE 1** AIC Modeling of the Sunspot Series

| | AR(9) | Subset of AR(9) | BL |
|---|---|---|---|
| $k$ | 9 | 3 | 15 |
| $\sigma_\epsilon^2$ | 194.43 | 199.2 | 143.86 |

**TABLE 2** MDL Modeling of the Sunspot Series

| | MDL | |
|---|---|---|
| Model | Lin | NL |
| $k$ | 3 | 2 |
| $\Delta RRV$ | 84.4% | 3.1% |
| $\sigma_\epsilon^2$ | 159.88 | |

sunspot series. In Table 1, we see that this nonlinear model reduces the MSE considerably. Using AIC with $c = 2$, 15 parameters have been retained.

$$y(n) = 2.23 + \underbrace{1.4480 * y(n-1)}_{67.10\%} \underbrace{-0.7745 * y(n-2)}_{15.29\%} + \underbrace{0.169 * y(n-9)}_{2.02\%}$$
$$\underbrace{-0.0045 * y(n-1) * y(n-1)}_{1.78\%} + \underbrace{0.0042 * y(n-2) * y(n-2) * y(n-2)}_{1.3\%} + \underbrace{\epsilon(n)}_{12.51\%} \quad (44)$$

Results using our selection algorithm with the complete criterion (see Eq. 39) are presented in Table 2, and the corresponding minimal model is given by Eq. 44. The most relevant terms are, as in Ref. 25, the three linear terms. Comparing the bilinear model in Ref. 25, the nonlinear terms are small compared with the linear ones but their RRV reduction (39.4% for both) is not negligible. The sunspot series (dotted line) and its prediction (solid line) are displayed in Figure 2. Table 2 summarizes the RRV reduction due to linear (Lin) and nonlinear (NL) terms and the overall MSE.

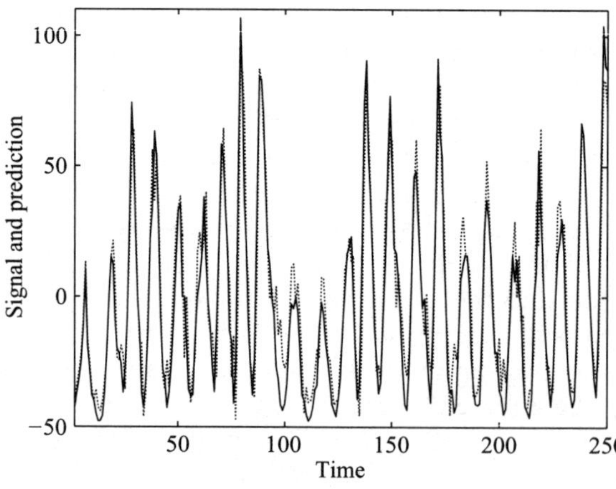


**Figure 2** Sunspot time series and its prediction.

### *3.4.3. RR-Interval Time Series*

In the previous examples, our modeling algorithm has been shown to be a very valuable method for nonlinear polynomial model selection. Before analyzing more closely signals of the cardiovascular system, three MDL criteria and two AICs are compared and validated on an ensemble of heart rate signals. RR intervals of 10 control subjects have been recorded (see Section 5.3.1). Each time series consists of 500 samples. We applied the polynomial search algorithm with $D_{max} = 3$ and $l_{max} = 10$ and used the SVD for the estimation part. Figure 3 shows the evolution averaged over the recordings of the five model selection criteria with the number of parameters.

Typically, the model selection criteria take high values for low numbers of parameters because of high values of the MSE. As the number of parameters grows, the MSE decreases and so do the model selection criteria. The values taken by the criteria should start to increase when the trade-off between model complexity and MSE is such that the model begins to overfit the data. The best model is located at the minimum.

As seen in Figure 3, the AIC curves are flatter than the MDL ones. The $c$ factor determines the curvature of these functions. The simplified $\mathrm{MDL_S}$(k) given by Eq. 39 is close to AIC with $c = 4$ but selects a reasonable number of parameters, i.e., $k \approx 7$ (see Section 5 for more details about RR prediction). Taking into account the parameter precisions $\delta_j$ and with $\gamma = 4$ or $\gamma = 32$, the MDL(k) functions 38 give a pronounced minimum for five parameters. The curvature of the MDL(k) curves at the minimum is conditioned by $\delta_j$ and $\gamma$. The larger the value of $\gamma$, the more pronounced the curvature, with a slight displacement in the location of the minimum.

In this section, the complete MDL(k) criterion was compared with the simplified one, which is asymptotically valid. Both criteria were compared with the AIC(k). Various synthetic and experimental simulations have shown that the MDL(k) criterion provides excellent results. $\mathrm{MDL_S}$(k) overestimates the true model order for *small* Gaussian data sets. The performance of the AIC(k), which systematically overfits the data, is not satisfactory. Finally, we want to give some advice to the reader about the use of the various model selection criteria:

- TheMDLs given by Eqs. 38 or 39, or AIC by Eq. 40, are valid only for Gaussian data, which is usually far from being the case for biomedical signals.
- The complete MDL(k) given by Eq. 38 depends on the $\gamma$ factor, which introduces an arbitrary component in the criterion.
- Although $\mathrm{MDL_S}$(k) given by Eq. 39 seems to be valid only for large data sets, it has the great advantage of alleviating the previously mentioned drawback.

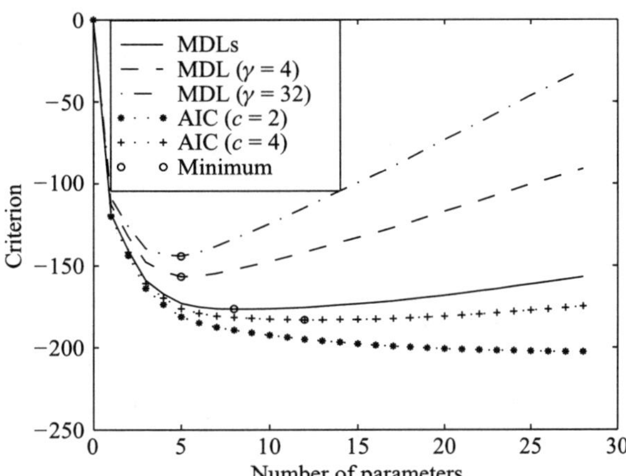


**Figure 3** Model selection criteria with RR intervals.

## 4. SUBBAND DECOMPOSITION

Subband signal processing was introduced as a powerful tool essentially for coding purposes [26]. Subsequent work [27] established that the subband decomposition approach outperforms classical techniques in terms of modeling and spectral estimation. In addition, the decomposition into different frequency bands allows one to take into account specific properties of the signal under investigation. Indeed, for a large number of physical signals, in particular biomedical signals, distinct regulation mechanisms are associated with different frequency bands. Consequently, a prior subband decomposition of this kind of signal may facilitate the comprehension of the underlying physical phenomena.

### 4.1. Subband Signal Processing

The basic idea behind subband signal processing is the use of a filter bank consisting of several band-pass filters. In the classical scheme, the source signal is first band-pass filtered through an *analysis filter bank* and then downsampled to obtain subsignals representing the source signal within different frequency bands. The advantages of such a decomposition are that all operations can be parallelized and that the downsampling performed on the subsignals yields a reduced amount of data to be processed. The source signal can eventually be reconstructed using a *synthesis filter bank*. An appropriate design of both filter banks (analysis and synthesis) guarantees a perfect reconstruction of the original signal. Figure 4 shows a general analysis–synthesis scheme.

The design of the filters may be achieved using a number of well-known signal processing techniques. However, careful selection of the design process can provide a filter bank with some specific desirable properties. One of the most popular implementations relies on wavelets. The main advantage is that a perfect reconstruction scheme can be obtained with a very straightforward implementation.

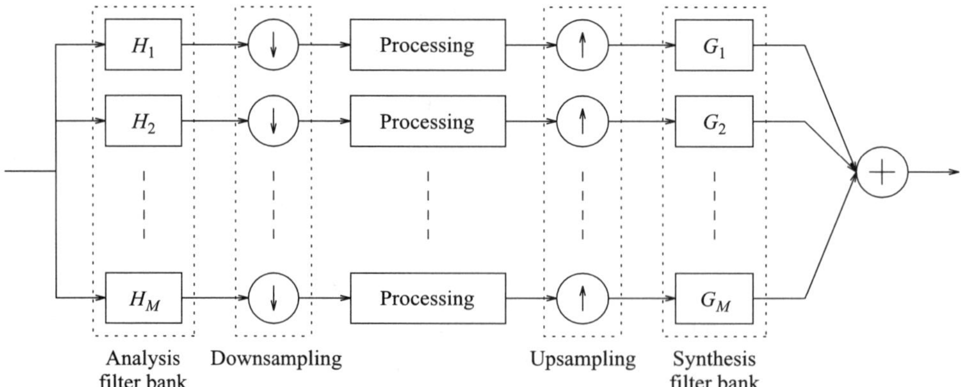


**Figure 4** Decomposition and reconstruction using an analysis–synthesis subband scheme.

#### 4.1.1. The Wavelet Transform

Our purpose is not to give a complete description of the subject because excellent reviews may be found in the literature [28, 29]. We will give here only the basic definitions required to understand the relationships between wavelets and filter banks.

The main motivation for the wavelet transform comes from the fundamental limitations of both the Fourier transform and the short-time Fourier transform. These transforms map the input signal from the time domain to a new domain where it is expressed as a linear combination of characteristic basis functions. For the Fourier transform, these functions are simply sines and cosines, excluding any time localization in the new domain (which corresponds to the frequency domain). In the case of the short-time Fourier transform, the basis functions are also sines and cosines but multiplied by a window so as to introduce time localization in the new domain and give rise to a *time–frequency representation*. However, because the window shape and length are fixed for the whole signal, the time–frequency resolution is constant.

The *wavelet transform* goes beyond the short-time Fourier transform by decomposing the signal into basis functions called *wavelets*, which are the translation and the dilation of a unique function, often referred to as the *mother wavelet* [30]. Both translations and dilations allow one to obtain a variable time–frequency resolution.

Given an input signal $x(t)$, the wavelet transform can be written

$$W_x(a, b) = \frac{1}{\sqrt{|a|}} \int_{-\infty}^{+\infty} x(t)\overline{\psi\left(\frac{t-b}{a}\right)}\, dt \tag{45}$$

where $a \in \mathcal{R}\backslash\{0\}$ is the dilation (or scaling) factor, $b \in \mathcal{R}$ is the translation factor, and the wide bar denotes complex conjugation. Equation 45 may also be read in the Fourier domain

$$W_x(a, b) = \sqrt{|a|} \int_{-\infty}^{+\infty} X(\omega)\overline{\Psi(a\omega)}e^{j\omega b}\, d\omega \tag{46}$$

where $X$ and $\Psi$ are, respectively, the Fourier transform of the signal and the Fourier transform of the wavelet.

Equations 45 and 46 show that the wavelet transform is fundamentally a time–frequency representation or, more properly, a time–scale representation (the scaling parameter $a$ behaves as the inverse of a frequency).

#### 4.1.2. Wavelets as a Filter Bank

If we denote the scaling and conjugation of $\psi(t)$ by

$$\psi_a^{\#}(t) = \frac{1}{\sqrt{|a|}}\overline{\psi\left(\frac{-t}{a}\right)} \tag{47}$$

it is immediate to see that the wavelet transform may also be expressed by a convolution product

$$W_x(a, b) = (x * \psi_a^{\#})(b) \tag{48}$$

The input signal $x(t)$ can be reconstructed from its wavelet transform if the Fourier transform $\Psi(\omega)$ of $\psi(t)$ satisfies the so-called admissibility condition

$$\int_{-\infty}^{+\infty} \frac{|\Psi(\omega)|^2}{|\omega|} \, d\omega < +\infty \tag{49}$$

This condition implies that $\Psi(0) = 0$ and that $\Psi(\omega)$ is small enough in the neighborhood of $\omega = 0$. Accordingly, the wavelet $\psi(t)$ may be viewed as the impulse response of a band-pass filter. Using expression 48, the frequency response of this filter is given by

$$\Psi_a(\omega) = \sqrt{|a|}\Psi(a\omega) \tag{50}$$

showing that the time–frequency resolution of the wavelet transform changes with the scaling factor $a$. If we evaluate 48 for a discrete set of scaling factors, we obtain what is called a *constant-Q filter bank*. The wavelet transform performs a subband decomposition of the signal with a constant relative bandwidth, i.e., $\Delta\omega/\omega = \text{const}$. The choice $a = 2^j$, $j \in \mathcal{Z}$, leads to a *dyadic* subband decomposition, and each frequency band is approximately one octave wide [31]. Figure 5 illustrates this type of filter bank.

If we denote by $H_1(\omega), \ldots, H_J(\omega)$ the discrete analysis filters associated with a $J$-channel dyadic wavelet decomposition, it is possible to obtain the corresponding reconstruction algorithm if the system acts globally as the identity, i.e., if the synthesis filters $G_1(\omega), \ldots, G_J(\omega)$ are chosen such that

$$\sum_{j=1}^{J} H_j(\omega)G_j(\omega) = 1 \tag{51}$$

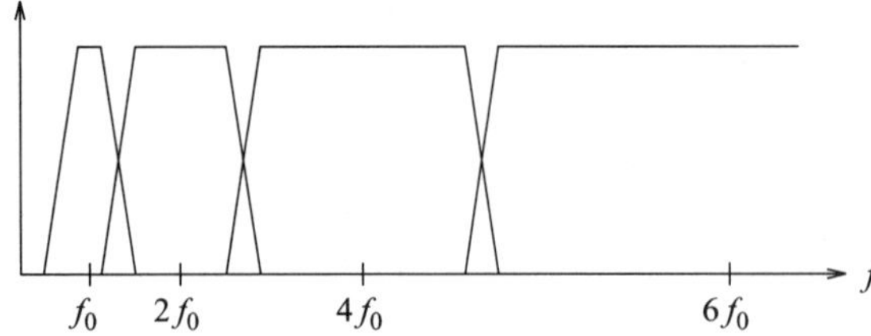


**Figure 5** Filter bank corresponding to a dyadic wavelet decomposition of the signal.

#### *4.1.3. Subband Spectral Estimation and Modeling*

The dyadic subband decomposition has given rise to the so-called *multiresolution approach* [31], widely employed for coding purposes. But other signal processing applications have emerged. The interest in subband decomposition for spectral estimation has been highlighted by some researchers [27]. Using information theory, important properties have been derived:

- The sum of the minimum prediction errors over the subbands is lower than the minimum prediction error achievable in the full band.
- The subband spectra are whiter than the full-band spectrum.
- The subband prediction error spectrum is whiter than the full-band prediction error spectrum.

A more general review of the superiority of subband decomposition for coding, linear prediction, or spectral estimation may be found in Ref. 32.

## 4.2. Frequency Decomposition of Physical Signals

### *4.2.1. Modeling of the Human Visual System*

In addition to the advantages in terms of estimation and modeling, the division into several frequency bands of interest corresponds to a large number of physical phenomena, especially in the biomedical field. The first attempt to characterize a natural phenomenon using this kind of scheme was probably the modeling of the human visual system [33]. Numerous efforts have been devoted in the study of psychophysiology to analyze the response of the human visual system to stimuli having particular orientations and frequency tunings. The first experiments [34] have shown that the retinal image seems to be processed in separate frequency channels. Later, some investigators [35] tried to characterize these filters more accurately and showed that their frequency bandwidth was around one octave. Accordingly, the retinal image is likely to be processed by a filter bank whose filter widths are approximately the same on a logarithmic scale, thus corresponding to a *constant-Q filter bank*. This finding has been an important motivation in the development of the multifrequency channel decomposition of images by Mallat [31].

### *4.2.2. Other Physiological Signals*

The division into frequency bands may be found in many biomedical signals, particularly those of neurophysiological or cardiovascular origin. Relevant frequency ranges corresponding to the various biological oscillations are large. For instance:

- **Electroencephalogram (EEG)**. The EEG is the main signal used to study the general functional states of the brain, such as the sleep stages, neurological disorders, or metabolic disturbances. The signal collected by standard recorders has a frequency range of 0.5 to 40 Hz and is usually described with the following frequency bands: $\delta$ band (below 4 Hz), $\theta$ band (from 4 to 8 Hz), $\alpha$ band (from 8 to 12 Hz), and high-rhythm $\beta$ band (above 12 Hz).
- **Electromyogram (EMG)**. The EMG is mainly used to assess the average muscular tone and can give more precise insights into the recognition of sleep stages. The useful frequency range is between 10 and 1000 Hz.
- **Electrogastrogram (EGG)**. Electrical activity also exists in the human stomach. The EGG is a noninvasive means of obtaining myoelectrical activity in the stomach by placing surface electrodes on the abdomen. The gastric electric signal may display normal slow waves (regular frequency of about 0.03 to 0.07 Hz), periods of tachygastria (frequency range of 0.07 to 0.15 Hz), or periods of bradygastria (frequency range of 0.008 to 0.03 Hz).

## 5. APPLICATION TO CARDIOVASCULAR SYSTEM MODELING

### 5.1. The Cardiovascular System

Spontaneous fluctuations in heart rate and blood pressure were described for the first time several centuries ago. In 1733, Hales reported respiration-related fluctuations in blood pressure. Later, Ludwig extended this observation to fluctuations in heart rate, and Mayer gave his name to blood pressure fluctuations that were slower and therefore appeared to be independent of those related to respiration. It is now well established that these variations in heart rate and blood pressure are mediated by the autonomic nervous system [36, 37] on a closed-loop basis in order to maintain cardiovascular homeostasis. Indeed, afferent signals from many peripheral receptors, from respiratory neurons, and from higher cerebral centers impinge on the neurons located in the cardiovascular centers of the brain stem and thereby modulate the activity of the sympathetic and parasympathetic nerves.

Let us describe briefly the regulatory processes involved in cardiovascular system function. The heart is composed of two pumps working in parallel and in series in a finite space (defined by the pericardium). The right side of the heart propels the desaturated venous blood to the lungs, and the left side of the heart propels the newly oxygenated blood from the lungs to the arterial systemic circulation. In addition, both sides of the heart interact with each other during the cardiac cycle. The systolic function of the left heart improves the right heart function (systolic interaction), and the diastolic state of the right heart modifies the left one (diastolic interaction). Inspiration increases the venous return to the right heart, increasing its end-diastolic volume and consequently reducing the left end-diastolic volume. According to the Frank–Starling law, the resulting left ventricular ejection volume and systolic blood pressure decrease, unloading pressure receptors located in the arterial tree (sinoaortic baroreceptors). Baroreceptor unloading results in a drop of inhibitory inputs to cardiovascular brain stem centers, inhibiting cardiac parasympathetic neuron activity. As a consequence, inspiration is responsible for a reflex increase in heart rate to compensate for the mechanically induced decrease in left ventricular ejection volume and systolic blood pressure (homeostatic process). In the frequency domain, parasympathetically mediated heart rate variations and mechanically induced systolic blood pressure fluctuations have the same high-frequency ($\geq 0.15$ Hz) location as the respiratory rate. We thus have a closed-loop homeostatic process located in the high-frequency domain.

In addition, fluctuations in diastolic blood pressure are sensed by the same baroreceptors. However, the homeostatic regulatory process responds with a delay estimated to be a few beats. This delay is responsible for the 0.1 Hz rhythm (resonance phenomenon) observed in neurocardiovascular signals. Diastolic blood pressure decrease is sensed by arterial baroreceptors, resulting in an increase in sympathetic vascular activity. Because of intrinsic delays in the feedback loop, the increases in vascular tone and diastolic blood pressure resulting from the activation of vascular sympathetic nerves are located in the low-frequency domain ($\leq 0.15$ Hz) of the blood pressure spectrum. The resulting blood pressure fluctuations are then sensed by arterial baroreceptors, which modulate the heart rate in order to limit the amplitude of diastolic blood pressure fluctuations. This analysis results in a standard division of cardiovascular signal spectra [38]: very low frequencies (VLFs) = 0–0.04 Hz, low frequencies (LFs) = 0.04–0.15 Hz, and high frequencies (HFs) = 0.15–0.5 Hz.

Since the 1970s, many authors have attempted to model the cardiovascular system. Kitney [39] was the first to apply spectral analysis techniques to heart rate time series and to confirm that spontaneous fluctuations in heart rate and blood pressure signals had distinct frequency peaks. Then, Akselrod et al. [40] reported the preponderant role played by the sympathetic and parasympathetic nervous systems as mediators of heart rate variations. Since then, several other studies summarized in Refs 41 and 42 have confirmed their results.

Different approaches have been used to assess the spontaneous variation of these signals:

- The continuous-time approach, in which signal fluctuations are considered as slowly varying analog signals [39, 43].
- The beat-to-beat approach, in which cardiac interbeat intervals and systolic and diastolic blood pressure are considered as nonuniformly sampled discrete time signals [44–46].
- The hybrid continuous/beat-to-beat approach [47].

Both respiration and baroreceptor activity are characterized by nonlinear behaviors [48, 39]. For simplicity, however, linear models were used first to analyze variability of and relationships between cardiovascular (CV) signals. The first approach in which the CV system was considered as a closed-loop system was performed by De Boer et al. [44]. The authors were the first to suggest that the 0.1 Hz rhythm present in the heart rate and blood pressure was caused by a resonance phenomenon [44] and not a central oscillator as reported initially in the literature. In 1988, Baselli et al. [45] developed a beat-to-beat closed-loop model of the CV system taking into account heartbeat intervals, systolic and diastolic blood pressures, and respiration signals. Causal models together with open- and closed-loop identification procedures were tested using white noise input signals and least-squares methods. Since then, closed-loop linear models have been used by several authors [43, 49, 50]. In 1991, Saul et al. [46] proposed a model based on transfer function analysis of the autonomic regulation in dogs that were excited with erratic rhythms. Using an open-loop spectral analysis of the CV system, they examined effects of random excitations of vagal, sympathetic, and respiratory sites.

Figure 6 displays the CV system considered as a closed-loop system with different autonomic components: the muscle sympathetic nerve activity (MSNA), the cardiac sympathetic nerve activity (CSNA), the cardiac vagal nerve activity (CVNA), the phrenic nerve (PN), and the main CV signals (RR, MBP, ILV).

Nonlinear neurocardiovascular models have not been thoroughly examined so far (see H. Seidel and H. Herzel in Ref. 51). The reason for such parsimonious work in this field may be related to the potential complexity of nonlinear methods. Several works can be mentioned in this direction [13, 52–56].

Until recent publications, models were validated using power spectral analysis, a procedure with important limitations. Although model selection has been the subject of numerous publications [10, 11, 16], it is only since 1996, with the work of Perrot and Cohen [57] and Celka et al. [55], that it has been applied to CV system modeling.

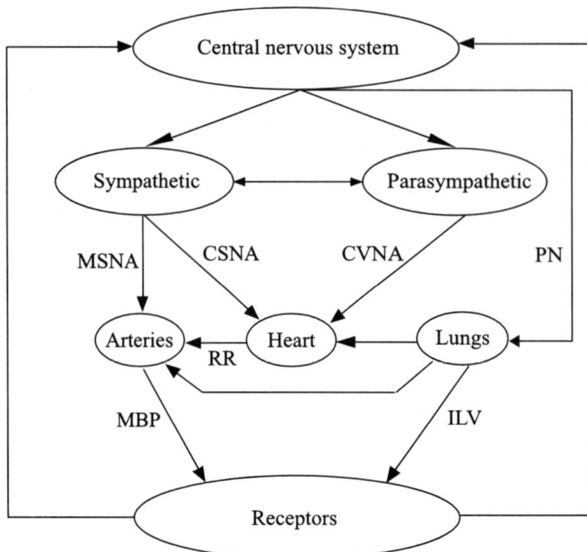


**Figure 6** Cardiovascular closed-loop system.

One of the major limitations of all current models is that they do not take into account the particular relationships of CV signals within specific frequency bands. It is thus important to investigate the various regulatory processes according to their location in the spectral domain and analyze CV signal variability and interactions within VLF, LF, and HF bands.

For all these reasons, in the following experiments, we first proceed to a subband decomposition of CV signals according to the three formerly mentioned frequency bands (Section 4). Then we apply polynomial NARMA modeling (Section 2) together with model selection (Section 3) to subband signals in order to assess their nonlinear relationships.

## 5.2. Subband Decomposition of the Cardiovascular Signals

### 5.2.1. Dyadic Filter Bank

In order to split CV signals according to the VLF, LF, and HF bands, we used a digital dyadic wavelet-based decomposition with orthogonal Daubechies [58] wavelets. This procedure has the advantage of compact support and a finite impulse response (FIR) representation with quadratic mirror filters [59, 60]. The perfect reconstruction property and zero phase lag at the synthesis level make this filter bank very attractive. The dyadic decomposition forced us to choose a sampling frequency of 1.2 Hz and to take $J = 5$ (Section 4) in order to have the best possible correspondence between physiological subbands and resulting dyadic frequency bands. The resulting bands are $\mathrm{SB}^1 = \mathrm{VLF} = 0$–$0.0375$ Hz, $\mathrm{SB}^2 = \mathrm{LF} = 0.0375$–$0.15$ Hz, and $\mathrm{SB}^3 = \mathrm{HF} = 0.15$–$0.3$ Hz. Figure 7 shows the three filters corresponding to the Daubechies wavelet with 32 coefficients and defining the subbands $SB^i$ ($i = 1, 2, 3$). The filters are overlapping due to the finite length of the wavelet $\psi(n)$.

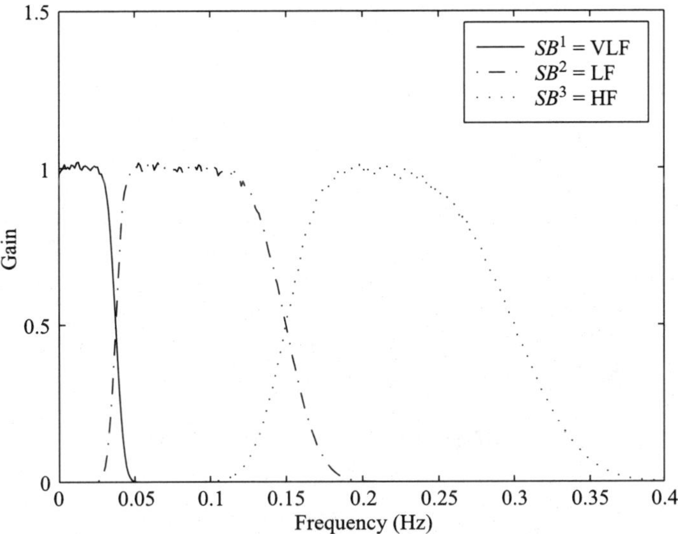


**Figure 7** Wavelet-based dyadic filter bank.

Prior to modeling CV signals, both input and output signals were decomposed into the three frequency bands: VLF, LF, and HF. Identification focused on the relationships, inside a particular frequency band, between the signals under consideration. Then we compared our subband approach with a "global" one in the complete frequency band 0–0.3 Hz. Due to the perfect reconstruction property of our filters, no residual error was added at the synthesis level.

A drawback of this scheme is that the frequency overlap observed in Figure 7 causes some of the information from adjacent frequency bands to be used to estimate the output signal in one frequency band.

**Remark**. Any signal corresponding the frequency band $SB^i$ is denoted by a superscript $i$ or by the frequency label (VLF, LF, HF). An analysis in the subbands is denoted by SB, and an analysis on the original signals (the global case) by $G$.

#### 5.2.2. Subband Identification

Figure 8 illustrates the general architecture of the subband identification scheme. The vector signal $x^i$ which corresponds to the subband $SB^i$, and which is composed of

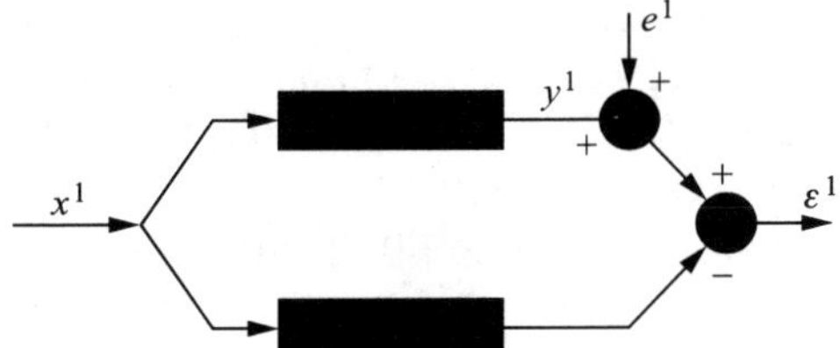


**Figure 8** Block diagram of the identification in $SB^i$.

the $M$ input signals (see Eq. 4), is used for identification together with the output signal $y^i$. Prior to model estimation, we perform a subsampling of the various signals in accordance with Shannon's theorem. We have used a sampling frequency corresponding to three times the upper cutoff frequency of the chosen subband. This resampling is quite important for any MDL criterion to work properly because the number of samples plays an active role in MDL computation.

The function $f(.)$ represents the subsystem characteristic function as introduced in Eq. 1. It is approximated using a polynomial NARMA model (Eq. 9). Observe that only output noise perturbation is assumed in the model. As input and output signals can be interchanged, with different subsystems being identified, input noise should be introduced as well. We do not take into account input noise perturbation because it augments drastically the complexity of the identification problem [61].

## 5.3. Recording Methods and Medical Protocols

### 5.3.1. Recording Methods

***General Procedure.*** Our study population consisted of six normal subjects (male gender, aged 25 ± 3 years) (mean ± SD). Recording conditions consisted of a cool room temperature (21°C) and subjects in supine resting position with a metronome-controlled respiratory rate fixed at 13 ± 2 breaths/min (mean ± SD). We obtained simultaneous measurements of surface ECG, instantaneous lung volume (ILV) (pneumobelt), arterial blood pressure (ABP) (Finapres Ohmeda, Englewood), and muscle sympathetic nerve activity (MSNA). RR interbeat interval time series were extracted from the ECG. The respiratory signal was expressed in arbitrary units. All these signals were digitized at 500 Hz on a personal computer coupled with a multichannel analog-to-digital acquisition board (National Instruments, AT-MIO-16E10) and preprocessed with LabView and MATLAB softwares.

***Muscle Sympathetic Nerve Activity.*** Multiunit recordings of MSNA were obtained with unipolar tungsten microelectrodes inserted selectively into muscle nerve fasicles of the peroneal nerve posterior to the fibular head by the microneurographic technique of Vallbo et al. [62]. The neural signal was amplified (by 20–50 × $10^3$), filtered (bandwidth 700–2000 Hz), rectified, and integrated (time constant 0.1 second) to obtain a mean voltage display of sympathetic activity. A recording of MSNA was considered acceptable when it revealed spontaneous, pulse synchronous bursts of neural activity that increased during the Valsalva maneuver but not during arousal stimuli such as loud noise.

***Data Preprocessing.*** All signals were acquired during 5 minutes, and simultaneously wide-sense stationary time intervals were extracted (mean time interval of 200 seconds) in order to use least-mean-squares techniques. This stationarity requirement had to be satisfied especially in the case of multi-input single–output biological modeling. All artifacts were removed from ABP, ILV, and ECG using adapted signal processing tools (linear and nonlinear filters). The ABP was filtered using a fourth-order Butterworth filter with a cutoff frequency of 0.5 Hz to obtain the mean arterial blood pressure (MBP).

During the acquisition process, some artifacts in MSNA were noted: (1) electrostatic discharges from the recording device, (2) slow fluctuations of the baseline due to variations in muscle tension, and (3) other noise sources such as parasitic electromagnetic fields. Artifact cancellation proceeded as follows: suppression of electrostatic discharges using a slope thresholding method followed by a linear interpolation and suppression of slow baseline fluctuations using morphological filters. The computation time was reduced by downsampling the MSNA at 20 Hz using an antialiasing fourth-order Chebyshev filter.

Finally, all signals were resampled at 1.2 Hz using spline interpolation for MBP, ILV, MSNA, and the method described by Berger et al. for the RR signal [46]. Note that, as the relevant frequency band goes from 0 to 0.5 Hz, it should theoretically be sufficient to undersample the data at 1 Hz, but the 1.2 Hz sampling rate is dictated by CV signal characteristics (see Section 4).

#### 5.3.2. Medical Protocols

All subjects had provided informed written consent. Three protocols were conducted:

1. *Baseline*: Supine resting position. This protocol is denoted *Ba*.
2. *Phenylephrine infusion*: Phenylephrine ($0$–$1.5\,\mu g\,kg^{-1}\,min^{-1}$) was infused for 15 minutes in order to increase the mean arterial pressure by 10 mm Hg. Phenylephrine infusion results in a marked decrease in both heart rate and MSNA in response to the resulting increase in vascular tone and resistance (through the unloading of arterial baroreceptors). This protocol is denoted *Pe*.
3. *Nipride infusion*: Sodium nitroprusside (Nipride) ($0$–$1.5\,\mu g\,kg^{-1}\,min^{-1}$) was infused for 15 minutes to decrease mean arterial pressure by 10 mm Hg. Nipride infusion markedly increased heart rate and MSNA in response to peripheral arterial vasodilation. This protocol is denoted *Ni*.

### 5.4. Results on Cardiovascular Modeling

#### 5.4.1. General Description

Let us summarize the main features of the experiments whose main results are presented below:

- We have performed parsimonious ARMA (linear) and NARMA (nonlinear) modeling (Sections 2 and 3) of the cardiovascular system in terms of both RR-interval prediction and estimation from MBP/MSNA and ILV. This allowed us to make a fair comparison of the respective performances of the linear and nonlinear approaches.
- A priori information led us to choose the values $D_{max} = 3$ for NARMA model selection (of course $D_{max} = 1$ for ARMA model selection) and $l_{max} = 6$ in both cases (Section 3.3). This corresponds to time lags of 6.6 s in the HF, 13 s in the LF, and 53 s in the VLF band. The simplified MDL criterion (3.8) was used in all cases. In the RR-interval nonlinear predictive model we have with Eq. 8 $N_{Par} = 84$, and in the nonlinear estimation model we have $N_{Par} = 680$.
- When solving the normal equation (23), we have observed that SVD is a much more robust approach than Gram-Schmidt, as already noted in Ref. 11. SVD was thus employed throughout the experiments.

- Three physiological conditions corresponding to various degrees of activation or inhibition of the sympathetic–parasympathetic systems (Subsection 5.3.2) were investigated. They are referred to as *Ba*, *Pe*, and *Ni*. For each experiment, we investigated the possibility of using the obtained model parameters for the classification of the three physiological states.

Results on linear modeling are presented in terms of relative residual variance (RRV) in the different frequency bands as well as globally. We used RRV as a validation parameter in different conditions: (1) global modeling in the full-band $RRV_G$, (2) subband modeling $RRV_{SB}$, (3) modeling in the VLF band $RRV_{VLF}$, (4) modeling in the LF band $RRV_{LF}$, and (5) modeling in the HF band $RRV_{HF}$. The $RRV_{SB}$ was computed after application of the reconstruction filter bank so that it can be compared with $RRV_G$.

The results on nonlinear modeling are expressed by the relative mean square error together with the number of MDL-selected parameters. The reason for this representation is that the selected parameters are usually different, depending on the frequency band and the patient, so that a mean parameter coefficient value cannot be computed. NARMA models comprise distinct linear and nonlinear parts. Thus, we can evaluate their relative weight in terms of modeling error reduction.

Note that the optimization algorithm (Section 3.3) enables us to select a reduced number of parameters, and only the most significant ones. For example, if the addition of nonlinear related terms does not improve the model with respect to our criterion, then none of these parameters are selected and the resulting model remains linear. This fact is important because the use of nonlinear modeling such as NARMA usually decreases the MSE but drastically increases the complexity of the model by selecting small amplitude higher order parameters.

Gains in RRV from one experiment to another—linear–nonlinear (*L-NL*), basal–nipride (*Ba-Ni*), basal–phenylephrine (*Ba-Pe*), phenylephrine–nipride (*Pe-Ni*)—are denoted by $\Delta RRV_i^j$ with $i = (\text{VLF}, \text{LF}, \text{HF}, \text{SB}, \text{G})$ and $j = (L - NL, Ba - Ni, Ba - Pe, Pe - Ni)$ and expressed in percentages.

Results are grouped as follows: Section 5.4.2 deals with RR-interval prediction, Section 5.4.3 shows RR-interval estimation from ILV and MBP, and Section 5.4.4 analyzes the effect of the replacement of MBP with MSNA.

***Statistical Analysis.*** Since the number of patients is rather small, results are presented as mean $\pm$ SEM where $\text{SEM} = \text{SD}/\sqrt{N}$, with $N$ the number of data. Significance testing was performed with ANOVA. Results with a $p$ value less than or equal to 0.05 are considered as statistically significant.

### 5.4.2. RR-Interval Prediction

Prediction of RR-interval time series using linear AR models has already been employed for spectral analysis purposes [63]. In this section, we perform the prediction of RR-interval time series with AR and NARMA models and analyze both model structure and classification performance. Figure 9 shows the RRV in all frequency bands for both linear and nonlinear models.

***Discussion of the Prediction Models.*** From Figure 9, we can observe that nonlinear prediction performs slightly better than linear prediction and that subband modeling

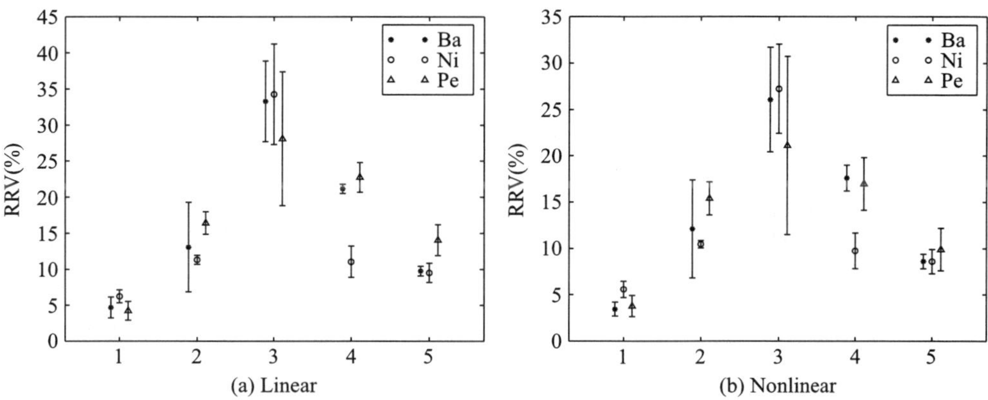


**Figure 9** RRV for RR-interval prediction: (1) VLF, (2) LF, (3) HF, (4) G, (5) SB.

yields slightly lower *RRV* than global modeling. The latter point confirms some general results of Rao and Pearlman [32]. Higher RRV is observed when passing from the VLF band to the HF band. In the HF band, the standard deviation on RRV in all physiological conditions is much higher than in the other frequency bands. This result is certainly due to the prevalent influence of respiration in this band. Indeed, respiratory sinus arrhythmia (RSA) has several and certainly interdependent origins (central medullary respiratory neuron activity, baroreflex stimulation via venous return, and afferent stimuli from stretch lung receptors) [37, 64–66] and hence cannot be predicted efficiently. Results of Section 5.4.3 confirm this observation.

Values of the parameters of the linear predictors averaged over all the patients, together with their SEM values, are displayed in Figure 10. The first two plots present the LF and HF subband terms, and the last one represents the global model terms. The corresponding frequency responses show a broad, ill-defined peak for the LF model and a well-defined peak for the HF one. An important remark is that exactly the same terms are selected first in the nonlinear models. One can also observe the very small variance of the global model parameters. This is the key to successful classification of physiological conditions based on these parameters (see Table 6).

***Discussion of the Nonlinearity.*** The number of nonlinear model parameters selected is consistent with the preliminary tests in Section 3. According to the MDL criterion and as shown in Figure 11, there exists nonlinearity for each of the physiological situations examined. However, the linear contribution is preponderant, i.e., more than 85% in LF, SB, and G. The nonlinear contribution is higher in the HF domain than in all the others when looking at $\Delta\mathrm{RRV}_{\mathrm{HF}}^{\mathrm{L-NL}}$ (see Figure 11b). Nevertheless, the variance over $\Delta\mathrm{RRV}_{\mathrm{HF}}^{\mathrm{L-NL}}$ is large. One can observe in the VLF band that, while MDL has selected substantially many more parameters than in the other frequency bands (Table 3), $\Delta\mathrm{RRV}_{\mathrm{VLF}}^{\mathrm{L-NL}}$ is very small ($\approx 1\%$). This may be due to the simplified MDL criterion, which tends to select a larger number of parameters than the MDL with parameter precision.

From Figure 11, $\Delta\mathrm{RRV}_{\mathrm{G,SB}}^{\mathrm{L-NL}}$ under *Ni* is lower than $\Delta\mathrm{RRV}_{\mathrm{G,SB}}^{\mathrm{L-NL}}$ under *Pe*. This indicates that the cardiovascular system tends to be more linear under sympathetic

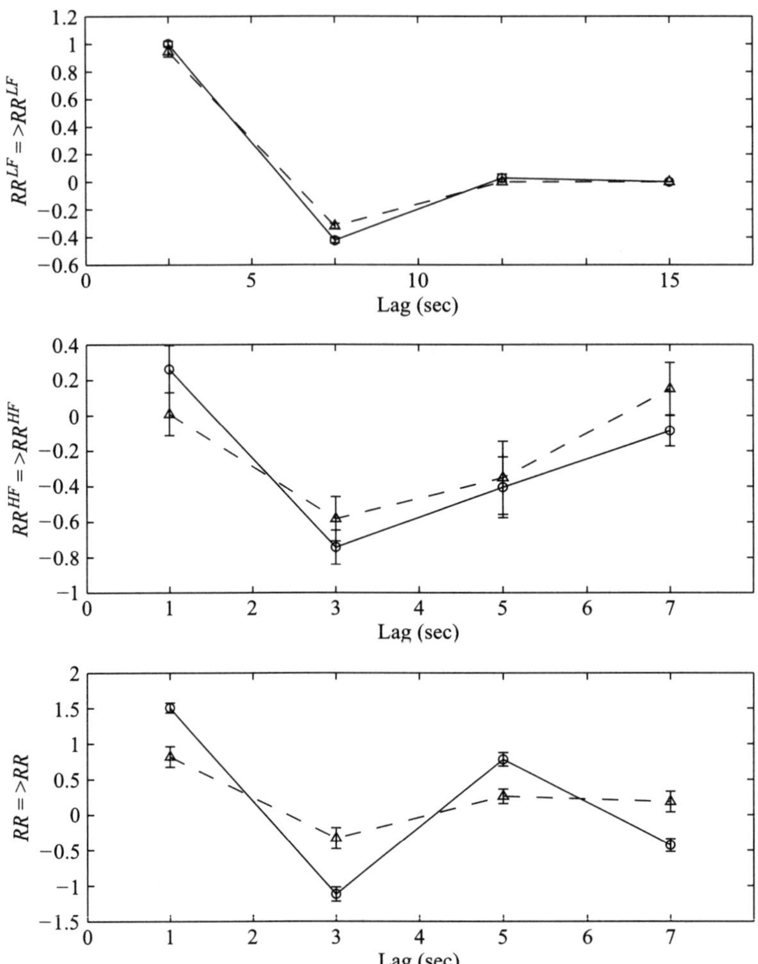


**Figure 10** Parameters of linear RR-interval prediction model: (○) *Ni*, (△) *Pe*.

activation and more nonlinear under sympathetic inhibition. However, statistical tests do not confirm the significance of this observation (i.e., in SB with $Ni - Pe$, $p = 0.1$, and in G, $p = 0.058$).

It thus seems that sympathetic stimulation has a direct influence on RR-interval predictability and the linearity of the relationships inside the cardiovascular system. With increasing sympathetic activity, the RR-interval signal appears to be more correlated. If the sympathetic influence is reduced, the resulting signal contains a more unpredictable structure. One can also compare the respective RRVs for *Ni* and *Pe* at the global level in Figure 9.

The number of parameters selected with the simplified MDL criterion is very small compared with the overall model parameter number $N_{\mathrm{Par}} = 680$. Going from linear to nonlinear models introduces one to five additional coefficients in the LF, HF, or G (see Table 4).

**TABLE 3** Number of Parameters in the Linear Models

| Linear | VLF | LF | HF | G |
|---|---|---|---|---|
| *Ba* | $2.8 \pm 0.7$ | $3.0 \pm 0.44$ | $4.0 \pm 0.68$ | $5.0 \pm 0.36$ |
| *Ni* | $2.6 \pm 0.66$ | $3.3 \pm 0.61$ | $4.0 \pm 0.51$ | $4.6 \pm 0.21$ |
| *Pe* | $1.5 \pm 0.34$ | $2.33 \pm 0.21$ | $2.83 \pm 0.65$ | $4.0 \pm 0.25$ |

**TABLE 4** Number of Parameters in the Nonlinear Models

| Nonlinear | VLF | LF | HF | G |
|---|---|---|---|---|
| *Ba* | $13 \pm 2.7$ | $4.1 \pm 0.5$ | $8.4 \pm 1.3$ | $7.7 \pm 0.86$ |
| *Ni* | $7.5 \pm 2.6$ | $4.7 \pm 0.5$ | $7.5 \pm 0.68$ | $6.4 \pm 0.36$ |
| *Pe* | $10.6 \pm 3.8$ | $7.33 \pm 2.38$ | $8.5 \pm 2.24$ | $7.5 \pm 1.23$ |

***Discussion of the Physiological Aspects.*** Table 5 summarizes the modeling results with respect to the RRV index. We observe that we can separate *Ni* and *Pe* in the LF band. Using the global model, *Ni* can be distinguished from *Pe* and *Ba*. These conclusions apply to both linear and nonlinear models.

In the subbands only one significant difference between the parameters of the linear model has been obtained: for $Ni - Pe$, in the LF band, for $\text{lag} = 2$ ($p = 0.0331$). Table 6 shows the results of ANOVA on the AR global model parameters when the three physiological situations are compared. They are almost all significant up to the third lag. From Figure 10, we observe a marked and significant difference between *Pe* and *Ni* in the global model, whereas only trends are observed in the LF and HF bands.

From Figure 10, we can observe that the global response of the system is similar for *Pe* and *Ni* up to a scaling factor. Table 7 lists the average Euclidean norms of the linear AR parameter vectors together with ANOVA results on these norms for the various physiological situations. This norm can be associated with gains of the feedback loop system, in particular the baroreflex gain. We observe that the norms in LF and G

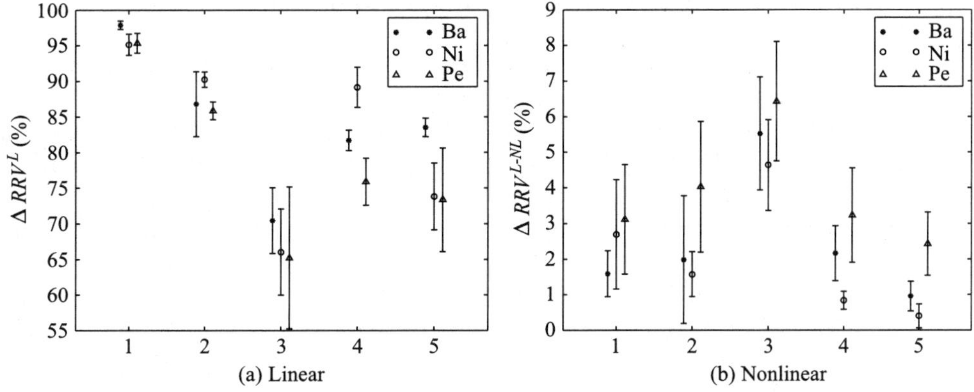


**Figure 11** ΔRRV for RR-interval prediction: (1) VLF, (2) LF, (3), HF, (4) G, (5) SB.

**TABLE 5** $p$ Values of ANOVA Tests on $RRV$ for Linear and Nonlinear Models

| | (a) Linear | | | | | (b) Nonlinear | | | | |
|---|---|---|---|---|---|---|---|---|---|---|
| Band | VLF | LF | HF | G | SB | VLF | LF | HF | G | SB |
| *Ni–Pe* | 0.25 | **0.022** | 0.61 | **0.0079** | 0.12 | 0.26 | **0.035** | 0.58 | **0.00795** | 0.64 |
| *Ba–Ni* | 0.39 | 0.79 | 0.91 | **0.0043** | 0.87 | 0.11 | 0.76 | 0.87 | **0.016** | 0.99 |
| *Ba–Pe* | 0.82 | 0.61 | 0.64 | 0.48 | 0.10 | 0.81 | 0.57 | 0.67 | 0.84 | 0.61 |

**TABLE 6** $p$ Values of ANOVA Tests on Global Linear Model Parameters

| Lag | 1 | 2 | 3 | 4 | 5 | 6 |
|---|---|---|---|---|---|---|
| $Ni - Pe$ | **0.0016** | **0.0014** | **0.0038** | **0.0059** | 0.089 | 0.68 |
| $Ba - Ni$ | **0.0044** | **0.021** | **0.026** | 0.0515 | 0.12 | 0.49 |
| $Ba - Pe$ | **0.03** | **0.0246** | 0.201 | 0.19 | 0.81 | 0.36 |

**TABLE 7** Norm of All Parameters of the Linear Model and $p$ Values of ANOVA Tests

| | (a) Norms | | | (b) p-values | | | |
|---|---|---|---|---|---|---|---|
| Band | LF | HF | Global | Band | LF | HF | Global |
| *Pe* | $\mathbf{0.989 \pm 0.036}$ | $1.16 \pm 0.16$ | $\mathbf{1.56 \pm 0.065}$ | $Ni - Pe$ | **0.05** | 0.57 | **0.0002** |
| *Ni* | $\mathbf{1.159 \pm 0.054}$ | $1.25 \pm 0.20$ | $\mathbf{2.03 \pm 0.117}$ | $Ba - Ni$ | **0.025** | 0.75 | **0.006** |
| *Ba* | $\mathbf{1.004 \pm 0.005}$ | $1.06 \pm 0.26$ | $\mathbf{1.13 \pm 0.11}$ | $Ba - Pe$ | 0.8 | 0.72 | **0.007** |

are well separated (except $Ba - Pe$ in LF) and correspond to physiologically different conditions. Under *Ni*, the sympathetic system is activated through vasodilation, and baroreflex gain is increased with respect to the basal condition. Under *Pe*, the opposite is observed. This confirms the results of Pagani et al. [67].

### 5.4.3. RR-Interval Estimation from ILV and MBP

This section describes results on multi-input single-output nonlinear moving average estimation of models in which the output signal $y$ corresponds to the RR intervals and the two input signals are $x_1 = \text{ILV}$ and $x_2 = \text{MBP}$. Some of this work has already been published [56], but results for the two physiological conditions (*Ni*, *Pe*) and their classification on the basis of the model parameters are new. Figure 12 displays one RR-interval time series and its nonlinear estimation in the frequency bands VLF, LF, and HF as well as in the full band (G).

One notices immediately the quality of the estimation in LF under *Ni* and in HF under *Pe*. This provides a first qualitative insight into the effect of nipride and phenylephrine on the autonomic nervous system and their respective influences on RR-interval fluctuations in the LF and HF frequency bands. These qualitative impressions are confirmed by Figure 13 and Table 9.

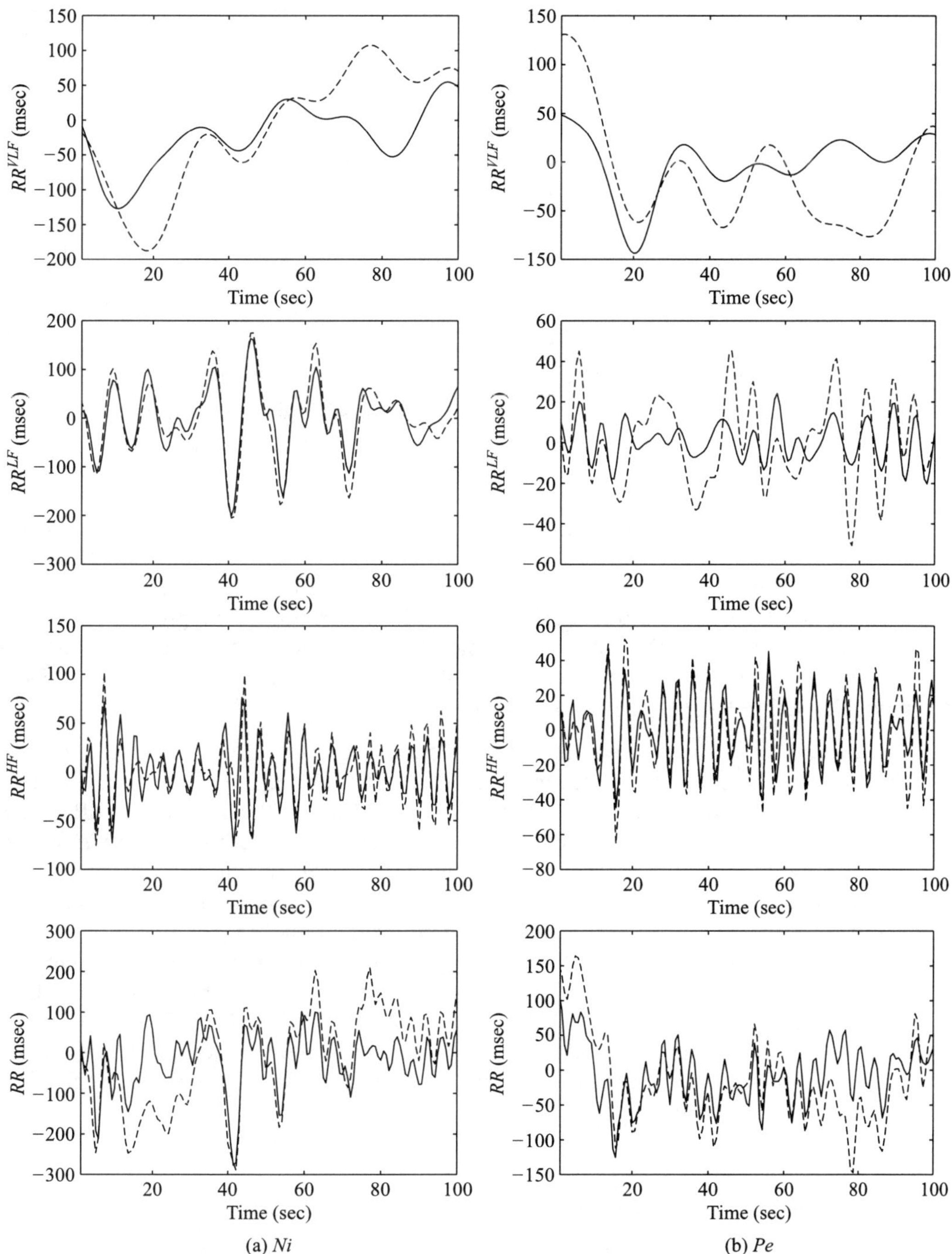


**Figure 12** RR-interval (dashed line) time series in the frequency bands and their nonlinear estimations (solid line) from ILV and MBP.

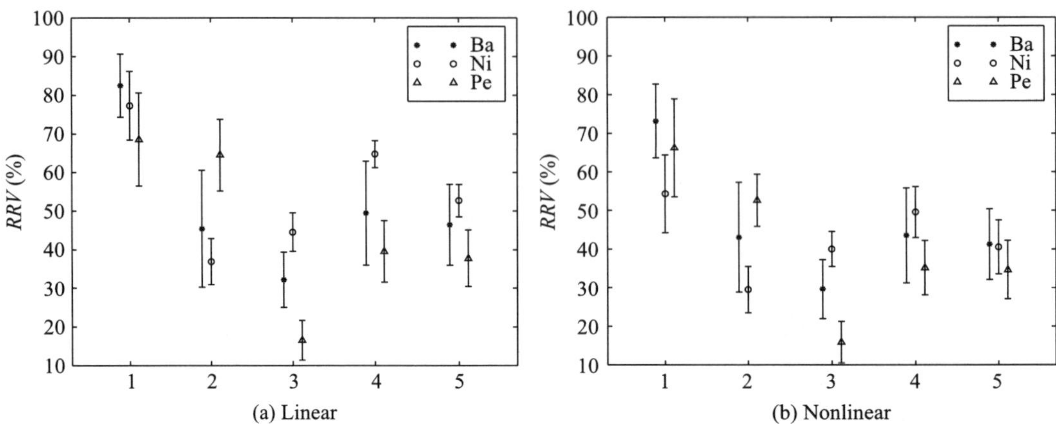


(a) Linear (b) Nonlinear

**Figure 13** RRV for RR-interval estimation from ILV and MBP: (1) VLF, (2) LF, (3) HF, (4) G, (5) SB.

***Discussion of the Estimation Models.*** Comparing Figure 13 and Figure 9, one can observe that the RRVs for RR-interval estimation are larger than for RR-interval prediction. This may be due to the fact that we introduce two more signals in the model and therefore additional noise sources. Moreover, finger Finapres measurements do not correspond to the true arterial blood pressure wave but only to a distorted version of it (see K. H. Wesseling in Ref. 42, p. 118). Figure 13 shows that subband modeling still outperforms global modeling.

***Discussion of the Nonlinearity.*** When comparing Table 8 and Table 3 and 4, we observe that the numbers of parameters in LF, HF, and G are almost the same for linear and nonlinear models. Also, Figure 14 indicates that the decrease in RRV when changing from a linear to a nonlinear model is quite modest, although more important than for RR prediction. Surprisingly, the number of parameters in the VLF band in the estimation of the RR interval is much lower than in the prediction model. This results in a much larger $RRV_{VLF}$ in the estimation than in the prediction.

In the LF band, Figure 14 shows that *Ni* has a more important linear content than *Pe* ($p = 0.047$). Nonlinear models can still separate *Ni* and *Pe* in the LF band ($p = 0.043$), but only linear models are successful in G ($p = 0.0275$). The increased nonlinearity of the cardiovascular system under *Ni* in the LF, found in this section, confirms results obtained with the RR-interval prediction model. Nonlinear content in HF remains very small.

**TABLE 8** Number of Parameters in the Linear and Nonlinear Models

| | (a) Linear | | | | (b) Nonlinear | | | |
|---|---|---|---|---|---|---|---|---|
| | VLF | LF | HF | Global | VLF | LF | HF | Global |
| *Ba* | $1.2 \pm 0.3$ | $3.7 \pm 0.8$ | $4.2 \pm 0.6$ | $4.5 \pm 0.5$ | $2.0 \pm 0.4$ | $4.2 \pm 0.7$ | $7.5 \pm 1.6$ | $7.3 \pm 1.2$ |
| *Ni* | $1.3 \pm 0.2$ | $3.3 \pm 0.3$ | $4.3 \pm 0.7$ | $4.4 \pm 0.4$ | $2.5 \pm 0.4$ | $5.1 \pm 1.4$ | $6.3 \pm 1.2$ | $7.3 \pm 1.1$ |
| *Pe* | $1.6 \pm 0.4$ | $3.2 \pm 0.6$ | $6.2 \pm 0.6$ | $4.2 \pm 0.6$ | $1.8 \pm 0.5$ | $4.8 \pm 0.9$ | $7.6 \pm 1.2$ | $6.2 \pm 0.6$ |

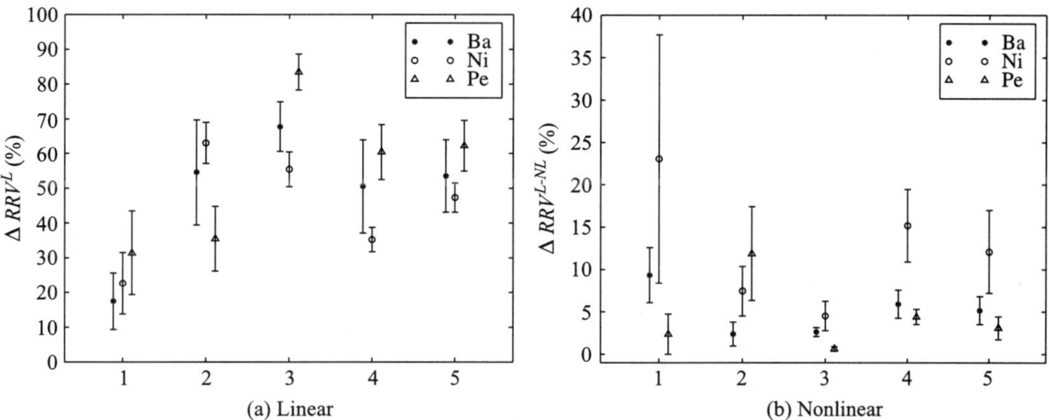


**Figure 14** ΔRRV for RR-interval estimation from ILV and MBP: (1) VLF, (2) LF, (3) HF, (4) G, (5) SB.

***Discussion of the Physiological Aspects.*** Mean RRV values in VLF and LF are much larger for estimation than for prediction. This can be explained as resulting from indirect measurements of blood pressure, which impair the estimation process in VLF and LF. On the other hand, in HF, the mean RRV value is much smaller for the estimation under *Pe* than for the other physiological conditions. In addition, Table 9 shows clearly that HF RR-interval estimation allows the discrimination of *Pe* and *Ni*, which was impossible in the prediction experiment (see Table 5). Indeed, in the HF band, respiration is the main factor responsible for RR-interval variations. It operates through mediation of the parasympathetic component of the baroreflex buffering respiration-induced fluctuations in blood pressure [66, 68] but also through central oscillatory and mechanical stimuli. We conclude that respiration should if possible be included in any model of this type, be it for interpretation or classification.

Figure 15 presents the impulse responses corresponding respectively to MBP and ILV in LF, HF, and the full band G. These results confirm those in Ref. 56 for subband modeling and in Ref. 57 for global modeling.

Note that all these impulse responses, which model the short-term interactions between the signals, are indicative of a high-pass behavior. The results presented here for the global modeling are consistent with those obtained by Perrott and Cohen [57] or Chon et al. [13].

**TABLE 9** *p* Values of ANOVA Tests on RRV for Linear and Nonlinear Models

| | (a) Linear | | | | | (b) Nonlinear | | | | |
|---|---|---|---|---|---|---|---|---|---|---|
| | VLF | LF | HF | G | SB | VLF | LF | HF | G | SB |
| *Ni* − *Pe* | 0.58 | **0.047** | **0.0079** | **0.0275** | 0.1279 | 0.48 | **0.043** | **0.0136** | 0.18 | 0.58 |
| *Ba* − *Ni* | 0.68 | 0.62 | 0.21 | 0.31 | 0.6 | 0.22 | 0.41 | 0.28 | 0.68 | 0.95 |
| *Ba* − *Pe* | 0.38 | 0.32 | 0.12 | 0.54 | 0.52 | 0.67 | 0.56 | 0.19 | 0.57 | 0.59 |

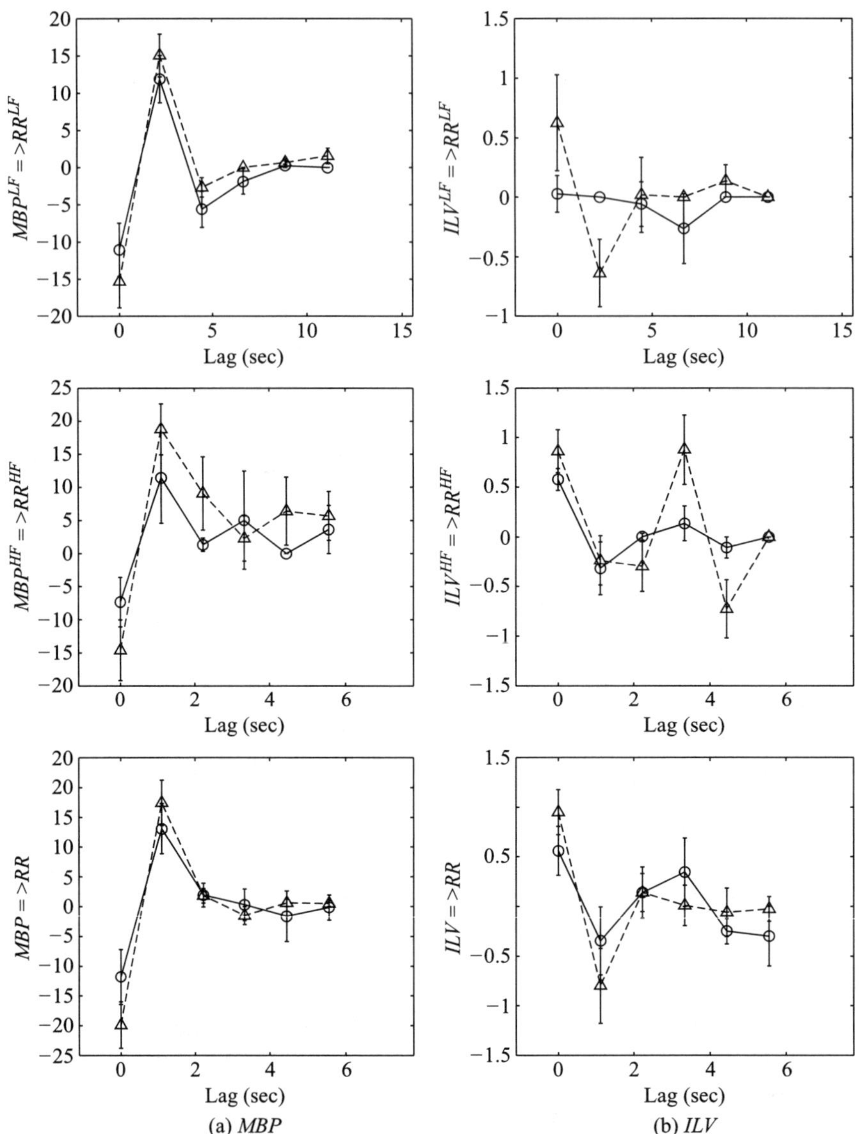


**Figure 15** Impulse responses in LF, HF, and G: (○) *Ni*, (△) *Pe*.

The shape of the impulse response ILV → RR intervals (Figure 15b) leads to physiological interpretations. The impulse response in the HF band under *Ni* is very similar to the impulse response in G, whereas in the LF band it displays a slow oscillation. This indicates that the HF band is responsible for short-term respiration-mediated RR-interval variations through parasympathetic activity, while the LF band responds more slowly and reflects the sympathetic action on the heart [69].

### 5.4.4. RR-Interval Estimation from ILV and MSNA

We consider again a multi-input NARMA model with the RR intervals as output signal, but this time the input signals are ILV and MSNA. It has been shown [70] that the cardiac sympathetic activity (CSNA) and the MSNA are correlated. This study is based on simultaneous measurements of MSNA and cardiac noradrenaline spillover and indicates that a common mechanism influences MSNA and CSNA. Thus, RR-interval estimation using MSNA instead of MBP allows us to bypass the arterial baroreceptors and the central nervous system. Here again, linear and nonlinear modeling results are presented separately. VLF analysis has been discarded because the preprocessing on MSNA (baseline removal) suppresses most of the VLF content.

Figure 16 shows that, as in the two previous experiments, linear or nonlinear modeling using subband decomposition is superior to modeling on the original signals.

In what concerns linear modeling, the RRV values are quite the same for all physiological conditions with the exception of the LF band. In this band, the amount of linearity in the relationship between MSNA and RR intervals is larger for *Ni* than for *Ba* and *Pe*. But, when comparing Figure 13a with Figure 16a for *Ni*, we observe that this amount of linearity is very close to that of RR-interval prediction ($\mathrm{RRV}_{\mathrm{LF}}^{Ni} \approx 35\%$). This confirms that MSNA LF fluctuations are well correlated with those of RR intervals and that this relationship is more linear under sympathetic activation (for linear modeling, $\mathrm{RRV}_{\mathrm{LF}}^{Ni-Pe}$ and $\mathrm{RRV}_{\mathrm{LF}}^{Ni-Ba}$ are significant respectively with $p = 0.012$, and $p = 0.017$). Moreover, in LF and for *Ni*, linear estimation of the RR intervals with MSNA gives the same results as with MBP.

In this experiment, nonlinear modeling outperforms linear modeling: the various RRVs are clearly smaller in Figure 16, with some disparity between the various physiological conditions. In particular, one finds statistically significant differences in the LF band: $\mathrm{RRV}_{\mathrm{LF}}^{Ni-Ba}$ with $p = 0.016$ and $\mathrm{RRV}_{\mathrm{LF}}^{Ni-Pe}$ with $p = 0.0001$.

Although one can observe in Figure 17b that under *Ni*, mean $\Delta\mathrm{RRV}_{\mathrm{HF,G}}^{Ni}$ is larger than mean $\Delta\mathrm{RRV}_{\mathrm{HF,G}}^{Ba,Pe}$, these differences are not significant enough to classify the physiological conditions. When comparing Figure 13b with Figure 16b or Figure 14b

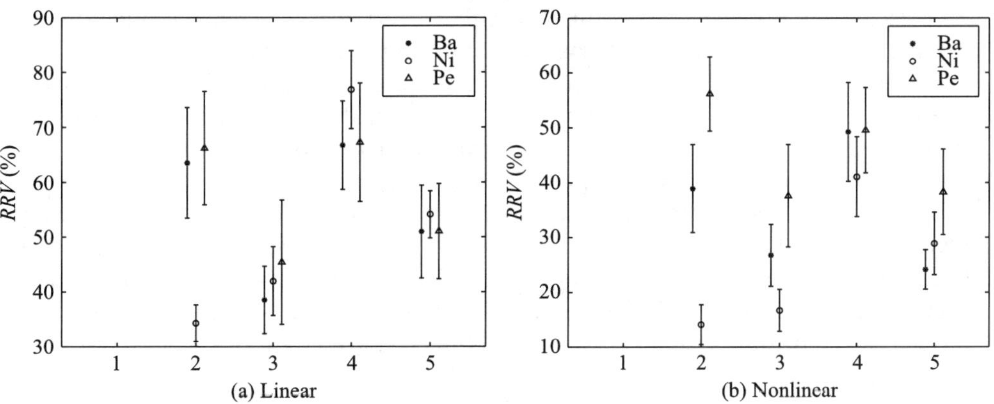


**Figure 16** RRV for RR-interval estimation from ILV and MSNA: (2) LF, (3) HF, (4) G, (5) SB.

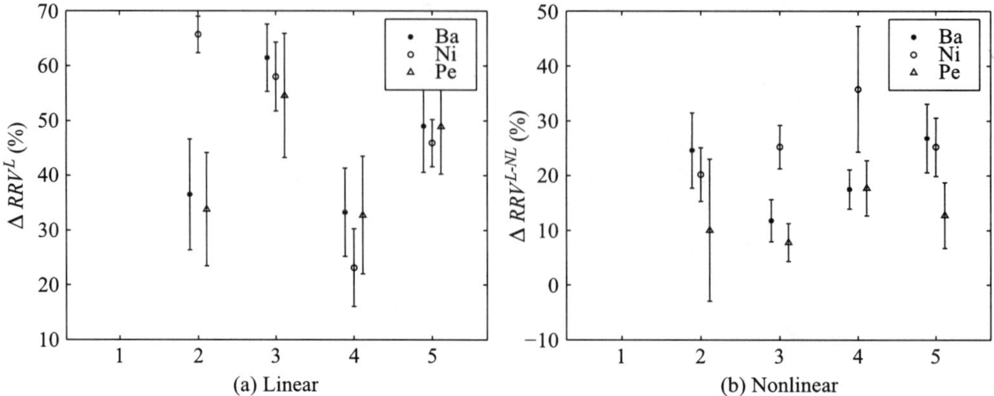


**Figure 17** $\Delta RRV$ for RR-interval estimation from $ILV$ and $MSNA$: (2) LF, (3) HF, (4) G, (5) SB.

with Figure 17b, one can observe that the amount of nonlinearity in (ILV, MSNA) $\rightarrow$ RR is larger than that in (ILV, MBP) $\rightarrow$ RR.

We first conclude that the muscle sympathetic nerve activity displays LF fluctuations related to those of the RR intervals. Second, this relationship is more linear under *Ni* but a substantial amount of nonlinearity is present. Third, the nonlinear model clearly outperforms the linear model in this experiment.

## 6. CONCLUSIONS

We have investigated the problem of nonlinear system identification and nonlinear time series prediction using NARMA models, which constitute a straightforward generalization of linear ARMA models.

We have briefly presented some NARMA model estimation techniques such as singular value decomposition and orthogonalization of the regression matrix. NARMA models are characterized by a huge number of terms, which leads unavoidably to *overfitting* if no term selection is exercised. Among the various model selection criteria proposed in the literature, the minimum description length (MDL) criterion has attracted increasing attention. We have demonstrated with numerical experiments that MDL is able to select the correct terms of various polynomial synthetic models. Such a criterion is thus able to provide objective and reliable solutions to modeling problems. A search strategy has been proposed in order to scan efficiently the parameter space and retain the essence of the system under study. This search method together with singular value decomposition of the regression matrix has given satisfactory and coherent results. For instance, when searching the best third-degree NARMA model, the algorithm first selects the correct linear part and adds higher degree terms if necessary.

We have also proposed to decompose the various signals according to specific frequency bands. This decomposition helps in separating phenomena taking place at different time scales. Subband filtering using wavelets has provided a simple and easy-to-implement solution, but it is probably not the most efficient one because of the lack of selectivity of the filters.

In what concerns cardiovascular system modeling, we now summarize our main results. We have demonstrated that, under the various physiological conditions under study, the cardiovascular system is predominantly linear. Indeed, nonlinear terms selected with our method do not improve model performance much, from both physiological interpretation and classification viewpoints. A notable exception is modeling the heart interbeat intervals with muscle sympathetic nerve activity, in which a substantial amount of nonlinearity is present. It is nonetheless important to keep in mind that these conclusions are valid only within the limits of the chosen model class.

NARMA modeling of the cardiovascular system has been shown to provide good insights into its neural autonomic regulation. Classification indices have been proposed such as relative residual error variance or its variation. Performing a physiologically based subband decomposition of the cardiovascular signals has led to a deeper understanding of the cardiovascular regulation process. Also, main results about already published works have been confirmed here:

- The importance of the baroreflex control loop and its manifestation in the low-frequency band
- The origin of the respiratory sinus arrhythmia and its relation to high-frequency heart interbeat interval fluctuations
- The correlation between cardiac and peripheral sympathetic nerve activities
- The correlation between low-frequency fluctuations of muscle sympathetic nerve activity and those of heart interbeat intervals

Finally, we would like to add that this chapter emphasizes parsimonious nonlinear system modeling from time series, which is quite different from system identification. In consequence, we do not claim to have identified some of the cardiovascular system components. Rather, we have provided modeling tools for the estimation and prediction of heart interbeat intervals. These models could serve for classification and help the clinician in diagnostic tasks. Although these signal processing techniques are very powerful and can provide some interesting new understanding about the system, they are still far from giving a realistic image of the complex phenomena appearing in biological systems.

## REFERENCES

[1] G. E. P. Box and G. M. Jenkins, *Time Series Analysis, Forecasting, and Control.* San Francisco: Holden-Day, 1970.

[2] H. Tong, *Nonlinear Time Series.* New York: Oxford University Press, 1990.

[3] C. W. J. Granger, Some properties of time series data and their use in econometric model specification. *J. Econometrics*, 22:121–130, 1981.

[4] S. Haykin, *Neural Networks—A Comprehensive Foundation.* New York: Macmillan, 1994.

[5] I. J. Leontaritis and S. A. Billings, Input–output parametric models for non-linear systems. Part I: Deterministic non-linear systems; Part II: Stochastic non-linear systems. *Int. J. Control* 41:303–344, 1985.

[6] J. Rissanen, *Stochastic Complexity in Statistical Inquiry.* Singapore: World Scientific, 1989.

[7] N. Kalouptsidis and S. Theodoridis, *Adaptive System Identification and Signal Processing Algorithms*. Englewood Cliffs, NJ: Prentice-Hall, 1993.

[8] S. A. Billings, Identification of nonlinear systems—A survey. *IEEE Proc.* 127:272–285, 1980.

[9] N. Wiener, *Nonlinear Problems in Random Theory*, New York: John Wiley & Sons, 1958.

[10] M. J. Korenberg, S. A. Billings, Y. P. Liu, and P. J. McIlroy, Orthogonal parameter estimation algorithm for non-linear stochastic systems. *Int. J. Control* 48:193–210, 1988.

[11] S. Chen, A. Billings, and W. Luo, Orthogonal least squares methods and their application to non-linear system identification. *Int. J. Control* 50:1873–1896, 1989.

[12] M. J. Korenberg and I. W. Hunter, The identification of nonlinear biological systems—Volterra kernel approaches. *Ann. Biomed. Eng.* 24:250–268, 1996.

[13] K. H. Chon, T. J. Mellen, and R. J. Cohen, A dual-input nonlinear system analysis of autonomic modulation of heart rate. *IEEE Trans. Biomed. Eng.* 43:530–544, 1996.

[14] K. Yana, J. P. Saul, R. D. Berger, M. H. Pert, and R. J. Cohen, A time domain approach for the fluctuation analysis of heart rate related to instantaneous lung volume. *IEEE Trans. Biomed. Eng.* 40:74–81, 1993.

[15] J. Rissanen, Minimum description length principle. In *Encyclopedia of Statistical Sciences*, S. Kotz and N. L. Johnson, 5:523–527, 1985.

[16] K. Judd and A. Mees, On selecting models for nonlinear time series. *Physica D* 82:426–444, 1995.

[17] R. E. Maine and K. W. Iliff, Formulation and implementation of a practical algorithm for parameter estimation with process and measurement noise. *SIAM J. Appl. Math.* 41:558–579, 1981.

[18] I. J. Leontaritis and S. A. Billings, Model selection and validation methods for nonlinear systems. *Int. J. Control* 45:318, 1987.

[19] M. Korenberg, S. A. Billings, Y. P. Liu, and P. J. McIlroy, Orthogonal parameter estimation algorithm for nonlinear stochastic systems. *Int J. Control* 48:193–210, 1988.

[20] S. A. Billings and W. S. F. Voon, Correlation based model validity tests for non-linear models. *Int. J. Control* 44:235–244, 1986.

[21] H. Akaike, A new look at statistical model identification. *IEEE Trans. Automat. Control* 19:716–723, 1974.

[22] H. H. Ku and S. Kullback, Approximating discrete probability distributions. *IEEE Trans. Inform. Theory* 15: 444–447, 1969.

[23] M. B. Priestley, *Spectral Analysis and Time Series*, Vols I and II. New York: Academic Press, 1981.

[24] S. M. Kay and S. L. Marple, Spectrum analysis—A modern perspective. *Proc. IEEE* 69:1380–1419, 1981.

[25] M. B. Priestley, *Nonlinear and Nonstationary Time Series Analysis*. New York: Academic Press, 1988.

[26] M. Vetterli, Multi-dimensional subband coding: Some theory and algorithms. *Signal Process* 6:97–112, April 1984.

[27] S. Rao and W. A. Pearlman, Spectral estimation from subbands. *Proceedings of the IEEE International Symposium on Time–Frequency and Time–Scale Analysis*, pp. 69–72, October 1992.

[28] I. Daubechies, *Ten Lectures on Wavelets*. Philadelphia: SIAM, 1992.

[29] A. Aldroubi and M. Unser, eds., *Wavelets in Medicine and Biology*. Boca Raton, FL: CRC Press, 1996.

[30] A Grossmann and J. Morlet, Decomposition of Hardy functions into square integrable wavelets of constant shape. *SIAM J. Math.* 15:723–736, 1984.

[31] S. G. Mallat, A theory for multiresolution signal decomposition: The wavelet representation. *IEEE Trans. Pattern Anal. Machine Intell.* 11:674–693, 1989.

[32] S. Rao and W. A. Pearlman, Analysis of linear prediction, coding, and spectral estimation from subbands. *IEEE Trans. Inform. Th.* 42:1160–1178, 1996.

[33] M. D. Levine, *Vision in Man and Machine*. New York: McGraw-Hill, 1985.

[34] F. Campbell and J. Robson, Application of Fourier analysis to the visibility of gratings. *J. Physiol.* 197:551–566, 1968.

[35] J. Nachmais and A. Weber, Discrimination of simple and complex gratings. *Vision Res.* 15:217–223, 1975.

[36] M. L. Appel, R. D. Berger, J. P. Saul, J. M. Smith, and R. J. Cohen, Beat-to-beat variability in cardiovascular variables: Noise or music? *JACC* 14:1139–1148, 1989.

[37] K. Toska and M. Eriksen, Respiration-synchronous fluctuations in stroke volume, heart rate and arterial pressure in humans. *J. Physiol.* 472:501–512, 1993.

[38] Task Force of the European Society of Cardiology and the North American Society of Pacing and Electrophysiology, Heart rate variability—Standards of measurement, physiological interpretation, and clinical use. *Circulation* 93:1043–1063, 1996.

[39] R. I. Kitney, A nonlinear model for studying oscillations in the blood pressure control system. *J. Biomed. Eng.* 1:89–99, 1979.

[40] S. Akselrod, D. Gordon, J. B. Madwed, N. C. Snidman, D. C. Shannon, and R. J. Cohen, Hemodynamic regulatory investigation by spectral analysis. *Am. J. Physiol.* 249:H867–H875, 1985.

[41] R. Hainsworth and A. L. Mark, *Cardiovascular Reflex Control in Health and Disease*. Philadelphia: W. B. Saunders, 1993.

[42] A. J. Man in't Veld, G. A. van Montfrans, Ir. G. J. Langewouters, K. I. Lie, and G. Mancia, *Measurement of Heart Rate and Blood Pressure Variability in Man*. Van Zuiden Communications, 1995.

[43] K. Toska, M. Eriksen, and L. Walloe, Short-term control of cardiovascular function: Estimation of control parameters in healthy humans. *Am. J. Physiol.* 270: H651–H660, 1996.

[44] R. W. De Boer, J. M. Karemaker, and J. Strackee, Hemodynamic fluctuations and baroreflex sensitivity in human: A beat-to-beat model. *Am. J. Physiol.* 253:H680–H689, 1987.

[45] G. Baselli, S. Cerutti, S. Civardi, A. Malliani, and M. Pagani, Cardiovascular variability signals: Towards the identification of a closed-loop model of the neural control mechanisms. *IEEE Trans. Biomed. Eng.* 35:1033–1045, 1988.

[46] J. P. Saul, R. D. Berger, P. Albrecht, S. P. Stein, M. H. Chen, and R. J. Cohen, Transfer function analysis of the circulation: Unique insight into cardiovascular regulation. *Am. J. Physiol.* 261:H1231–H1245, 1991.

[47] B. J. Ten Voorde, Modeling the baroreflex. PhD thesis, Free University, Amsterdam, 1992.

[48] J. P. Saul, D. T. Kaplan, and R. I. Kitney, Nonlinear interactions between respiration and heart rate: A phenomenon common to multiple physiologic states. *Comput. Cardiol.* 15:12–23, 1988.

[49] M. L. Appel, J. P. Saul, R. D. Berger, and R. J. Cohen, Closed-loop identification of cardiovascular regulatory mechanisms. *Comput. Cardiol.* 16:3–8, 1990.

[50] S. Cavalcanti and E. Belardinelli, Modeling of cardiovascular variability using a differential delay equation. *IEEE Trans. Biomed. Eng.* 43:982–989, 1996.

[51] E. Mosekilde and O. G. Nouritsen, *Modeling the Dynamics of Biological Systems*. New York: Springer, 1995.

[52] K. H. Chon, M. J. Korenberg, and N. H. Holstein-Rathlou, Application of fast orthogonal search to linear and nonlinear stochastic systems. *Ann. Biomed. Eng.* 25:793–801, 1997.

[53] P. Celka, R. Vetter, U. Scherrer, E. Pruvot, and M. Karrakchou, Analysis of the relationship between muscle sympathetic nerve activity and blood pressure in humans using linear and nonlinear modelization. *Med. Biol. Eng. Comput.* 34:425–426, 1996.

[54] P. Celka, R. Vetter, U. Scherrer, and E. Pruvot. Closed loop heart beat interval modelization in humans using mean blood pressure, instantaneous lung volume and muscle sympathetic nerve activity. *IEEE Annual International Conference of Engineering in Medicine and Biological Society (EMBS)*, Amsterdam, Vol. 1, 1996.

[55] P. Celka, R. Vetter, U. Scherrer, E. Pruvot, and M. Karrakchou, Analysis of the relationship between muscle sympathetic nerve activity and blood pressure signals using subband multi-input modeling. *J. Biomed. Eng. Appl. Basis Commun.* 8:518–528, 1996.
[56] R. Vetter, P. Celka, J.-M. Vesin, G. Thonet, E. Pruvot, M. Fromer, U. Scherrer, and L. Bernardi, Subband modeling of the human cardiovascular system: New insights into cardiovascular regulation. *Ann. Biomed. Eng.* 26:293–307, 1998.
[57] M. H. Perrott and R. J. Cohen, An efficient approach to ARMA modeling of biological systems with multiple inputs and delays. *IEEE Trans. Biomed. Eng.* 43:1–14, 1996.
[58] I. Daubechies, The wavelet transform, time–frequency localization and signal analysis. *IEEE Trans. Inform. Th.* 36:961–1005, 1990.
[59] M. Akay, Wavelets in biomedical engineering. *Ann. Biomed. Eng.* 23:531–542, 1995.
[60] A. Malliani, M. Pagani, F. Lombardi, and S. Cerutti, Cardiovascular neural regulation explored in the frequency domain. *Circulation* 84:482–492, 1991.
[61] N. J. Bershad, P. Celka, and J. M. Vesin, Stochastic analysis of gradient adaptive identification of nonlinear systems with memory. *ICASSP'98*, Seattle, Vol. 1, 1998.
[62] A. B. Vallbo, K. E. Hagbarth, H. E. Torebjörk, and B. G. Wallin, Somatosensory, proprioseptive, and sympathetic activity in human peripheral nerves. *Physiol. Rev.* 59:919–957, 1979.
[63] A. M. Bianchi, L. T. Mainardi, C. Meloni, S. Chierchia, and S. Cerutti, Continuous monitoring of the sympatho-vagal balance through spectral analysis. *IEEE EMBS Mag.* 16:64–73, 1997.
[64] J. R. Haselton and P. G. Guyenet, Central respiratory modulation of medulllary sympathoexcitatory neurons in rat. *Am. J. Physiol.* 256:R739–R750, 1989.
[65] D. L. Eckberg, C. Nerhed, and B. G. Wallin, Respiratory modulation of muscle sympathetic and vagal cardiac outflow in man. *J. Physiol.* 365:181–196, 1985.
[66] M. Piepoli, P. Sleight, S. Leuzzi, F. Valle, G. Spadacini, C. Passino, J. Johnston, and L. Bernardi, Origin of respiratory sinus arrhythmia in conscious humans. *Circulation* 95:1813–1821, 1997.
[67] M. Pagani, N. Montano, A. Porta, A. Malliani, F. M. Abboud, C. Birkett, and V. K. Somers, Relationship between spectral components of cardiovascular variabilities and direct measures of muscle sympathetic nerve activity in humans. *Circulation.* 95:1441–1448, 1997.
[68] J. A. Taylor and D. L. Eckberg, Fundamental relations between short-term *RR* interval and arterial pressure oscillations in humans. *Circulation* 93:1527–1532, 1996.
[69] J. K. Triedman, M. H. Perrott, R. J. Cohen, and J. P. Saul, Respiratory sinus arrhythmia: Time domain characterization using autoregressive moving average analysis. *Am. J. Physiol.* 208:H2232–H2238, 1995.
[70] G. Wallin, M. Esler, P. Dorward, G. Eisenhofer, C. Ferrier, R. Westerman, and G. Jennings, Simultaneous measurements of cardiac nonadrenaline spillover and sympathetic outflow to skeletal muscle in humans. *J. Physiol.* 453:45–58, 1992.

Chapter 5

# NONLINEAR BEHAVIOR OF HEART RATE VARIABILITY AS REGISTERED AFTER HEART TRANSPLANTATION

C. Maier, H. Dickhaus, L. M. Khadra, and T. Maayah

The heart rate variability registered from heart transplant recipients is investigated for possible nonlinear dynamics. The method used performs a local prediction of the underlying dynamical process. The algorithm compares the short-term predictability for a given time series with that of an ensemble of random data, called surrogate data, that have the same Fourier spectrum as the original time series. The short-term prediction error is computed for the given time series as well as for the surrogate data, and a discriminating statistic for performing statistical hypothesis testing is derived. This method is applied to simulated data to test its reliability and robustness. The results show that shortly after heart transplantation the heart rate variability of nearly all patients has stochastic behavior with linear dynamics. However, with increasing time after transplantation we find in most of the patients nonlinear characteristics, or chaotic behavior, as known for healthy subjects.

## 1. INTRODUCTION

Qualitative and quantitative measures of heart rate variability (HRV) are applied increasingly in investigations of central regulation of autonomic state, physiological functions, and psychological behavior and particularly in assessment of clinical risk.

Changes or reduction of HRV have been investigated in regard to different pathological or pathophysiological phenomena in cardiac disorders such as myocardial infarction and coronary artery disease. HRV is even discussed as a predictor of mortality or as a marker for endangered patients indicating an impaired cardiac system or at least deficiencies of the autonomous nervous control of the heart [1–4].

Although the nature of HRV is not yet fully understood, the basic idea behind these investigations and suggestions implies a model of controlled complexity of heart dynamics with interacting subsystems operating on different time scales. Autonomic as well as nonautonomic influences controlling the heart are well studied and dynamic and steady-state characteristics of the cardiac response to vagal and sympathetic activity have been reported by many groups [4–6].

In addition to its autonomous control, the heart rate is modulated by humoral factors. These are findings that humoral influences on HRV are apparent even after cardiac denervation [7]. Berntson et al. report in their review [6] on low-frequency HRV of cardiac transplant patients that it is probably attributable primarily to humoral effects and small residual higher frequency fluctuations that are probably due to respiratory-related mechanical stretch of the sinoatrial node.

Our investigation was carried out on electrocardiographic (ECG) data recorded from heart transplant recipients at different times after the cardiac transplantation.

We know that immediately after transplantation, the heart shows a quite periodically paced rhythm with low variability. With increasing time after transplantation HRV becomes more and more apparent. This fact can be explained in terms of the background of increasing autonomous reinnervation. Therefore, it is of special interest to investigate quantitatively the time course of HRV properties, which should reflect the increasing trend to characteristics that are usually established for healthy subjects.

Various approaches for describing HRV with emphasis on different aspects have been suggested, and numerous studies employ different methods for expressing HRV numerically.

Global descriptive statistics are applied to characterize the distribution of heart periods in the time domain. For example, the standard deviation of differences between adjacent RR intervals or the so-called triangular index and the suggested PNN50 are clinical measures frequently used to quantify short- or long-term variability [4, 8, 9].

Spectral methods for calculating fast Fourier transform (FFT) estimates from RR interval time series, respectively power spectra and related measures, are applied because characteristic frequency bands—very low frequency (VLF, 0–0.05 Hz), low frequency (LF, 0.05–0.15 Hz), and high frequency (HF 0.15–0.4 Hz)—correspond to the main streams of autonomous nervous activations [10]. In this context, the autoregressive models should be mentioned [11]. Some authors prefer to apply time and frequency approaches together for HRV characterization because these methods are to some extent complementary to each other [12, 13]. The inherent assumption for these methods is a model of heart rate generation that is characterized by properties of a linear system [14].

However, precise analysis of the time behavior of the recorded heart rate data and the consideration of the regulating mechanisms of heart rate suggest many nonlinearities and nonstationary system properties. This motivated some authors to introduce nonstationary methods such as the frequently applied time–frequency approaches [15] or complex demodulation techniques [16].

Well-understood physiological mechanisms correspond directly to nonlinearities, for example, the adaptation and saturation of the baroreceptors, changes in gain of feedback loops, and delays in conduction time [14]. These facts are reason enough to describe the dynamics of the heart rate, particularly its variability, in terms of nonlinear system analysis, with the help of chaos theory [17–20].

Nonlinear systems have been used to model biological systems that are some of the most chaotic systems imaginable. In terms of this theory, the system controlling heart rate is characterized by a deterministic behavior that is hidden by random fluctuations. It is important to reveal and characterize the underlying order of the apparently random fluctuations to gain more insight into the heart rate dynamics provided by the theory of nonlinear dynamic systems. For this reason, distinguishing nonlinearity or deterministic chaos from stochastic behavior has become an important task in the analysis of HRV.

Essential and powerful tools of the methodological framework are introduced in the literature [21, 22] as well as in this book (see Chapter 1). As far as necessary, some facts and relations will be roughly described in the context of our evaluation in the next section.

## 2. DETECTION OF NONLINEAR DYNAMICS IN TIME SERIES

The problem of identifying nonlinearities or even chaos is not easy. Both chaotic and random processes can have similar broadband spectra. Basically, there are two approaches for distinguishing chaos from random noise: the dimension estimating approach and the nonlinear prediction approach.

The first approach is based on the fact that randomness arises from an infinite-dimensional attractor, whereas finite-dimensional attractors indicate chaos. An example of this approach is the Grassberger–Procaccia algorithm [23]. There are biases, however, in this approach. For example, a random process with power spectra $1/f$ can fool the dimension algorithm into detecting chaos where there is none [24]. Although the algorithm is relatively simple to compute, it is sensitive to variations in its parameters such as the number of data points, embedding dimension, reconstruction delay, and initial conditions. Moreover, it is very sensitive to noise. Hence, the practical significance of this algorithm is questionable. It is well known that a positive Lyapunov exponent is an indicator of chaotic dynamics. Therefore, the largest Lyapunov exponent is regarded as a good indicator for chaos [25]. The algorithm estimates lag and mean period using Fourier transformation (FT) and the attractor is reconstructed using the method of time delay embedding. The nearest neighbors are found and the average separation of neighbors is measured. This method is robust to noise and variations in data length but may misclassify high-dimensional chaos. Moreover, some types of composite signals may fool it.

The second approach is based on a fundamental property of time series showing dynamical nonlinearities: as opposed to random time series, their values are predictable in the short-term sense. The analysis of the prediction error allows discrimination of stochastic and nonlinear deterministic behavior [26]. Sugihara and May [27] used this approach to distinguish between deterministic chaos and random noise added to periodic signals. They showed that for chaotic signals, the correlation coefficient between predicted and actual values from the time series will decrease as the prediction time increases, whereas in the case of random data, the correlation will not change as the prediction time is changed. However, this technique for chaos identification will fail when presented with colored noise. Tsonis and Elsner [28] solved this problem with a particular class of colored noise, the random fractal sequences. They showed that by the scaling properties of the prediction errors as a function of the prediction time, nonlinear prediction can be used to distinguish chaotic and random fractal sequences. The approach used in this paper solves the problem for the more general class of colored random noise [24, 26, 29]. This approach is described in the language of statistical hypothesis testing, where the prediction error of the time series is compared with the prediction error of an ensemble of surrogate random time series that share the same Fourier spectrum and amplitude distribution with the observed data but are otherwise random.

## 3. THE DETECTION SCHEME FOR NONLINEARITIES

The general idea of the detection scheme is a simple statistical test: Because we would like to show that the time series we measured is generated by a nonlinear deterministic process, the null hypothesis H0 that we have to pose (and would like to reject) is: The time series results from a random process passing through a linear system.

The principle of a statistical test is the comparison of a parameter measured on a given realization to the distribution of the same parameter under presumption of the null hypothesis. In case the measured parameter does not match this distribution at a certain level of significance, we reject the null hypothesis.

One way to establish a suitable distribution is given by the concept of surrogate data. We generate artificial data sets that are consistent with the null hypothesis with respect to the following two properties:

- They are random.
- They pass the same hypothetical linear system as our measured time series.

We then calculate the parameter on this data ensemble. If we assume that the dynamics underlying the test data are stationary, then we have evidence that there are some dynamical nonlinearities in the test data in case of rejection of the null hypothesis. Chaos is one example of a dynamical nonlinearity, but this is not the only possibility. Other possibilities include nonlinear stochastic dynamics such as amplitude or frequency modulation with a stochastic modulating input [17].

Fortunately, we do not have to tie ourself down to choose a particular linear system to generate surrogate data. Because an optimal linear model of our original data is fully determined by its autocorrelation function, this property is all we have to preserve in the surrogate. However, the linear optimality of the model holds only in the case of Gaussian random input. In order to avoid problems with comparing non-Gaussian test data to Gaussian surrogate data, we perform a histogram transformation that renders any given amplitude distribution to a Gaussian distribution as proposed by Theiler et al. [24]. Nonnormal distributions can, for example, be generated by nonlinear measurement functions; that is, there exists an underlying Gaussian time series $y(n)$ and an observed time series $x(n)$ given by

$$x(n) = H\{y(n)\} \tag{1}$$

where $H\{\}$ is an invertible nonlinear transformation or observation function. The whole surrogate generation process is explained in detail later.

In principle, a variety of parameters can be chosen as a discriminating statistic. The only requirement is that they may not be derivable from the autocorrelation function or the amplitude histogram. In our case, we perform a nonlinear prediction and compare the prediction error of the original time series to that of the surrogate data. A block diagram of the detection approach is given in Figure 1. The different steps of surrogate data generation, nonlinear forecasting, calculation of the prediction error, and hypothesis testing are explained in the following.

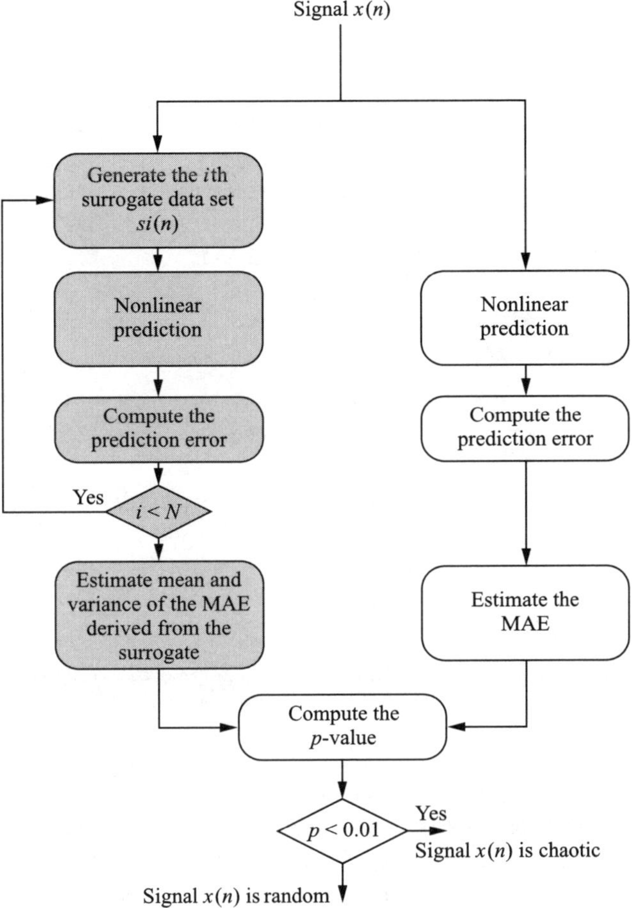


**Figure 1** Flowchart of the detection scheme for nonlinearities.

## 3.1. Surrogate Data Generation

The surrogate data sets are constructed according to the so-called amplitude-adjusted Fourier transform method, which combines the Fourier transform method with the histogram transformation as explained in Figure 2.

Since the only requirement for generating different surrogates is to maintain the autocorrelation function, which is linked to the power spectrum by Wiener's theorem, we perform the surrogate construction in the frequency domain. To implement this idea, the FT of $x'(n)$ is computed, and then the phase is randomized by multiplying the complex spectrum $X'$ at each frequency by $e^{j\phi}$, where $\phi \in [0, 2\pi]$ is a uniformly distributed random variable and is independently chosen at each frequency. For the inverse FT to be real, the phases must be symmetrical. The inverse FT is then calculated, which will result in a surrogate time series $s'(n)$.

The adjusted amplitude Fourier transform method is used to generate an ensemble of surrogate data to be consistent with the null hypothesis of a nonlinear transformation of a linear Gaussian process. To construct a surrogate time series, first the original

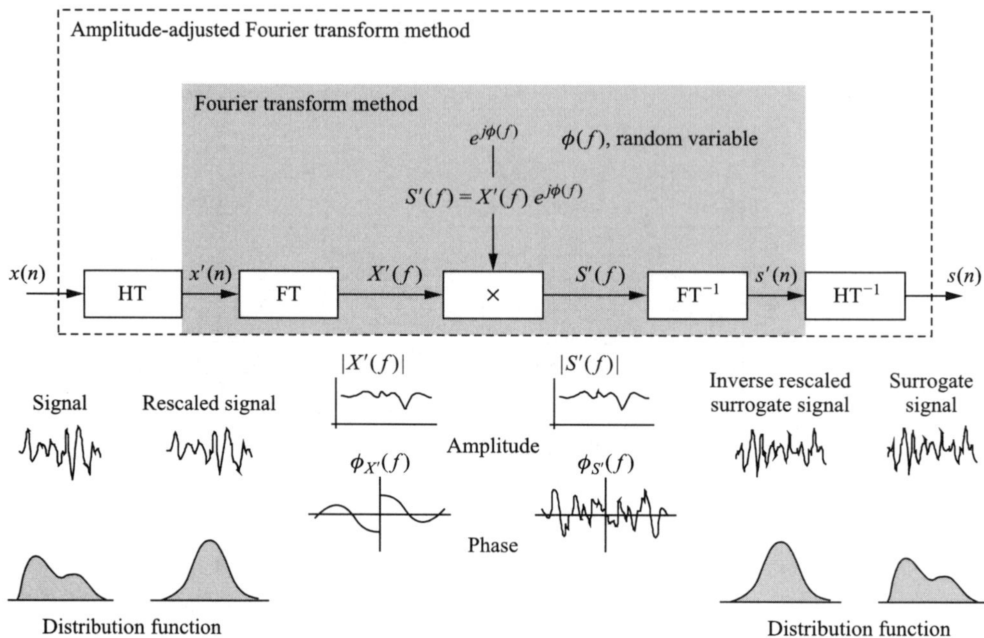


**Figure 2** Block diagram of the surrogate data generation process by the method of amplitude-adjusted Fourier transform.

time series $x(n)$ is nonlinearly rescaled by a process known as the histogram transformation (HT). The rescaled time series $x'(n)$ will have nearly the same shape as $x(n)$ but its amplitude distribution will be Gaussian. This transformation will preserve the linear correlation of the original time series and thus preserve the dynamics of the system and its fractal dimension. Then the Fourier transform method as explained earlier is applied on the rescaled time series to generate a Gaussian time series $s'(n)$. This time series is then rescaled using the inverse histogram transformation so that the resulting time series will follow the shape of the Gaussian time series $s'(n)$ and its amplitude distribution will be similar to that of the original time series $x'(n)$. The HT is carried out by generating a set of Gaussian random numbers with the same size as the given time series. Then the time order of the Gaussian data set as well as that of the original time series is shuffled according to the amplitude rank order. A look-up table is constructed from the shuffled sets, and each value in the original time series is replaced by the Gaussian random number of equal rank. The resulting series has the same shape as the original data set. Note that if the original time series is random, so will be the transformed one, however, with a Gaussian distribution.

### 3.2. Nonlinear Prediction

In the second phase of the algorithm, a predictive model is built from the time series, and its predictive performance is tested. This involves two steps: state space reconstruction and nonlinear function approximation. To fully describe a deterministic system at time $n$, the state vector must be known. Using this state vector, the future states of the system can be easily predicted. In many situations the observed time series

$x(n)$ is the only information available, which may represent one state variable. The goal of state space reconstruction is to reconstruct the system state using the past behavior of this time series, which bears the marks of all other variables participating in the dynamics. To reconstruct the state, we perform *time-delay embedding* [26], that is, we embed the time series in a state space with a state vector $\vec{x}(n)$, called the delay vector, as

$$\vec{x}(n) = \begin{pmatrix} x_1(n) \\ x_2(n) \\ \vdots \\ x_d(n) \end{pmatrix} = \begin{pmatrix} x(n) \\ x(n-\tau) \\ \vdots \\ x(n-(d-1)\tau) \end{pmatrix} \tag{2}$$

where $d$ is the embedding dimension and $n$ is an integer in the range $d \leq n \leq N$ with $N$ the number of points in the time series. The choice of $\tau$ is somewhat arbitrary according to Ref. 26 and we choose it to be 1, so that all data points of the measured time series can be included. The next step is splitting the data points in the time series into a fitting set $N_{\mathrm{f}}$ and a testing set $N_{\mathrm{t}}$ each containing half of the data points. Then we perform local approximation. This yields better accurate predictions than any other approximation method [26, 30]. To predict $x(n+T)$ of the test set, we fit a function between the previous nearby states $\{\vec{x}(n_1), \ldots, \vec{x}(n_k)\}$, $n_i \leq N/2$, of the fitting set and the points they evolve into a time index $T = 1$ later. The simplest form of local predictor is the approximation by only one nearest neighbor, called a zeroth-order approximation. That is, to predict $x(n+1)$, it is necessary only to search for the nearest neighbor $x(n_1)$ and to set the predicted future value of $x(n)$ as

$$x_{\mathrm{pred}}(n+1) = x(n_1+1) \tag{3}$$

This kind of local approximation was implemented by Kennel and Isabelle [29] for chaos detection in time series. A superior approach would be the first-order approximation. This kind of local approximation is used in our implementation, where a linear polynomial is fitted to the $k$ pairs of nearby states and their scalar future iterates. Thus, after searching for the $k$ neighboring states $\{\vec{x}(n_1), \ldots, \vec{x}(n_k)\}$, a linear predictor with parameters $\vec{a}$ and $b$ is fitted and used to compute the future iterate of $\vec{x}(n_i)$ as

$$x_{\mathrm{pred}}(n_i+1) = \vec{a}^T \cdot \vec{x}(n_i) + b \qquad \forall i = 1, \cdots, k \tag{4}$$

so that

$$\sum_{i=1}^{k} \left(x_{\mathrm{pred}}(n_i+1) - x(n_i+1)\right)^2 = \min \tag{5}$$

The fit is made using least squares by LU decomposition. The parameters derived by this procedure are used to estimate the value of the time series $x(n+1)$ according to

$$x_{\mathrm{pred}}(n+1) = \vec{a}^T \cdot \vec{x}(n) + b \tag{6}$$

The forecasting error

$$\epsilon(n) = x_{\text{pred}}(n+1) - x(n+1) \tag{7}$$

is calculated for all values $n \geq N/2$. The statistic quantifying the predictive performance is established by calculating the median of the absolute value of the forecasting error ($MAE$):

$$MAE = \text{median}\{|\epsilon(n)|\} \qquad n = \frac{N}{2}, \cdots, N-1 \tag{8}$$

## 3.3. Testing the Hypothesis

As outlined before, the scheme for detecting nonlinearities is a statistical test based on the null hypothesis that the observed time series is a realization of random noise passing through a linear system. Testing the hypothesis means evaluating the extent to which the null hypothesis is suitable to explain the data measured by comparing the statistic—the $MAE$—to the distribution of the $MAE$ derived by prediction of the surrogate data.

The comparison is made using a parametric approach. Assuming a Gaussian distribution of the $MAE$ for our surrogate data, we standardize our statistic by calculating

$$Z = \frac{MAE_{\text{test}} - \tilde{\mu}_{\text{surr}}}{\tilde{\sigma}_{\text{surr}}} \tag{9}$$

where $MAE_{\text{test}}$ denotes the value of the statistic for the test data set and $\tilde{\mu}_{\text{surr}}$ and $\tilde{\sigma}_{\text{surr}}$ are the sample mean and standard deviation of the $MAE$ of the surrogate data ensemble. With $Z$ we can estimate the significance of differences between test and surrogate data according to

$$p = \text{erfc}\left(\frac{|Z|}{\sqrt{2}}\right) \tag{10}$$

where erfc stands for the complementary error function. Provided that the null hypothesis is true, the $p$ value quantifies the probability of finding a $Z$ as extreme as that measured from the test data. So, a small $p$ value of 0.01 indicates that on the average only 1% of the realizations will result in an $MAE$ of the same value as the $MAE$ of our test data. Because it is very unlikely that our test series represents that one case out of a hundred, we prefer to reject the null hypothesis in cases in which the $p$ value is $< 0.01$. This is the maximum probability of rejecting a true null hypothesis or the level of significance. If $p < 0.01$, we assume that the test data reflect the dynamics of a nonlinear process. Moreover, we call these cases chaotic, knowing that we do not have evidence for this property in the strict sense.

## 3.4. Realization Aspects

The whole detection scheme as just described is realized in three modules that are documented in the Appendix. The main program is a MATLAB function (nl_test.m), which calls two external programs written in Pascal that implement the histogram transform (hstgram.pas) and the prediction (forecast.pas). The phase randomization is done within nl_test.m.

## 4. SIMULATIONS

We apply the proposed detection scheme to chaotic data generated from the Henon map:

$$\begin{aligned} x(n+1) &= \alpha + \beta \cdot y(n) + x^2(n) \\ y(n+1) &= x(n) \end{aligned} \tag{11}$$

When $\alpha = 1.4$ and $\beta = 0.3$, the map exhibits chaos [21]. In Figure 3 we see that the input time series $x'(n)$ to the surrogate data generation algorithm follows the original time series $x(n)$ but is otherwise Gaussian as its histogram shows. This nonlinearly scaled version of $x(n)$ is obtained by the histogram transformation. The resulting time series from the surrogate data generation process $s'(n)$ is Gaussian as shown in the histogram. The Gaussian surrogate time series $s'(n)$ is rescaled resulting in signal $s(n)$ by means of the inverse histogram transform as demonstrated in Figure 3. The $s(n)$ has the same amplitude distribution as the original time series $x(n)$.

Applying the detection scheme to the Henon time series $x(n)$, where the values of the statistic ($MAE$) are computed for an ensemble of 64 surrogate data sets, we get the results listed in Table 1. Figure 4 shows the histogram of the $MAE$. As displayed, the mean value of the $MAE$ for the surrogate data ensemble is about 0.6, which is very large compared with the prediction error of the actual Henon data. This will result in a large statistic value Z leading to a $p$ value $< 0.01$ and a detection of low-dimensional chaos. To test the hypothesis H0, the statistic is computed for the input signal $x(n)$ and compared with the distribution of the surrogate statistics. The resulting $MAE\{x(n)\} = 2.3395 \times 10^{-5}$ is very far from the mean of the surrogate statistics as demonstrated in the histogram in Figure 4. Testing the hypothesis in a more systematic way can be performed as described in Section 3, by calculating the $p$ value. For $N$ (number of data

**TABLE 1** Sensitivity of the Detection Scheme to Simulated Increasing Noise Levels $\sigma$[a]

| $\sigma$ | $p$ value | Detection |
|---|---|---|
| 0 | $\sim 0$ | Nonlinear |
| 10% | $\sim 0$ | Nonlinear |
| 20% | 5.03E-14 | Nonlinear |
| 30% | 1.59E-06 | Nonlinear |
| 40% | 1.11E-05 | Nonlinear |
| 50% | 0.0002 | Nonlinear |
| 60% | 0.0813 | Random |
| 70% | 0.4946 | Random |
| 80% | 0.5007 | Random |
| 90% | 0.5669 | Random |
| 100% | 0.9253 | Random |

[a] The correct detection holds to a noise level of 50% for a threshold $p$ value of 0.01.

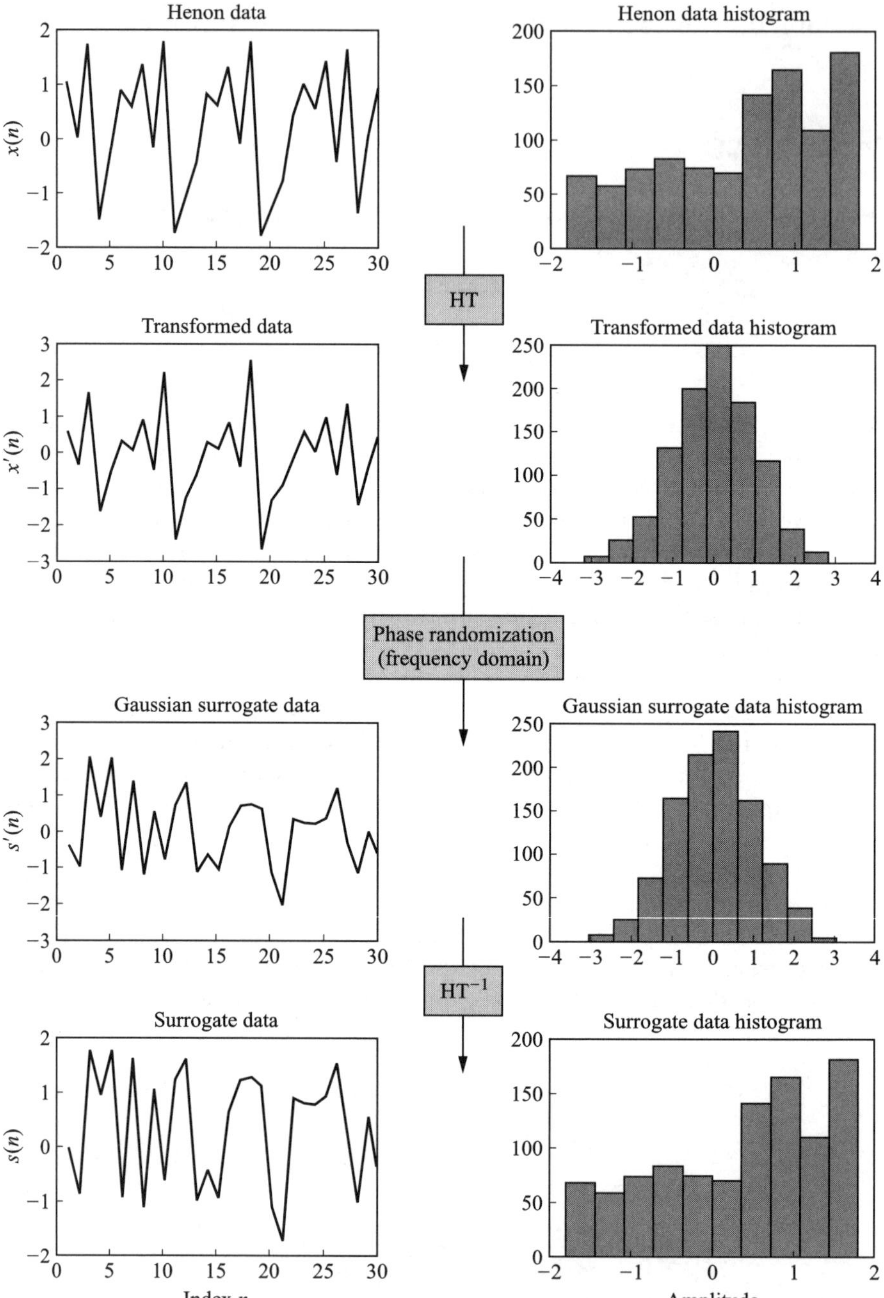


**Figure 3** Process of generating surrogate data $s(n)$ from the input Henon data $x(n)$ by the amplitude–adjusted Fourier transform method. Initial histogram transformation as well as inverse histogram transformation after phase randomization is demonstrated by the time series and their corresponding histograms. For details see text.

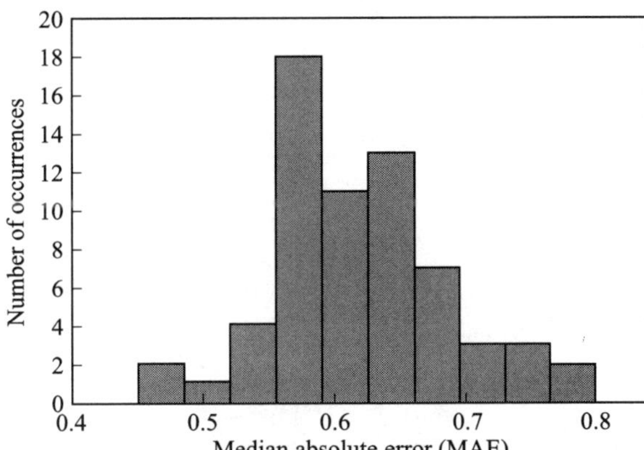


**Figure 4** Histogram of the median absolute error (*MAE*) derived from the Henon surrogate data.

points in the tested time series) = 1024, $N_s$ (number of generated surrogate time series) = 64, $k = 5$ and $d = 2$, we obtain a large value of $Z = 12.037$ ($p$ value $\approx 0$), which indicates that the data are chaotic.

## 4.1. The Choice of Modeling Parameter

The prediction time $T$ is set to 1. This is because we are distinguishing nonlinear from linear stochastic dynamics based on the capability of the chaotic time series to be predicted in the short-term sense. That is, as $T$ becomes larger the prediction accuracy decreases.

The selection of the embedding dimension $d$ depends on the dimension $D$ of the supposedly underlying strange attractor of the input time series $x(n)$. A condition on this parameter is to have $d \geq D$ as shown in Ref. 26. However, the true attractor dimension $D$ is not available. Moreover, the choice of the number of nearest neighbors $k$ is also important. Generally, it is chosen to be $k \geq 2d$ [30], but this choice is not always optimal. So we sweep the parameter $d$ from 1 to 10 and the corresponding $k$ from 2 to 20. Then the prediction error of the input time series is computed at each set of values. The lowest value of the median absolute error will correspond to the best fitting possible and its corresponding values of $d$ and $k$ are chosen for the detection scheme. Figure 5 illustrates the simulation result of applying this technique to the Henon time series using a number of neighbors $k = 5$. From our results it is apparent that the values $d = 2$ and $k = 5$ yield the best fitting. Hence, they are used in the detection scheme for the Henon map. This is in accordance with the theory, for we know that the Henon attractor is of fractal dimension 1.26 and is embedded in a plane.

## 4.2. Algorithm Robustness

To test the robustness of the detection scheme against noise contamination, noise is added to the Henon input signal $x(n)$ at different levels and the detection scheme is applied. The noise signal $\zeta(n)$ is generated in the same way as the surrogate data.

$$x_{\text{noise}}(n) = \sigma \cdot \zeta(n) = (1 - \sigma) \cdot x(n) \qquad 0 \leq \sigma \leq 1 \tag{12}$$

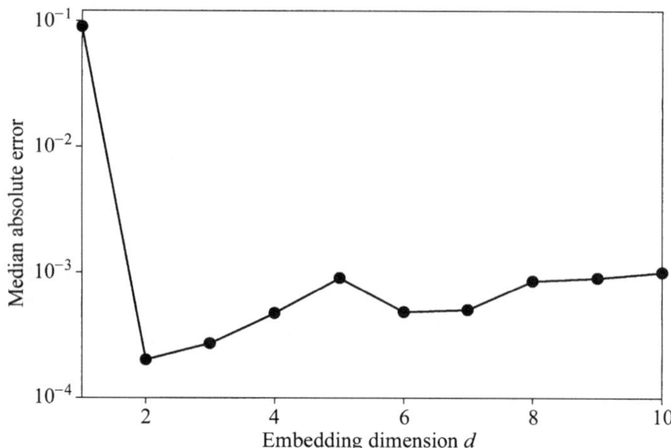


**Figure 5** Median absolute error ($MAE$) derived from the Henon data as a function of the embedding dimension $d$. The number of nearest neighbors in the prediction was set to $k = 5$.

where $\sigma$ denotes the noise level. The surrogate method of noise generation will result in noise $\zeta(n)$ that has the same power spectrum as the input time series $x(n)$. Table 1 summarizes the results of applying the detection scheme on $x_{\text{noise}}(n)$ for a range of noise levels $\sigma$ increasing from 0 to 100%. The limit of the detection is reached at a noise level of $\sigma = 50\%$.

Moreover, we apply the detection scheme to random data generated from the following autoregressive random process:

$$x(n+1) = 0.9 \cdot x(n) + \zeta(n) \tag{13}$$

For $n = 1, 2, \ldots, 1024$, $\zeta(n)$ are white Gaussian random numbers with zero mean and unity variance. We first determine the optimal values of $d$ and $k$. Figure 6 shows the results of this simulation. As expected, because the input time series is random noise, the prediction error will not improve upon varying the parameters. Applying the noisy time series $x(n)$ to the detection algorithm, we get a $p$ value of 0.4908 for $k = 8$ and $d = 4$.

## 5. CLINICAL STUDY

At the Heidelberg University Hospital, ECG data for eight heart-transplanted patients were recorded at different times after transplantation. Our data sets range from at least five successive measurements for two patients to at most nine different recordings for one patient. The individual observation time differs for the eight patients from minimally 70 days since the surgery to maximally more than 1 year for one patient. The time intervals between successive recordings also differ considerably. In general, the recording interval between two measurements increased with increasing time after transplantation. Because of the individual clinical status of each patient, a fixed protocol for recording was difficult to realize.

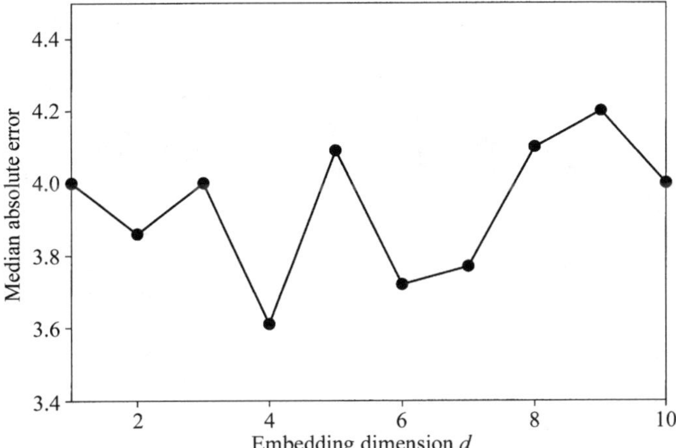


**Figure 6** Median absolute error ($MAE$) for the simulated colored Gaussian noise as function of the embedding dimension $d$. The number of nearest neighbors in the prediction was set to $k = 8$.

The registration of the ECGs was performed with the Predictor II System from Corazonix. The signal was sampled at 1000 Hz. In order to derive the time series of RR intervals between successive QRS complexes, the signal was triggered using a correlation-based algorithm. The resulting time series of RR intervals was visually checked for contamination with noise and artifacts. From each record with appropriate signal-to-noise ratio, we extracted more than 1000 successive RR intervals corresponding to about 10 minutes of recording time. This data set was used for the detection procedure. Figure 7 shows an example of four different time series of RR intervals from patient 3, recorded 28, 35, 47, and 70 days after transplantation.

For the different data sets of all patients, the $p$ values of the hypothesis test were calculated by the method described earlier. We first determined the optimal values for $k$ and $d$ with $k = 6$ and $d = 3$ as outlined in a previous paper [31]. Table 2 presents the results for the eight patients and their different recordings after transplantation. For each data set, the corresponding $p$ value and the detected kind of dynamics is presented. Figure 8 illustrates these results for the eight patients. Each diagram is for one patient and demonstrates the log scaled $p$ value as a function of the days elapsed since transplantation. The threshold for discriminating random behavior of the time series from a nonlinear or chaotic one was chosen to be 0.01, indicated as a horizontal line in each diagram. For all $p$ values above this line, the dynamic of the corresponding heart rate is considered as random; in other words, the time series of RR intervals results from a random process passing through a linear system. It is obvious that for nearly all patients the $p$ values start with high numbers and show a more or less monotonic decrease. Two patients, no. 5 and no. 8, have a small $p$ value for the second recording after 12 and 80 days, respectively. Only one patient, no. 4, shows an interesting inverse time course for the early recordings after transplantation. We do not have any reasonable explanation for this behavior. For all patients, the $p$ value dropped below our threshold at the final measurement. This is in accordance with the hypothesis that HRV of normal subjects is characterized by nonlinear or chaotic behavior as pointed out in an earlier study [31].

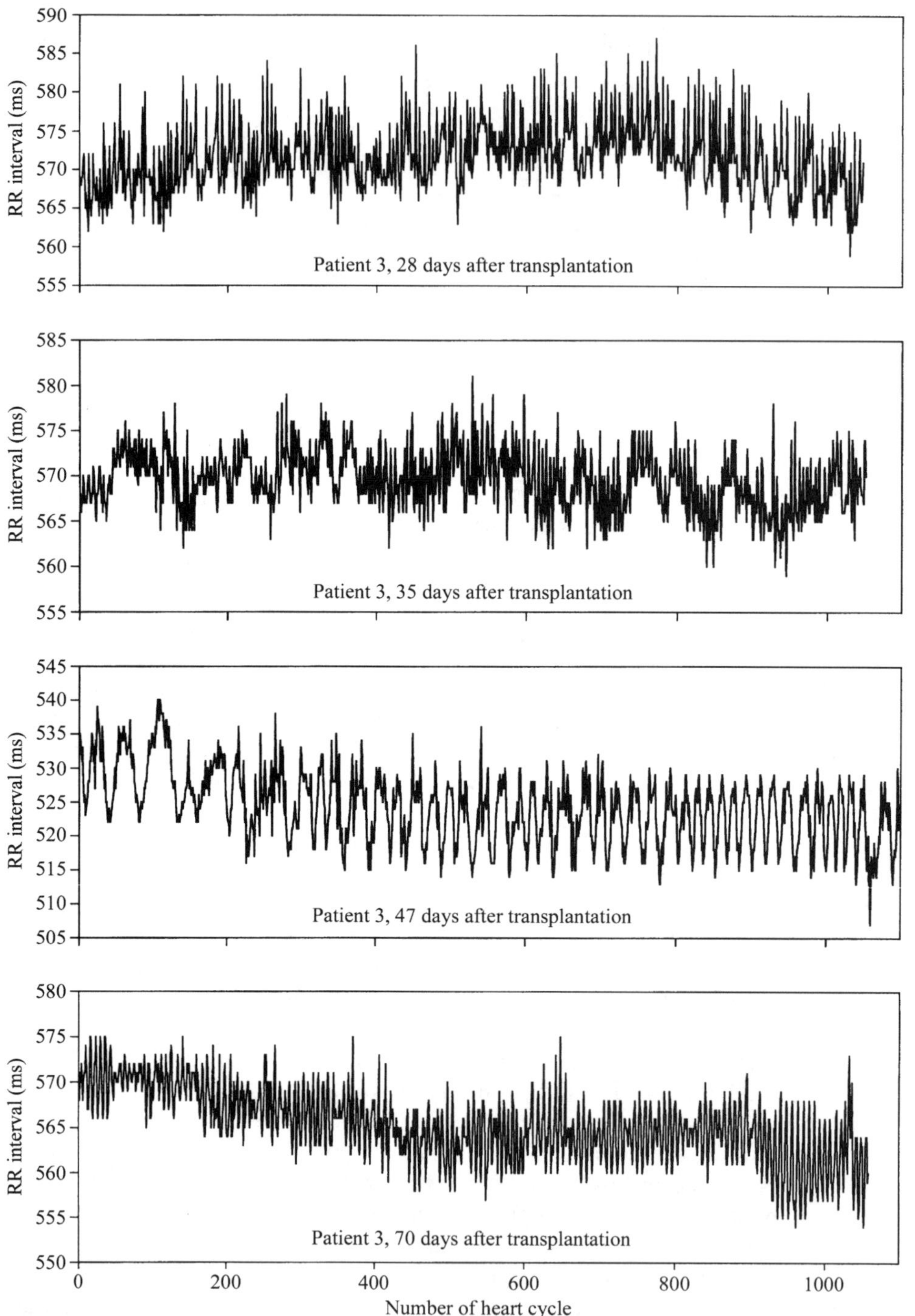


**Figure 7** RR interval time series recorded from patient 3 at 28, 35, 47, and 70 days after transplantation.

**TABLE 2** *P* Values Derived from the HRV Signals of Eight Patients at Different Times after Transplantation[a]

| Pat-ID | Days elapsed since transplantation | *P* value | Detection | Pat-ID | Days elapsed since transplantation | *P* value | Detection |
|---|---|---|---|---|---|---|---|
| 1 | 52 | 0.204 | Random | 5 | 2 | 0.807 | Random |
| | 61 | 0.01565 | Random | | 12 | > 0.00001 | Nonlinear |
| | 75 | 0.0043 | Nonlinear | | 26 | 0.1225 | Random |
| | 120 | 0.00503 | Nonlinear | | 56 | 0.0021 | Nonlinear |
| | 171 | 0.00091 | Nonlinear | | 70 | 0.0026 | Nonlinear |
| 2 | 27 | 0.075 | Random | 6 | 52 | 0.093 | Random |
| | 35 | 0.0431 | Random | | 68 | 0.1015 | Random |
| | 56 | 0.0109 | Random | | 94 | 0.0069 | Nonlinear |
| | 138 | 0.00005 | Nonlinear | | 157 | > 0.00001 | Nonlinear |
| | 166 | 0.00086 | Nonlinear | | 192 | > 0.00001 | Nonlinear |
| | 196 | 0.0001 | Nonlinear | | 276 | 0.00502 | Nonlinear |
| 3 | 28 | 0.3305 | Random | 7 | 27 | 0.014 | Random |
| | 35 | 0.0125 | Random | | 39 | 0.002 | Nonlinear |
| | 47 | 0.0053 | Nonlinear | | 48 | 0.001 | Nonlinear |
| | 56 | 0.0007 | Nonlinear | | 55 | > 0.00001 | Nonlinear |
| | 70 | 0.0005 | Nonlinear | | 67 | > 0.00001 | Nonlinear |
| | 82 | 0.0006 | Nonlinear | 8 | 47 | 0.22 | Random |
| | 112 | 0.0009 | Nonlinear | | 75 | 0.00026 | Nonlinear |
| 4 | 21 | > 0.00001 | Nonlinear | | 87 | 0.311 | Random |
| | 28 | 0.0029 | Nonlinear | | 131 | 0.953 | Random |
| | 40 | 0.0097 | Nonlinear | | 157 | 0.00014 | Nonlinear |
| | 54 | 0.0511 | Random | | 199 | 0.00083 | Nonlinear |
| | 63 | 0.9498 | Random | | 241 | 0.00062 | Nonlinear |
| | 75 | > 0.00001 | Nonlinear | | 299 | 0.00005 | Nonlinear |
| | 131 | 0.007 | Nonlinear | | 370 | > 0.00001 | Nonlinear |
| | 160 | 0.00069 | Nonlinear | | | | |

[a] The corresponding detection result is indicated in the last column.

## 6. DISCUSSION

In the scope of this research we studied the fundamental concept of nonlinear dynamics and chaos theory, in particular the problem of distinguishing between nonlinearity and random noise in time series. A detection scheme for nonlinearities was digitally implemented and applied to both simulated data and real HRV signals. The method was able to distinguish chaos from random data, even when applied to correlated random data.

Our results, derived from the small sample size of only eight patients, suggest that autonomic reinnervation of transplanted hearts follows a specific physiological time course. This behavior seems to be reflected by the $p$ values of the statistical hypothesis test procedure. ECG signals recorded shortly after heart transplantation are typically less complex in their interbeat dynamics, whereas the heart rate of healthy subjects shows more variability. The HRV of normal subjects seems to be actually nonlinear and not an accidental result of some random signals in the body. Normal subjects, even those at rest, show a high degree of heart rate variability that is not subjectively

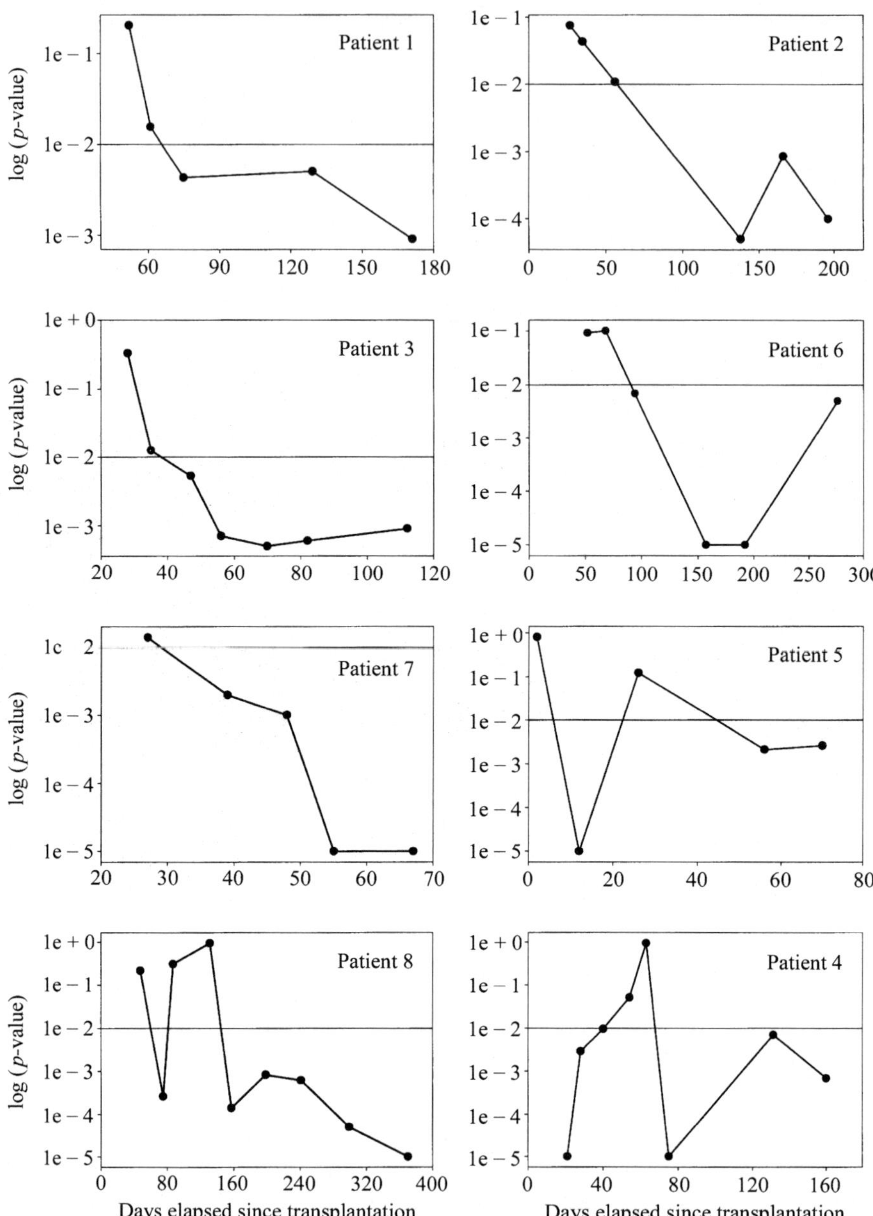


**Figure 8** Time courses of the log-scaled $p$ value for eight different patients. The decision threshold was set to 0.01. Values below that threshold indicate nonlinear dynamics in contrast to the random behavior for $p$ values above the threshold.

perceptible. The underlying order may be revealed and characterized by the theory of nonlinear dynamics. It permits a quantitative description of the complexity of the system. The interbeat interval fluctuations of the healthy heart may in part be due to the intrinsic variability of the autonomic nervous system and its control function.

Although, from the small sample size in our study, we cannot generalize a time course of the increasing physiological control by the autonomic nervous system in absolute time constants, we find this characteristic of adaptation principally in all transplant recipients. It seems to be that after 3 months in nearly all patients the normal characteristic of HRV is restored. Of course, there might be individual variations associated with the physiological or pathophysiological individuality of each subject. For instance, complications after transplantation such as infections or even acute rejections as well as the medication can delay or modulate this process of reinnervation. Further investigations including more patients will be necessary

However, our reported results confirm our previous findings [31]. In addition, it is interesting to note that other researchers reached similar conclusions indicating possible chaotic behavior of HRV signals [32].

## ACKNOWLEDGMENTS

The authors are pleased to acknowledge support from Dr. Daniela Gawliczek and Dr. Thomas Hilbel, Department of Cardiology at the Heidelberg University Hospital

## REFERENCES

[1] R. E. Kleiger, J. P. Miller, J. T. Bigger, A. J. Moss, and the Multicenter Post-Infarction Research Group, Decreased heart rate variability and its association with increased mortality after acute myocardial infarction. *Am. J. Cardiol.* 59:256–262, 1987.

[2] F. Lombardi, G. Sandrone, S. Pernpruner, R. Sala, M. Garimoldi, S. Cerutti, G. Baselli, M. Pagani, and A. Malliani, Heart rate variability as an index of sympathetic interaction after acute myocardial infarction. *Am. J. Cardiol.* 60:1239–1245, 1987.

[3] T. G. Farrell, Y. Bashir, T. Cripps, M. Malik, J. Poloniecky, E. D. Bennett, D. E. Ward, and F. Camm, Risk stratification for arryhthmic events in postinfarction patients based on heart rate variability, ambulatory electrocardiographic variables and the signal-averaged electrocardiogram. *J. Am. Coll. Cardiol.* 18:687–697, 1991.

[4] Task Force of the European Society of Cardiology and NASPE, Heart rate variability: Standards of measurement, physiological interpretation, and clinical use. *Circulation* 93:1043–1065, 1996.

[5] J. P. Saul, Y. Arai, R. D. Berger, L. S. Lilly, W. S. Colucci, and R. J. Cohen, Assessment of autonomic regulation in chronic congestive heart failure by heart rate spectral analysis. *Am. J. Cardiol.* 61:1292–1299, 1988.

[6] G. G. Berntson, T. Bigger, D. L. Eckberg, P. Grossmann, P. G. Kaufmann, M. Malik, H. N. Nagaraja, S. W. Porges, J. P. Saul, P. H. Stone, and M. W. Van der Molen, Heart rate variability: Origins, methods, and interpretative caveat. *Psychophysiology* 34:623–648, 1997.

[7] L. Bernardi, F. Salvucci, R. Suardi, P. L. Solda, A. Calciati, S. Perlini, C. Falcone, and L. Ricciardi, Evidence for an intrinsic mechanism regulating heart rate variability in the transplanted and the intact heart during submaximal dynamic exercise? *Cardiovasc. Res.* 24:969–981, 1990.

[8] M. Malik and A. J. Camm, *Heart Rate Variability*. Armonk, NY: Futura Publishing, 1995.

[9] D. J. Ewing, J. M. M. Neilson, and P. Travis, New methods for assessing cardiac parasympathetic activity using 24 hour electrocardiograms. *Br. Heart J.* 52:396, 1984.

[10] S. Akselrod, D. Gordon, F. A. Ubel, D. C. Shannon, A. C. Barger, and R. J. Cohen, Power spectrum analysis of heart rate fluctuations: A quantitative probe of beat-to-beat cardiovascular control. *Science* 213:220–222, 1981.

[11] S. Cerutti, A. M. Bianchi, and L. T. Mainardi, Spectral analysis of the heart rate variability signal. In *Heart Rate Variability*, M. Malik and A. J. Camm, eds., pp. 63–74, Armonk, NY: Futura Publishing, 1995.

[12] D. A. Litvack, T. F. Oberlander, L. H. Carney, and J. P. Saul, Time and frequency domain methods for heart rate variability analysis: A methodological comparison. *Psychophysiology* 32:492–504, 1995.

[13] M. W. F. Schweizer, J. Brachmann, U. Kirchner, I. Walter-Sack, H. Dickhaus, C. Metze, and W. Kübler, Heart rate variability in time and frequency domains: effects of gallopamil, nifedipine, and metoprolol compared with placebo. *Br. Heart J.* 70:252–258, 1993.

[14] D. Kaplan, Nonlinear mechanisms of cardiovascular control: We know they're there, but can we see them? *Proceedings 19th International Conference IEEE/EMBS*, Chicago, pp. 2704–2709, October 30–November 2, 1997.

[15] S. Pola, A. Macerata, M. Emdin, and C. Marchesi, Estimation of the power spectral density in nonstationary cardiovascular time series: Assessing the role of the time–frequency representations (TFR). *IEEE Trans. Biomed. Eng.* 43:46–59, 1996.

[16] S. J. Shin, W. N. Tapp, S. S. Reisman, and B. H. Natelson, Assessment of autonomic regulation of heart rate variability by the method of complex demodulation. *IEEE Trans. Biomed. Eng.* 36:274–283, 1989.

[17] D. T. Kaplan, Nonlinearity and nonstationarity: The use of surrogate data in interpreting fluctuations. *Proceedings of the 3rd Annual Workshop on Computer Applications of Blood Pressure and Heart Rate Signals*, Florence, Italy, May 4–5, 1995.

[18] A. Goldberger, Fractal mechanisms in the electrophysiology of the heart. *IEEE EMB Mag.* 11:47–52, 1992.

[19] A Garfinkel, M. L. Spano, W. L. Ditto, and J. N. Weiss, Controlling cardiac chaos. *Science* 257:1230–1235, 1992.

[20] M. E. Cohen, D. L. Hudson, and P. C. Deedwania, Applying continuous chaotic modeling to cardiac signal analysis. *IEEE EMB Mag.* 15:97–102, 1996.

[21] D. Kaplan and L. Glass, *Understanding Nonlinear Dynamics*. New York: Springer-Verlag, 1995.

[22] H. Kantz and T. Schreiber, *Nonlinear Time Series Analysis*. Cambridge, UK: Cambridge University Press, 1997.

[23] P. Grassberger and I. Procaccia, Measuring the strangeness of strange attractors. *Physica D* 9:189–208, 1983.

[24] J. Theiler, S. Eubank, A. Longtin, B. Galdrikian, and J. D. Farmer, Testing for nonlinearity in time series: The method of surrogate data. *Physica D* 58:77–94, 1992.

[25] M. Rosenstein, J. Collins, and C. Deluca, A practical method for calculating largest Lyapunov exponents for small data sets. *Physica D* 65:117–134, 1993.

[26] J. Farmer and J. Sidorowich, Predicting chaotic time series. *Phys. Rev. Lett.* 59:845–848, 1987.

[27] G. Sugihara and R. May, Nonlinear forecasting as a way of distinguishing chaos from measurement errors. *Nature* 344:734–741, 1990.

[28] A. Tsonis and J. Elsner, Nonlinear prediction as a way of distinguishing chaos from random fractal sequences. *Nature* 358:217–220, 1992.

[29] M. Kennel and S. Isabelle, Method of distinguishing possible chaos from colored noise and to determine embedding parameters. *Phys. Rev. A* 46:3111–3118, 1992.

[30] M. Casadagli, Nonlinear prediction of chaotic time series. *Physica D* 35:335–365, 1989.

[31] L. Khadra, T. Maayah and H. Dickhaus, Detecting chaos in HRV signals in human cardiac transplant recipients. *Comp. Biomed. Res.* 30:188–199, 1997.

[32] A. Babloyantz and A. Destexhe, Is the normal heart a periodic oscillator? *Biol. Cybern.* 58:203–211, 1988.

[33] A. L. Goldberger, Is normal heartbeat chaotic or homostatic? *News Physiol. Sci.* 6:87–91, 1991.

## APPENDIX

This is the source code of the MATLAB function nl_test.m, which implements the detection scheme as described earlier. For details please note the comments in the source code.

```
function p = nl_test(data, NumSurr, EmbDim, NumNeig);
%
% function p = nl_test(data, NumSurr [, EmbDim, NumNeig]);
%
% This function performs a test for nonlinearity in a time series by means of surrogate data.
% Surrogate calculation is carried out by phase randomization according to the method
% of amplitude adjusted fourier transform.
%
% Input parameters:
%
% data:       time series to be tested
% NumSurr:    number of surrogates to be calculated
% EmbDim:     embedding dimension
% NumNeig:    number of nearest neighbors to be included into calculation of predicted values
%
% Return parameter
% p-value
%
% This value should be less than 0.01 to claim that the time series under study is
% not of linear random origin
%
% Note that for HRV signals, after extensive study the optimal values for the embedding
% dimension and nearest neighbors were found to be 3 and 6 respectively

% nl_test is the main Matlab procedure that calls two Pascal programs
% Histgram.exe (which performs the histogram-transform) and
% Forecast.exe (which calculates the prediction error)
%
% Data exchange between Matlab and the external programs is handled through several
% files, see comments in the source code below.
% nl_test itself generates an ensemble of surrogate time series from the rescaled original
% time series, calculates the distribution of our test statistics - the MAE - for the
% surrogates, and compares it to the MAE of the original data set.

[row, col]= size(data);
if col > row
     data = data';
end
% now the original data is contained in a column-vector and we can save
% it to the file HRV.dat which is read by the program hstgram.exe
save HRV.dat data /ascii
RecordSize = length(data);
if (nargin == 2)
     EmbDim = 3;
     NumNeig = 6;
end
if (nargin == 3)
     NumNeig = 6;
end

i=sqrt (-1);

%We have to generate a call-String to invoke the external program forecast
CallForecastStr = sprintf('!Forecast %d %d ToPred.dat pred_err.dat', EmbDim, NumNeig);
% and initialize the random number generators
rand('seed', sum(100*clock));
randn('seed', 50000*rand);

% the following loop will generate an array of the Median Absolute Errors (MAE)
% for an ensemble of surrogate time series

for h=1:NumSurr
```

```
      % generate a gaussian time series with the same size as the HRV signal to be
      % tested and store it in an array G which is saved to the File Gauss.dat
      G = randn(RecordSize, 1);
      save Gauss.dat G /ascii

      % call the histogram transformation program to transform Gauss.dat time series to
      % have the same shape as HRV.dat time series but still have a gaussian PDF. The
      % transformed rescaled time series is stored in a file TGauss.dat ...

      !Hstgram HRV.dat Gauss.dat TGauss.dat

      % ... from which it is read into a local variable tgauss
      load TGauss.dat;

      % carry out phase randomization
      S=fft(tgauss, RecordSize);
      for 1=2:RecordSize/2
         r=2*pi*rand;
         % modify phase of spectral component no 1
         S(1) = real (S(1))*cos(r)+imag (S(1)*sin(r)*i;
         % we want the ifft to yield real valued signals, so the phases have to be symmetric
         S (RecordSize+2-1) = conj(S(1));
      end;
      SurrData=real(ifft(S));

      % Save the generated Surrogate data in a file named sg.dat
      save sg.dat SurrData /ascii

      % Call the inverse histogram transformation and store the transformed time series in
      % the file ToPred.dat
      !Hstgram sg.dat HRV.dat ToPred.dat

      % Call the forecasting algorithm to predict the surrogate time series and store the
      % absolute values of the prediction errors in the file named pred_err.dat
      eval (CallForecastStr);
      load pred_err.dat

      % Store the Median Absolute Error of the prediction error time series in the next
      % item in the array e
      e(h) = median(abs(pred_err));
end
% Call the forecasting algorithm to predict the original time series under study called
% HRV.dat and store the absolute values of the prediction errors in pred_err.dat
CallForecastStr =sprintf('!Forecast %d %d HRV.dat pred_err.dat', EmbDim, NumNeig);
eval(CallForecastStr);

load pred_err.dat;
% Calculate the Median Absolute Error (MAE)
E = median(abs(pred_err));

% Compute the p-value
p = erfc(abs((E-mean(e))/std(e))/sqrt(2));
```

This is the Pascal source code of the program hstgram.pas, which implements the histogram transform and is called from n1_pred.m. It reads three parameters from the command line, two input file names, and one output file name. All files are ASCII text files. The first input file is rescaled according to the values of the second input file, and the resulting rescaled time series is written to the output file.

```
{
   This program implements the Histogram and Inverse Histogram transformations
   The way this transformation is implemented is by storing the two time series
   in an array LIST where each element in the array consists of four values

      valx : is the data point in the first time series
      indx : is the time index of the point in valx
      valg : is the data point in the second time series
      indg : is the time index of the point in valg
```

```
   the program reads the two time series and store them in LIST, indx is equal indg

   The first step:
   is to sort the first time series in valx using the quik sort algorithm.

   this sorting will preserve the indices of the data points, that is when we swap two data
   point in valx we swap the indices in indx as well.

   The second step:
   is sort the second time series in valg and preserving the indices as done in step one.
   This will result in having both time series sorted out in descending order and their
   corresponding indices are stored in indx and indg.

   The third step:
   Now if we sort the LIST array using indx values, that is if we sort the values in indx in des-
   cending order and when we swap values, we swap the whole item; that is the
complete
   record (indx, valx, valg, indg). This will result in the first time series restored in
valx
   in terms of its original shape and the second time series sorted in valg being sorted
   out in such a way to mimic the shape of the first time series. That is the second
   time series is a non-linearly rescaled version of the first time series.

   If we simply swap the inputs passed to the program, we are simply doing the Inverse
   Histogram transformation.
}
program hstgram;
uses crt;
type
      element=^node;
      node=record
            indx:real;
            valx:real;
            indg:real;
            valg:real;
            end;
      arraytype=array[1..1024] of element;
      indextype=0..1024;
var   list:arraytype;
      number,i:integer;
      fp,fp2:text;
      temp:element;
      filename:string[20];

procedure inter_swap(n:integer);
begin
      for i:=1 to number do
      begin
         temp^.valx:=list[i]^.valx;
         temp^.indx:=list[i]^.indx;
         if n=1 then
         begin
            list[i]^.valx:=list[i]^.valg;
            list[i]^.indx:=list[i]^.indg;
         end
         else
         begin
            list[i]^.valx:=list[i]^.indg;
            list[i]^.indx:=list[i]^.valg;
         end;
         list[i]^.valg:=temp^.valx;
         list[i]^.indg:=temp^.indx;
      end;

end
procedure swap (var x,y:element;n:integer);
begin
      temp^.indx:=x^.indx;temp^.valx:=x^.valx;
      if n=2 then begin temp^.indg:=x^.indg;temp^.valg:=x^.valg;end;
      x^.indx:=y^.indx;x^.valx:=y^.valx;
      if n=2 then begin x^.indg:=y^.indg;x^.valg:=y^.valg;end;
```

```
      y^.indx:=temp^.indx;y^.valx:=temp^.valx;
      if n=2 then begin y^.indg:=temp^.indg;y^.valg:=temp^.valg; end;
end;

( This procedure is called recursively by the quicksort procedure qsort )

procedure split(var list:arraytype;first,last:indextype;
                  var splitpt1,splitpt2:indextype;n:integer);
var right,left:indextype;
      v:real;
begin
      v:=list[((first+last)div 2)]^.valx;
      right:=first;
      left:=last;
      repeat
         while list[right]^.valx< v do
            right:=right+1;
         while list[left]^.valx> v do
            left:=left-1;
         if right<=left then
         begin
            swap(list[right],list[left],n);
            inc(right);
            dec(left);
         end
      until right>left;
      splitpt1:=right;
      splitpt2:=left;
end;

(  This procedure implements the quicksort recursive algorithm and uses the split procedure )

procedure qsort (var list: arraytype;first,last:indextype;n:integer);
var splitpt1,splitpt2:indextype;
begin
      if first<last then
      begin
         split(list,first,last,splitpt1,splitpt2,n);
         if splitpt1<last then
            qsort(list,splitpt1,last,n);
         if first<splitpt2 then
            qsort(list,first,splitpt2,n);
      end;
end;

begin
      if ParamCount <> 3 then
      begin
         writeln('');
         writeln('ERROR - wrong number of parameters');
         writeln('Correct call:');
         writeln(' Hstgram InFile1 InFile2 OutFile')'
         exit
      end;
      clrscr;
      assign(fp, ParamStr(1));
      reset(fp);
      assign(fp2, ParamStr (2));
      reset(fp2);
      new(temp);
      (  Read the two time series into the array LIST - each item in the array
         stores the index of the time series and the two corresponding data points)
      number := 0;
      while not eof(fp) do
      begin
         inc(number);
         new(list[number]);
         readln(fp,list[number]^.valx);
         list[number]^.indx:=number;
         list[number]^.indg:=number;
         readln(fp2,list[number]^.valg);
      end;
```

```
        close(fp);
        close(fp2);
        (  sort the first time series in descending order - the data points and keep track of the
           indices )
        qsort(list,1,number,1);
        (  swap the two time series )
        inter_swap(1);

        (  sort the second time series in descending order - the data points and keep track of the
           indices )
        qsort(list,1,number,1);

        (  swap the indices )
        inter_swap(2);

        (  sort the indices )
        qsort(list,1,number,2);
        assign(fp, ParamStr(3));
        rewrite(fp);
        for i:=1 to number do
             writeln(fp.list[i]^.valg);
        close(fp);
end.
```

This is the Pascal source code of the program forecast.pas, which implements the nonlinear prediction algorithm and is called from nl_pred.m. It takes four command line parameters: three input parameters—the embedding dimension, the number of neighbors to be used for the local model, and the name of the file containing the values of the time series to be predicted as ASCII text—and one output parameter, the name of the file to which the absolute values of the prediction errors are written.

```
(  This program implements the non-linear prediction algorithm )

program forecast;
uses crt;
const
        dmax=50;
        tiny=1e-20;
type
        x_arr=array[1..511] of double;
        int=array [1..dmax] of integer;
        one=array [1..dmax] of double;
        two=array [1..dmax,1..dmax] of double;
var
        filename:string[20];
        ConvErr, number,n,w,dim,kv,i,j,k,d,ii:integer;
        b:one;
        temp:double;
        indx,ndex:int;
        a:two;
        fpl,fp:text;
        x:array [0..1024] of double;
        deff.x_pred:x_arr;

(  The source code of the following two procedures is omitted due to copyright reasons
   It can be found in the book

   W. H. Press, S. A. Teukolsky, W. T. Vetterling, B. P. Flannery
   NUMERICAL RECIPES IN PASCAL
   1st Edition
   Cambridge University Press

   These procedures perform an L/U-decomposition of a matrix.
)

procedure ludcmp(var a:two;n:integer;var indx:int;var d:integer);
procedure lubksb(var a:two;n:integer;var inx:int;var b:one);
```

```
( The following procedure is used to select the k nearest neighbors to
the predicted point using the Euclidean Norm as basis for comparison )

procedure selectsort(var deff:x_arr;n:integer;var ndex:int);
var
     m,k,mindex:integer;
begin
     for k:=1 to kv do
     begin
        m:=1;
        while m<=n do
        begin
           if deff[m]<deff[mindex] then mindex:=m;
           m:=m+1;
        end;

        ndex[k]:=mindex;
        deff[mindex]:=300000;
     end;
end;
begin
if ParamCount <> 4 then
begin
     writeln('');
     writeln('ERROR - wrong number of parameters');
     writeln('Correct call:');
     writeln(' ForeCast EmbDim NumNeig PredFile PredErrFile');
     writeln(' Parameters:');
     writeln(' EmbDim      embedding dimension');
     writeln(' NumNeig     number of neighbors to include in linear fit');
     writeln(' PredFile input ASCII-File containing values of time series to be predicted');
     writeln(' PredErrFile output ASCII-File containing prediction error');
     exit;
end;
clrscr;
val(ParamStr(1), dim, ConvErr);
if ConvErr <> 0 then
begin
     writeln('Cannot evaluate expression for command line argument EmbDim');
     exit
end;
val(ParamStr(2), kv, ConvErr);
if ConvErr <> 0 then
begin
     writeln('Cannot evaluate expression for command line argument NumNeig');
     exit
end;
(fp points to the file containing the time series to be predicted)
assign(fp,ParamStr(3));
reset(fp);
(fpl points to the prediction-error file )
assign(fpl.ParamStr(4));
rewrite(fpl);
n:=1;
number := 0;
while not eof (fp) do
begin
     readln(fp,x[number]);
     inc(number);
end;
close(fp);
for j:=(number div 2 -1) to number-1 do
begin
     for i:=1 to dim-1 do deff[i]:=30000;
     for i:=dim to (number div 2-2) do
     begin
        temp:=0;
        for ii:=0 to dim-1 do temp:=temp+sqr(x[j-ii]-x[i-ii]);
        deff[i]:=sqrt(temp);
     end;

     selectsort(deff, (number div 2-2),ndex);
```

```
        for i:=1 to dim do
        begin
           for k:=1 to dim do
           begin
              a[i,k]:=0
              for ii:=1 to kv do a[i,k]:=a[i,k]+x[ndex[ii]-k+1]*x[ndex[ii]-i+1];
           end;
           a[i,dim+1]:=0;
           for ii:=1 to kv do a[i,dim+1]:=a[i,dim+1]+x[ndex[ii]-i+1];
        end;

        for k:=1 to dim do
        begin
           a[dim+1,k]:=0
           for ii:=1 to kv do a[dim+1,k]:=a[dim+1,k]+x[ndex[ii]-k+1];
        end;
        a[dim+1,dim+1]:=kv;

        for i:=1 to dim do
        begin
           b[i]:=0
           for ii:=1 to kv do b[i]:=b[i]+x[ndex[ii]+1]*x[ndex[ii]-i+1];
        end;
        b[dim+1]:=0;
        for ii:=1 to kv do b[dim+1]:=b[dim+1]+x[ndex[ii]+1];
        ludcmp(a,dim+1,indx,d);
        lubksb(a,dim+1,indx,b);
        x_pred[n]:-0;
        for ii:=1 to dim do
        x_pred[n]:=x_pred[n]+x[j-ii+1]*b[ii];
        x_pred[n]:=x_pred[n]+b[dim+1];
        writeln(fpl,abs(x_pred[n]-x[j+1]));
        inc(n);
  end;
  end.
```

Chapter 6

# HEART RATE VARIABILITY: MEASURES AND MODELS

Malvin C. Teich, Steven B. Lowen, Bradley M. Jost, Karin Vibe-Rheymer, and Conor Heneghan

## 1. INTRODUCTION

The human heart generates the quintessential biological signal: the heartbeat. A recording of the cardiac-induced skin potentials at the body's surface, an electrocardiogram (ECG), reveals information about atrial and ventricular electrical activity. Abnormalities in the temporal durations of the segments between deflections, or of the intervals between waves in the ECG, as well as their relative heights, serve to expose and distinguish cardiac dysfunction. Because the electrical activity of the human heart is influenced by many physiological mechanisms, electrocardiography has become an invaluable tool for the diagnosis of a variety of pathologies that affect the cardiovascular system [1]. Electrocardiologists have come to excel at visually interpreting the detailed form of the ECG wave pattern and have become adept at differential diagnoses.

Readily recognizable features of the ECG wave pattern are designated by the letters P-QRS-T; the wave itself is often referred to as the *QRS complex*. Aside from the significance of various features of the QRS complex, the timing of the *sequence* of QRS complexes over tens, hundreds, and thousands of heartbeats is also significant. These inter-complex times are readily measured by recording the occurrences of the peaks of the large R waves, which are, perhaps, the most distinctive feature of the normal ECG.

In this chapter we focus on various measures of the fluctuations of this sequence of interbeat intervals and how such fluctuations can be used to assess the presence or likelihood of cardiovascular disease [2]. This approach has come to be called *heart rate variability* (HRV) analysis [3, 4] even when it is the time *intervals* whose fluctuations are studied (heart rate has units of inverse time rather than time). HRV analysis serves as a marker for cardiovascular disease because cardiac dysfunction is often manifested by systematic changes in the variability of the RR-interval sequence relative to that of normal controls [1, 3, 5, 6]. A whole host of HRV measures, some scale dependent and others scale independent, have been developed and examined over the years in an effort to develop readily available, inexpensive, and noninvasive measures of cardiovascular function.

We examine 16 HRV measures and their suitability for correctly classifying ECG records of various lengths as normal or revealing the presence of cardiac dysfunction. Particular attention is devoted to HRV measures that are useful for discriminating patients with congestive heart failure from normal subjects. Using receiver operating characteristic (ROC) analysis, we demonstrate that scale-dependent HRV measures (e.g., wavelet and spectral measures) are substantially superior to scale-independent measures (such as wavelet and spectral fractal exponents) for discriminating these two classes of data over a broad range of record lengths. The wavelet-transform standard deviation at a scale near 32 heartbeat intervals and its spectral counterpart near 1/32 cycles/interval turn out to provide reliable results using ECG records just minutes long.

A long-standing issue of importance in cardiac physiology is the determination of whether the normal RR sequence arises from a chaotic attractor or has an underlying stochastic origin [6]. We present a phase-space analysis in which *differences* between adjacent RR intervals are embedded. This has the salutary effect of removing most of the correlation in the time series, which is well known to be deleterious to the detection of underlying deterministic dynamics. We demonstrate that RR sequences, from normal subjects and from patients with cardiac dysfunction alike, have stochastic rather than deterministic origins, in accord with our earlier conclusions [7, 8].

Finally, we develop a mathematical point process that emulates the human heartbeat time series for both normal subjects and patients with heart failure. Using simulations, we show that a jittered integrate-and-fire model built around a fractal-Gaussian-noise kernel provides a realistic, although not perfect, simulation of real heartbeat sequences. A construct of this kind may well be useful in a number of venues, including pacemaker excitation.

## 2. METHODS AND MEASURES

### 2.1. The Heartbeat Sequence as a Point Process

The statistical behavior of the sequence of heartbeats can be studied by replacing the complex waveform of an individual heartbeat recorded in the ECG (an entire QRS complex) with the time of occurrence of the contraction (the time of the peak of the R phase), which is a single number [8, 9]. In mathematical terms, the heartbeat sequence is then modeled as an unmarked point process. This simplification greatly reduces the computational complexity of the problem and permits us to use the substantial methodology that exists for point processes [10–12].

The occurrence of a contraction at time $t_i$ is therefore simply represented by an impulse $\delta(t - t_i)$ at that time, where $\delta$ is the Dirac delta function, so that the sequence of heartbeats is represented by

$$h(t) = \sum_i \delta(t - t_i) \tag{1}$$

A realization of a point process is specified by the set of occurrence times $\{t_i\}$ of the events. A single realization of the data is often all that is available to the observer, so the identification of the point process, and the elucidation of the mechanisms that underlie it, must be gleaned from this one realization.

One way in which the information in an experimental point process can be made more digestible is to reduce the data into a statistic that emphasizes a particular aspect of the data (at the expense of other features). These statistics fall into two broad classes that derive from the sequence of interevent intervals and the sequence of counts, as illustrated in Figure 1 [10, 13].

Figure 1 illustrates how an ECG may be analyzed to obtain the sequence of interbeat intervals as well as the sequence of counts. Figure 1a illustrates an ECG (sequence of QRS complexes) recorded from a patient. The R waves are schematically represented by a sequence of vertical lines, as shown in Figure 1b. The time between the first two R waves is $\tau_1$, the first RR (or interbeat) interval as indicated by the horizontal arrows in this figure. The time between the second and third R waves is $\tau_2$, and so forth. In Figure 1c, the time axis is divided into equally spaced, contiguous time windows, each of duration $T$ seconds, and the (integer) number of R waves that fall in the $i$th window is counted and denoted $N_i$. This sequence $\{N_i\}$ forms a discrete-time random counting process of nonnegative integers. Varying the duration $T$ yields a family of sequences $\{N_i\}(T)$. The RR intervals $\{\tau_i\}$ themselves also form a sequence of positive real-valued random numbers, which is shown schematically in Figure 1d. Here the abscissa is the interval number, which is not a simple function of time.

In this section we examine several statistical measures (including some that are novel) to characterize these stochastic processes; the development is assisted by an understanding of point processes.

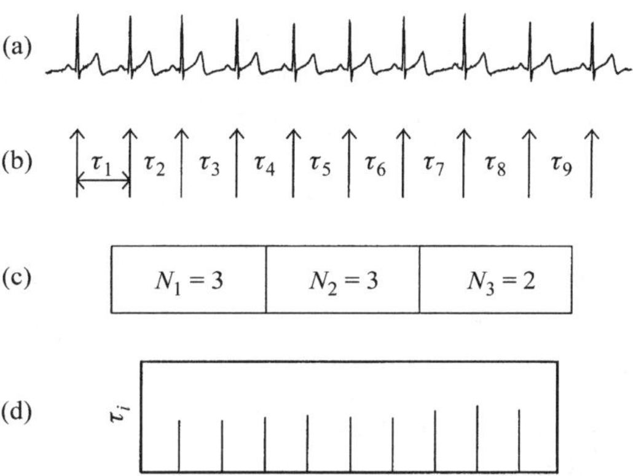


**Figure 1** In HRV analysis the electrocardiogram (ECG) schematized in (a) is represented by a sequence of times of the R phases that form an unmarked point process [vertical arrows in (b)]. This sequence may be analyzed as a sequence of counts $\{N_i\}(T)$ in a predetermined time interval $T$ as shown in (c) or as a sequence of interbeat (RR) intervals $\{\tau_i\}$ as shown in (d). The sequence of counts forms a discrete-time random counting process of nonnegative integers, whereas the sequence of intervals forms a sequence of positive real-valued random numbers.

### 2.1.1. Conventional Point Processes

The homogeneous Poisson point process, perhaps the simplest of all stochastic point processes, is described by a single parameter, the rate $\lambda$. This point process is memoryless: the occurrence of an event at any time $t_0$ is independent of the presence (or absence) of events at other times $t \neq t_0$. Because of this property, both the intervals $\{\tau_i\}$ and counts $\{N_i\}$ form sequences of independent, identically distributed random variables. The homogeneous Poisson point process is therefore completely characterized by the interevent-interval distribution (also referred to as the interbeat-interval histogram), which is exponential, or the event-number distribution (also referred to as the counting distribution), which is Poisson, together with the property of being independent. This process serves as a benchmark against which other point processes are measured; it therefore plays the role that the white Gaussian process enjoys in the realm of continuous-time stochastic processes.

A related point process is the nonparalyzable fixed-dead-time-modified Poisson point process, a close cousin of the homogeneous Poisson point process that differs only by the imposition of a dead-time (refractory) interval after the occurrence of each event, during which other events are prohibited from occurring [10, 14]. Another cousin is the gamma-$r$ renewal process which, for integer $r$, is generated from a homogeneous Poisson point process by permitting every $r$th event to survive while deleting all intermediate events [10, 15]. Both the dead-time-modified Poisson point process and the gamma-$r$ renewal process require two parameters for their description.

Some point process exhibit no dependences among their interevent intervals at the outset, in which case the sequence of interevent intervals forms a sequence of identically distributed random variables and the point process is completely specified by its interevent-interval histogram, i.e., its first-order statistic. Such a process is called a renewal process [10], a definition motivated by the replacement of failed parts, each replacement of which forms a renewal of the point process. Both examples of point processes presented above belong to the class of renewal point processes.

The interevent-interval histogram is, perhaps, the most commonly used of all statistical measures of point processes in the life sciences. The interevent-interval histogram estimates the interevent-interval probability density function $p_\tau(\tau)$ by computing the relative frequency of occurrence of interevent intervals as a function of interval size. Its construction involves the loss of interval ordering and therefore of information about dependences among intervals; a reordering of the sequence does not alter the interevent-interval histogram because the order plays no role in the relative frequency of occurrence.

The interevent-interval probability density function for the homogeneous Poisson point process assumes the exponential form

$$p_\tau(\tau) = \lambda \exp(-\lambda\tau) \tag{2}$$

where $\lambda$ is the mean number of events per unit time. The interevent-interval mean and variance are readily calculated to be $\mathrm{E}[\tau] = \int_0^\infty \tau p_\tau(\tau)\, d\tau = 1/\lambda$ and $\mathrm{Var}(\tau) = \mathrm{E}[\tau^2] - \mathrm{E}^2[\tau] = 1/\lambda^2$, respectively, where $\mathrm{E}[\cdot]$ represents expectation over the quantity inside the brackets. The interevent-interval probability density function for the dead-time-modified Poisson point process exhibits the same exponential form as for the homogeneous Poisson point process but is truncated at short interevent intervals as a result of the dead time [10]:

$$p_\tau(\tau) = \begin{cases} 0 & \tau < \tau_\mathrm{d} \\ \lambda \exp[-\lambda(\tau - \tau_\mathrm{d})] & \tau \geq \tau_\mathrm{d} \end{cases} \tag{3}$$

Here $\tau_\mathrm{d}$ is the dead time and $\lambda$ is the rate of the process before dead time is imposed.

If a process is nonrenewal, so that dependencies exist among its interevent intervals, then the interevent-interval histogram does not completely characterize the process [13]. In this case, measures that reveal the nature of the dependencies provide information that is complementary to that contained in the interevent-interval histogram. The heartbeat time series is such a nonrenewal process.

### 2.1.2. Fractal and Fractal-Rate Point Processes

The complete characterization of a stochastic process involves a description of all possible joint probabilities of the various events occurring in the process. Different statistics provide complementary views of the process; no single statistic can in general describe a stochastic process completely. Fractal stochastic processes exhibit scaling in their statistics. Such scaling leads naturally to power-law behavior, as demonstrated in the following. Consider a statistic $w$, such as the Allan factor for long counting times (see Section 2.5.1), which depends continuously on the scale $x$ over which measurements are taken [16, 17]. Suppose changing the scale by any factor $a$ effectively scales the statistic by some other factor $g(a)$, related to the factor but independent of the original scale:

$$w(ax) = g(a)w(x) \tag{4}$$

The only nontrivial solution of this scaling equation, for real functions and arguments, that is independent of $a$ and $x$ is

$$w(x) = bg(x) \quad \text{with} \quad g(x) = x^c \tag{5}$$

for some constants $b$ and $c$ [16–18]. Thus, statistics with power-law forms are closely related to the concept of a fractal [19–21]. The particular case of fixed $a$ admits a more general solution [22]:

$$g(x; a) = x^c \cos[2\pi \ln(x)/\ln(a)] \tag{6}$$

Consider once again, for example, the interevent-interval histogram. This statistic highlights the behavior of the times between adjacent events but reveals none of the information contained in the relationships among these times, such as correlation between adjacent time intervals. If the interevent-interval probability density function follows the form of Eq. 5 so that $p(\tau) \sim \tau^c$ over a certain range of $\tau$ where $c < -1$, the process is known as a fractal renewal point process [17, 19], a form of fractal stochastic process.

A number of statistics may be used to describe a fractal stochastic point process, and each statistic that scales will in general have a different scaling exponent $c$. Each of these exponents can be simply related to a more general parameter $\alpha$, the fractal exponent, where the exact relation between these two exponents will depend upon the statistic in question. For example, the exponent $c$ of the interevent-interval probability density function defined above is related to the fractal exponent $\alpha$ by $c = -(1 + \alpha)$. As the fractal exponent is a constant that describes the overall scaling behavior of a statistic, it does not depend on the particular scale and is therefore *scale independent*. Scale-independent measures are discussed in Sections 2.6 and 3.6.1.

Sample functions of the fractal renewal point process are true fractals; the expected value of their generalized dimensions assumes a nonintegral value between the topological dimension (zero) and the Euclidean dimension (unity) [19].

The sequence of unitary events observed in many biological and physical systems, such as the heartbeat sequence, do not exhibit power-law distributed interevent-interval histograms but nevertheless exhibit scaling in other statistics. These processes therefore have integral generalized dimensions and are consequently not true fractals. They may nevertheless be endowed with rate functions that are either fractals or their increments: fractal Brownian motion, fractal Gaussian noise, or other related processes. Therefore, such point processes are more properly termed fractal-*rate* stochastic point processes [17]. It can be shown by surrogate data methods, e.g., shuffling the order of the intervals (see Section 5.1.3), that it is the ordering and not the relative interval sizes that distinguish these point processes [8].

## 2.2. Standard Frequency-Domain Measures

A number of HRV measures have been used as standards in cardiology, both for purposes of physiological interpretation and for clinical diagnostic applications [3]. We briefly describe some of the more commonly used measures that we include in this chapter for comparison with several novel measures that have been recently developed.

Fourier transform techniques provide a method for quantifying the correlation properties of a stochastic process through spectral analysis. Two definitions of power spectral density have been used in the analysis of HRV [9]. A *rate*-based power spectral density $S_\lambda(f)$ is obtained by deriving an underlying random continuous process $\lambda(t)$, the heart rate, based on a transformation of the observed RR interbeat intervals. The power spectral density of this random process is well defined, and standard techniques may be used for estimating the power spectral density from a single observation of $\lambda(t)$. An advantage of this technique is that the power spectral density thus calculated has temporal frequency as the independent variable, so that spectral components can be interpreted in terms of underlying physiological processes with known time scales. The power spectral density itself is usually expressed in units of $\text{sec}^{-1}$. However, the choice of how to calculate $\lambda(t)$, the underlying rate function, may influence the calculated power spectral density.

The second spectral measure that is more widely used is an *interval*-based power spectral density $S_\tau(f)$ that is directly calculated from measured RR interbeat intervals without transformation [9]. In this case, the intervals are treated as discrete-index samples of an underlying random process, and there is no intermediate calculation of an underlying rate function. The power spectral density in this case has cycles/interval as the independent variable and therefore has units of $\text{sec}^2$/interval.

The two types of power spectral densities are easily confused and care must be taken in their interpretation. For example, one could mistakenly interpret the abscissa of an interval-based power spectral density plot as being equivalent to temporal frequency (e.g., cycles/s). Although this is generally incorrect, for point processes whose interevent-interval coefficient of variation is relatively small [9], the interval-based and rate-based power spectral density plots can be made approximately equivalent by converting the interval-based frequency $f_{\text{int}}$ (in cycles/interval) to the time-based frequency $f_{\text{time}}$ (in cycles/s) using

$$f_{\text{time}} = f_{\text{int}}/\text{E}[\tau] \tag{7}$$

For typical interbeat-interval sequences, the coefficient of variation is indeed relatively small and this conversion can be carried out without the introduction of significant error [9]. In the remainder of this chapter we work principally with the interval-based power-spectral density. We use the notation $f \equiv f_{\text{int}}$ for the interval-based frequency (cycles/interval) and retain the notation $f_{\text{time}}$ for temporal frequency (cycles/s).

We make use of nonparametric technique for estimating the spectral density. A simple reliable method for estimating the power spectral density of a process from a set of discrete samples $\{\tau_i\}$ is to calculate the averaged periodogram [23–25]. The data are first divided into $K$ nonoverlapping blocks of $L$ samples. After the optional use of a Hanning window, the discrete Fourier transform of each block is calculated and squared. The results are then averaged to form the estimate

$$\hat{S}_\tau(f) \equiv \frac{1}{K} \sum_{k=1}^{K} |\tilde{\tau}_k(f)|^2 \tag{8}$$

Here $\tilde{\tau}_k(f)$ is the discrete Fourier transform of the $k$th block of data and the hat explicitly indicates that we are dealing with an *estimate* of $S_\tau(f)$, which is called an averaged periodogram.

The periodogram covers a broad range of frequencies that can be divided into bands that are relevant to the presence of various cardiac pathologies. The power within a band is calculated by integrating the power spectral density over the associated frequency range. Some commonly used measures in HRV are [3]:

**VLF**. The power in the very-low-frequency range: 0.003–0.04 cycles/interval. Physiological correlates of the VLF band have not been specifically identified [3].

**LF**. The power in the low-frequency range: 0.04–0.15 cycles/interval. The LF band may reflect both sympathetic and vagal activity but its interpretation is controversial [3].

**HF**. The power in the high-frequency range: 0.15–0.4 cycles/interval. Efferent vagal activity is a major contributor to the HF band [26–28].

**LF/HF**. The ratio of the power in the low-frequency range to that in the high-frequency range. This ratio may either mirror sympathovagal balance or reflect sympathetic modulations [3].

## 2.3. Standard Time-Domain Measures

We consider three time-domain measures commonly used in HRV analysis. The first and last are highly correlated with each other inasmuch as they estimate the high-frequency variations in the heart rate [3]. They are:

**pNN50**. The relative proportion of successive NN intervals (normal-to-normal intervals, i.e., all intervals between adjacent QRS complexes resulting from sinus node depolarizations [3]) with interval differences greater than 50 ms.

**SDANN**. The Standard Deviation of the Average NN interval calculated in 5-minute segments. It is often calculated over a 24-hour period. This measure estimates fluctuations over frequencies smaller than 0.003 cycles/sec.

**SDNN ($\sigma_{\text{int}}$)**. The Standard Deviation of the NN interval set $\{\tau_i\}$ specified in units of seconds. This measure is one of the more venerable among the many scale-dependent measures that have long been used for HRV analysis [3, 5, 29, 30].

## 2.4. Other Standard Measures

There are several other well-known measures that have been considered for HRV analysis. For completeness, we briefly mention two of them here: the event-number histogram and the Fano factor [7, 8]. Just as the interevent-interval histogram provides an estimate of the probability density function of interevent-interval magnitude, the event-number histogram provides an estimate of the probability mass function of the number of events. Construction of the event-number histogram, like that of the interevent-interval histogram, involves loss of information, in this case the ordering of the counts. However, whereas the time scale of information contained in the interevent-interval histogram is the mean interevent interval, which is intrinsic to the process under consideration, the event-number histogram reflects behavior occurring on the adjustable time scale of the counting window $T$. The Fano factor, which is the variance of the number of events in a specified counting time $T$ divided by the mean number of events in that counting time, is a measure of correlation over different time scales $T$. This measure is sometimes called the index of dispersion of counts [31]. In terms of the sequence of counts illustrated in Figure 1, the Fano factor is simply the variance of $\{N_i\}$ divided by the mean of $\{N_i\}$.

## 2.5. Novel Scale-Dependent Measures

The previous standard measures are all well-established scale-dependent measures. We now describe a set of recently devised scale-dependent measures whose performance we evaluate. Throughout this chapter, when referring to intervals, we denote the fixed scale as $m$; when referring to time, we employ $T$.

### 2.5.1. Allan Factor [A(T)]

In this section we present a measure we first defined in 1996 [32] and called the Allan factor. We quickly found that this quantity was a useful measure of HRV [8]. The Allan factor is the ratio of the event-number Allan variance to twice the mean:

$$A(T) \equiv \frac{\mathrm{E}\left\{\left[N_{i+1}(T) - N_i(T)\right]^2\right\}}{2\mathrm{E}\left\{N_{i+1}(T)\right\}} \tag{9}$$

The Allan variance, as opposed to the ordinary variance, is defined in terms of the variability of *successive* counts [17, 33, 34]. As such, it is a measure based on the Haar wavelet. The Allan variance was first introduced in connection with the stability of atomic-based clocks [33]. Because the Allan factor functions as a derivative, it has the salutary effect of mitigating against linear nonstationarities.

The Allan factor of a point process generally varies as a function of the counting time $T$; the exception is the homogeneous Poisson point process. For a homogeneous Poisson point process, $A(T) = 1$ for any counting time $T$. Any deviation from unity in the value of $A(T)$ therefore indicates that the point process in question is not Poisson in nature. An excess above unity reveals that a sequence is less ordered than a homogeneous Poisson point process, whereas values below unity signify sequences which are more ordered. For a point process without overlapping events, the Allan factor approaches unity as $T$ approaches zero.

A more complex wavelet Allan factor can be constructed to eliminate polynomial trends [35–37]. The Allan variance, $\mathrm{E}[(N_{i+1} - N_i)^2]$, may be recast as the variance of the integral of the point process under study multiplied by the following function:

$$\psi_{\text{Haar}}(t) = \begin{cases} -1 & \text{for } -T < t < 0 \\ +1 & \text{for } 0 < t < T \\ 0 & \text{otherwise} \end{cases} \tag{10}$$

Equation 10 defines a scaled wavelet function, specifically the Haar wavelet. This can be generalized to any admissible wavelet $\psi(t)$; when suitably normalized the result is a wavelet Allan factor [36, 38].

### *2.5.2. Wavelet-Transform Standard Deviation [$\sigma_{\text{wav}}(m)$]*

Wavelet analysis has proved to be a useful technique for analyzing signals at multiple scales [39–44]. It permits the time and frequency characteristics of a signal to be simultaneously examined and has the advantage of naturally removing polynomial nonstationarities [36, 37, 45]. The Allan factor served in this capacity for the counting process $\{N_i\}$, as discussed earlier. Wavelets similarly find use in the analysis of RR-interval series. They are attractive because they mitigate against the nonstationarities and slow variations inherent in the interbeat-interval sequence. These arise, in part, from the changing activity level of the subject during the course of a 24-hour period.

Wavelet analysis simultaneously gives rise to both scale-dependent and scale-independent measures [46], affording the experimenter an opportunity to compare the two approaches. In this latter capacity wavelet analysis provides an estimate of the wavelet-transform fractal (scaling) exponent $\alpha_W$ [46, 47], as discussed in the context of HRV in Sections 2.6.2 and 3.6.1. As a result of these salutary properties, we devote particular attention to the wavelet analysis of HRV in this chapter.

A dyadic discrete wavelet transform for the RR-interval sequence $\{\tau_i\}$ may be defined as [41–43]

$$W_{m,n}(m) = \frac{1}{\sqrt{m}} \sum_{i=0}^{L-1} \tau_i \psi(i/m - n) \tag{11}$$

The quantity $\psi$ is the wavelet basis function, and $L$ is the number of RR intervals in the set $\{\tau_i\}$. The scale $m$ is related to the scale index $j$ by $m = 2^j$. Both $j$ and the translation variable $n$ are nonnegative integers. The term dyadic refers to the use of scales that are integer powers of 2. This is an arbitrary choice; the wavelet transform could be calculated at arbitrary scale values, although the dyadic scale enjoys a number of convenient mathematical properties [41, 42].

The dyadic discrete wavelet transform calculated according to this prescription generates a three-dimensional space from a two-dimensional signal graph. One axis is time or, in our case, the RR-interval number $i$; the second axis is the scale $m$; and the third axis is the strength of the wavelet component. Pictorially speaking, the transform gives rise to a landscape whose longitude and latitude are RR-interval number and scale of observation, while the altitude is the value of the discrete wavelet transform at the interval $i$ and the scale $m$.

Figure 2 provides an example of such a wavelet transform, where $\psi(x)$ is the simple Haar wavelet. Figure 2a illustrates the original wavelet, a function that is by definition $\psi(x) = 1$ for $x$ between 0 and 0.5; $\psi(x) = -1$ for $x$ between 0.5 and 1; and $\psi(x) = 0$ elsewhere. Figure 2b illustrates the wavelet scaled by the factor $m = 16$, which causes it to last for 16 samples rather than 1; and delayed by a factor of $n = 3$ times the length of the wavelet, so that it begins at $x = nm = 48$. Figure 2c shows a sequence of interbeat-interval values multiplied by the scaled and shifted wavelet (the summand in Eq. 11). The abscissa is labeled $i$ rather than $x$ to indicate that we have a discrete-time process composed of the sequence $\{\tau_i\}$. In this particular example, only values of $\tau_i$ between $i =$ 48 and 63 survive. Adding them (with the appropriate sign) provides the wavelet transform beginning at interval number $i = 48$ at a scale of $m = 16$.

For the Haar wavelet the calculation of the wavelet transform is therefore tantamount to adding the eight RR intervals between intervals 48 and 55 inclusive and then subtracting the eight subsequent RR intervals between intervals 56 and 63 inclusive, as illustrated in Figure 2c. Moving this window across interval number allows us to see how the wavelet transform evolves with interval number, whereas varying the scale of the window permits this variation to be observed over a range of resolutions, from fine to coarse (smaller scales allow the observation of more rapid variations, i.e., higher frequencies).

A simple measure that can be devised from the wavelet transformation is the standard deviation of the wavelet transform as a function of scale [46–49]:

$$\sigma_{\text{wav}}(m) = \left[\mathrm{E}\left\{\left|W_{m,n}(m) - \mathrm{E}\left[W_{m,n}(m)\right]\right|^2\right\}\right]^{1/2} \tag{12}$$

where the expectation is taken over the process of RR intervals and is independent of $n$. It is readily shown that $\mathrm{E}[W_{m,n}(m)] = 0$ for all values of $m$ so that a simplified form for the wavelet-transform standard deviation emerges:

$$\sigma_{\text{wav}}(m) = \left\{\mathrm{E}\left[\left|W_{m,n}(m)\right|^2\right]\right\}^{1/2} \tag{13}$$

This quantity has been shown to be quite valuable for HRV analysis [46–50]. The special case obtained by using the Haar-wavelet basis and evaluating Eq. 13 at $m = 1$ yields the standard deviation of the difference between pairs of consecutive interbeat intervals. This special case is therefore identical to the well-known HRV measure

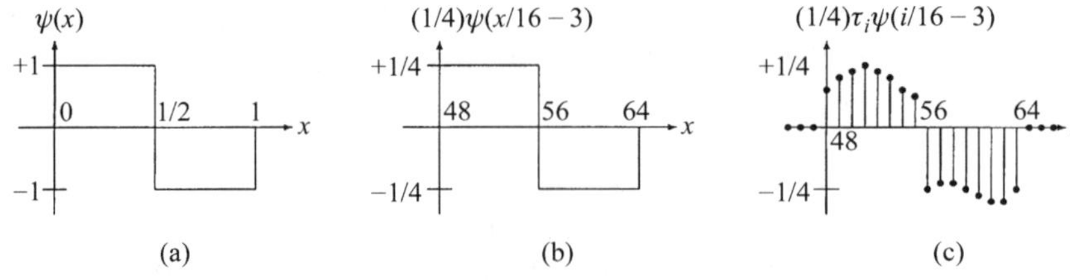


**Figure 2** Estimating the wavelet transform using the Haar wavelet: (a) original Haar wavelet, (b) delayed and scaled version of the wavelet ($m = 16$, $n = 3$), and (c) time series multiplied by this wavelet.

referred to as **RMSSD** [3], an abbreviation for Root-Mean-Square of Successive-interval Differences.

Figure 3 provides an example in which the discrete wavelet transform is calculated using an RR-interval data set. In Figure 3a, the original RR interbeat-interval series is shown, and Figure 3b shows the dyadic discrete wavelet transform at three different scales as a function of RR-interval number. It is important and interesting to note that the trends and baseline variations present in the original time series have been removed by the transform. As illustrated in Figure 3b, the wavelet-transform standard deviation $\sigma_{\mathrm{wav}}$ typically increases with the scale $m$. When plotted versus scale, this quantity provides information about the behavior of the signal at all scales. In Section 3 we show how this measure can be effectively used to separate patients with heart failure from normal subjects.

### 2.5.3. Relationship of Wavelet [$\sigma_{\mathrm{wav}}(m)$] and Spectral Measures [$S_\tau(f)$]

Is there a spectral measure equivalent to the wavelet-transform standard deviation? We proceed to show that the wavelet-transform standard deviation $\sigma_{\mathrm{wav}}(m)$ and the interval-based power spectral density $S_\tau(f)$ are isomorphic [49], so the answer is yes under conditions of stationarity. Although their equivalence is most easily analyzed in the continuous domain, the results are readily translated to the discrete domain by interpreting the discrete wavelet transform as a discretized version of a continuous wavelet transform.

The continuous wavelet transform of a signal $\tau(t)$ is defined as

$$W_\tau(s, r) = \frac{1}{\sqrt{s}} \int_{-\infty}^{\infty} \tau(t)\psi^*\left(\frac{t-r}{s}\right) dt \tag{14}$$

where $s$ and $r$ are continuous-valued scale and translation parameters, respectively, $\psi$ is a wavelet basis function, and $*$ denotes complex conjugation. Since $\mathrm{E}[W_\tau] = 0$, the variance of $W_\tau$ at scale $s$ is

$$D(s) = \sigma_{\mathrm{wav}}^2(s) = \mathrm{E}\left[\left|W_\tau(s, r)\right|^2\right] \tag{15}$$

which can be written explicitly as

$$D(s) = \mathrm{E}\left[\frac{1}{\sqrt{s}} \int_{-\infty}^{\infty} \tau(t)\psi^*\left(\frac{t-r}{s}\right) dt \, \frac{1}{\sqrt{s}} \int_{-\infty}^{\infty} \tau^*(t')\psi\left(\frac{t'-r}{s}\right) dt'\right] \tag{16}$$

For a wide-sense stationary signal the variance can be written as

$$D(s) = \frac{1}{s} \int_{-\infty}^{\infty} \int_{-\infty}^{\infty} R(t - t')\psi^*\left(\frac{t-r}{s}\right)\psi\left(\frac{t'-r}{s}\right) dt\, dt' \tag{17}$$

where $R$ is the autocorrelation function of $\tau$. Routine algebraic manipulation then leads to

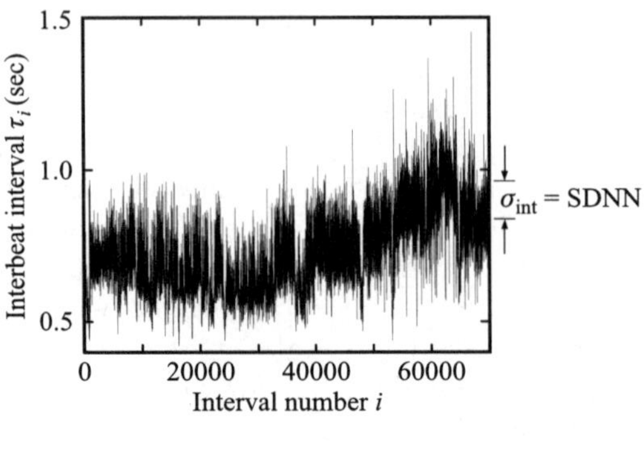


(a)

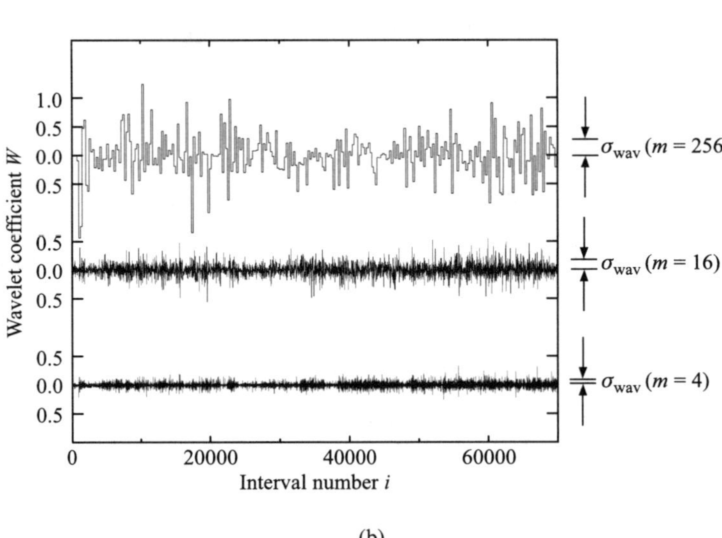


(b)

**Figure 3** (a) Series of interbeat intervals $\tau_i$ versus interval number $i$ for a typical normal patient (data set n16265). (Adjacent values of the interbeat interval are connected by straight lines to facilitate viewing.) Substantial trends are evident. The interbeat-interval standard deviation $\sigma_{\text{int}} \equiv$ SDNN is indicated. (b) Wavelet transform $W_{m,n}(m)$ (calculated using a Daubechies 10-tap analyzing wavelet) at three scales ($m = 4$, 16, 256) versus interval number $i$ ($= mn$) for the RR-interval sequence shown in (a). The trends in the original interbeat-interval time series are removed by the wavelet transformation. The wavelet-transform standard deviation $\sigma_{\text{wav}}(m)$ for this data set is seen to increase with the scale $m$.

$$D(s) = s\int_{-\infty}^{\infty} R(sy)W_{\psi}(1, y)\,dy \tag{18}$$

or, alternatively,

$$D(s) = s\int_{f=-\infty}^{\infty} S_{\tau}(f)\left[\int_{y=-\infty}^{\infty} W_{\psi}(1, y)\,\exp(j2\pi fsy)\,dy\right]df \tag{19}$$

where $W_{\psi}(1, y)$ is the wavelet transform of the wavelet itself (termed the wavelet kernel), and $S_{\tau}(f)$ is the power spectral density of the signal.

For the dyadic discrete wavelet transform that we have used, Eq. 19 becomes

$$\begin{aligned} D(m) = \sigma^2_{\text{wav}}(m) &= m \int_{f=-\infty}^{\infty} S_{\tau}(f) \left[ \int_{y=-\infty}^{\infty} W_{\psi}(1, y) \exp(j2\pi fmy)\, dy \right] df \\ &= m \int_{-\infty}^{\infty} S_{\tau}(f) H(mf)\, df \end{aligned} \tag{20}$$

We conclude that for stationary signals the interval-based power spectral density $S_{\tau}(f)$ is directly related to the wavelet-transform standard deviation $\sigma_{\text{wav}}(m)$ through an integral transform. This important result has a simple interpretation: the factor in square brackets in Eq. 20 represents a bandpass filter $H(mf)$ that passes only spectral components in a bandwidth surrounding the frequency $f_m$ that corresponds to the scale $m$. This is because the Fourier transform of a wavelet kernel is constrained to be bandpass in nature. For a discrete-index sequence, the sampling "time" can be arbitrarily set to unity so that a frequency $f_m$ corresponds to $1/m$. We conclude that information obtained from a $D(m)$-based statistic is also accessible through interval-based power spectral density measures. In Section 3 we explicitly show that the two measures are comparable in their abilities to discriminate patients with heart failure from normal subjects.

#### 2.5.4. Detrended Fluctuation Analysis [DFA(m)]

Detrended fluctuation analysis (DFA) was originally proposed as a technique for quantifying the nature of long-range correlations in a time series [51–53]. As implied by its name, it was conceived as a method for detrending variability in a sequence of events. The DFA computation involves the calculation of the summed series

$$y(k) = \sum_{i=1}^{k} \{\tau_i - \mathrm{E}[\tau]\} \tag{21}$$

where $y(k)$ is the $k$th value of the summed series and $\mathrm{E}[\tau]$ denotes the average over the set $\{\tau_i\}$. The summed series is then divided into segments of length $m$ and a least-squares fit is performed on each of the data segments, providing the trends for the individual segments. Detrending is carried out by subtracting the local trend $y_m(k)$ in each segment. The root-mean-square fluctuation of the resulting series is then

$$F(m) = \left\{ \frac{1}{L} \sum_{k=1}^{L} [y(k) - y_m(k)]^2 \right\}^{1/2} \tag{22}$$

The functional dependence of $F(m)$ is obtained by evaluations over all segment sizes $m$.

Although detrended fluctuation analysis was originally proposed as a method for estimating the scale-independent fractal exponent of a time series [51], as discussed in Section 2.6.1, we consider its merits as a scale-*dependent* measure. As will be demonstrated in Section 3, a plot of $F(m)$ versus $m$ reveals a window of separation between patients with congestive heart failure and normal subjects over a limited range of scales, much as that provided by the other scale-dependent measures discussed in this section. Because DFA is an ad hoc measure that involves nonlinear computations, it is difficult to relate it to other scale-dependent measures in the spirit of Eq. 20. Furthermore, as will become clear in Section 3.6.2, relative to other measures DFA is highly time intensive from a computational point of view.

## 2.6. Scale-Independent Measures

Scale-independent measures are designed to estimate fractal exponents that characterize scaling behavior in one or more statistics of a sequence of events, as discussed in Section 2.1.2. The canonical example of a scale-independent measure in HRV is the fractal exponent $\alpha_S$ of the interbeat-interval power spectrum, associated with the decreasing power-law form of the spectrum at sufficiently low frequencies $f$: $S_\tau(f) \propto f^{-\alpha_S}$ [3, 30, 54]. Other scale-independent measures have been examined by us [7, 8, 16, 17, 46] and by others [6, 51, 55, 56] in connection with HRV analysis. For exponent values encountered in HRV and infinite data length, all measures should in principle lead to a unique fractal exponent. In practice, however, finite data length and other factors introduce bias and variance, so that different measures give rise to different results. The performance of scale-independent measures has been compared with that of scale-dependent measures for assessing cardiac dysfunction [46, 47].

### *2.6.1. Detrended-Fluctuation-Analysis Power-Law Exponent ($\alpha_D$)*

The DFA technique, and its use as a scale-dependent measure, has been described in Section 2.5.4. A number of studies [51, 53, 56] have considered the extraction of power-law exponents from DFA and their use in HRV. As originally proposed [51], $\log[F(m)]$ is plotted against $\log(m)$ and scaling exponents are obtained by fitting straight lines to sections of the resulting curve—the exponents are simply the slopes of the linearly fitted segments on this doubly logarithmic plot. The relationship between the scaling exponents has been proposed as a means of differentiating normal from pathological subjects [51, 55, 56].

### *2.6.2. Wavelet-Transform Power-Law Exponent ($\alpha_W$)*

The use of the wavelet transform as a scale-dependent measure was considered in Section 2.5.2. It was pointed out that a scale-independent measure also emerges from the wavelet-transform standard deviation. The wavelet-transform fractal exponent $\alpha_W$ is estimated directly from the wavelet transform as twice the slope of the curve $\log[\sigma_{\text{wav}}(m)]$ versus $\log(m)$, measured at large values of $m$ [46]. The factor of 2 is present because the fractal exponent is related to variance rather than to standard deviation.

### *2.6.3. Periodogram Power-Law Exponent ($\alpha_S$)*

The description of the periodogram as a scale-dependent measure was provided in Section 2.2. The periodogram fractal exponent $\alpha_S$ [17, 54] is obtained as the least-square-fit slope of the spectrum when plotted on doubly logarithmic coordinates. The range of low frequencies over which the slope is estimated stretches between $10/L$ and $100/L$ where $L$ is the length of the data set [17].

#### *2.6.4. Allan-Factor Power-Law Exponent ($\alpha_A$)*

The use of the Allan factor as a scale-dependent measure was considered in Section 2.5.1. The Allan factor fractal exponent $\alpha_A$ [8, 17] is obtained by determining the slope of the best-fitting straight line, at large values of $T$, to the Allan factor curve (Eq. 9) plotted on doubly logarithmic coordinates. Estimates of $\alpha$ obtained from the Allan factor can range up to a value of 3 [32]. The use of wavelets more complex than the Haar enables an increased range of fractal exponents to be accessed, at the cost of a reduction in the range of counting time over which the wavelet Allan factor varies as $T^{\alpha_A}$. In general, for a particular wavelet with regularity (number of vanishing moments) $R$, fractal exponents $\alpha < 2R + 1$ can be reliably estimated [36, 38]. For the Haar basis, $R = 1$, whereas all other wavelet bases have $R > 1$. A wavelet Allan factor making use of bases other than the Haar is therefore required for fractal-rate stochastic point processes for which $\alpha \geq 3$. For processes with $\alpha < 3$, however, the Allan factor appears to be the best choice [36, 38].

#### *2.6.5. Rescaled-Range-Analysis Power-Law Exponent ($\alpha_R$)*

Rescaled range analysis [19, 57–59] provides information about correlations among blocks of interevent intervals. For a block of $k$ interevent intervals, the difference between each interval and the mean interevent interval is obtained and successively added to a cumulative sum. The normalized range $R(k)$ is the difference between the maximum and minimum values that the cumulative sum attains, divided by the standard deviation of the interval size. $R(k)$ is plotted against $k$. Information about the nature and the degree of correlation in the process is obtained by fitting $R(k)$ to the function $k^H$, where $H$ is the so-called Hurst exponent [57]. For $H > 0.5$ positive correlation exists among the intervals, whereas $H < 0.5$ indicates the presence of negative correlation; $H = 0.5$ obtains for intervals with no correlation. Renewal processes yield $H = 0.5$. For negatively correlated intervals, an interval that is larger than the mean tends, on average, to be preceded or followed by one smaller than the mean.

The Hurst exponent $H$ is generally assumed to be well suited to processes that exhibit long-term correlation or have a large variance [19, 57–59], but there are limits to its robustness because it exhibits large systematic errors and highly variable estimates for some fractal sequences [6, 60, 61]. Nevertheless, it provides a useful indication of correlation in a sequence of events arising from the ordering of the interevent intervals alone.

The exponent $\alpha_R$ is ambiguously related to the Hurst exponent $H$, because some authors have used the quantity $H$ to index fractal Gaussian noise whereas others have used the same value of $H$ to index the integral of fractal Gaussian noise (which is fractional Brownian motion). The relationship between the quantities is $\alpha_R = 2H - 1$ for fractal Gaussian noise and $\alpha_R = 2H + 1$ for fractal Brownian motion. In the context of this work, the former relationship holds.

## 2.7. Estimating the Performance of a Measure

We have, to this point, outlined a variety of candidate measures for use in HRV analysis. The task now is to determine the relative value of these measures from a clinical perspective. We achieve this by turning to estimation theory [62].

A statistical measure obtained from a finite set of actual data is characterized by an estimator. The fidelity with which the estimator can approximate the true value of the measure is determined by its bias and variance. The bias is the deviation of the expected value of the estimator from its true underlying value (assuming that this exists), whereas the variance indicates the expected deviation from the mean. An ideal estimator has zero bias and zero variance, but this is not achievable with a finite set of data. For any unbiased estimator the Cramér–Rao bound provides a lower bound for the estimator variance; measures that achieve the Cramér–Rao bound are called efficient estimators. The estimator bias and variance play a role in establishing the overall statistical significance of conclusions based on the value returned by the estimator.

### 2.7.1. Statistical Significance: p, d′, h, and d

The concept of statistical significance extends the basic properties of bias and variance [63]. It provides a probabilistic interpretation of how likely it is that a particular value of the estimator might occur by chance alone, arising from both random fluctuations in the data and the inherent properties of the estimator itself.

A frequently used standard of statistical significance is the $p$ value, the calculation of which almost always implicitly assumes a Gaussian-distributed dependent variable. A lower $p$ value indicates greater statistical significance, and a measure is said to be statistically significant to a value of $p_0$ when $p < p_0$. The distributions obtained from HRV measurements are generally not Gaussian, however, so the usual method for estimating the $p$ value cannot be used with confidence. Because other methods for estimating the $p$ value require more data than is available, we do not consider this quantity further.

Another often-used distribution-dependent standard is the $d'$ value. It serves to indicate the degree of separation between two distributions and has been widely used in signal detection theory and psychophysics, where the two distributions represent noise and signal plus noise [64]. The most common definition of $d'$ is the difference in the means of two Gaussian distributions divided by their common standard deviation. Two closely related distribution-dependent cousins of $d'$ are the detection distance $h$, defined as the difference in the means of the two Gaussian distributions divided by the square root of the sum of their variances, and the detection distance $d$, defined as the difference in the means of the two Gaussian distributions divided by the sum of their standard deviations. Larger values of $d'$, $h$, and $d$ indicate improved separation between the two distributions and therefore reduced error in assigning an outcome to one or the other of the hypotheses.

Because HRV measures are intended to provide diagnostic information in a clinical setting and do not return Gaussian statistics, the evaluation of their performance using distribution-independent means is preferred. Two techniques for achieving this, positive and negative predictive values and receiver operating characteristic (ROC) analysis, are described next. Neither requires knowledge of the statistical distribution of the measured quantities and both are useful.

#### 2.7.2. Positive and Negative Predictive Values

The performance of the various HRV measures discussed previously can be effectively compared using positive predictive values and negative predictive values, the proportion of correct positive and negative identifications, respectively. When there is no false positive (or negative) detection, the predictive value is equal to unity and there is perfect assignment. Furthermore, when the individual values of a measure for normal subjects and patients do not overlap, the predictive value curves are typically monotonic, either increasing or decreasing, with the threshold. A detailed discussion of positive and negative predictive values is provided in Section 3.4.

#### 2.7.3. Receiver Operating Characteristic (ROC) Analysis

ROC analysis [62, 64] is an objective and highly effective technique for assessing the performance of a measure when it is used in binary hypothesis testing. This format provides that a data sample be assigned to one of two hypotheses or classes (e.g., pathologic or normal) depending on the value of some measured statistic relative to a threshold value. The efficacy of a measure is then judged on the basis of its sensitivity (the proportion of pathologic patients correctly identified) and its specificity (the proportion of normal subjects correctly identified). The ROC curve is a graphical presentation of sensitivity versus 1 − specificity as a threshold parameter is swept. Note that sensitivity and specificity relate to the status of the patients (pathologic and normal), whereas predictive values relate to the status of the identifications (positive and negative).

The area under the ROC curve serves as a well-established index of diagnostic accuracy [64]; the maximum value of 1.0 corresponds to perfect assignment (unity sensitivity for all values of specificity), whereas a value of 0.5 arises from assignment to a class by pure chance (areas $< 0.5$ arise when the sense of comparison is reversed). ROC analysis can be used to choose the best of a host of different candidate diagnostic measures by comparing their ROC areas, or to establish for a single measure the tradeoff between data length and misidentifications (misses and false positives) by examining ROC area as a function of record length. A minimum record length can then be specified to achieve acceptable classification accuracy.

As pointed out before, ROC analysis relies on no implicit assumptions about the statistical nature of the data set [62], so it is generally more suitable [47] for analyzing non-Gaussian time series than are measures of statistical significance such as $p$ value, $h$, and $d$. Another important feature of ROC curves is that they are insensitive to the units employed (e.g., spectral magnitude, magnitude squared, or log magnitude); ROC curves for a measure $M$ are identical to those for any monotonic transformation thereof such as $M^x$ or $\log(M)$. In contrast, the values of $d'$, $h$, and $d$ are generally modified by such transformation, as demonstrated in Section 3.5.1.

## 3. DISCRIMINATING HEART-FAILURE PATIENTS FROM NORMAL SUBJECTS

We now proceed to examine the relative merits of various HRV measures for discriminating patients with congestive heart failure (CHF) from normal subjects.

Specifically, we contrast and compare the performance of the 16 measures set forth in Section 2: VLF, LF, HF, LF/HF, pNN50, SDANN, SDNN ($\sigma_{\text{int}}$), $A(T)$, $\sigma_{\text{wav}}(m)$, $S_\tau(f)$, DFA($m$), $\alpha_D$, $\alpha_W$, $\alpha_S$, $\alpha_A$, and $\alpha_R$.

After discussing the selection of an appropriate scale $m$, we use predictive value plots and ROC curves to select a particular subset of HRV markers that appears to be promising for discerning the presence of heart failure in a patient population.

### 3.1. Database

The RR recordings analyzed in this section were drawn from the Beth-Israel Hospital (Boston, MA) Heart-Failure Database, which includes 12 records from normal subjects (age 29–64 years, mean 44 years) and 12 records from patients with severe CHF (age 22–71 years, mean 56 years). The recordings were made with a Holter monitor digitized at a fixed value of 250 samples/s. Also included in this database are 3 RR records for CHF patients who also suffered from atrial fibrillation (AF); these records are analyzed as a separate class. All records contain both diurnal and nocturnal segments. The data were originally provided to us in 1992 by D. Rigney and A. L. Goldberger.

A detailed characterization of each of the records is presented in Table 1 of Ref. 8; some statistical details are provided in Table A1. Of the 27 recordings, the shortest contained $L_{\max} = 75{,}821$ RR intervals; the remaining 26 recordings were truncated to this length before calculating the 16 HRV measures.

### 3.2. Selecting a Scale

A value for the scale $m$ that suitably discriminates heart-failure patients from normal subjects can be inferred from our recent wavelet studies of the CHF and normal records from the same database as discussed in Section 3.1 [46, 47]. With the help of the wavelet-transform standard deviation $\sigma_{\text{wav}}(m)$ discussed in detail in Section 2.5.2, we discovered a critical scale window near $m = 32$ interbeat intervals over which the normal subjects exhibited greater fluctuations than those afflicted with heart failure. For these particular long data sets, we found that it was possible to discriminate perfectly between the two groups [46, 47, 48].

The results are displayed in Figure 4, where $\sigma_{\text{wav}}(m)$ is plotted against wavelet scale $m$ for the 12 normal subjects (+), the 12 CHF patients without atrial fibrillation ($\times$), and the 3 CHF patients with atrial fibrillation ($\triangle$), using Haar-wavelet analysis. The AF patients ($\triangle$) typically fell near the high end of the non-AF patients ($\times$), indicating greater RR fluctuations, particularly at small scales. This results from the presence of nonsinus beats. Nevertheless it is evident from Figure 4 that the wavelet measure $\sigma_{\text{wav}}$ completely separates the normal subjects from the heart-failure patients (both without and with AF) at scales of 16 and 32 heartbeat intervals, as reported in Ref. 46. One can do no better. This conclusion persists for a broad range of analyzing wavelets, from Daubechies 2-tap (Haar) to Daubechies 20-tap [46].

The importance of this scale window has been confirmed in an Israeli–Danish study of diabetic patients who had not yet developed clinical signs of cardiovascular disease [50]. The reduction in the value of the wavelet-transform standard deviation $\sigma_{\text{wav}}(32)$ that leads to the scale window occurs not only for CHF (both with and without AF) and diabetic patients but also for heart-transplant patients [48, 50] and also in records preceding sudden cardiac death [46, 48]. The depression of $\sigma_{\text{wav}}(32)$ at these scales is prob-

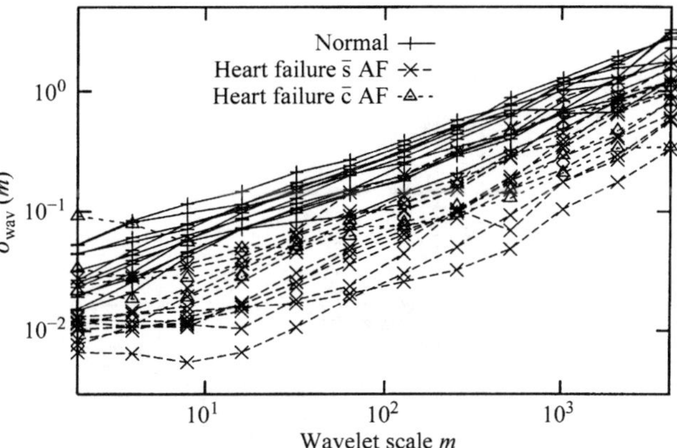


**Figure 4** Haar-wavelet-transform standard deviation $\sigma_{wav}(m)$ vs scale $m$ for the 12 normal subjects (+), 12 CHF patients without (s̄) atrial fibrillation (×), and 3 CHF patients with (c̄) atrial fibrillation (△). Each data set comprises the first 75,821 RR intervals of a recording drawn from the Beth-Israel Hospital heart-failure database. Complete separation of the normal subjects and heart-failure patients is achieved at scales $m = 16$ and 32 interbeat intervals.

ably associated with the impairment of autonomic nervous system function. Baroreflex modulations of the sympathetic or parasympathetic tone typically lie in the range 0.04–0.09 cycles/s (11–25 s), which corresponds to the scale where $\sigma_{wav}(m)$ is reduced.

These studies, in conjunction with our earlier investigations that revealed a similar critical scale window in the *counting* statistics of the heartbeat [8, 35] (as opposed to the time-*interval* statistics under discussion), lead to the recognition that scales in the vicinity of $m = 32$ enjoy a special status. The measures that depend on a particular scale are therefore evaluated at $m = 32$ and $f = 1/32$ in the expectation that these values maximize discriminability in the more usual situation in which the two classes of data cannot be fully separated.

### 3.3. Individual Value Plots

Having devised a suitable scale value $m$, we now proceed to evaluate the 16 measures for all 27 normal and CHF data sets, each comprising 75821 RR intervals. The results are presented in Figure 5, where each of the 16 panels represents a different measure. For each measure the individual values for normal subjects (+), CHF patients without AF (×), and CHF patients with AF (△) comprise the left three columns, respectively. Values in the right four columns correspond to other cardiovascular pathologies and will be discussed in Section 4.

To illustrate how particular measures succeed (or fail to succeed) in distinguishing between CHF patients and normal subjects, we focus in detail on two measures: VLF power and pNN50. For this particular collection of patients and record lengths, the normal subjects all exhibit larger values of VLF power than the CHF patients; indeed a horizontal line drawn at VLF = 0.000600 completely separates the two classes. On the other hand, for pNN50, although the normals still have larger values on average, there is a region of overlap of CHF patients and normal subjects near 0.05, indicating that the two classes of patients cannot be entirely separated using this measure. Thus, for the full data set, comprising 75,821 RR intervals, VLF succeeds in completely distinguishing CHF patients and normal subjects whereas pNN50 does not.

Examining all 16 panels, we find that 6 measures manage to completely separate the normal subjects (first column) from the heart-failure patients (second and third columns) while the remaining 10 fail to do so. The six successful measures are highlighted by boldface font in Figure 5: **VLF, LF, $A$(10), $\sigma_{wav}$(32), $S_\tau$(1/32)**, and **DFA(32)**.

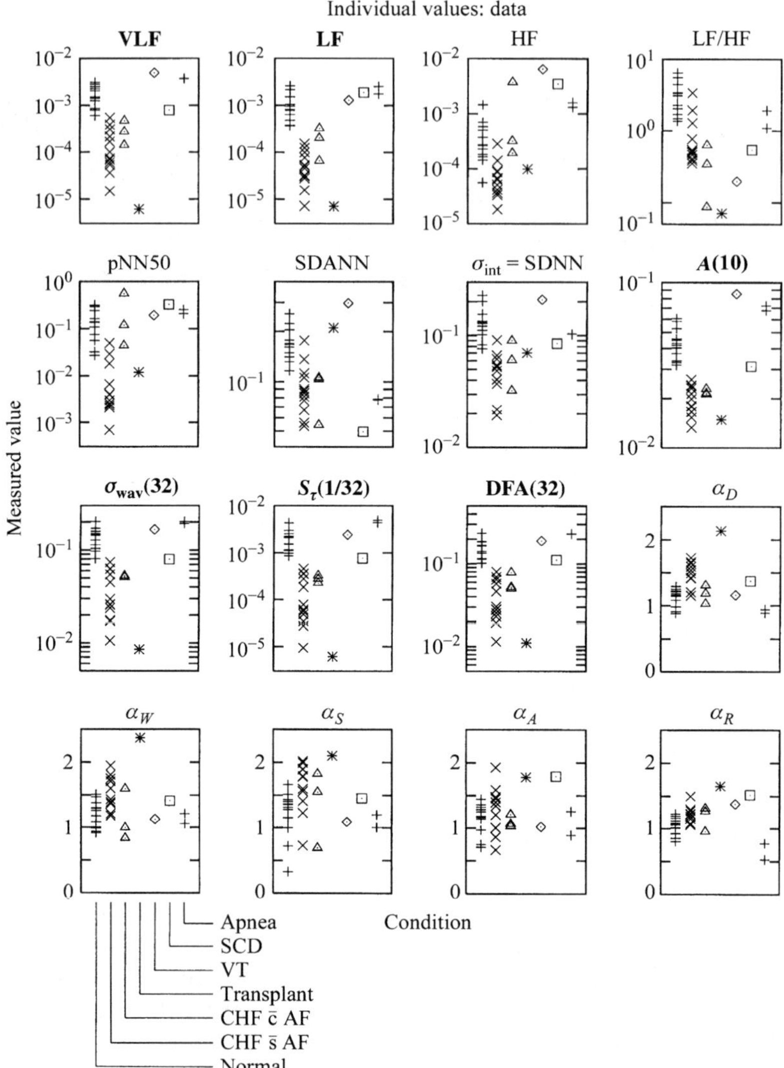


**Figure 5** Individual value plots (data) for the 16 measures. Each panel corresponds to a different HRV measure. The seven columns in each panel, from left to right, comprise data for (1) 12 normal subjects (+), (2) 12 CHF patients s̄ AF (×), (3) 3 CHF patients c̄ AF (△), (4) 1 heart-transplant patient (∗), (5) 1 ventricular-tachycardia patient (◇), (6) 1 sudden-cardiac-death patient (□), and (7) 2 sleep-apnea patients (+). Each data set comprises 75,821 RR intervals except for the two sleep-apnea data sets, which comprise 16,874 and 15,751 RR intervals, respectively. The six measures highlighted in boldface font succeed in completely separating normal subjects and CHF patients (s̄ and c̄ atrial fibrillation) **VLF**, **LF**, ***A*(10)**, $\boldsymbol{\sigma_{wav}(32)}$, $\boldsymbol{S_\tau(1/32)}$, and **DFA(32)**.

### 3.4. Predictive Value Plots

How can the ability of a measure to separate two classes of subjects be quantified? Returning to the VLF panel in Figure 5, we place a threshold level $\theta$ at an arbitrary position on the ordinate and consider only the leftmost two columns: normal subjects and heart-failure patients who do not suffer from atrial fibrillation. We then classify all subjects for whom the VLF values are $< \theta$ as CHF patients (positive) and for those for whom the VLF values are $> \theta$ as normal (negative). (Measures that yield smaller results for normal patients, on average, obey a reversed decision criterion.)

If a subject labeled as a CHF patient is indeed so afflicted, then this situation is referred to as a true positive ($P_T$); a normal subject erroneously labeled as a CHF patient is referred to as a false positive ($P_F$). We define negative outcomes that are true ($N_T$) and false ($N_F$) in an analogous manner. As pointed out in Section 2.7.2, the positive predictive value $V_P = P_T/(P_T + P_F)$ and negative predictive value $V_N = N_T/(N_T + N_F)$ represent the proportion of positives and negatives, respectively, that are correctly identified. This determination is carried out for many values of the threshold $\theta$.

Figure 6 shows the positive (solid curves) and negative (dotted curves) predictive values for all 16 measures, plotted against the threshold $\theta$, each in its own panel. These curves are constructed using the 12 normal and 12 heart-failure (without AF) records that constitute the CHF database discussed in Section 3.1. For the VLF measure, both predictive values are simultaneously unity in the immediate vicinity of $\theta = 0.000600$. This occurs because $P_F$ and $N_F$ are both zero at this particular value of $\theta$ and reconfirms that the two classes of data separate perfectly in the VLF panel of Figure 5 at this threshold.

For threshold values outside the range $0.000544 < \theta < 0.000603$, some of the patients will be incorrectly identified by the VLF measure. If we set $\theta = 0.000100$ for example, 6 of the 12 CHF patients will be incorrectly identified as normal subjects, which is confirmed by examining the VLF panel in Figure 5. This yields $V_N = N_T/(N_T + N_F) = 12/(12 + 6) \doteq 0.67 < 1$, which is the magnitude of the negative predictive value (dotted curve) in the VLF panel in Figure 6. At this value of the threshold ($\theta = 0.000100$), the positive predictive value remains unity because $P_F$ remains zero.

The pNN50 panel in Figure 5, in contrast, reveals a range of overlap in the individual values of the normal subjects and CHF patients. Consequently, as $\theta$ increases into the overlap region, $V_P$ decreases below unity and this happens before $V_N$ attains unity value. Thus, there is no threshold value, or range of threshold values, for which the positive and negative predictive values in Figure 6 are both unity. The best threshold for this measure lies in the range $0.026 < \theta < 0.050$, with the choice depending on the relative benefit of being able to predict accurately the presence or absence of CHF in a patient.

There are six measures in Figure 6 (indicated in boldface font) for which the positive and negative predictive values are both unity over the same range of threshold values. These measures are, of course, the same six measures for which the normal subjects and heart-failure patients fall into disjoint sets in Figure 5.

### 3.5. ROC Curves

Two other important clinically relevant quantities that depend on the threshold $\theta$ are the sensitivity, the proportion of heart-failure patients that are properly identified $[P_T/(P_T + N_F)]$, and the specificity, the proportion of normal subjects that are properly

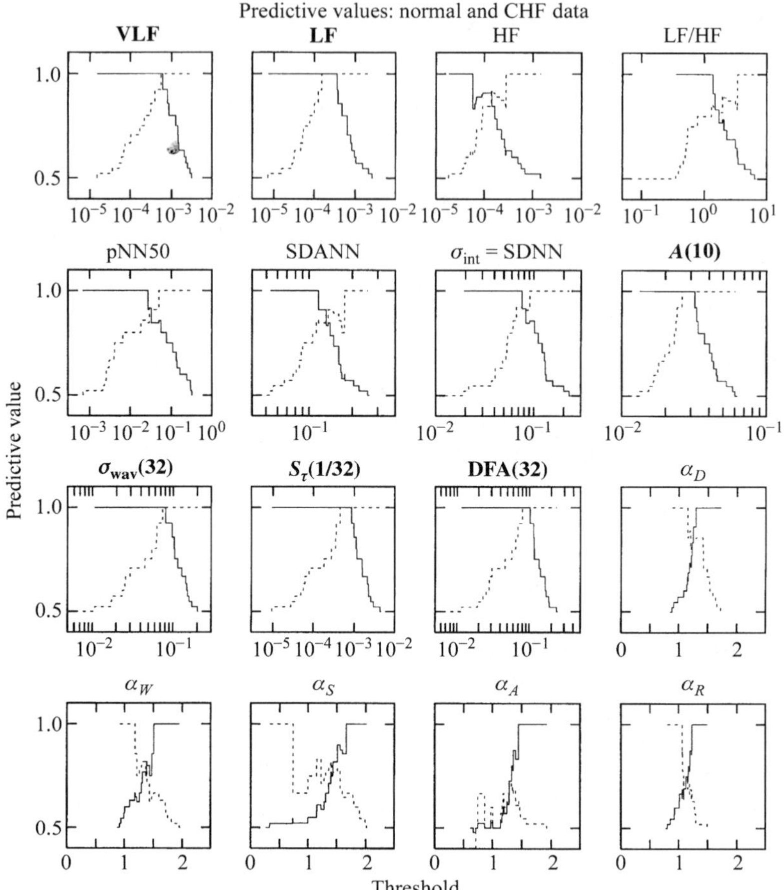


**Figure 6** Positive (solid curves) and negative (dotted curves) predictive values for all 16 HRV measures plotted against the threshold $\theta$, each in its own panel. These curves are constructed using the 12 normal and 12 heart-failure ($\bar{s}$ AF) records drawn from the CHF database, each of which has been truncated to 75,821 RR intervals. The six measures highlighted in boldface font exhibit threshold regions for which both the positive and negative predictive values are unity: **VLF**, **LF**, ***A*(10)**, $\boldsymbol{\sigma_{wav}}$**(32)**, $\boldsymbol{S_\tau}$**(1/32)**, and **DFA(32)**. This indicates that the normal subjects and CHF ($\bar{s}$ AF) patients can be completely distinguished by these six measures, in accordance with the results established in Figure 5.

identified $[N_T/(N_T + P_F)]$. As pointed out in Sections 2.7.2 and 2.7.3, sensitivity and specificity are related to patient status (pathologic and normal, respectively), whereas the positive and negative predictive values are related to identification status (positive and negative, respectively). Sensitivity and specificity are both monotonic functions of the threshold, but this is not generally true for the predictive values. The monotonicity property is salutary in that it facilitates the use of a parametric plot which permits these quantities to be represented in compact form. A plot of sensitivity versus 1 − specificity, traced out various values of the threshold $\theta$, forms the ROC curve (see Section 2.7.3).

ROC curves are presented in Figure 7 for all 16 measures, again using the same 12 normal and 12 heart-failure (without AF) records that constitute the CHF database discussed in Section 3.1. Because of the complete separation between the two classes of patients (leftmost two columns of the VLF panel in Figure 5) near $\theta = 0.000600$, the VLF ROC curve in Figure 7 simultaneously achieves unity (100%) sensitivity and unity (100%) specificity (the point at upper left corner of the ROC curve). For the pNN50 statistic, in contrast, the overlap evident in Figure 5 prevents this, so that the upper left corner of the pNN50 ROC curve in Figure 7 instead reveals smaller simultaneous values of sensitivity and specificity.

Six measures in Figure 7 simultaneously exhibit unity sensitivity and specificity; these are indicated by boldface font and have ROC curves that are perfectly square. They are clearly the same measures for which the normal subjects and heart-failure

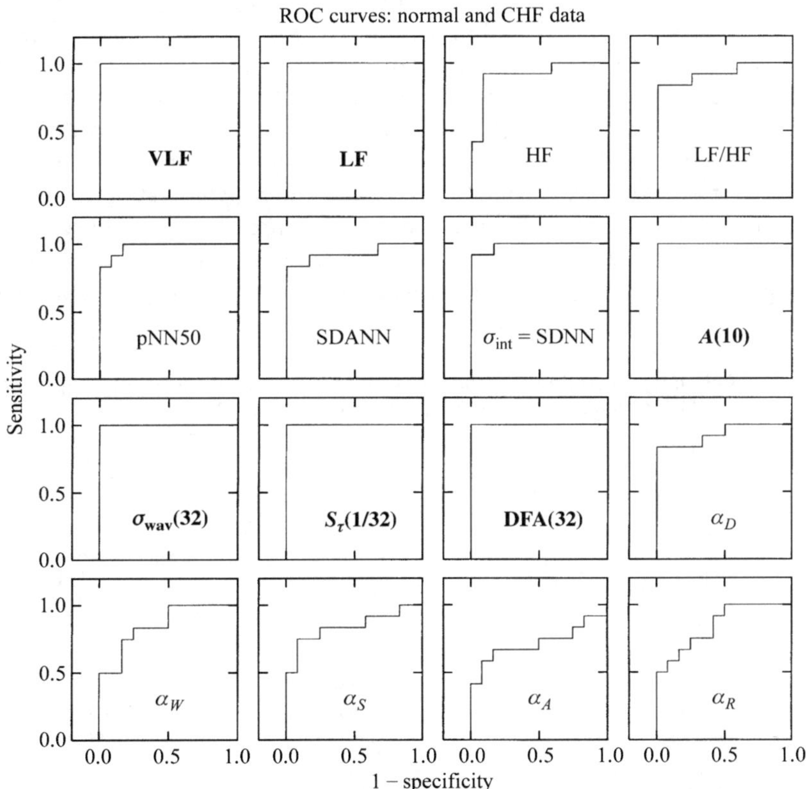


**Figure 7** ROC curves (sensitivity vs. 1 − specificity) for all 16 HRV measures, each in its own panel. These curves are constructed using the 12 normal and 12 heart-failure ($\bar{s}$ AF) records drawn from the CHF database, each of which has been truncated to 75,821 RR intervals. The six measures highlighted in boldface font simultaneously achieve 100% sensitivity and 100% specificity so that the ROC curve is perfectly square: **VLF**, **LF**, ***A*(10)**, $\boldsymbol{\sigma_{wav}}$**(32)**, $\boldsymbol{S_\tau}$**(1/32)**, and **DFA(32)**. This indicates that the normal subjects and CHF ($\bar{s}$ AF) patients can be completely distinguished by these six measures, in accordance with the results established in Figures 5 and 6.

patients fall into disjoint sets in Figure 5 and for which simultaneous positive and negative predictive values of unity are observed in Figure 6.

#### 3.5.1. Comparison with Detection-Distance Measures

For didactic purposes we compare the ROC results presented immediately above with those obtained using detection-distance analysis. As we indicated in Section 2.7.1, care must be exercised when using these techniques for anything other than Gaussian-distributed quantities. The calculations were carried out using the same 12 normal-subject and 12 CHF-patient records, each comprising $L_{max} = 75{,}821$ intervals. In Table 1 we provide the detection distances $h$ and $d$, in order of descending value of $h$, for all 16 measures. Large values are best because they indicate that the two distributions are well separated.

Five of the six measures that entirely separate the CHF patients and normal subjects using ROC analysis fall in the top five positions in Table 1. The sixth measure, LF, falls in the ninth position. This confirms that detection-distance analysis applied to these long recordings provides results that qualitatively agree with those obtained using ROC analysis. However, detection-distance analysis does not provide any indication of how many (or indeed whether any) of the measures at the top of the list completely separate the two classes of patients, nor does it provide estimates of sensitivity and specificity. Moreover, the rankings according to $d$ differ form those according to $h$.

Finally, the detection distance for a particular measure, as well as the relative ranking of the measure, depends on what appear to be insignificant details about the

**TABLE 1** Detection Distances $h$ and $d$ for All 16 HRV Measures Applied to the 12 Normal and 12 CHF Records of Length $L_{max}$[a]

| Measure | $h$ | $d$ |
|---|---|---|
| DFA(32) | 2.48253 | 1.81831 |
| $\sigma_{wav}(32)$ | 2.33614 | 1.70153 |
| $A(10)$ | 2.32522 | 1.77482 |
| VLF | 1.84285 | 1.56551 |
| $S_\tau(1/32)$ | 1.77422 | 1.55200 |
| $\sigma_{int}$ | 1.74750 | 1.32475 |
| $\sigma^2_{wav}(32)$ | 1.71343 | 1.47165 |
| $\alpha_D$ | 1.64883 | 1.17679 |
| SDANN | 1.46943 | 1.04079 |
| LF | 1.36580 | 1.28686 |
| pNN50 | 1.36476 | 1.20896 |
| LF/HF | 1.24507 | 0.91444 |
| $\alpha_W$ | 1.09916 | 0.77800 |
| $\alpha_R$ | 1.02367 | 0.72463 |
| HF | 0.85361 | 0.73077 |
| $\alpha_S$ | 0.82125 | 0.58071 |
| $\alpha_A$ | 0.38778 | 0.27895 |

[a]The measures are listed in order of descending value of $h$. The rankings according to $d$ differ from those according to $h$. The wavelet-transform standard deviation $\sigma_{wav}(32)$ and variance $\sigma^2_{wav}(32)$, although related by a simple monotonic transformation, yield different values of $h$ and have different rankings.

specific form in which the measure is cast. For example the $h$ values for $\sigma_{\mathrm{wav}}(32)$ and its square $\sigma^2_{\mathrm{wav}}(32)$ are substantially different, and so are their rankings in Table 1. As discussed in Section 2.7.3, ROC analysis is invariant to monotonic transformations of the measure and therefore does not suffer from this disadvantage.

### 3.5.2. ROC-Area Curves

Perfectly square ROC curves, associated with the group of six boldface-labeled measures in Figures 5–8, exhibit unit area. These ROCs represent 100% sensitivity for all values of specificity, indicating that every patient is properly assigned to the appropriate status: heart failure or normal. Although the perfect separation achieved by these six measures endorses them as useful diagnostic statistics, the results of most studies are seldom so clear-cut. ROC area will surely decrease as increasing numbers of out-of-sample records are added to the database, because increased population size means increased variability [46].

ROC area also decreases with diminishing data length; shorter records yield less information about patient condition and these patients are therefore more likely to be misclassified [47, 48]. The ROC curves in Figure 7 have been constructed from Holter-monitor records that contain many hours of data (75,821 RR intervals). It would be useful in a clinical setting to be able to draw inferences from HRV measures recorded over shorter times, say minutes rather than hours. It is therefore important to examine the performance of the 16 HRV measures as data length is decreased. As indicated in Section 2.7.3, ROC analysis provides an ideal method for carrying out this task [47, 48].

In Figure 8 we present ROC-area curves as a function of the number of RR intervals (data length) $L$ analyzed ($64 \leq L \leq 75{,}821$). The ROC areas for the full-length records ($L_{\mathrm{max}} = 75{,}821$), which correspond to the areas under the ROC curves presented in Figure 7, are the rightmost points in the ROC-area curves shown in Figure 8. Results for shorter records were obtained by dividing the 12 normal and 12 heart-failure (without AF) records that comprise the CHF database (Section 3.1) into smaller segments of length $L$. The area under the ROC curve for that data length $L$ was computed for the first such segment for all 16 measures, and then for the second segment, and so on, for all segments of length $L$ (remainders of length $< L$ of the original files were not used for the ROC-area calculations). From the $L_{\mathrm{max}}/L$ values of the ROC area, the mean and standard deviation were computed and are plotted in Figure 8. The lengths $L$ examined ranged from $L = 2^6 = 64$ to $L = 2^{16} = 65{,}536$ RR intervals, in powers of two, in addition to the entire record of $L_{\mathrm{max}} = 75{,}821$ intervals.

To illustrate the information provided by these curves, we direct our attention to the VLF and pNN50 panels in Figure 8. For the full-length records the rightmost point in the VLF panel reveals unit area while that for pNN50 lies somewhat lower, as expected from the corresponding ROC curves in Figure 7. VLF clearly outperforms pNN50. As the data length analyzed decreases, so too do the ROC areas for both measures while their variances increase, also as expected. However, when the data length dips to 256 or fewer RR intervals, the performance of the two measures reverses so that pNN50 outperforms VLF. There is an important point to be drawn from this example. Not only does the performance of a measure depend on data length, but so too does the relative performance of different measures.

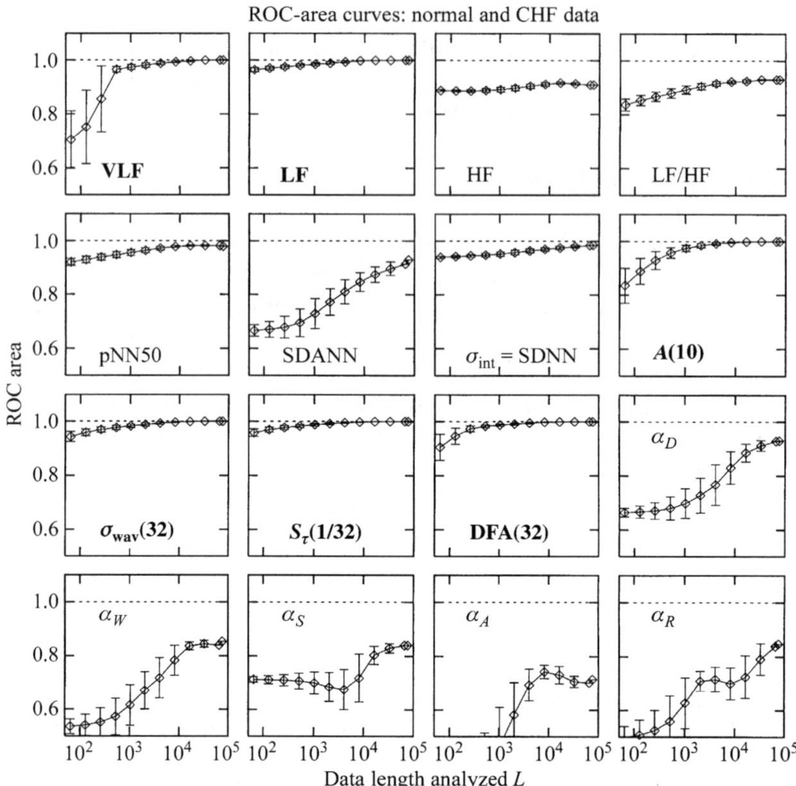


**Figure 8** Diagnostic accuracy (area under ROC curve) as a function of record length analyzed $L$ (number of RR intervals) for the 16 HRV measures (mean $\pm$ 1 SD). An area of unity corresponds to perfect separability of the two classes of patients. The six measures highlighted in boldface font—**VLF, LF, $A(10)$, $\sigma_{wav}(32)$, $S_\tau(1/32)$, and DFA(32)**—provide such perfect separability at the longest record lengths, in accordance with the results in Figure 7. As the record length decreases, performance degrades at a different rate for each measure. The five scale-independent measures, $\alpha_D$, $\alpha_W$, $\alpha_S$, $\alpha_A$, and $\alpha_R$, perform poorly at all data lengths.

## 3.6. Comparing the Measures for Various Data Lengths

Based on their overall ability to distinguish between CHF patients and normal subjects over a range of data lengths, the 16 measures shown in Figure 8 divide roughly into three classes. The six measures that fall in the first class, VLF, LF, $A(10)$, $\sigma_{wav}(32)$, $S_\tau(1/32)$, and DFA(32), succeed in completely separating the two classes of patients for data lengths down to $L = 2^{15} = 32{,}768$ RR intervals. These six measures share a dependence on a single scale, or small range of scales, near 32 heartbeat intervals. For this collection of data sets, this scale appears to yield the best performance. Members of this class outperform the other 10 measures at nearly all data lengths. Apparently, the scale value itself is far more important than the measure used to evaluate it.

The second class, consisting of HF, the ratio LF/HF, pNN50, and $\sigma_{int}$, fail to achieve complete separation for any data size examined. Nevertheless, the members of this class are not devoid of value in separating CHF patients from normal subjects. Interestingly, all but LF/HF provide better results than $A(10)$, a member of the first class, for the shortest data lengths. Results for these four measures varied relatively little with data size, thus exhibiting a form of robustness.

Members of the third class, consisting of SDANN and the five scale-independent measures $\alpha_{[\cdot]}$, exhibit poor performance at all data lengths. These six measures require long sequences of RR intervals to make available the long-term fluctuations required for accurate estimation of the fractal exponent. Data lengths $L < 5000$ RR intervals lead to large variance and (negative) bias and are not likely to be meaningful. As an example of the kind of peculiarity that can emerge when attempting to apply scale-independent measures to short records, the $\alpha_A$ ROC area decreases below 0.5 when the data size falls below 2048 intervals (reversing the sense of the comparison only for these data sizes increases the ROC area, although not above 0.7; however this clearly violates the spirit of the method). SDANN requires several 5-minute segments to determine the standard deviation accurately.

### 3.6.1. Scale-Independent versus Scale-Dependent Measures

As indicated in the previous subsection, all five scale-independent measures ($\alpha_D$, $\alpha_W$, $\alpha_S$, $\alpha_A$, and $\alpha_R$) perform poorly at all data lengths. These fractal-exponent estimators return widely differing results as is plainly evident in Figure 5. This suggests that there is little merit in the concept of a single exponent for characterizing the human heartbeat sequence, no less a "universal" one as some have proposed [51, 55, 56].

A variation on this theme is the possibility that pairs of fractal exponents can provide a useful HRV measure. At small scales $m$, Figure 4 reveals that heart-failure patients exhibit smaller values of the wavelet-transform standard-deviation slope than do normal subjects. Following Peng et al. [51], who constructed a measure based on differences of DFA scaling exponents in different scaling regions in an attempt to discriminate CHF patients from normal subjects, Thurner et al. [46] constructed a measure based on differences in the wavelet-transform standard-deviation slope at different scales. However, the outcome was found to be unsatisfactory when compared with other available measures; we concluded the same about the results obtained by Peng et al. [51]. Using ROC analysis, as described in Section 2.7.3, we determined that the ROC area for the measure described by Thurner et al. [46] was sufficiently small (0.917 for $m = 4$, 16, and 256) that we abandoned this construct.

Four of the techniques we have discussed in this chapter (spectral, wavelet, detrended fluctuation analysis, and Allan factor) yield both scale-independent and scale-dependent measures and therefore afford us the opportunity of directly comparing these two classes of measures in individual calculations: $\alpha_W \leftrightarrow \sigma_{\text{wav}}(32)$; $\alpha_S \leftrightarrow S_\tau(1/32)$; $\alpha_D \leftrightarrow \text{DFA}(32)$; $\alpha_A \leftrightarrow A(10)$. In each of these cases the fixed-scale measure is found to greatly outperform the fractal-exponent measure for all data sizes examined, as we previously illustrated for the pairs $\alpha_W \leftrightarrow \sigma_{\text{wav}}(32)$ and $\alpha_S \leftrightarrow S_\tau(1/32)$ [47]. These results were recently confirmed in a follow-up Israeli–Danish study [65]. Moreover, in contrast to the substantial variability returned in fractal-exponent estimates, results for the different scale-dependent measures at $m = 32$ intervals bear reasonable similarity to each other.

Nunes Amaral et al. [56] concluded exactly the opposite, namely that scaling exponents provide superior performance to scale-dependent measures. This may be because they relied exclusively on the distribution-dependent measures $\eta \equiv h^2$ and $d^2$ (See Sections 2.7.1. and 3.5.1) rather than on distribution-independent ROC analysis. The same authors [56] also purport to glean information from higher moments of the wavelet coefficients, but the reliability of such information is questionable because estimator variance increases with moment order [17].

### 3.6.2. Computation Times of the Various Measures

The computation times for the 16 measures considered in this chapter are provided in Table 2. All measures were run 10 times and averaged except for the two DFA measures, which, because of their long execution time, were run only once. These long execution times are associated with the suggested method for computing DFA [66], which is an $N^2$ process. DFA computation times therefore increase as the square of the number of intervals, whereas all 14 other methods, in contrast, are either of order $N$ or $N \log(N)$.

Based on computation time, we can rank order the six scale-dependent measures that fall into the first class from fastest to slowest: $\sigma_{wav}(32)$, $S_\tau(1/32)$, $A(10)$, LF, VLF, and DFA(32). Because of its computational simplicity, the wavelet-transform standard deviation $\sigma_{wav}(32)$ computes more rapidly than any of the other measures. It is 3 times faster than its nearest competitor $S_\tau(1/32)$, 16.5 times faster than LF, and 32,500 times faster than DFA(32).

### 3.6.3. Comparing the Most Effective Measures

In Sections 3.3–3.5, we established that six measures succeeded in completely separating the normal subjects from the CHF patients in our database for data lengths $L \geq 2^{15} = 32{,}768$ RR intervals: VLF, LF, $A(10)$, $\sigma_{wav}(32)$, $S_\tau(1/32)$, and DFA(32). We

**TABLE 2** Computation Times (to the nearest 10 msec) for the 16 HRV Measures for Data Sets Comprising 75,821 RR Intervals[a]

| Measure | Execution time (ms) |
|---|---|
| VLF, LF, HF, and LF/HF | 330 |
| pNN50 | 40 |
| SDANN | 160 |
| $\sigma_{int}$ | 190 |
| $A(10)$ | 160 |
| $\sigma_{wav}(32)$ | 20 |
| $S_\tau(1/32)$ | 60 |
| DFA(32) | 650,090 |
| $\alpha_D$ | 650,110 |
| $\alpha_W$ | 220 |
| $\alpha_S$ | 920 |
| $\alpha_A$ | 610 |
| $\alpha_R$ | 570 |

[a]The long execution times for the two DFA measures result from the fact that it is an $N^2$ process whereas the 14 other methods are either $N$ or $N \log(N)$.

now demonstrate from a fundamental point of view that these measures can all be viewed as transformations of the interval-based power spectral density and that all are therefore closely related.

Consider first the relationships of VLF and LF to $S_\tau(1/32)$. At a frequency $f = 1/32 \approx 0.031$ cycles/interval, $S_\tau(f)$ separates CHF patients from normal subjects; however, a range of values of $f$ provide varying degrees of separability. For the data sets we analyzed, separation extends from $f = 0.02$ to $f = 0.07$ cycles/interval [49]. Recall that VLF and LF are simply integrals of $S_\tau(f)$ over particular ranges of $f$: from $f = 0.003$ to $f = 0.04$ cycles/interval for VLF and from $f = 0.04$ to $f = 0.15$ cycles/interval for LF. Because these ranges overlap with those for which the power spectral density itself provides separability, it is not surprising that these spectral integrals also exhibit this property. This illustrates the close relationship among VLF, LF, and $S_\tau(1/32)$.

We next turn to the relation between $\sigma_{\text{wav}}(32)$ and $S_\tau(1/32)$. In Section 2.5.3 the relationship between these two measures was established as

$$\sigma^2_{\text{wav}}(m) = m \int_{-\infty}^{\infty} S_\tau(f) H(mf)\, df \tag{23}$$

revealing that $\sigma^2_{\text{wav}}(m)$ is simply a bandpass-filtered version of the interval-based power spectral density for stationary signals. For the Haar wavelet, the bandpass filter $H(mf)$ has a center frequency near $1/m$ and a bandwidth about that center frequency also near $1/m$ [49]. Accordingly, $\sigma^2_{\text{wav}}(32)$ is equivalent to a weighted integral of $S_\tau(f)$ centered about $f = 1/32 \approx 0.031$, with an approximate range of integration of $0.031 \pm 0.016$ (0.016 to 0.047). Thus, the separability of $\sigma^2_{\text{wav}}(32)$ is expected to resemble that of $S_\tau(1/32)$ and therefore that of VLF and LF as well.

Now consider the count-based Allan factor evaluated at a counting time of 10 seconds, $A(10)$. Since the Allan factor is directly related to the variance of the Haar wavelet transform of the *counts* [36], $A(10)$ is proportional to the wavelet transform of the counting process evaluated over a wavelet duration of $T = 20$ seconds. The count-based and interval-based measures are approximately related via Eq. 7 so that $A(10)$ should be compared with $\sigma_{\text{wav}}(20/\mathrm{E}[\tau]) \approx \sigma_{\text{wav}}(25)$, whose value, in turn, is determined by $S_\tau(1/25 = 0.04)$ and spectral components at nearby frequencies. Since this range of $S_\tau(f)$ is known to exhibit separability, so too should $A(10)$.

The relationship between detrended fluctuation analysis and the other measures proves more complex than those already delineated. Nevertheless, DFA(32) can also be broadly interpreted in terms of an underlying interval-based power spectral density. To forge this link, we revisit the calculation of $F(m)$ (see Section 2.5.4). The mean is first subtracted from the sequence of RR intervals $\{\tau_i\}$, resulting in a new sequence that has zero mean. This new random process has a power spectral density everywhere equal to $S_\tau(f)$, except at $f = 0$, where it assumes a value of zero. The summation of this zero-mean sequence generates the series $y(k)$, as shown in Eq. 21. Since summation is represented by a transfer function $1/j2\pi f$ for small frequencies $f$, with $j = \sqrt{-1}$, the power spectral density of $y(k)$ is approximately given by

$$S_y(f) \approx \begin{cases} 0 & f = 0 \\ \dfrac{S_\tau(f)}{4\pi^2 f^2} & f \neq 0 \end{cases} \tag{24}$$

Next, the sequence $y(k)$ is divided into segments of length $m$. For the first segment, $k = 1$ to $m$, the best fitting linear trend $y_m(k) = \beta + \gamma k$ is obtained; the residuals $y(k) - y_m(k)$ therefore have the minimum variance over all choices of $\beta$ and $\gamma$. Over the duration of the entire sequence $y(k)$, the local trend function $y_m(k)$ will consist of a series of such linear segments and is therefore piecewise linear. Since $y_m(k)$ changes behavior over a time scale of $m$ intervals, its power will be concentrated at frequencies below $1/m$. Because the residuals $y(k) - y_m(k)$ have minimum variance, the spectrum of $y_m(k)$ will resemble that of $y(k)$ as much as possible and in particular at frequencies below $1/m$, where most of its power is concentrated. Thus

$$S_{ym}(f) \approx \begin{cases} S_y(f) & |f| < 1/m \\ 0 & |f| > 1/m \end{cases} \tag{25}$$

Equation 22 shows the mean-square fluctuation $F^2(m)$ to be an estimate of the variance of the residuals $y(k) - y_m(k)$. by Parseval's theorem, this variance equals the integral of the spectrum of the residuals over frequency. Therefore

$$F^2(m) \approx 2 \int_{0^+}^{\infty} \left[S_y(f) - S_{ym}(f)\right] df = (2\pi^2)^{-1} \int_{1/m}^{\infty} S_\tau(f) f^{-2}\, df \tag{26}$$

so that the mean-square fluctuation $F^2(m)$ can be represented as a weighted integral of the RR-interval spectrum over the appropriate frequency range.

We conclude that all six measures that best distinguish CHF patients from normal subjects are related in terms of transformations of the interval-based power spectral density.

### 3.6.4. Selecting the Best Measures

In the previous section we showed that the six most effective measures are interrelated from a fundamental point of view. This conclusion is confirmed by the overall similarity of the ROC-area curves for these measures, as can be seen in Figure 8.

We now proceed to compare these ROC areas more carefully in Figure 9. This figure presents 1 − ROC area plotted against the data length analyzed $L$ on a doubly logarithmic plot. This is in contrast to the usual presentation (such as Figure 8) in which ROC area is plotted against $L$ on a semilogarithmic plot. The unconventional form of the plot in Figure 9 is designed to allow small deviations from unit area to be highlighted. Clearly, small values of the quantity 1 − ROC area are desirable. Of course, as the data length $L$ diminishes, ROC area decreases (and therefore 1 − ROC area increases) for all six measures.

The trends exhibited by all six measures displayed in Figure 9 [VLF, $A(10)$, DFA(32), $\sigma_{wav}(32)$, $S_\tau(1/32)$, and LF] are indeed broadly similar, as we observed in Figure 8. For data sets that are sufficiently long, the six measures provide identical diagnostic capabilities (unity ROC area). At shorter data lengths, however, VLF, $A(10)$, and DFA(32) exhibit more degradation than do the other three measures. In accordance with the analysis provided in Section 3.6.3, VLF deviates the most at small values of $L$ because it incorporates information from values of $S_\tau(f)$ well below the frequency range of separability. Similarly, DFA(32) favors low-frequency contributions by virtue

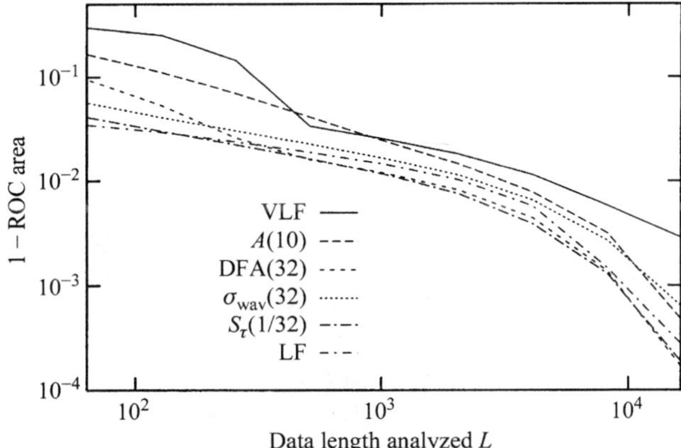


**Figure 9** Doubly logarithmic plot of 1 – ROC area as a function of data length $L$ for the six most effective measures: VLF, $A(10)$, DFA(32), $\sigma_{wav}(32)$, $S_\tau(1/32)$, and LF. The unconventional form of this ROC plot allows small deviations from unity area to be highlighted. The trends exhibited by all six measures are broadly similar, but VLF, $A(10)$, and DFA(32) exhibit more degradation at shorter data lengths than do the other three measures. Thus three measures emerge as superior: **LF**, $\boldsymbol{\sigma_{wav}(32)}$, and $\boldsymbol{S_\tau(1/32)}$. When computation time is taken into account in accordance with the results provided in Table 2, $\boldsymbol{\sigma_{wav}(32)}$ and $\boldsymbol{S_\tau(1/32)}$ are the two preferred measures.

of the $1/f^2$ weighting function in Eq. 26. Finally, the performance of $A(10)$ is suppressed because of the confounding effect of values of $S_\tau(f)$ for $f > 0.07$, as well as from possible errors deriving from our assumptions regarding the equivalence of the count-based and interval-based wavelet-transform variances.

The remaining three measures therefore emerge as superior: LF, $\sigma_{wav}(32)$, and $S_\tau(1/32)$. Judged on the basis of both performance (Figure 9) *and* computation time (Table 2), we conclude that two of these are best: $\boldsymbol{\sigma_{wav}(32)}$ and $\boldsymbol{S_\tau(1/32)}$.

It is desirable to confirm these conclusions for other records, comprising out-of-sample data sets from CHF patients and normal subjects. It will also be of interest to examine the correlation of ROC area with severity of cardiac dysfunction as judged, for example, by the New York Heart Association (NYHA) clinical scale. It is also important to conduct comparisons with other CHF studies that make use of HRV measures [3].

A sufficiently large database of normal and CHF records would make it possible to examine the detailed distinctions between the two classes of patients over a range of scales surrounding $m = 32$. A superior measure that draws on these distinctions in an optimal way could then be designed; for example, the weighting function in an integral over $S_\tau(f)$ could be customized. Other measures that simultaneously incorporate properties inherent in collections of the individual measures considered here could also be developed.

## 4. MARKERS FOR OTHER CARDIAC PATHOLOGIES

Returning to the individual-value plots in Figure 5, we briefly examine the behavior of the 16 measures for several other cardiovascular disorders for which these measures may hold promise. Among these are three CHF patients who also suffer from atrial fibrillation (CHF with AF, third column, △), one heart-transplant patient (TRANSPLANT, fourth column, ∗), one ventricular-tachycardia patient (VT, fifth column, ◇), one sudden-cardiac-death patient (SCD, sixth column, □), and two sleep-apnea patients (APNEA, last column, +). A summary of the more salient statistics of the original RR recordings for these patients is presented in Table A1. Prior to analysis, the RR records were truncated at $L_{max} = 75{,}821$ RR intervals to match the lengths of the records of CHF patients and normal subjects. The two sleep-apnea data records were drawn from the MIT-BIH Polysomnographic Database (Harvard–MIT Division of Health Sciences and Technology) and comprised 16,874 and 15,751 RR intervals, respectively. Because of the limited duration of the sleep period, records containing larger numbers of RR intervals are not available for this disorder.

Too few patient recordings are available to us to reasonably evaluate the potential of these HRV measures for the diagnosis of these forms of cardiovascular disease, but the results presented here might suggest which measures would prove useful given large numbers of records.

We first note that for most of the 16 measures, the values for the 3 CHF patients with atrial fibrillation (△) tend to fall within the range set by the other 12 CHF patients without atrial fibrillation. In particular, the same six measures that completely separate normal subjects from CHF patients without AF continue to do so when CHF patients with AF are included. The inclusion of the AF patients reduces the separability only for 3 of the 16 measures: pNN50, HF, and $\alpha_W$.

Results for the heart-transplant patient (∗) appear toward the pathological end of the CHF range for many measures, and for some measures the transplant value extends beyond the range of all of the CHF patients. Conversely, the sleep apnea values (+) typically fall at the end of, or beyond, the range spanned by the normal subjects. Indeed, three thresholds can be chosen for $S_\tau(1/32)$ (i.e., 0.000008, 0.0006, and 0.00436) that completely separate four classes of patients: sleep apnea [largest values of $S_\tau(1/32)$], normal subjects (next largest), CHF with or without atrial fibrillation (next largest), and heart transplant (smallest). Although such separation will no doubt disappear as large numbers of additional patient records are included, we nevertheless expect the largest variability (measured at 32 intervals as well as at other scales) to be associated with sleep-apnea patients because their fluctuating breathing patterns induce greater fluctuations in the heart rate. Conversely, heart-transplant patients (especially in the absence of reinnervation) are expected to exhibit the lowest values of variability because the autonomic nervous system, which contributes to the modulation of heart rate at these time scales, is subfunctional.

Results for the ventricular-tachycardia patient (◇) lie within the normal range for most measures, although the value for LF/HF is well beyond the pathological end of the range for CHF patients. Interestingly, four measures [VLF, HF, SDANN, and $A(10)$] yield the largest values for the ventricular-tachycardia patient of the 32 data sets examined. Perhaps $A(10)$, which exhibits the largest separation between the value for this patient and those for the other 31 patients, will prove useful in detecting ventricular tachycardia.

Finally, we consider results for the sudden-cardiac-death patient (□), which lie between the ranges of normal subjects and CHF patients for all but one (LF) of the six measures that separate these two patient classes. The pNN50 and HF results, on the other hand, reveal values beyond the normal set, thereby indicating the presence of greater variability on a short time scale than for other patients. The value of SDANN for the SCD patient is greater than that for any of the other 31 patients, so it is possible that SDANN could prove useful in predicting sudden cardiac death.

## 5. DOES DETERMINISTIC CHAOS PLAY A ROLE IN HEART RATE VARIABILITY?

The question of whether normal HRV arises from a low-dimensional attractor associated with a deterministic dynamical system or has stochastic underpinnings has continued to entice researchers [61]. Previous studies devoted to this question have come to conflicting conclusions. The notion that the normal human-heartbeat sequence is chaotic appears to have been first set forth by Goldberger and West [67]. Papers in support of this notion [68], as well as in opposition to it [69, 70], have appeared recently. An important factor that pertains to this issue is the recognition that correlation, rather than nonlinear dynamics, can lie behind these seemingly chaotic recordings [8].

We address this question in this section and conclude, for both normal subjects and patients with several forms of cardiac dysfunction, that the sequence of heartbeat intervals does not exhibit chaos.

### 5.1. Methods

#### 5.1.1. Phase-Space Reconstruction

One way of approaching the question of chaos in heartbeat recordings is in terms of a phase-space reconstruction and the use of an algorithm to estimate one or more of its fractal dimensions $D_q$ [6, 59, 71]. The box-counting algorithm [72] provides a method for estimating the capacity (or box-counting) dimension $D_0$ of the attractor [6, 8, 71, 73].

The phase-space reconstruction approach operates on the basis that the topological properties of the attractor for a dynamical system can be determined from the time series of a single observable [74–76]. A $p$-dimensional vector $\vec{T} \equiv \{\tau_i, \tau_{i+1}, \ldots, \tau_{i+(p-1)l}\}$ is formed from the sequence of RR intervals $\{\tau_i\}$. The parameter $p$ is the embedding dimension, and $l$ is the lag, usually taken to be the location of the first zero crossing of the autocorrelation function of the time series under study. As the time index $i$ progresses, the vector $\vec{T}$ traces out a trajectory in the $p$-dimensional embedding space.

The box-counting algorithm estimates the capacity dimension of this trajectory. Specifically, the negative slope of the logarithm of the number of full boxes versus the logarithm of the width of the boxes, over a range of $4/[\max(\tau) - \min(\tau)]$ to $32/[\max(\tau) - \min(\tau)]$ boxes, in powers of two, provides an estimate of $D_0$. For convenience in the generation of phase-randomized surrogates, data sets were limited to powers of two: $2^{16} = 65{,}536$ intervals for all but the two sleep-apnea data sets, for which $2^{14} = 16{,}384$ and $2^{13} = 8192$ intervals were used respectively (the sleep-apnea data sets are much shorter than the others as indicated in Table A1). For uncorrelated noise, the capacity dimension $D_0$ continues to increase as the embedding dimension $p$ increases. For an attractor in a deterministic system, in contrast, $D_0$ saturates as $p$ becomes larger than $2D_0 + 1$ [73]. Such saturation, however, is not a definitive signature of deterministic dynamics; temporal correlation in a stochastic process can also be responsible for what appears to be underlying deterministic dynamics [8, 77–79]. The distribution of the data can also play a role.

Surrogate data analysis, discussed in Section 5.1.3, is useful in establishing the underlying cause, by proving or disproving various null hypotheses that the measured behavior arises from other, nonchaotic causes.

### 5.1.2. Removing Correlations in the Data

Adjacent intervals in heartbeat sequences prove to exhibit large degrees of correlation. The serial correlation coefficient for RR intervals

$$\rho(\tau) \equiv \frac{\mathrm{E}[\tau_{i+1}\tau_i] - \mathrm{E}^2[\tau_i]}{\mathrm{E}[\tau_i^2] - \mathrm{E}^2[\tau_i]}$$

exceeds 0.9 for two thirds of the data sets examined, with an average value of 0.83; the theoretical maximum is unity, with zero representing a lack of serial correlation. Since such large correlation is known to interfere with the detection of deterministic dynamics, we instead employ the first difference of the RR intervals: $v_i \equiv \tau_i - \tau_{i+1}$. Such a simple transformation does not change the topological properties of dynamical system behavior, and a related but more involved procedure is known to reduce error in estimating other fractal dimensions [80].

The serial correlation coefficient for the differenced sequence $\{v_i\}$ turns out to have a mean value of $-0.084$, and none exceeds 0.6 in magnitude. This sequence will therefore be used to generate the embedding vector $\overrightarrow{T} \equiv \{v_i, v_{i+1}, \ldots, v_{i+p-1}\}$, with the lag $l$ set to unity since $\rho(v) \approx 0$.

### 5.1.3. Surrogate Data Analysis

Information about the nature of an interval or segment series may be obtained by applying various statistical measures to surrogate data sets. These are processes constructed from the original data in ways designed to preserve certain characteristics of the original data while eliminating (or modifying) others. Surrogate data analysis provides a way of determining whether a given result arises from a particular property of the data set.

We make use of two kinds of surrogate data sets: shuffled intervals and randomized phases. In particular, we compare statistical measures calculated from both the original data and its surrogates to distinguish those properties of the data set that arise from correlations (such as from long-term rate fluctuations) from those properties inherent in the form of the original data.

***Shuffled Intervals.*** The first class of surrogate data is formed by shuffling (randomly reordering) the sequence of RR intervals of the original data set. Such random reordering destroys dependencies among the intervals, and therefore the correlation properties of the data, while exactly preserving the interval histogram. As a refinement, it is possible to divide the data into blocks and shuffle intervals within each block while keeping each block separate. In this case dependencies remain over durations much larger than the block size, while being destroyed for durations smaller than it.

***Randomized Phases.*** The other class of surrogate data we consider is obtained by Fourier transforming the original interval data and then randomizing the phases while leaving the spectral magnitudes intact. The modified function is then inverse transformed to return to a time-domain representation of the intervals. This technique exactly preserves the second-order correlation properties of the interval sequence while removing other temporal structure, for example that needed for phase-space reconstruction. The interval histogram typically becomes Gaussian for this surrogate as a result of the central limit theorem.

## 5.2. Absence of Chaos

In Figure 10 we present plots of the capacity dimension $D_0$ against the embedding dimension $p$ for the 12 normal subjects. Similar plots are presented in Figure 11 for CHF patients without atrial fibrillation and in Figure 12 for patients with other cardiac pathologies. For most of the panels in Figure 10, the $D_0$ estimate for the original data (solid curve) is indeed seen to rise with increasing embedding dimension $p$ and shows evidence of approaching an asymptotic value of about 2. The phase-randomized data (dotted curve) yield somewhat higher results, consistent with the presence of a deterministic attractor.

However, estimates of $D_0$ for the shuffled data (dashed curve) nearly coincide with those of the original data. This surrogate precisely maintains the distribution of the relative sizes of the differenced RR intervals but destroys any correlation or other

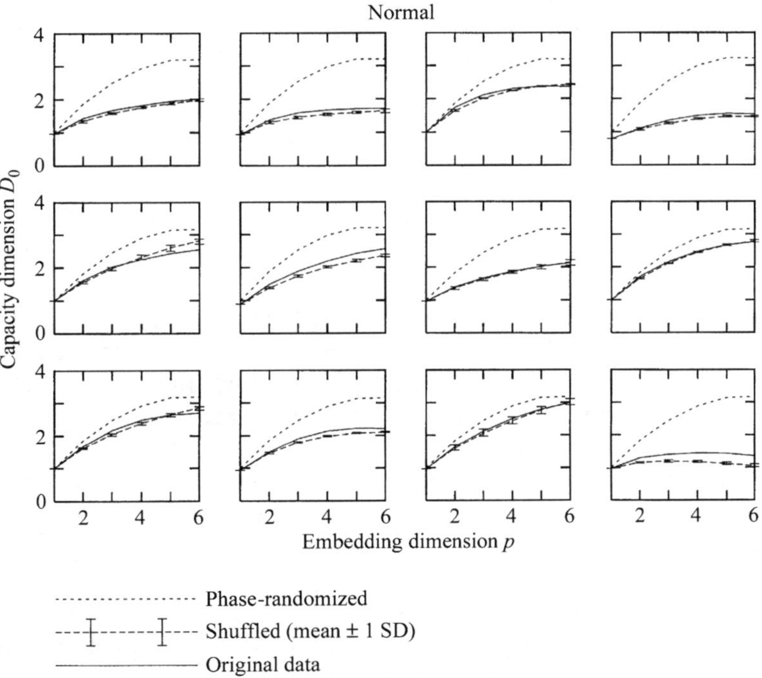


**Figure 10** Capacity dimension $D_0$ as a function of embedding dimension $p$ for the 12 normal subjects. The three curves shown for each subject correspond to the original differenced intervals (solid curves), the shuffled surrogates (dashed curves), and the phase-randomized surrogates (dotted curves). The dashed curves are quite similar to the solid curves in each panel, while the 12 dotted curves closely resemble each other.

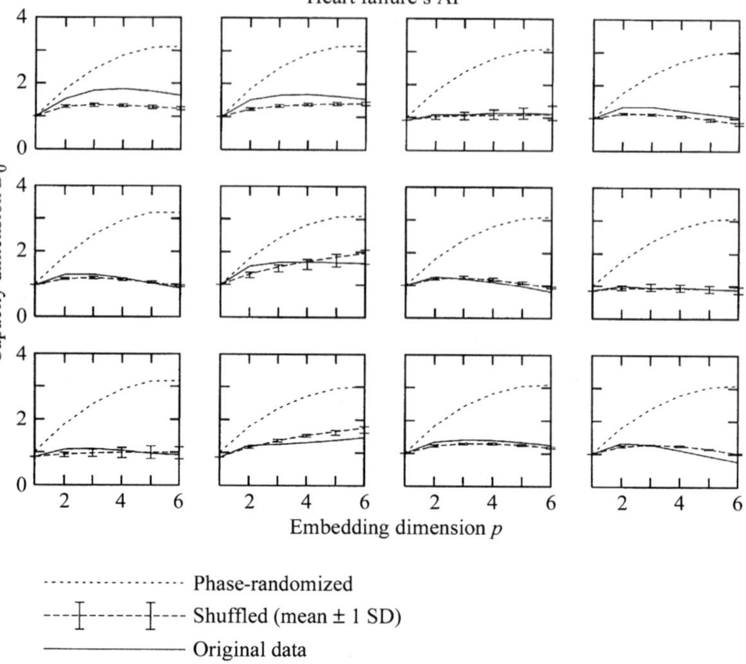


**Figure 11** Capacity dimension $D_0$ as a function of embedding dimension $p$ for the 12 CHF patients without atrial fibrillation. The three curves shown for each patient correspond to the original differenced intervals (solid curves), the shuffled surrogates (dashed curves), and the phase-randomized surrogates (dotted curves). The dashed curves are reasonably similar to the solid curves in most panels, while the 12 dotted curves closely resemble each other.

dependencies in the data. Specifically, no deterministic structure remains in the shuffled surrogate. Therefore, the apparent asymptote in $D_0$ must derive from the particular form of the RR-interval distribution and not from a nonlinear dynamical attractor. To quantify the closeness between results for original and shuffled data, the shuffling process was performed 10 times with 10 different random seeds. The average of these 10 independent shufflings forms the dashed curve, with error bars representing the largest and smallest estimates obtained for $D_0$ at each corresponding embedding dimension. Results for the original data indeed lie within the error bars. Therefore, the original data do not differ from the shuffled data at the $[(10-1)/(10+1)]100\% = 82\%$ significance level.

Results for a good number of the other 31 data sets, normal and pathological alike, are nearly the same; the original data lie within the error bars of the shuffled surrogates and consequently do not exhibit significant evidence of deterministic dynamics. However, for some records the original data exceed the limits set by the 10 shufflings. But attributing this behavior to low-dimensional chaos proves difficult for almost all the data records for three reasons. First, the surrogates often yield *smaller* results for $D_0$ than the original data, whereas shuffling data from a true chaotic system will theoretically yield *larger* $D_0$ estimates. Second, of the data sets for which the original results

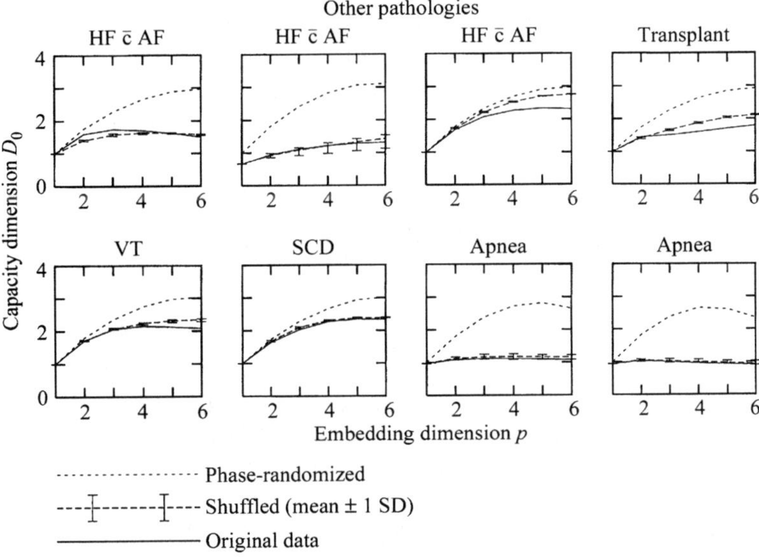


**Figure 12** Capacity dimension $D_0$ as a function of embedding dimension $p$ for the eight patients with other cardiac pathologies. The three curves shown for each patient correspond to the original differenced intervals (solid curves), the shuffled surrogates (dashed curves), and the phase-randomized surrogates (dotted curves). The dashed curves are reasonably similar to the solid curves in most panels while the eight dotted curves closely resemble each other.

fall below those of their shuffled surrogates, few achieve an asymptotic value for $D_0$, but rather continue to climb steadily with increasing embedding dimension. Finally, for the remaining data sets, the original and shuffled data yield curves that do not differ nearly as much as do those from simulated systems known to be chaotic, such as the Hénon attractor (not shown) [81]. Taken together, these observations suggest that finite data length, residual correlations, and/or other unknown effects in these RR-interval recordings give rise to behavior in the phase-space reconstruction that can masquerade as chaos.

Results for the phase-randomized surrogates (dotted curves in Figures 10–12), while exhibiting a putative capacity dimension in excess of the original data and of the shuffled surrogates for every patient, exhibit a remarkable similarity to each other. This occurs because the differenced-interval sequence $\{v_i\}$ displays little correlation and the phase-randomization process preserves this lack of correlation while imposing a Gaussian differenced-interval distribution. The result is nearly white Gaussian noise regardless of the original data set, and indeed results for independently generated white Gaussian noise closely follow these curves (not shown).

Time series measured from the heart under distinctly nonnormal operating conditions such as ventricular fibrillation [82], or those obtained under nonphysiological conditions such as from excised hearts [83], may exhibit chaotic behavior under some circumstances. This is precisely the situation for cellular vibrations in the mammalian cochlea [84–86]. When the exciting sound pressure level (SPL) is in the normal physiological regime, the cellular-vibration velocities are nonlinear but not chaotic. When the

SPL is increased beyond the normal range, however, routes to chaos emerge. It has been suggested that since chaos-control methods prove effective for removing such behavior, the heartbeat sequence must be chaotic. This is an unwarranted conclusion because such methods often work equally well for purely stochastic systems, which cannot be chaotic [87].

## 6. MATHEMATICAL MODELS FOR HEART RATE VARIABILITY

The emphasis in this chapter has been on HRV analysis. The diagnostic capabilities of various measures have been presented and compared. However, the results that emerge from these studies also serve to expose the mathematical–statistical character of the RR time series. Such information can be of benefit in the development of mathematical models that can be used to simulate RR sequences. Such models may prove useful for generating realistic heartbeat sequences in pacemakers, for example, and for developing improved physiological models of the cardiovascular system. In this section we evaluate the suitability of several integrate-and-fire constructs for modeling heart rate variability.

### 6.1. Integrate-and-Fire Model

Integrate-and-fire models are widely used in the neurosciences [88] as well as in cardiology [8, 89–92]. These models are attractive in part because they capture known physiology in a simple way. The integration of a rate process can represent the cumulative effect of neurotransmitter on the postsynaptic membrane of a neuron or the currents responsible for the pacemaker potential in the sino-atrial node of the heart. The crossing of a preset threshold by the integrated rate then gives rise to an action potential or a heart contraction.

The sequence of discrete events constituting the human heartbeat can be viewed as a point process deriving from a continuous rate function. The integrate-and-fire method is perhaps the simplest means for generating a point process from a rate process [8, 17]. In this model, illustrated schematically in Figure 13, the rate function $\lambda(t)$ is integrated until it reaches a fixed threshold $\theta$, whereupon a point event is generated and the integrator is reset to zero. Thus the occurrence time of the $(i+1)$st event is implicitly obtained from the first occurrence of

$$\int_{t_i}^{t_{i+1}} \lambda(t)\, dt = \theta \tag{27}$$

The mean heart rate $\lambda$ (beats/s) is given by $\lambda = \mathrm{E}[\lambda(t)] = 1/\mathrm{E}[\tau]$. Because the conversion from the rate process to the point process is so direct, most statistics of the point

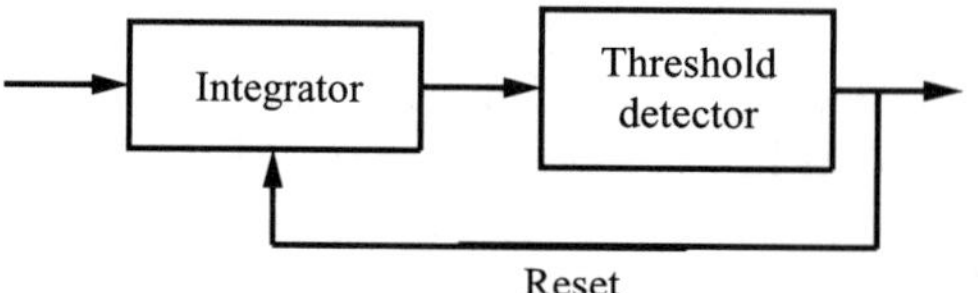


**Figure 13** Simple integrate-and-fire model often used in cardiology and neurophysiology. A rate function $\lambda(t)$ is integrated until it reaches a preset threshold $\theta$, whereupon a point event is generated and the integrator is reset to zero.

process closely mimic those of the underlying rate process $\lambda(t)$. In particular, for frequencies substantially lower than the mean heart rate, the theoretical power spectral densities of the point process and the underlying rate coincide [17].

## 6.2. Kernel of the Integrate-and-Fire Model

The detailed statistical behavior of the point process generated by the integrate-and-fire construct depends on the statistical character of the rate function in the integrand of Eq. 27 [8]. The fractal nature of the heartbeat sequence requires that the rate function $\lambda(t)$ itself be fractal. Assuming that the underlying rate is stochastic noise, there are three plausible candidates for $\lambda(t)$ [17]: fractal Gaussian noise, fractal lognormal noise, and fractal binomial noise. We briefly discuss these three forms of noise in turn.

### 6.2.1. Fractal Gaussian Noise

Fractal Gaussian noise served admirably in our initial efforts to develop a suitable integrate-and-fire model for the heartbeat sequence [8]. A Gaussian process has values that, at any set of times, form a jointly Gaussian random vector. This property, along with its mean and spectrum, completely defines the process. We consider a stationary rate process with a mean equal to the expected heart rate and a $1/f$-type *rate*-based spectrum that takes the form [17]

$$S_{\lambda}(f_{\text{time}}) = \lambda^2 \delta(f_{\text{time}}) + \lambda[1 + (f_{\text{time}}/f_0)^{-\alpha}] \tag{28}$$

Here, $\lambda$ (beats/s) and $\alpha$ are the previously defined heart rate and fractal exponent, respectively, $\delta(\cdot)$ is again the Dirac delta function, and $f_0$ is the cutoff frequency (cycles/s). Mathematically, cutoffs at low or high frequencies (or sometimes both) are required to ensure that $\tau$ has a finite variance. In practice, the finite duration of the simulation guarantees the former cutoff, and the finite time resolution of the simulated samples of $\tau$ guarantees the latter. It is to be noted that the designation fractal Gaussian noise properly applies only for $\alpha < 1$; the range $1 < \alpha < 3$ is generated by fractal Brownian motion [19]. In the interest of simplicity, however, we employ the term fractal Gaussian noise for all values of $\alpha$.

There are a number of recipes in the literature for generating fractal Gaussian noise [93–96]. All typically result in a sampled version of fractal Gaussian noise, with equal spacing between samples. A continuous version of the signal is obtained by using a zeroth-order interpolation between the samples.

The use of a fractal Gaussian noise kernel in the integrate-and-fire construct results in the so-called fractal-Gaussian-noise driven integrate-and-fire model [8, 17]. The mean of the rate process is generally required to be much larger than its standard deviation, in part so that the times when the rate is negative (during which no events may be generated) remain small.

The *interval*-based spectrum is obtained by applying Eq. 7 to Eq. 28 (see Section 2.2 and Ref. 9). The following approximate result, in terms of interval-based frequency $f$, is obtained:

$$S_{\tau}(f) \approx \lambda^{-2}[\delta(f) + 1 + (\lambda f/f_0)^{-\alpha}] \tag{29}$$

We proceed to discuss two other physiologically based noise inputs [17]; both reduce to fractal Gaussian noise in appropriate limits.

### 6.2.2. Fractal Lognormal Noise

A related process results from passing fractal Gaussian noise through a memoryless exponential transform. This process plays a role in neurotransmitter exocytosis [97]. Since the exponential of a Gaussian is lognormal, we refer to this process as fractal lognormal noise [17, 97]. If $X(t)$ denotes a fractal Gaussian noise process with mean $\mathrm{E}[X]$, variance $\mathrm{Var}[X]$, and autocovariance function $K_X(\tau)$, then $\lambda(t) \equiv \exp[X(t)]$ is a fractal lognormal noise process with moments $\mathrm{E}[\lambda^n] = \exp(n\mathrm{E}[X] + n^2\mathrm{Var}[X]/2)$ and autocovariance function $K_\lambda(\tau) = \mathrm{E}^2[\lambda](\exp[K_X(\tau)] - 1)$ [97].

By virtue of the exponential transform, the autocorrelation functions of the Gaussian process $X(t)$ and the lognormal process $\lambda(t)$ differ; thus their spectra differ. In particular, the power spectral density of the resulting point process does not follow the exact form of Eq. 28, although for small values of $\mathrm{Var}[X]$ the experimental periodogram closely resembles this ideal form [97]. In the limit of very small values of $\mathrm{Var}[X]$, the exponential operation approaches a linear transform, and the rate process reduces to fractal Gaussian noise.

### 6.2.3. Fractal Binomial Noise

A third possible kernel for the integrate-and-fire model is fractal binomial noise. This process results from the addition of a number of independent identical alternating fractal renewal processes. The sum is a binomial process with the same fractal exponent as each of the individual alternating fractal processes [98, 99]. This binomial process can serve as a rate function for an integrate-and-fire process; the result is the fractal-binomial-noise driven integrate-and-fire model [17]. This construct was initially designed to model the superposition of alternating currents from a number of ion channels with fractal behavior [98–101]. As the number of constituent processes increases, fractal binomial noise converges to fractal Gaussian noise with the same fractal exponent therefore leading to the fractal-Gaussian-noise driven integrate-and-fire point process [17].

## 6.3. Jittered Integrate-and-Fire Model

Conventional integrate-and-fire constructs have only a single source of randomness, the rate process $\lambda(t)$. It is sometimes useful to incorporate a second source of randomness into such models. One way to carry this out is to impose random jitter on the interevent times generated in the integrate-and-fire process [17].

The procedure used for imposing this jitter is as follows. After generating the time of the $(i+1)$st event $t_{i+1}$ in accordance with Eq. 27, the $(i+1)$st interevent time $\tau_{i+1} = t_{i+1} - t_i$ is multiplied by a dimensionless Gaussian-distributed random variable with unit mean and variance $\sigma^2$. This results in the replacement of $t_{i+1}$ by

$$t_{i+1} + \sigma(t_{i+1} - t_i)\mathcal{N}_i \tag{30}$$

before the simulation of subsequent events $(t_{i+2}, t_{i+3}, \ldots)$. The quantity $\{\mathcal{N}_i\}$ represents a sequence of zero-mean, unity-variance independent Gaussian random variables and the standard deviation $\sigma$ is a free parameter that controls the strength of the proportional jitter. In accordance with the construction of the model, events at subsequent times experience jitter imposed by all previous events (see Figure 14).

The overall result is the jittered integrate-and-fire model. The jitter serves to introduce a source of scale-independent noise into the point process. Depending on the character of the input rate function, this model can give rise to the fractal-Gaussian-noise, fractal-lognormal-noise, or fractal-binomial-noise driven jittered integrate-and-fire processes, among others. Two limiting behaviors readily emerge from this model. When $\sigma \to 0$, the jitter disappears and the jittered integrate-and-fire model reduces to the ordinary integrate-and-fire model. At the other extreme, when $\sigma \to \infty$, the jitter dominates and the result is a homogeneous Poisson process. The fractal behavior present in the rate function $\lambda(t)$ then disappears and $\lambda(t)$ behaves as if it were constant. Between these two limits, as $\sigma$ increases the fractal onset time $1/f_0$ of the resulting point process increases and the fractal characteristics of the point process are progressively lost, first at higher frequencies (shorter times) and subsequently at lower frequencies (longer times) [17].

### 6.3.1. Simulating the Jittered Integrate-and-Fire Point Process

We use a fractal Gaussian noise kernel in the jittered integrate-and-fire construct to simulate the human heartbeat. Fractal Gaussian noise with the appropriate mean, standard deviation, and fractal exponent $\alpha_S$ is generated using Fourier-transform methods. This noise serves as the kernel in an integrate-and-fire element (Eq. 27), the output of which is jittered in accordance with Eq. 30. For each of the 12 normal subjects and 15 CHF (without and with AF) patients, the model parameters were adjusted iteratively to obtain a simulated data set that matched the patient recording as closely as possible. In particular, each of the 27 simulations contained exactly $L_{\max} = 75{,}821$ RR intervals, displayed mean heart rates that fell within 1% of the corresponding data recordings, and exhibited wavelet-transform standard deviations that were nearly identical to those of the data. The simulation results for CHF patients with and without atrial fibrillation were very similar.

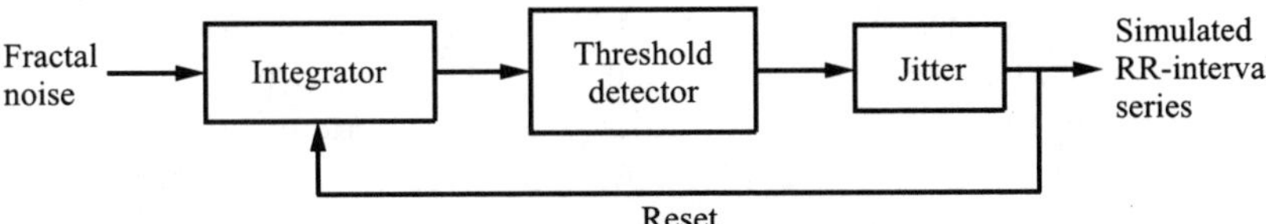


**Figure 14** Schematic representation of the jittered integrate-and-fire model for generating a simulated RR-interval series. An underlying rate process $\lambda(t)$, assumed to be band-limited fractal Gaussian noise, is integrated until it reaches a fixed threshold $\theta$, whereupon a point event is generated. The occurrence time of the point event is jittered in accordance with a Gaussian distribution and the integrator is then reset. The continuous rate process is thereby converted into a point process representing the sequence of R phases in the human heartbeat.

### 6.3.2. Statistics of the Simulated Point Process for Normal Subjects and CHF Patients

Having developed the jittered integrate-and-fire model and simulated sequences of RR intervals, we proceed to examine some of the statistical features of the resulting point processes.

Figure 15 displays representative results in which the data (solid curves) and simulations (dotted curves) are compared for a single normal subject (left column) and a single CHF (without AF) patient (right column). Figure 15a illustrates that the RR-interval sequences obtained from the model and the data are qualitatively similar for both the normal and heart-failure cases. Furthermore, the agreement between the simulated and actual wavelet-transform standard-deviation curves is excellent in both cases, as is evident in Figure 15b. Even the gentle increase of $\sigma_{wav}$ for the heart-failure patient at small scale values is reproduced, by virtue of the jitter introduced into the model. It is apparent in Figure 15c, however, that the simulated RR-interval histogram for the heart-failure patient does not fit the actual histogram well; it is too broad, which indicates that $\sigma_{int}$ of the simulation is too large. Finally, Figure 15d shows that the simulated spectra of the RR intervals match the experimental spectra quite nicely, including the whitening of the heart-failure spectrum at high frequencies. This latter effect is the counterpart of the flattening of $\sigma_{wav}$ at small scales and results from the white noise introduced by the jitter at high frequencies.

### 6.3.3. Simulated Individual Value Plots and ROC-Area Curves

To examine how well the jittered integrate-and-fire model mimics the collection of data from a global perspective, we constructed individual-value plots and ROC-area curves using the 27 simulated data sets. The results are presented in Figures 16 and 17, respectively. Direct comparison should be made with Figures 5 and 8, respectively, which provide the same plots for the actual heart-failure patient and normal-subject data.

Comparing Figure 16 with Figure 5 and Figure 17 with Figure 8, we see that in most cases the simulated and actual results are remarkably similar. Based on their overall ability to distinguish between CHF and normal simulations for the full set of 75,821 RR intervals, the 16 measures portrayed in Figure 16 roughly fall into three classes. A similar division was observed for the actual data, as discussed in Section 3.6.

As is evident in Figure 16, five measures (indicated by boldface font) succeed in fully separating the two kinds of simulations and, by definition, fall into the first class: VLF, LF, $\sigma_{wav}(32)$, $S_\tau(1/32)$, and DFA(32). These measures share a dependence on a single scale, or small range of scales near 32 heartbeat intervals. However, of these five measures only two (LF and VLF) successfully separate the simulated results for the next smaller number of RR intervals ($L = 65,536$) and none succeeds in doing so for yet smaller values of $L$.

The same five measures also fully distinguish the normal subjects from the CHF patients (see Figure 5); however, a sixth measure, $A(10)$, also achieves this. For the actual data all six of these measures successfully separate the actual heart-failure patients from the normal subjects for data lengths down to $L = 32,768$ RR intervals.

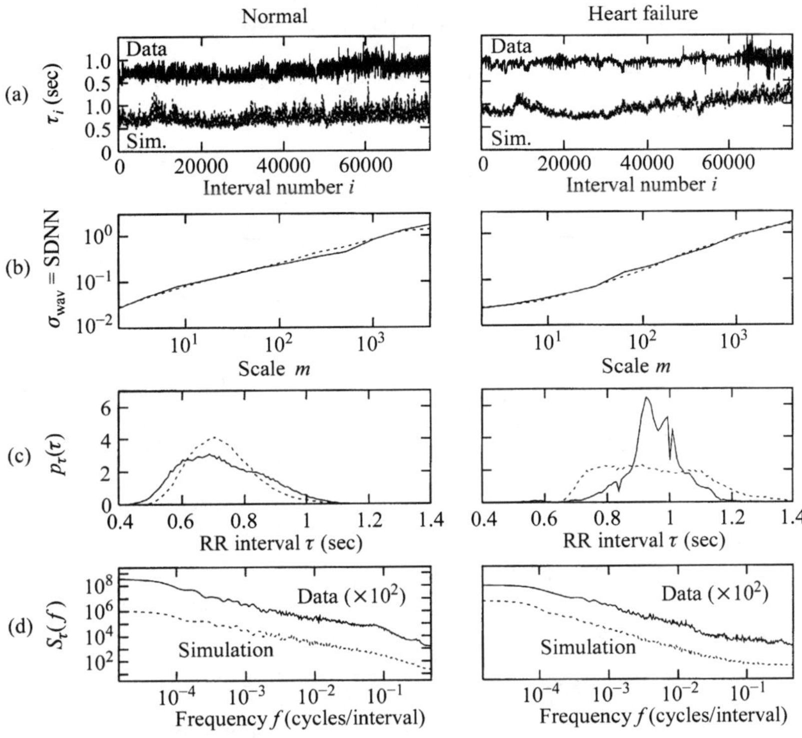


**Figure 15** Comparison of data (solid curves) and simulations (dotted curves) for a single normal subject (left column, data set n16265) and a single CHF $\bar{s}$ AF patient (right column, data set a6796). The parameters $E[\tau_i]$ (representing the mean interbeat interval) and $\alpha_W$ (representing the fractal exponent) used in the simulations were derived from the data for the two individuals. The normal-subject simulation parameters were $1/\lambda = 0.74\,s$, $\alpha_S = 1.0$, $f_0 = 0.0012$ cycles/s, and $\sigma = 0.022$; the CHF-patient simulation parameters were $1/\lambda = 0.99\,s$, $\alpha_S = 1.5$, $f_0 = 0.00055$ cycles/s, and $\sigma = 0.025$. The lengths of the simulated data sets were identical to those of the experimental records analyzed. (a) RR-interval sequence over the entire data set. Qualitative agreement is apparent in both the normal and heart-failure panels. (b) Wavelet-transform standard deviation versus scale. The simulations mimic the scaling properties of the data in both cases, as well as the gentle flattening of $\sigma_{wav}$ for the heart-failure patient at small scale values. (c) Interbeat-interval histogram. The model is satisfactory for the normal subject but fails to capture the narrowing of the histogram (reduction of $\sigma_{int}$) for the heart-failure patient. (d) Spectrum of the sequence of RR intervals (the data are displaced upward by a factor of $10^2$ for clarity). The simulations capture the subtleties in the spectra quite well, including the whitening of the heart-failure spectrum at high frequencies.

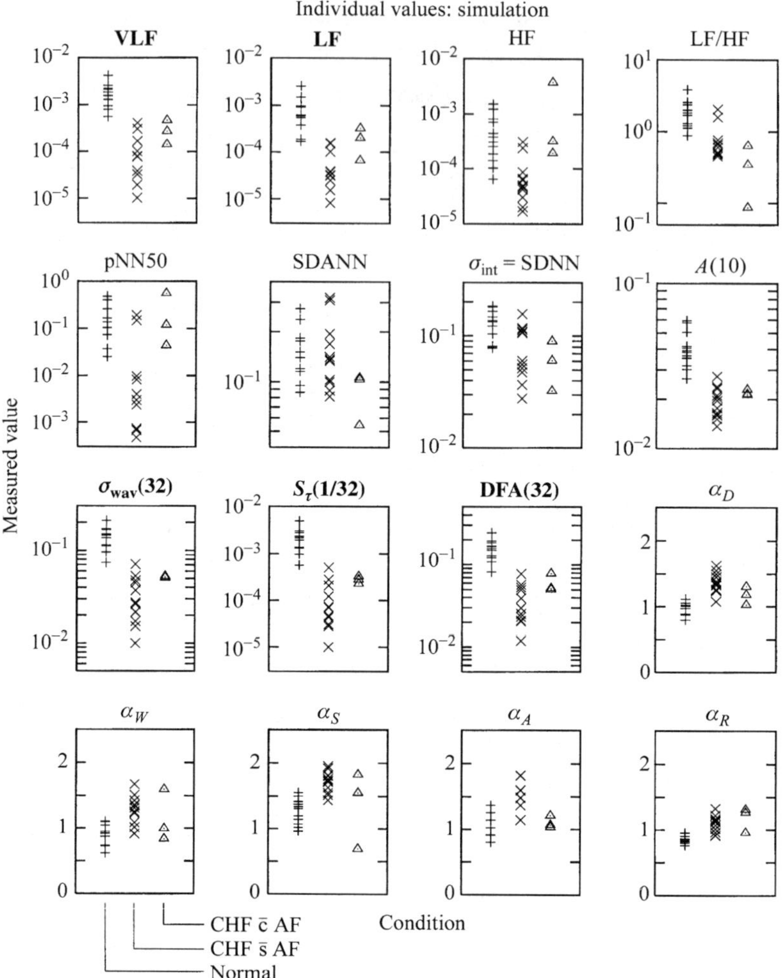


**Figure 16** Individual value plots (jittered integrate-and-fire-simulations) for the 16 measures. Each panel corresponds to a different HRV measure. The three columns in each panel, from left to right, comprise data for (1) 12 simulations using normal-subject parameters (+), (2) 12 simulations using CHF-patient (s̄ AF) parameters (×), and (3) 3 simulations using CHF-patient (c̄ AF) parameters (△). Each simulation comprises 75,821 RR intervals and is carried out using parameters drawn from a single actual data set. The five measures highlighted in boldface font succeed in completely separating normal-subject and CHF-patient (s̄ and c̄ atrial fibrillation) simulations: **VLF, LF, $\sigma_{wav}(32)$, $S_\tau(1/32)$,** and **DFA(32)**. Each panel should be compared with the corresponding panel for the actual normal and CHF data in Figure 5 (leftmost three columns).

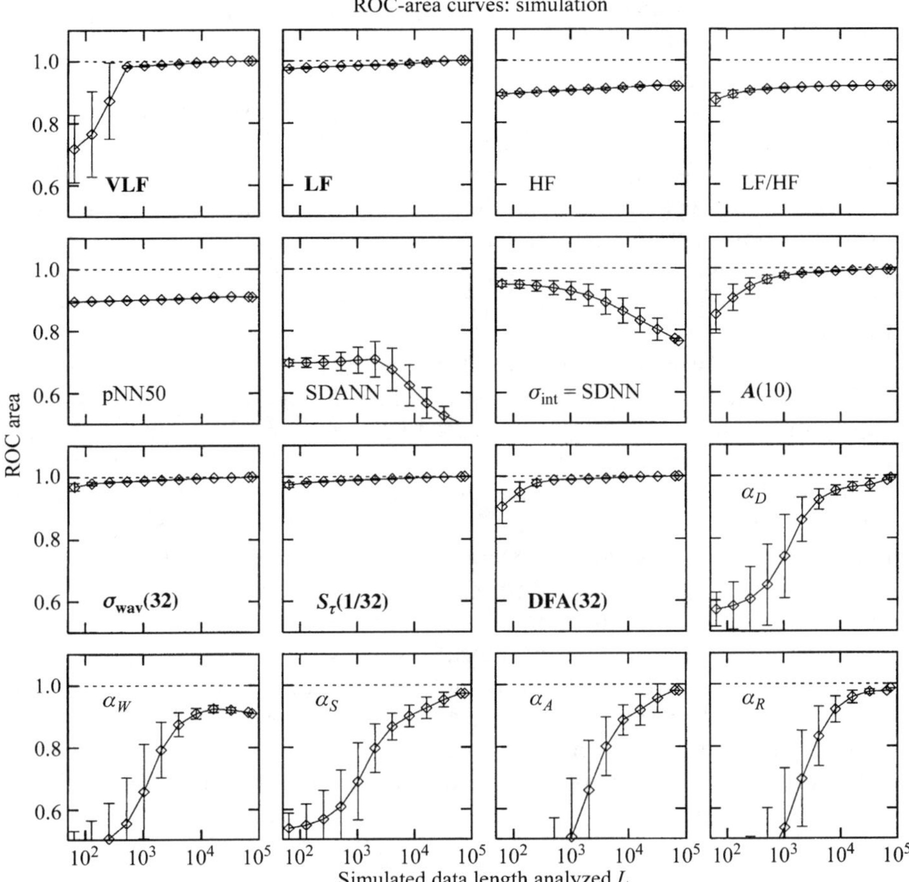


**Figure 17** Area under the ROC curve (jittered integrate-and-fire simulations) as a function of number of RR intervals analyzed ($L$) for the 16 HRV measures (mean $\pm$ 1 SD). An area of unity corresponds to perfect separability of the two classes of simulations. The five measures highlighted in boldface font [**VLF, LF,** $\boldsymbol{\sigma_{wav}}$**(32)**, $\boldsymbol{S_\tau}$**(1/32)**, and **DFA(32)**] provide such perfect separability but only for the largest number of RR intervals analyzed. Each panel should be compared with the corresponding panel for the actual normal and CHF data in Figure 8. The ROC-area simulations differ from those for the data in two principal respects: the performance of the two interval measures SDANN and $\sigma_{int}$ severely degrades as the number of RR intervals increases, and the performance of the five fractal-exponent measures is substantially enhanced as the number of RR intervals increases.

The simulated ROC-area curves presented in Figure 17 resemble the curves for the normal and CHF data shown in Figure 8, but there are a few notable distinctions. The simulated and actual ROC areas (Figures 17 and 8, respectively) for the interval measures SDANN and $\sigma_{int}$ are just about the same for $L = 64$ RR intervals. However, both of these measures exhibit *decreasing* ROC areas as the number of simulated RR intervals ($L$) increases (Figure 17). In contrast, the ROC areas based on the actual data increase with increasing data length $L$ in normal fashion (Figure 8). The decrease in ROC area observed in Figure 17 means that performance is degraded as the simulated data length increases. This paradoxical result probably arises from a deficiency of the model; it proved impossible to fit simultaneously the wavelet-transform standard deviation at all scales, the mean heart rate, and the RR-interval standard deviation. Choosing to fit the two former quantities closely to the data rendered inaccurate the simulated interval standard deviation. Since shorter files have insufficient length to exhibit the longer term fluctuations that dominate fractal activity, the interval standard deviation for small values of $L$ is not influenced by these fluctuations. Apparently, the manner in which the long-term effects are expressed differs in the original data and in the simulation.

The most surprising distinction between the actual and simulated results, perhaps, is the dramatic putative improvement in the ability of the five scale-independent fractal-exponent measures to separate simulated heart-failure and normal data as the number of RR intervals increases. The Allan factor exponent $\alpha_A$ appears to offer the greatest "improvement," as seen by comparing Figures 17 and 8. All of the exponents except $\alpha_W$ attain an ROC area exceeding 0.97 for $L_{\max} = 75{,}821$ RR intervals. Nevertheless, the improved separation yields performance that remains inferior to that of the fixed-scale measures. The apparent improved separation in the simulated data may therefore arise from a nonlinear interaction in the model between the fractal behavior and the signal component near a scale of 32 heartbeat intervals. Or it may be that some aspect of the data, not reliably detected by fractal-exponent measures applied to the original data, emerges more clearly following the modeling and simulation. This could mean that another measure of the fractal exponent of the actual data, perhaps incorporating in some way the processing provided by the modeling procedure, might yield better results than the more straightforward fractal-exponent measures that we have used to this point.

### 6.3.4. Limitations of the Jittered Integrate-and-Fire Model

Figure 18 presents a direct comparison between four measures derived from the original data and from the corresponding simulations. Each panel contains 12 normal results (+), 12 CHF without AF results (×), and 3 CHF with AF results (△). The diagonal lines correspond to perfect agreement.

The results for $S_\tau(1/32)$ and $\sigma_{\text{wav}}(32)$ are excellent, illustrating the remarkable ability of the model to mimic the data at a time scale of 32 interbeat intervals. The correlation coefficients for these two measures lie very close to unity, indicating almost perfect correspondence between the original and simulated data.

The upper left panel presents values for $\sigma_{\text{int}}$, the standard deviation of the RR intervals. Agreement between the simulation and original data is only fair, with simulated values of $\sigma_{\text{int}}$ generally somewhat higher than desired. The correlation coefficient $\rho = 0.71$ for all 27 simulations indicates significant correlation ($p < 0.0003$ that 27 random pairs of numbers with the same distribution would achieve a magnitude of $\rho \geq 0.71$) but not extremely close agreement between simulated and original RR-interval sequences.

This disagreement highlights a problem with the simple jittered integrate-and-fire model. The fractal-Gaussian-noise integrate-and-fire model *without* jitter yields sub-

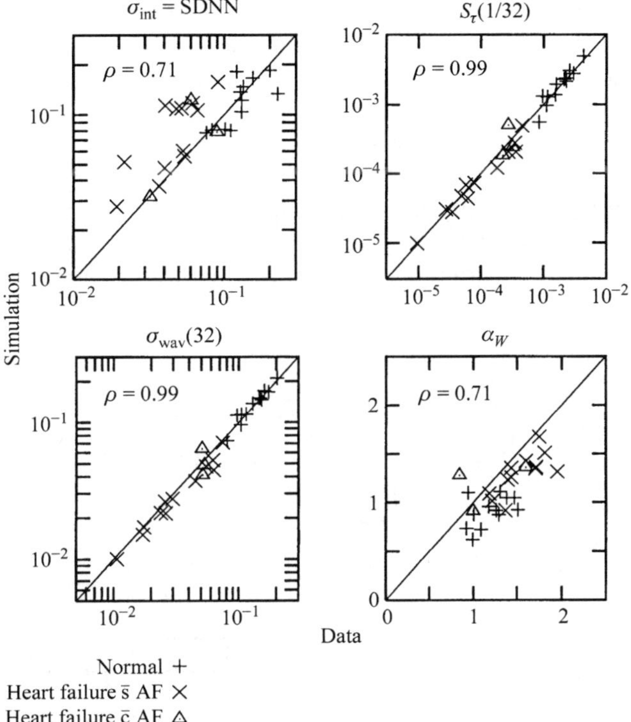


**Figure 18** Simulation accuracy for four measures with their correlation coefficients $\rho$. The jittered integrate-and-fire model performs remarkably well for the measures $S_\tau(1/32)$ and $\sigma_{wav}(32)$; however, it does not perform nearly as well for the measures $\sigma_{int}$ and $\alpha_W$. These disagreements highlight problems with the simple jittered integrate-and-fire model.

stantially closer agreement between the data and simulation for this statistic [8]. Although the introduction of jitter leads to improved agreement for $\sigma_{wav}$, it brings the model results farther away from the data as measured by $\sigma_{int}$. Apparently, a method other than jitter must be devised to forge increased agreement with $\sigma_{wav}$ while not degrading the agreement with $\sigma_{int}$. One possibility is the reordering of the individual RR intervals over short time scales.

The simulated exponent $\alpha_W$ also fails to show close agreement with the data, suggesting that there are other features of the model that are less than ideal. However, since this measure is of little use in separating CHF patients from normal subjects, this disagreement takes on reduced importance.

### 6.4. Toward an Improved Model of Heart Rate Variability

Agreement between the simulations and data would most likely be improved by adding other inputs to the model aside from fractal noise, as illustrated schematically in Figure 19. Input signals related to respiration and blood pressure are two likely candidates because it is well known that variations in these physiological functions influence

HRV. Experience also teaches us that there is a measure of white Gaussian noise present, possibly associated with the measurement process.

As indicated in Section 6.3.4, the introduction of jitter in the integrate-and-fire construct improves the agreement of the model with certain features of the data but degrades it for others. An adaptive reset will provide a more flexible alternative to the jitter. Moreover, the threshold $\theta$ could be converted into a stochastic process as a way of incorporating variability in other elements of the system [17].

All things considered, the jittered integrate-and-fire construct does a rather good job of mimicking the actual data, although the introduction of a more physiologically based model would be a welcome addition.

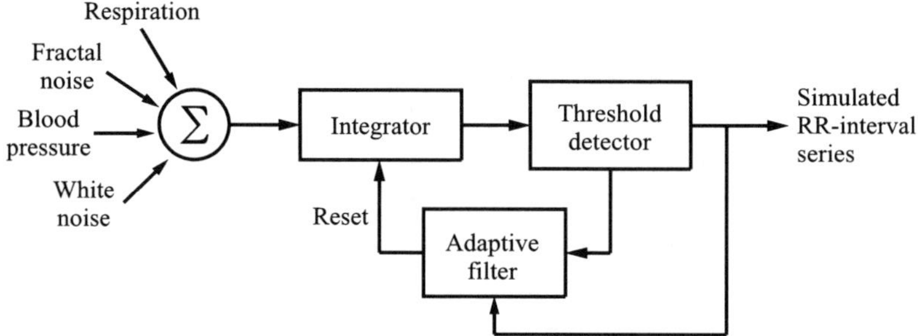


**Figure 19** Potential modifications to the simple integrate-and-fire model. Multiple physiologically based inputs and adaptive feedback could serve to produce more realistic RR-interval simulations.

## 7. CONCLUSION

For the purposes of HRV analysis, the occurrence times of the sequence of R phases in the electrocardiogram can be represented as a fractal-rate stochastic point process. Using a collection of standard and novel HRV measures, we have examined the sequence of times between the R phases, namely the RR time series, of ECGs recorded from normal subjects and patients with congestive heart failure, as well as from several patients with other cardiovascular disorders. CHF patients who also suffered from atrial fibrillation were treated as a separate class. We examined 16 HRV measures comprising frequency-domain, time-domain, scaling-exponent, Allan-factor, detrended-fluctuation-analysis, and wavelet-transform methods.

ROC analysis was used to compare and contrast the performance of these 16 measures with respect to their abilities to distinguish properly heart-failure patients from normal subjects over a broad range of record lengths (minutes to hours). Scale-dependent measures rendered performance that was substantially superior to that of scale-independent ones. Judged on the basis of performance and computation time, 2 measures of the 16 emerged at the top of the list: the wavelet-transform standard deviation $\sigma_{\text{wav}}(32)$ and the RR interval–based power spectral density $S_{\tau}(1/32)$. They share in common the dependence on a single scale, or small range of scales, near 32 heartbeat intervals. The behavior of the ECG at this particular scale, corresponding to about 25 seconds, turns out to be a significant marker for the presence of cardiac dysfunction. The scale value itself is far more important than the details of the measures used to examine it.

Application of these techniques to a large database of normal and CHF records will make it possible to uncover just how the ECGs of pathological patients differ from those of normal subjects in the vicinity of this special scale. This would facilitate the development of an optimal diagnostic measure, which might take the form of a specialized weighting function over the interval-based power spectral density or over some combination of elementary measures.

We also addressed an issue of fundamental importance in cardiac physiology: the determination of whether the RR time series arises from a chaotic attractor or has an underlying stochastic origin. Using nonlinear-dynamics theory, together with surrogate-data analysis, we established that the RR sequences from both normal subjects and pathological patients have stochastic, rather than deterministic, origins. The use of a special embedding of the *differences* between adjacent RR intervals enabled us to reach this conclusion. This technique substantially reduced the natural correlations inherent in the interbeat-interval time series, which can masquerade as chaos and confound simpler analyses.

Finally, we developed a jittered integrate-and-fire model built around a fractal-Gaussian-noise kernel. Simulations based on this model provided realistic, although not perfect, replicas of real heartbeat sequences. The model was least successful in predicting interbeat-interval statistics and fractal-exponent values. Nevertheless, it could find use in some applications such as pacemaker excitation.

To confirm the observations we have reported, it will be desirable to carry out continuation studies using large numbers of out-of-sample records. It will also be useful to examine how the performance of the various measures correlates with the severity of cardiac dysfunction as judged, for example, by the NYHA clinical scale. Comparisons with existing CHF studies that make use of HRV measures should also be conducted.

## REFERENCES


[1] J. R. Levick, *An Introduction to Cardiovascular Physiology.* London: Butterworths, 1991.

[2] R. Kitney, D. Linkens, A. Selman, and A. McDonald, The interaction between heart rate and respiration: Part II—Nonlinear analysis based on computer modeling. *Automedica* 4:141–153, 1982.

[3] M. Malik, J. T. Bigger, A. J. Camm, R. E. Kleiger, A. Malliani, A. J. Moss, P. J. Schwartz, and the Task Force of the European Society of Cardiology and the North American Society of Pacing and Electrophysiology, Heart rate variability—Standards of measurement, physiological interpretation, and clinical use. *Eur. Heart J.* 17:354–381, 1996. Also published in *Circulation* 93:1043–1065, 1996.

[4] E. H. Hon and S. T. Lee, Electronic evaluations of the fetal heart rate patterns preceding fetal death, further observations. *Am. J. Obstet. Gynec.* 87:814–826, 1965.

[5] R. E. Kleiger, J. P. Miller, J. T. Bigger, A. J. Moss, and the Multicenter Post-Infarction Research Group, Decreased heart rate variability and its association with increased mortality after acute myocardial infarction. *Am. J. Cardiol.* 59:256–262, 1987.

[6] J. B. Bassingthwaighte and G. M. Raymond, Evaluating rescaled range analysis for time series. *Ann. Biomed. Eng.* 22:432–444, 1994.

[7] R. G. Turcott and M. C. Teich, Long-duration correlation and attractor topology of the heartbeat rate differ for healthy patients and those with heart failure. *Proc. SPIE* 2036:22–39, 1993.

[8] R. G. Turcott and M. C. Teich, Fractal character of the electrocardiogram: Distinguishing heart-failure and normal patients. *Ann. Biomed. Eng.* 24:269–293, 1996.

[9] R. W. DeBoer, J. M. Karemaker, and J. Strackee, Comparing spectra of a series of point events particularly for heart rate variability data. *IEEE Trans. Biomed. Eng.* BME-31:384–387, 1984.

[10] D. R. Cox and P. A. W. Lewis, *The Statistical Analysis of Series of Events.* London: Methuen, 1966.

[11] P. A. W. Lewis, ed., *Stochastic Point Processes: Statistical Analysis, Theory, and Applications.* New York: Wiley-Interscience, 1972.

[12] B. E. A. Saleh, *Photoelectron Statistics.* New York: Springer-Verlag, 1978.

[13] M. C. Teich and S. M. Khanna, Pulse-number distribution for the neural spike train in the cat's auditory nerve. *J. Acoust. Soc. Am.* 77:1110–1128, 1985.

[14] P. R. Prucnal and M. C. Teich, Refractory effects in neural counting processes with exponentially decaying rates. *IEEE Trans. Syst. Man Cybern.* SMC-13:1028–1033, 1983.

[15] M. C. Teich, C. Heneghan, S. B. Lowen, T. Ozaki, and E. Kaplan, Fractal character of the neural spike train in the visual system of the cat, *J. Opt. Soc. Am. A* 14:529–546, 1997.

[16] S. B. Lowen and M. C. Teich, Estimation and simulation of fractal stochastic point processes. *Fractals* 3:183–210, 1995.

[17] S. Thurner, S. B. Lowen, M. C. Feurstein, C. Heneghan, H. G. Feichtinger, and M. C. Teich, Analysis, synthesis, and estimation of fractal-rate stochastic point processes. *Fractals* 5:565–595, 1997.

[18] W. Rudin, *Principles of Mathematical Analysis*, 3rd ed. New York: McGraw-Hill, 1976.

[19] B. B. Mandelbrot, *The Fractal Geometry of Nature*, 2nd ed. New York: W. H. Freeman, 1983.

[20] J. Feder, *Fractals.* New York: Plenum, 1988.

[21] L. S. Liebovitch, *Fractals and Chaos Simplified for the Life Sciences.* New York: Oxford, 1998.

[22] M. F. Shlesinger and B. J. West, Complex fractal dimension of the bronchial tree. *Phys. Rev. Lett.* 67:2106–2108, 1991.

[23] A. Papoulis, *Probability, Random Variables, and Stochastic Processes*, 3rd ed. New York: McGraw-Hill, 1991.

[24] A. V. Oppenheim and R. W. Schafer, *Discrete-Time Signal Processing.* Englewood Cliffs, NJ: Prentice-Hall, 1989.

[25] W. H. Press, S. A. Teukolsky, W. T. Vetterling, and B. P. Flannery, *Numerical Recipes in C*, 2nd ed. New York: Cambridge University Press, 1992.

[26] S. Akselrod, D. Gordon, F. A. Ubel, D. C. Shannon, A. C. Barger, and R. J. Cohen, Power spectrum analysis of heart rate fluctuation: a quantitative probe of beat to beat cardiovascular control. *Science* 213:220–222, 1981.

[27] B. Pomeranz, R. J. Macaulay, M. A. Caudill, I. Kutz, D. Adam, D. Gordon, K. M. Kilborn, A. C. Barger, D. C. Shannon, and R. J. Cohen, Assessment of autonomic function in humans by heart rate spectral analysis. *Am. J. Physiol.* 248:H151–H153, 1985.

[28] A. Malliani, M. Pagani, F. Lombardi, and S. Cerutti, Cardiovascular neural regulation explored in the frequency domain. *Circulation* 84:482–492, 1991.

[29] M. M. Wolf, G. A. Varigos, D. Hunt, and J. G. Sloman, Sinus arrhythmia in acute myocardial infarction. *Med. J. Australia.* 2:52–53, 1978.

[30] J. P. Saul, P. Albrecht, R. D. Berger, and R. J. Cohen, Analysis of long term heart rate variability: Methods, $1/f$ scaling and implications. In *Computers in Cardiology 1987*, pp. 419–422. Washington: IEEE Computer Society Press, 1988.

[31] D. R. Cox and V. Isham, *Point Processes*. London: Chapman & Hall, 1980.

[32] S. B. Lowen and M. C. Teich, The periodogram and Allan variance reveal fractal exponents greater than unity in auditory-nerve spike trains. *J. Acoust. Soc. Am.* 99:3585–3591, 1996.

[33] D. W. Allan, Statistics of atomic frequency standards. *Proc. IEEE* 54:221–230, 1966.

[34] J. A. Barnes and D. W. Allan, A statistical model of flicker noise. *Proc. IEEE* 54:176–178, 1966.

[35] M. C. Teich, Fractal behavior of the electrocardiogram: Distinguishing heart-failure and normal patients using wavelet analysis. *Proc. Int. Conf. IEEE Eng. Med. Biol. Soc.* 18:1128–1129, 1996.

[36] M. C. Teich, C. Heneghan, S. B. Lowen, and R. G. Turcott, Estimating the fractal exponent of point processes in biological systems using wavelet- and Fourier-transform methods. In *Wavelets in Medicine and Biology*, A. Aldroubi and M. Unser, eds., pp. 383–412. Boca Raton, FL: CRC Press, 1996.

[37] P. Abry and P. Flandrin, Point processes, long-range dependence and wavelets. In *Wavelets in Medicine and Biology*, A. Aldroubi and M. Unser, eds., pp. 413–437. Boca Raton, FL: CRC Press, 1996.

[38] C. Heneghan, S. B. Lowen, and M. C. Teich, Wavelet analysis for estimating the fractal properties of neural firing patterns. In *Computational Neuroscience: Trends in Research 1995*, J. M. Bower, ed., pp. 441–446. San Diego: Academic Press, 1996.

[39] S. Mallat, A theory for multiresolution signal decomposition: The wavelet representation. *IEEE Trans. Pattern Anal. Mach. Intel.* 11:674–693, 1989.

[40] Y. Meyer, *Ondelettes et opérateurs*. Paris: Hermann, 1990.

[41] I. Daubechies, *Ten Lectures on Wavelets*. Philadelphia: Society for Industrial and Applied Mathematics, 1992.

[42] A. Aldroubi and M. Unser, eds., *Wavelets in Medicine and Biology*. Boca Raton, FL: CRC Press, 1996.

[43] M. Akay, ed., *Time Frequency and Wavelets in Biomedical Signal Processing*. New York: IEEE Press, 1997.

[44] M. Akay, Wavelet applications in medicine. *IEEE Spectrum* 34(5):50–56, 1997.

[45] A. Arneodo, G. Grasseau, and M. Holschneider, Wavelet transform of multifractals. *Phys. Rev. Lett.* 61:2281–2284, 1988.

[46] S. Thurner, M. C. Feurstein, and M. C. Teich, Multiresolution wavelet analysis of heartbeat intervals discriminates healthy patients from those with cardiac pathology. *Phys. Rev. Lett.* 80:1544–1547, 1988.

[47] S. Thurner, M. C. Feurstein, S. B. Lowen, and M. C. Teich, Receiver-operating-characteristic analysis reveals superiority of scale-dependent wavelet and spectral measures for assessing cardiac dysfunction. *Phys. Rev. Lett.* 81:5688–5691, 1998.

[48] M. C. Teich, Multiresolution wavelet analysis of heart rate variability for heart-failure and heart-transplant patients. *Proc. Int. Conf. IEEE Eng. Med. Biol. Soc.* 20:1136–1141, 1998.

[49] C. Heneghan, S. B. Lowen, and M. C. Teich, Analysis of spectral and wavelet-based measures used to assess cardiac pathology. *Proceedings 1999 IEEE International Conference on Acoustics, Speech, and Signal Processing (ICASSP)*, 1999, paper no. SPTM-8.2.

[50] Y. Ashkenazy, M. Lewkowicz, J. Levitan, H. Moelgaard, P. E. Bloch Thomsen, and K. Saermark, Discrimination of the healthy and sick cardiac autonomic nervous system by a new wavelet analysis of heartbeat intervals. *Fractals* 6:197–203, 1998.

[51] C.-K. Peng, S. Havlin, H. E. Stanley, and A. L. Goldberger, Quantification of scaling exponents and crossover phenomena in nonstationary heartbeat time series. *Chaos* 5:82–87, 1995.

[52] C.-K. Peng, S. Havlin, J. M. Hausdorff, J. Mietus, H. E. Stanley, and A. L. Goldberger, Fractal mechanisms and heart rate dynamics: Long-range correlations and their breakdown with disease. *J. Electrocardiol.* 28:59–65, 1996.

[53] K. K. L. Ho, G. B. Moody, C.-K. Peng, J. E. Mietus, M. G. Larson, D. Levy, and A. L. Goldberger, Predicting survival in heart failure case and control subjects by use of fully automated methods for deriving nonlinear and conventional indices of heart rate dynamics. *Circulation* 96:842–848, 1997.

[54] M. Kobayashi and T. Musha, $1/f$ fluctuations of heartbeat period. *IEEE Trans. Biomed. Eng.* BME-29:456–457, 1982.

[55] C.-K. Peng, J. Mietus, J. M. Hausdorff, S. Havlin, H. E. Stanley, and A. L. Goldberger, Long-range anticorrelations and non-Gaussian behavior of the heartbeat. *Phys. Rev. Lett.* 70:1343–1346, 1993.

[56] L. A. Nunes Amaral, A. L. Goldberger, P. Ch. Ivanov, and H. E. Stanley, Scale-independent measures and pathologic cardiac dynamics. *Phys. Rev. Lett.* 81:2388–2391, 1998.

[57] H. E. Hurst, Long-term storage capacity of reservoirs. *Trans. Am. Soc. Civ. Eng.* 116:770–808, 1951.

[58] W. Feller, The asymptotic distribution of the range of sums of independent random variables. *Ann. Math. Stat.* 22:427–432, 1951.

[59] H. E. Schepers, J. H. G. M. van Beek, and J. B. Bassingthwaighte, Comparison of four methods to estimate the fractal dimension of self-affine signals. *IEEE Eng. Med. Biol. Mag.* 11:57–64, 1992.

[60] J. Beran, *Statistics for Long-Memory Processes.* New York: Chapman & Hall, 1994.

[61] J. B. Bassingthwaighte, L. S. Liebovitch, and B. J. West, *Fractal Physiology.* New York: Oxford, 1994.

[62] H. L. Van Trees, *Detection Estimation, and Modulation Theory*, Part I. New York: Wiley, 1968.

[63] A. Hald, *Statistical Theory with Engineering Applications.* New York: Wiley, 1952.

[64] D. M. Green and J. A. Swets, *Signal Detection Theory and Psychophysics.* Los Altos, CA: Peninsula Press, 1988.

[65] Y. Ashkenazy, M. Lewkowicz, J. Levitan, S. Havlin, K. Saermark, H. Moelgaard, P. E. Bloch Thomsen, M. Moller, U. Hintze, and H. Huikuri, Scale specific and scale independent measures of heart rate variability as risk indicators. e-print http://xxx.lanl.gov/abs/physics/9909029.

[66] S. V. Buldyrev, A. L. Goldberger, S. Havlin, R. N. Mantegna, M. E. Matsa, C.-K. Peng, M. Simons, and H. E. Stanley, Long-range correlation properties of coding and noncoding DNA sequences: GenBank analysis. *Phys. Rev. E* 51:5084–5091, 1995.

[67] A. L. Goldberger and B. J. West, Applications of nonlinear dynamics to clinical cardiology. *Ann. N. Y. Acad. Sci.* 504:195–213, 1987.

[68] C.-S. Poon and C. K. Merrill, Decrease of cardiac chaos in congestive heart failure. *Nature* 389:492–495, 1997.

[69] J. K. Kanters, N.-H. Holstein-Rathlou, and E. Agner, Lack of evidence for low-dimensional chaos in heart rate variability. *J. Cardiovasc. Electrophysiol.* 5:591–601, 1994.

[70] D. E. Roach and R. S. Sheldon, Information scaling properties of heart rate variability. *Am. J. Physiol.* 274:H1970–H1978, 1998.

[71] J. Theiler, Estimating fractal dimension. *J. Opt. Soc. Am. A* 7:1055–1073, 1990.

[72] L. S. Liebovitch and T. Tóth, A fast algorithm to determine fractal dimensions by box counting. *Phys. Lett. A* 141:386–390, 1989.

[73] M. Ding, C. Grebogi, E. Ott, T. Sauer, and J. A. Yorke, Plateau onset for correlation dimension: When does it occur? *Phys. Rev. Lett.* 70:3872–3875, 1993.

[74] P. Grassberger and I. Procaccia, Measuring the strangeness of strange attractors. *Physica D* 9:189–208, 1983.

[75] L. S. Liebovitch, Introduction to the properties and analysis of fractal objects, processes, and data. In *Advanced Methods of Physiological System Modeling*, V. Z. Marmarelis, ed., Vol. 2, pp. 225–239. New York: Plenum, 1989.

[76] A. Wolf, J. B. Swift, H. L. Swinney, and J. A. Vastano, Determining Lyapunov exponents from a time series. *Physica D* 16:285–317, 1985.

[77] A. R. Osborne and A. Provenzale, Finite correlation dimension for stochastic systems with power-law spectra. *Physica D* 35:357–381, 1989.

[78] S. J. Schiff and T. Chang, Differentiation of linearly correlated noise from chaos in a biologic system using surrogate data. *Biol. Cybern.* 67:387–393, 1992.

[79] J. Theiler, S. Eubank, A. Longtin, B. Galdrikian, and J. D. Farmer, Testing for nonlinearity in time series: The method of surrogate data. *Physica D* 58:77–94, 1992.

[80] A. M. Albano, J. Muench, C. Schwartz, A. I. Mees, and P. E. Rapp, Singular-value decomposition and the Grassberger–Procaccia algorithm. *Phys. Rev. A* 38:3017–3026, 1988.

[81] C. Grebogi, E. Ott, and J. A. Yorke, Chaos, strange attractors, and fractal basin boundaries in nonlinear dynamics. *Science* 238:632–638, 1987.

[82] F. X. Witkowski, K. M. Kavanagh, P. A. Penkoske, R. Plonsey, M. L. Spano, W. L. Ditto, and D. T. Kaplan, Evidence for determinism in ventricular fibrillation. *Phys. Rev. Lett.* 75:1230–1233, 1995.

[83] A. Garfinkel, J. N. Weiss, W. L. Ditto, and M. L. Spano, Chaos control of cardiac arrhythmias. *Trends Cardiovasc. Med.* 5:76–80, 1995.

[84] C. Heneghan, S. M. Khanna, Å. Flock, M. Ulfendahl, L. Brundin, and M. C. Teich, Investigating the nonlinear dynamics of cellular motion in the inner ear using the short-time Fourier and continuous wavelet transforms. *IEEE Trans. Signal Process.* 42:3335–3352, 1994.

[85] M. C. Teich, C. Heneghan, S. M. Khanna, Å. Flock, M. Ulfendahl, and L. Brundin, Investigating routes to chaos in the guinea-pig cochlea using the continuous wavelet transform and the short-time Fourier transform. *Ann. Biomed. Eng.* 23:583–607, 1995.

[86] M. C. Teich, C. Heneghan, and S. M. Khanna, Analysis of cellular vibrations in the living cochlea using the continuous wavelet transform and the short-time Fourier transform. In *Time Frequency and Wavelets in Biomedical Engineering*, M. Akay, ed., pp. 243–269. New York: IEEE Press, 1997.

[87] D. J. Christini and J. J. Collins, Controlling nonchaotic neuronal noise using chaos control techniques. *Phys. Rev. Lett.* 75: 2782–2785, 1995.

[88] H. C. Tuckwell, *Stochastic Processes in the Neurosciences*. Philadelphia: Society for Industrial and Applied Mathematics, 1989.

[89] R. D. Berger, S. Akselrod, D. Gordon, and R. J. Cohen, An efficient algorithm for spectral analysis of heart rate variability. *IEEE Trans. Biomed. Eng.* BME-33:900–904, 1986.

[90] B. W. Hyndman and R. K. Mohn, A pulse modulator model for pacemaker activity. *Digest 10th Int. Conf. Med. Biol. Eng.*, Dresden, Germany, p. 223, 1973.

[91] B. W. Hyndman and R. K. Mohn, Model of the cardiac pacemaker and its use in decoding the information content of cardiac intervals. *Automedica* 1:239–252, 1975.

[92] O. Rompelman, J. B. I. M. Snijders, and C. J. van Spronsen, The measurement of heart rate variability spectra with the help of a personal computer. *IEEE Trans. Biomed. Eng.* BME-29:503–510, 1982.

[93] P. Flandrin, Wavelet analysis and synthesis of fractional Brownian motion. *IEEE Trans. Inform. Theory* 38:910–917, 1992.

[94] B. B. Mandelbrot, A fast fractional Gaussian noise generator. *Water Resour. Res.* 7:543–553, 1971.

[95] T. Lundahl, W. J. Ohley, S. M. Kay, and R. Siffert, Fractional Brownian motion: A maximum likelihood estimator and its application to image texture. *IEEE Trans. Med. Imaging* 5:152–161, 1986.

[96] M. A. Stoksik, R. G. Lane, and D. T. Nguyen, Accurate synthesis of fractional Brownian motion using wavelets. *Electron. Lett.* 30:383–384, 1994.

[97] S. B. Lowen, S. S. Cash, M.-M. Poo, and M. C. Teich, Quantal neurotransmitter secretion rate exhibits fractal behavior. *J. Neurosci.* 17:5666–5677, 1997.

[98] S. B. Lowen and M. C. Teich, Fractal renewal processes generate $1/f$ noise. *Phys. Rev. E* 47:992–1001, 1993.

[99] S. B. Lowen and M. C. Teich, Fractal renewal processes, *IEEE Trans. Inform. Theory* 39: 1669–1671, 1993.

[100] S. B. Lowen and M. C. Teich, Fractal auditory-nerve firing patterns may derive from fractal switching in sensory hair-cell ion channels. In *Noise in Physical Systems and $1/f$ Fluctuations*, P. H. Handel and A. L. Chung, eds., pp. 781–784. New York: American Institute of Physics, 1993. AIP Conference Proceedings 285.

[101] L. S. Liebovitch and T. I. Tóth, Using fractals to understand the opening and closing of ion channels. *Ann. Biomed. Eng.* 18:177–194, 1990.

# APPENDIX A

Table A1 provides a summary of some significant statistics of the RR records analyzed in this chapter, before truncation to 75,821 intervals.

**TABLE A1** Elementary Statistics of the Original RR-Interval Recordings before Truncation to 75,821 RR Intervals

| File name | Condition | Number of RR intervals | Recording duration (s) | RR-interval mean (s) | Mean rate $(s)^{-1}$ | RR interval SD (s) |
|---|---|---|---|---|---|---|
| n16265 | Normal | 100,460 | 80,061 | 0.797 | 1.255 | 0.171 |
| n16272 | Normal | 93,177 | 84,396 | 0.906 | 1.104 | 0.142 |
| n16273 | Normal | 89,846 | 74,348 | 0.828 | 1.208 | 0.146 |
| n16420 | Normal | 102,081 | 77,761 | 0.762 | 1.313 | 0.101 |
| n16483 | Normal | 104,338 | 76,099 | 0.729 | 1.371 | 0.089 |
| n16539 | Normal | 108,331 | 84,669 | 0.782 | 1.279 | 0.150 |
| n16773 | Normal | 82,160 | 78,141 | 0.951 | 1.051 | 0.245 |
| n16786 | Normal | 101,630 | 84,052 | 0.827 | 1.209 | 0.116 |
| n16795 | Normal | 87,061 | 74,735 | 0.858 | 1.165 | 0.212 |
| n17052 | Normal | 87,548 | 76,400 | 0.873 | 1.146 | 0.159 |
| n17453 | Normal | 100,674 | 74,482 | 0.740 | 1.352 | 0.103 |
| nc4 | Normal | 88,140 | 71,396 | 0.810 | 1.235 | 0.132 |
| a6796 | HF s̄ AF | 75,821 | 71,934 | 0.949 | 1.054 | 0.091 |
| a8519 | HF s̄ AF | 80,878 | 71,948 | 0.890 | 1.124 | 0.062 |
| a8679 | HF s̄ AF | 119,094 | 71,180 | 0.598 | 1.673 | 0.051 |
| a9049 | HF s̄ AF | 92,497 | 71,959 | 0.778 | 1.285 | 0.058 |
| a9377 | HF s̄ AF | 90,644 | 71,952 | 0.794 | 1.260 | 0.060 |
| a9435 | HF s̄ AF | 114,959 | 71,158 | 0.619 | 1.616 | 0.034 |
| a9643 | HF s̄ AF | 148,111 | 71,958 | 0.486 | 2.058 | 0.024 |
| a9674 | HF s̄ AF | 115,542 | 71,968 | 0.623 | 1.605 | 0.084 |
| a9706 | HF s̄ AF | 115,064 | 71,339 | 0.620 | 1.613 | 0.100 |
| a9723 | HF s̄ AF | 115,597 | 71,956 | 0.622 | 1.607 | 0.027 |
| a9778 | HF s̄ AF | 93,607 | 71,955 | 0.769 | 1.301 | 0.070 |
| a9837 | HF s̄ AF | 115,205 | 71,944 | 0.624 | 1.601 | 0.066 |
| a7257 | HF c̄ AF | 118,376 | 71,140 | 0.601 | 1.664 | 0.039 |
| a8552 | HF c̄ AF | 111,826 | 71,833 | 0.642 | 1.557 | 0.062 |
| a8988 | HF c̄ AF | 118,058 | 71,131 | 0.603 | 1.660 | 0.091 |
| tp987 | Transplant | 106,394 | 67,217 | 0.632 | 1.583 | 0.083 |
| vt973 | VT | 86,992 | 70,044 | 0.805 | 1.242 | 0.202 |
| sd984 | SCD | 77,511 | 62,786 | 0.810 | 1.234 | 0.089 |
| slp59 | Apnea | 16,874 | 14,399 | 0.853 | 1.172 | 0.103 |
| slp66 | Apnea | 15,751 | 13,199 | 0.838 | 1.193 | 0.103 |

Chapter 7

# VENTRICULOARTERIAL INTERACTION AFTER ACUTE INCREASE OF THE AORTIC INPUT IMPEDANCE: DESCRIPTION USING RECURRENCE PLOT ANALYSIS

Stephan Schulz, Robert Bauernschmitt, Andreas Schwarzhaupt, C. F. Vahl, and Uwe Kiencke

## 1. ABOUT CARDIOVASCULAR MECHANICS AND REGULATION

The heart cycle is the result of sequentially contracting atria and ventricles resulting in a cycle of pressure and volume changes. The cardiac cycle has four phases. Plotting ventricular pressure versus its volume forms a closed loop representing a cardiac cycle. Pressure–volume loops reveal a ventricular filling period (0.5 s), an isovolumetric contraction period (0.05 s), a systolic ejection period (0.3 s), and an isovolumetric relaxation period (0.08 s).

During ventricular filling blood flows, supported by the sucking effect of the relaxing ventricle, passively from the great veins through the atria and the open inlet valves into the ventricle until the ventricle reaches it natural volume. Now the rate of filling slows down, the ventricle starts to expand, and the ventricular pressure rises. After the atrial contraction forces some additional blood into the ventricle, the end-diastolic volume and pressure are reached. In the short isovolumetric contraction period the atrioventricular valves close, forced by the reversed pressure gradient between the atrium and the ventricle. Now the contracting ventricle is a closed chamber and pressure rises steeply.

At the moment ventricular pressure exceeds arterial pressure, the outflow valves are compelled to open and the systolic ejection period starts. During the first phase the blood is not able to escape fast enough out of the arterial tree and is stored in the large elastic arteries. The arterial pressure reaches its maximum. As systolic ejection diminishes, the blood volume flowing away through the arterial system exceeds the ejection rate and the pressure begins to decrease. Even after the active ejection phase, the blood flow continues owing to the inertia of the blood mass. The upcoming inversion of the ventriculoarterial pressure gradient reduces the outflow until a brief backflow closes the outflow valve. An end-systolic volume remains in the ventricle and allows the stroke volume to increase during exercise.

After the closing of the outflow valves, the ventricles are closed chambers again and the isovolumetric relaxation period begins. When ventricular pressure falls under the atrial pressure, the atrioventricular valves open and blood floods in from the atria, which have been refilled during the ventricular systole.

Examining a ventriculoarterial interaction, the influences of different filling volumes or pressure gradients on cardiac mechanics are of interest. Depending on the filling of the heart (preload) or an increased aortic pressure (afterload), there are influences on the cardiac mechanics. One important intracardiac mechanism is the so-called Frank–Starling mechanism. Compared with an initial condition, an increased venous supply results in increased end-diastolic filling of the ventricle (preload). As a consequence, the more stretched heart muscle fibers can perform a stronger shortening and the stroke volume increases. This mechanism plays a role in short-time volume balance and in the coordination of the stroke volumes of the left and right sides of the heart. Increased aortic pressure, for example, after increased peripheral resistance, results in an increased pressure course in the ventricle. The reduced ejection volume after an increase in diastolic aortic pressure causes a higher end-systolic volume and, because of the unchanged venous reflow, increased diastolic filling for the next heart action. Again, increased contractility is the result.

The behavior of this mechanism makes it possible to derive static parameters from the pressure–volume loops. The slope of a line fitted to the end-systolic pressure points in the pressure–volume loops after different preloads is called the left ventricular end-systolic elastance ($E_{es}$). The slope of the arterial end-systolic pressure–volume relationship is the arterial elastance ($E_a$). The interaction between heart and subsequent arterial system—usually termed "ventriculoarterial coupling"—is determined by the ratio of arterial ($E_a$) and left ventricular end-systolic elastance ($E_{es}$). This classical method provides static descriptions of the actual cardiac mechanics [8]. The cardiovascular system, however, is characterized by dynamical processes continuously updated by nonlinear feed-forward and feedback inputs [12]. Interacting output parameters vary depending on other system components and physiological state changes [12]. Other involved extracardiac adaptation mechanisms such as activation of the sympathetic tone, which influences the contractility and the heart rate, or the short-time pressure regulation of the baroreceptor reflex could not be described by only a static method. Depending on the number of involved parameters and control circuits, a physiological system shows a more or less deterministic structure and changing information content.

Physiologic systems usually include both linear and nonlinear properties. Still, it is unclear whether the cardiovascular system prefers linear or nonlinear regulation during healthy states, although there is at least strong experimental and clinical evidence that it shows nonlinear chaotic behavior [2, 4, 9, 10]. Furthermore, it could be of interest which changes of linear or nonlinear dynamics, as far as the degree of determinism and the information content (entropy) of the system are concerned, take place in response to pathophysiologic events. Recurrence plot analysis of dynamic signals was originally introduced to detect recurring patterns and nonstationarities [12]. Applied to physiologic signals, this technique could provide insight into regulation mechanisms during health and disease.

The determination of time correlation patterns and their information content assumes the introduction of a state-space representation for certain variables characterizing ventriculoarterial interaction. In particular, the question was posed whether a strong disturbance of ventriculoarterial dynamics—total occlusion of the descending aorta—leads to changes in the degree of system determinism or the number of involved control circuits, described by the system entropy, of the cardiocirculatory system.

## 2. BASICS BEFORE APPLYING RECURRENCE PLOTS ON AORTIC AND VENTRICULAR HEMODYNAMIC SIGNALS

### 2.1. Definition of a Ventriculoarterial State Space Representation

Very often the parameters of the $M$-dimensional system of interest are unknown and only one parameter can be accessed by measurements. The usual procedure is to reconstruct the $M$-dimensional state space from one time series by the method of time delays [3]. Using several dependent variables, it can be shown that the reconstruction of the dynamics from one $M$-dimensionally embedded time series shows the same behavior as the whole system [3]. In accordance with the classical methods, where the left ventricular pressure (LVP) and the aortic pressure (AOP) and flow (AOF) are used to calculate the ventricular and arterial elastances, time series of these signals are directly used to construct a three-dimensional orbit, the ventriculoarterial orbit (VAO) (Figure 3). The informational content of the LVP stands primarily for the left ventricular regulatory system and the AOP for the information content of the baroreceptor-controlled peripheral arterial system. The middle- and long-term mechanisms of blood pressure control are not examined in this study. The AOF is the result of the pressure gradient and keeps the informational content of the "Windkessel" function

$$\mathrm{VAO} = [\mathrm{LVP}(t),\ \mathrm{AOP}(t + \Delta t),\ \mathrm{AOF}(t + \Delta t)]$$

The time delay $\Delta t$ between the LVP and the aortic signals reflects the short delay between the heart action and the resulting waves in the arterial tree. A time lag of 0.01 second for $\Delta t$ corresponds to a distance of 5 cm between the sensor locations and a pulse wave velocity of 5 m/s.

Whereas LVP is the signal of the source, AOP and AOF represent the response of the arterial system. Every change in the ventriculoarterial interaction is supposed to influence the VAO and its parameters.

### 2.2. Visualization of the Ventriculoarterial Dynamics Using Recurrence Plot Strategies

The recurrence plot (Figure 4) is a graphical tool for visualizing the tendency of the traces of a dynamic system to follow an attractor and to revisit certain zones [3]. A sphere with radius $r$ is chosen and a maximum of 10 nearest neighboring points within this sphere is calculated. A fixed radius of 10 units has been used for our calculations. The recurrence map of the VAO($i$) ($i = 1, \ldots, N$, $N =$ number of points) can be obtained by plotting the index $i$ of every single point on the orbit ($x$ axis) against the indices of its maximum 10 nearest neighbors ($y$ axis) in distance. Periodic signals show diagonal straight lines in the recurrence plot (Figure 4). The distance between two consecutive points is the sample period. So the number of points of an orthogonal line between two neighbors is the time period of the sampled signals.

### 2.3. Quantitative Description of Recurrence Plots

Webber and Zbilut [12] proposed parameters for quantifying recurrence plots. The percentage of a plot occupied by recurrent points is quantified by the *percent recurrence*. This parameter distinguishes between aperiodic and periodic systems. Aperiodic systems are characterized by smaller percent recurrence values than periodic systems. Depending on the total number of points, the recurrence values can become very small.

$$\text{percent recurrence} = \frac{\text{recurrentpoints} \times 100}{(\text{all points})^2}$$

Recurrent points that form upward diagonal line segments are an indicator of self-repeating strings of vectors farther down the dynamics. This separates dispersed recurrence plots from those that are organized in certain diagonal structures. The percentage of these points compared with all recurrent points is called *percent determinism*. The larger this value, the more deterministic is a system.

$$\text{percent determinism} = \frac{\text{diagonalpoints} \times 100}{\text{recurrent points}}$$

A system is deterministic if its behavior is ruled by known equations and initial conditions and therefore predictable. Its time evolution is completely determined by its actual state and its history. There are no random components in a deterministic system. In contrast, chaos is an aperiodic, seemingly random behavior in a deterministic system that exhibits sensitive dependence on initial conditions [2]. The information contained in a signal can be measured using entropy.

Calculating with base 2 logarithms, the *entropy* has the unit of bits and bits are the unit of information The higher the entropy value, the more bits are used to encode the deterministic system dynamics and the more complex is the deterministic system.

$$\text{Entropy} = -\sum P_i \log_2(P_i)$$

### 2.4. Experimental Setting

We recorded our data from nine anesthetized and thoracotomized 20-kg pigs with a sample frequency of 512 Hz during a period of 6 seconds using piezoelectric pressure sensors and an ultrasonic Doppler flowmeter. The sensors are placed in the ascending aorta (pressure and flow at the same location) and in the left ventricle (pressure) with a distance of about 5 cm. The control condition is compared with total occlusion of the descending aorta.

# 3. APPLICATION OF RECURRENCE PLOTS ON HEMODYNAMIC PRESSURE AND FLOW SIGNALS TO DESCRIBE PATHOPHYSIOLOGIC STATES

## 3.1. Hemodynamic Measurements and Ventriculoarterial Orbits

Representative examples for LVP, AOF, and the corresponding three-dimensional orbits are given in Figures 1, 2, and 3. Occluding the descending aorta leads to an increase in pressure (Figure 1b), a slight decrease in peak aortic flow (Figure 2b), and a decrease in heart frequency from 70/min to 55/min. Forced by the endeavor of every dynamic system to reach its optimal energetic state, the ventriculoarterial system shows the behavior of an attractor (Figure 3a). The attractor corresponding to aortic occlusion (Figure 3b) was compared with control conditions (Figure 3a).

## 3.2. Application of Recurrence Plots on the Ventriculoarterial Orbit

Flow and pressure curves for the patient with one occlusion show a higher pressure amplitude (Figure 1b) and reduced AOF (Figure 2b) caused by total aortic occlusion, but there is no clearly visible effect on dynamic properties, for example, on periodicity. So it is necessary to extract information from the sequence of points on the orbit. In general, the recurrence plot of the VAO (Figure 4) shows periodic crossing points and forbidden zones that identify the nonlinear character of the chosen orbit. The dominating structures of all plots are diagonals as a result of using periodic signals to embed the ventriculoarterial dynamics. Diagonal structures reflect vector strings repeating themselves. The plots have no "line of identity" because the information that points are nearest neighbors of themselves is trivial, so they are neglected.

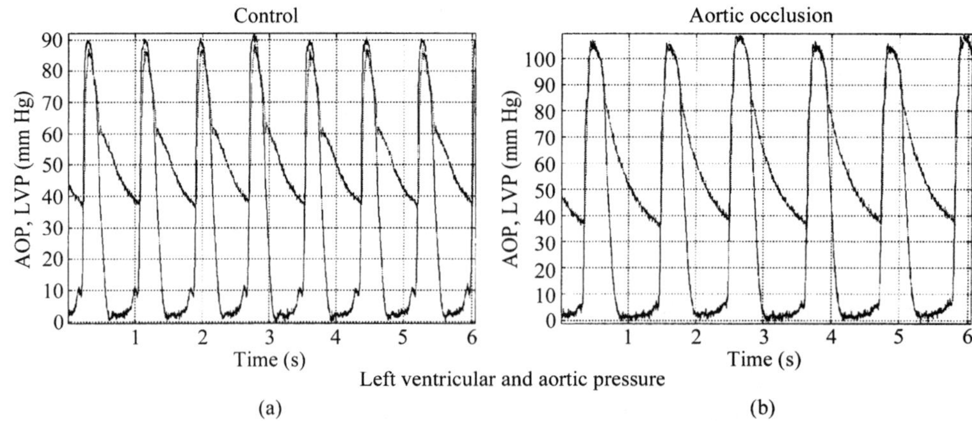


**Figure 1** From experiment 1: Compared with the control conditions (a) the pressure signals show increased amplitude during aortic occlusion (b). LVP, left ventricular pressure; AOP, aortic pressure.

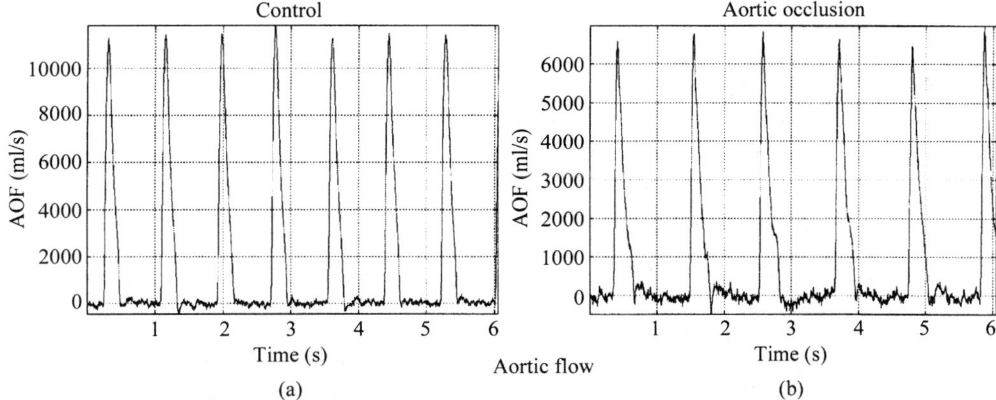


**Figure 2** From experiment 1: Aortic occlusion causes a decreased aortic peak flow (b) compared with the control condition (a). AOF, aortic flow.

The recurrence plot under control conditions (Figure 4) shows clear diagonal structures broken by small white spaces and a forbidden zone between the diagonals representing the period. Crosses in the diagonals mark sharp U-like curves in the attractor. In this case, a trajectory passes the reference point, turns, and comes back to the same region. Thickenings in the recurrence representation indicate a slowed-down velocity of the trajectory in this section of the attractor, so more nearest neighbors can be found. White or thin diagonal sections visualize rapid changes in time along the trajectory.

The recurrence plot patterns change after total aortic occlusion (Figure 5). Crossing points have almost completely disappeared. The sharp line structures in the control plot appear more smeared in the plot of aortic occlusion and there are less recurrent points in the spaces between the diagonals. The main diagonal appears more prominent in the occlusion plot.

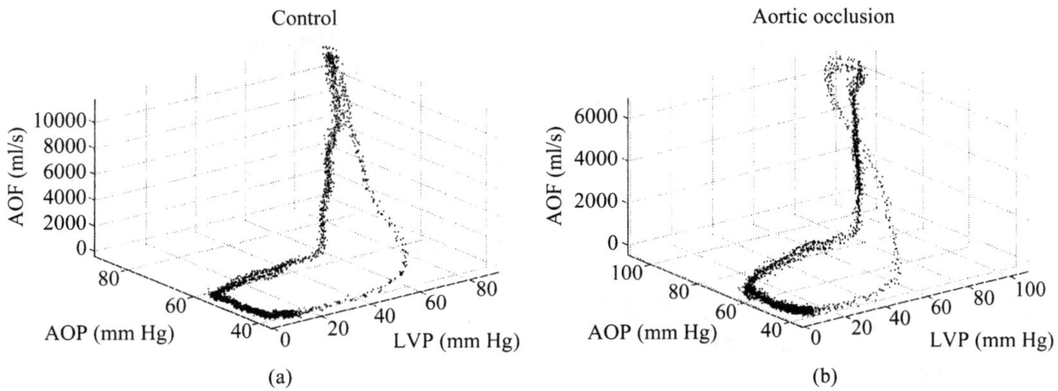


**Figure 3** From experiment 1: Example of a ventriculoarterial orbit VAO = [LVP($t$), AOP($t + \Delta t$), AOF($t + \Delta t$)] before (a) and after (b) aortic occlusion. $\Delta t$ is the time delay between the heart action and the arterial response. LVP, left ventricular pressure; AOP, aortic pressure; AOF, aortic flow.

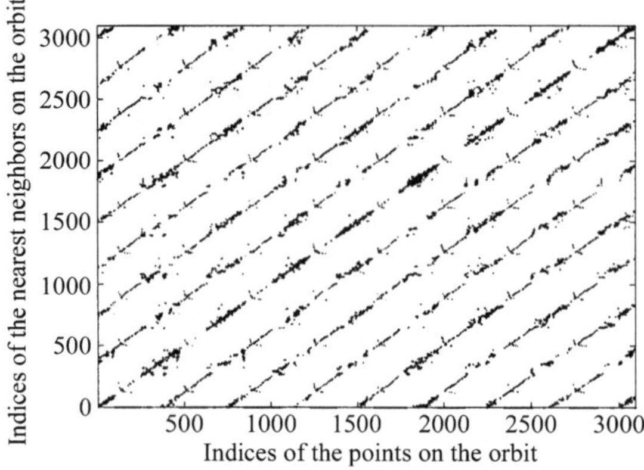


**Figure 4** Recurrence plot under control conditions.

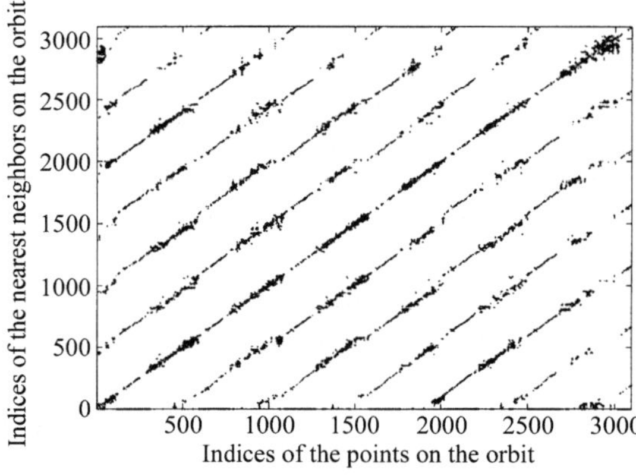


**Figure 5** Recurrence plot after total occlusion of the descending aorta.

### 3.3. Quantification of Ventriculoarterial Dynamics in the Recurrence Plot

From the recurrence plot, the nonlinear parameters percent recurrence, percent determinism, and entropy for each experiment were calculated (Figure 6).

Percent determinism and entropy showed significantly ($p < 0.003$) higher values. On average, percent determinism increases by 24% and the entropy by 2.3% as compared with the control conditions. The percent recurrence shows a small decrease (not statistically significant).

## 4. WHY NONLINEAR RECURRENCE PLOT ANALYSIS FOR DESCRIBING VENTRICULOARTERIAL INTERACTION

The understanding of "coupling" mechanisms between the left ventricle and the arterial system is essential for any description of cardiovascular action. The classical method for describing these interactions is based on the calculation of the ratio of ventricular and

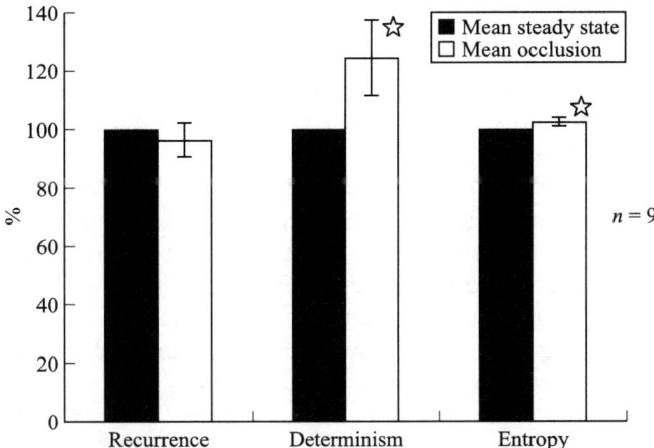


**Figure 6** Proportional changes of the parameters after aortic occlusion (white column). During aortic occlusion, the system shows a tendency to become less periodic. Rising determinism (24%) indicates more strings of vectors reoccurring farther down the dynamics. The complexity of the deterministic system (entropy) rises slightly (2.3%). (Number of experiments = 9, $p_{\mathrm{determinism}}$ and $p_{\mathrm{entropy}} < 0.003$.)

arterial elastance at end systole ($E_a/E_{es}$), providing a beat-by-beat analysis of the physical properties of ventricle and arteries [8].

Although yielding important information about cardiovascular action, this method has one serious drawback: it cannot provide statements about the dynamic nature of cardiovascular action and it completely neglects the nonlinear character of circulatory regulation. In general, physiologists are used to registering signals as one-dimensional functions of time or performing mathematical transformations yielding more information about the system (such as spectral analyses). Yet, even these transformations are still one-dimensional. By use of nonlinear methods, higher dimensional perspectives can be obtained from different but dependent physiological variables [11, 12].

In the present study, we suggest a nonlinear method to describe the interaction of ventricular and arterial variables by recurrence plot analysis. This method captures the nonlinear dynamics and depends on a few variables of the circulation, which are easily accessible even in vivo.

The parameters left ventricular pressure, aortic pressure, and aortic flow were chosen to construct a three-dimensional attractor. These parameters comprise information concerning the function of the source of energy (ventricle) as well as functional state and physical properties of the subsequent arterial system. Fundamentally, a description of any dynamic process is possible by less parameters, but any reduction of parameters may be associated with loss of information [12].

The chosen variables form attractors according to other dynamic systems described in the literature [1, 6]. The attractors show typical nonlinear behavior with forbidden zones and crossing points [12] and give a graphic description of the cardiocirculatory interaction. The plots are similar to those described by Eckmann, Mestivier,

Wagner and Webber [3, 5, 11, 12], but with clearer diagonal structures caused by the relatively high periodicity of flow and pressure signals.

Using these nonlinear methods, it should be possible not only to describe nonlinear properties but also to discriminate between different states of the cardiovascular system in response to pathologic disturbances. For this purpose, a simple but severe disturbance was introduced into the experimental model. By cross-clamping the descending aorta, a sudden increase in aortic input impedance was induced and the response of cardiac and vascular interaction was compared with control conditions.

An increased time period of the signals is manifested in less diagonals in the "occlusion" plot. The disappearance of crossing points is generated by broader pressure signals at the turning point of the flow peak. More smeared line structures in the plot of aortic occlusion and almost no recurrence points in the spaces between the diagonals reflect less complex dynamics with increased variability. This could also be a reason for a more prominent main diagonal in the occlusion plot.

When nonlinear parameters were calculated from the plots, no statistical difference was found between control conditions and occlusion of the descending aorta as far as percent recurrence is concerned. Thus, periodicity seems not to be altered by the intervention. Significant increases of percent determinism and entropy were found. This marks a more complex dynamic system coming up with a sudden increase of ventricular afterload, which may activate several mechanisms of local and systemic regulation [5, 6, 7]. Obviously, these mechanisms can cause a higher degree of determinism and complexity (as reflected by the increase of entropy) than those working during normal physiologic circumstances.

Although surprising at first sight, these findings partially reproduce the results of other groups, indicating that the healthy cardiovascular regulatory system follows nonlinear chaotic dynamics [2, 4, 9, 10]. Analyzing the interaction of aortic pressure and renal flow, Wagner et al. [9] demonstrated the nonlinear regulation of cardiovascular action. More deterministic behavior of the regulatory system, on the other hand, may indicate a disturbed state of this system, similar to findings in patients after heart transplantation [4] or after elimination of short-term control of blood pressure by baroreceptor denervation [10].

We conclude that cardiovascular action can be described using nonlinear techniques. Because recurrence plots impose no rigid constraints on data set size, stationarity, or statistical distribution [12], we hypothesize that this technique might be useful to discriminate different states of the ventriculoarterial system with respect to its dynamic structure and nonlinear behavior. To the best of our knowledge, this is the first attempt to describe ventriculoarterial interaction in physiologic and pathophysiologic states by nonlinear methods. This method is not intended to be competitive with the classical method of calculating $E_a/E_{es}$ but to be complementary. It not only yields information at a selected point of time during the circulatory cycle but also respects the dynamic nature of cardiovascular action. This may be of special interest to describe changes taking place when the system shifts from physiologic to pathologic conditions.

For further evaluation of these methods, larger studies are planned to investigate the influence of additional parameters of circulation as well as the influence of other pathophysiologic conditions in the experimental and clinical setup.

## ACKNOWLEDGMENTS

This research was performed at the Department of Cardiac Surgery, Prof. S. Hagl, and the Institute of Industrial Information Technique, Prof. U. Kiencke. The work is funded by the German Research Foundation (Deutsche Forschungsgemeinschaft) and is part of the special research program "Information Technology in Medicine"—"Computer and Sensor Supported Surgery."

Portions of this chapter were based on material previously published in: S. Shulz, R. Bauernschmitt, A. Schwarzhaupt, C. F. Vahl, U. Kiencke, "Description of the ventriculoarterial interaction dynamics using recurrence plot strategies." Biomedical Scientific Instruments 1998; 34: pp 269–74. 

## REFERENCES

[1] T. A. Denton and G. A. Diamond, Can the analytic techniques of nonlinear dynamics distinguish periodic, random and chaotic signals? *Comput. Biol. Med.* 21:243–263, 1991.

[2] T. A. Denton, G. A. Diamond, R. H. Helfant, S. Khan, and H. Karagueuzian, Fascinating rhythm: A primer on chaos theory and its application to cardiology. *Am. Heart J.* 120:1419–1440, 1990.

[3] J. P. Eckmann, S. O. Kamphorst, and D. Ruelle, Recurrence plots of dynamical systems. *Europhys. Lett.* 4:973–977, 1987.

[4] S. Guzzetti, M. G. Signorini, C. Cogliati, S. Mezzetti, A. Porta, S. Cerutti, and A. Malliani, Non-linear dynamics and chaotic indices in heart rate variability of normal subjects and heart-transplanted patients. *Cardiovasc. Res.* 31:441–446, 1996.

[5] D. Mestivier, N. P. Chau, X. Chanudet, B. Bauduceau, and P. Larroque, Relationship between diabetic autonomic dysfunction and heart rate variability assessed by recurrence plot. *Am. J. Physiol.* 272:H1094–H1099, 1997.

[6] H. O. Peitgen, H. Jürgens, and D. Saupe, *Fractals for the Classroom.* New York: Springer-Verlag, 1994.

[7] S. M. Pincus, Approximate entropy as a measure of system complexity. *Proc. Natl. Acad. Sci. USA* 88:2297–2301, 1991.

[8] K. Sunagawa, M. Sugimachi, K. Todaka, T. Kobota, K. Hayashida, R. Itaya, A. Chishaki, and A. Takeshita, Optimal coupling of the left ventricle with the arterial system. *Basic Res. Cardiol.* 88(Suppl 2):75–90, 1993.

[9] C. D. Wagner, B. Nafz, and P. B. Persson, Chaos in blood pressure control. *Cardiovasc. Res.* 31:380–387, 1996.

[10] C. D. Wagner and P. B. Persson, Nonlinear chaotic dynamics of arterial blood pressure and renal blood flow. *Am. J. Physiol.* 268:H621–H627, 1995.

[11] C. L. Webber Jr., M. A. Schmidt, and J. M. Walsh, Influences of isometric loading on biceps EMG dynamics as assessed by linear and nonlinear tools. *J. Appl. Physiol.* 78:814–822, 1995.

[12] C. L. Webber Jr. and J. P. Zbilut, Dynamical assessment of physiological systems and states using recurrence plot strategies. *J. Appl. Physiol.* 76:965–973, 1994.

# APPENDIX: MATLAB SOURCE CODES

## Calculate and Visualize Recurrence Plot

```
clear;

unumber=input('number of the first experiment ');
vnumber=input('number of the last experiment ');
date=input('date ');

for 1=unumber:vnumber

   clear map;
   clear mapall;
   clear points;
   clear distance;

name=['v'num2str(1)'.asc'];
file=['v'num2str(1)];

load(name);
h=eval(file);

c=1:length(h); %number of sample points
fs=512;%sample rate in Hz
Ts=1/fs;     %sample period in s
t=c*Ts;%define time axis

%time lag between LVP and AOP
lag=5;

%calculate recurrence plot
orbit=[h(1:length(h)-lag,1),h(lag+1:length(h),2),h(lag +1:length(h),4)];
neighbourcount=ones(length(orbit),1); %counts found nearest neighbours
distance=zeros(length(orbit),1);%avoids program interrupt if first or last point has
no nearest neighbours
points=zeros(length(orbit),1);%initialisation
maxpunkt=10;           %for placing the next nearest neighbour right
r=10;

disp('search nearest neighbours and calculate sorted nearest neighbour matrix...-
please wait ca. 15 min')
for i=1:(length(orbit)-1) %-1 because otherwise i+1>length(orbit) in the j loop
   for j=1+i:length(orbit)    %1+i avoids double comparison of distances e.g. 2
   with 3 and 3 with 2
      vector=orbit(j,:)-orbit(i,:);    %calculate difference vector
      betrag=sqrt(vector(1)^2+vector(2)^2+vector(3)^2); %amount of difference vector
      if betrag <=r  %if nearest neighbour found
         distance(i,neighbourcount(i))=betrag;     %store amount
         points(i,neighbourcount(i))=j;     %store point number
         distance(j,neighbourcount(j))=betrag;     %vice versa it is the same:
         distance 4->5=distance 5->4
         points(j,neighbourcount(j))=i;
         neighbourcount(i)=neighbourcount(i)+1; %increment counting pointer
         neighbourcount(j)=neighbourcount(j)+1;
      end
   end
   [hilf,index]=sort(nonzeros(distance(i,:)));
   index=index';
   Zwischen=points(i,index);
   mapall(i,1:length(Zwischen))=Zwischen;
end

%even respect the last row
i=length(orbit);
[hilf,index]=sort(nonzeros(distance(i,:)));
```

```
index=index';
Zwischen=points(i,index);
mapall(i,1:length(Zwischen))=Zwischen;
disp('finished!')

save([file,'allneig'],'mapall');

%exclude possibility that there is no point with 10 neighbours
[Zeilen,Spalten]=size(mapall);
if Spalten>=maxpunkt
   map=mapall(:,I:maxpunkt);%here only the e.g. 10 nearest neighbours are stored
else
   map=mapall(:,1:Spalten);
   disp('There is no point with 10 nearest neighbours');
end

save([file,'nearneig'],'map');

%end of calculation

set(0,'DefaultFigurePaperType','A4');

figure(1)

subplot(221)
plot(t,h(:,1),t,h(:,2));
axis tight
xlabel('time s');
grid
ylabel('AOPprox,LVP mmHg');
subplot(222)
plot(t,h(:,4));
axis tight
xlabel('time s');
grid
ylabel('AOF ml/s');

subplot(223)
plot3(h(1:length(h)-lag,1),h(lag+1:length(h),2),h(lag +1:length(h),4),'.);
axis tight
grid;
xlabel(['LVP mmHg - lag =',num2str(lag)]);
ylabel('AOP mmHg');
zlabel('AOF ml/s');
title([date,'pig experiment',num2str(1),'nn=',num2str (maxpunkt)]);

subplot(224)
plot(1:length(map),map,'.')
axis tight
print-dwin-f1;
print ('-dps',-f1',[file 'f1'])
print ('-dbitmap','-f1,[file'f1'])

figure(2)
plot(1:length(map),map,'.')
axis tight
title([date,'pig experiment',num2str(1),'lag=', num2str(lag),'nn=',
num2str(maxpunkt)]);
xlabel('Indices of the points on the orbit','FontSize',[20]);
ylabel('Indices of the nearest neighbours on the orbit','FontSize',[20]);
set(gca,'FontSize'.[20])

print-dwin-f2;
print('-dps','-f2';[file 'f2'])
print ('-dbitmap','-f2',[file'f2'])

end
```

## Calculate Parameters

```
clear

unumber=input('number of the first experiment ');
vnumber=input('number of the last experiment ');
date=input('date ');

for l=unumber:vnumber

name=['v'num2str(l)'_512hznearneig.mat'];
file=['v'num2str(l)];

load(name);

[rows,columns]=size(map)

points=0;
percent=0;

points=nnz(map);
percent=(points*100)/(rows^2);

disp([points'num2str(points)]);
disp(['density-percent recurrence'num2str(percent)]);

%search diagonals

histo=zeros(1,rows-1);
count=1;%if a neighbour is found the point itself counts too, therefore 1.
maxcount=0; %store the value of the largest string
altcount=0; %that the variables become not equal at the first comparison
i=1;
j=1;%in the first row is only the information that the point is a neighbour of
itself - so drop
while i<=rows
   while map(i,j)~=0
      k=1
      while count ~=altcount %stops if there is no new neighbour or to large point
      indexes (if i+/-k...)
         altcount=count; %if there are no new diagonal points stop while loop
         if i+k<=rows %search diagonal upwards
            for l=1:columns
               if map(i+k,l)==map(i,j)+k
                  count=count+1;
                  map(i+k,l:columns-1)=map(i+k,l+1:columns);
                  map(i+k,columns)=0;%erse used neighbour points, otherwise multiple
                  counting
                  break
               elseif map(i+k,l)==0 %in the matrix is a 0 if there are no neighbours
               any more
                  break
               end
            end
         end

         if i-k>=1 % search diagonal downwards
            for l=1:columns
               if(map(i-k,l)==map(i,j)-k) & (map(i,j)-k~=0) %if map(i,j)-k=1
                  count=count+1;
                  map(i-k,l:columns-1)=map(i-k,l+1:columns);
                  map(i-k,columns)=0;
                  break
               elseif map(i-k,l)==0
                  break
               end
            end
```

```
        end
        k=k+1; %after successful search of a neighbour the next diagonal point is
        searched
      end
   if count >1
      histo(count)=histo(count)+1;
      if count>maxcount
         maxcount=count;
      end
         count=1; %reset counter
         altcount=0;
      else
         altcount=0; %reset counter so count~=altcount
      end
      if j<columns %test if end of row is reached
         j=j+1
      else
         break
      end
   end
   j=1, %reset column counter
   i=i+1;
end

%percent determinism - percentage of points in diagonals compared to all points

detpoint=0;

for i=1:length(histo)
   detpoint=detpoint+(histo(i)*i);
end

percdet=(detpoint*100)/points;
disp(['density - per cent determinism 'num2str(percdet)]);

%calculate entropy

entropy=0;

for i=1:length(histo)
   if histo(i)~=0
      entropy=entropy - ((histo(i)/detpoint)*log2(histo(i)/detpoint));
   end
end

disp(['entropy' num2str(entropy)]);

%print histogram

set(0,'DefaultFigurePaperType','A4');
figure(3)
histo=histo(1:maxcount);
bar(histo);
axis tight
title(['date:'num2str(date)' experiment:'num2str(file)' percent recurrence:'-
num2str(percent)' percent
determinism: 'num2str(percdet)' entropy: 'num2str(entropy)])
xlabel('number of points on the diagonal')
ylabel('number of diagonals')

print-dwin-f3

end
```

Chapter 8

# NONLINEAR ESTIMATION OF RESPIRATORY-INDUCED HEART MOVEMENTS AND ITS APPLICATION IN ECG/VCG SIGNAL PROCESSING

Leif Sörnmo, Magnus Åström, Elena Carro, Martin Stridh, and Pablo Laguna

## 1. INTRODUCTION

Electrocardiographic (ECG) measurements from the body surface are often undesirably influenced by the presence of respiration-induced movements of the heart. Measures that quantify beat-to-beat variations in QRS morphology are particularly susceptible to this influence and special attention must therefore be given to this problem. The analysis of a vectorcardiographic (VCG) lead configuration has been found to reduce this problem. An important reason is that changes in the orientation of the electrical axis, caused by, for example, respiration, can to a certain degree be compensated for by VCG loop rotation. Such rotation can also improve the performance of serial VCG–ECG analysis in which two loops, recorded on different occasions, are compared in order to find pathological changes associated with, for example, myocardial infarction [1, 2].

In this chapter, a statistical signal model is described that compensates for heart movements by means of scaling and rotation in relation to a "reference" VCG loop [3]. Temporal loop misalignment is also parameterized within the model framework. The maximum likelihood (ML) estimator of the parameters describing these transformations is found to possess a nonlinear structure. The optimal parameter estimates can be determined without the need for iterative optimization techniques. Although the model initially assumes that two loops are to be aligned, the method can easily be extended to the case of multiple loop alignment.

The performance of the ML estimation method is assessed in the presence of noise and for different VCG loop morphologies. The results show that loop alignment can be done accurately at low to moderate noise levels. At high noise levels the estimation of rotation parameters breaks down in an abrupt manner. Furthermore, it is shown that the performance is strongly dependent on loop morphology; a planar loop is more difficult to align than a nonplanar loop. The issue of measuring morphologic variability in combination with loop alignment has been investigated [4]. Using an ECG simulation model based on propagation of action potentials in cardiac tissue, the ability of the method to separate morphologic variability of physiological origin from respiratory activity was studied. The results showed that the separation of these two activities can be done accurately up to moderate noise levels.

One application of the ML loop alignment is that of QRST complex cancellation for the analysis of atrial fibrillation in the surface ECG [5]. Again, shifts in the electrical axis of the heart sometimes cause the use of methods based on average beat subtraction to produce large QRST-related residuals. Using the loop alignment technique, residuals with a substantially lower amplitude were obtained and thereby the resulting residual ECG is much better suited for time–frequency analysis of atrial fibrillation. The new method for QRST complex cancellation is briefly reviewed here and an example illustrates the improved performance compared with that of the average beat subtraction method.

## 2. MAXIMUM LIKELIHOOD VCG LOOP ALIGNMENT

This section presents the essentials of the method for spatiotemporal alignment of VCG loops [3]. A statistical model is introduced in which a VCG loop is related to a reference loop by certain geometric transformations (Section 2.1). ML estimation is then investigated for finding the parameter values of the transformations that provide the optimal fit between the two loops (Section 2.2).

### 2.1. Model for Respiratory-Induced Heart Movements

The signal model is based on the assumption that an observed VCG loop of the QRS complex, $\mathbf{Y}$, derives from a reference loop, $\mathbf{Y}_\mathrm{R}$, but has been altered through a series of transformations. The matrix $\mathbf{Y} = [\mathbf{y}_1\ \mathbf{y}_2\ \mathbf{y}_3]$ contains column vectors $y_l$ with $N$ samples for the $l$th VCG lead. The reference loop $\mathbf{Y}_\mathrm{R}$ is $(N + 2\Delta)$-by-3 and includes $2\Delta$ additional samples in order to allow observations that constitute different consecutive subsets of $N$ samples from $\mathbf{Y}_\mathrm{R}$. The following transformations are considered:

**Amplitude Scaling**: Loop expansion or contraction is modeled by the positive-valued, scalar parameter $\alpha$ and represents, in a simplistic way, the effect of variations in locations of the heart, conductivity of the surrounding tissue, and so forth. When considering the problem of QRST cancellation (see Section 5) the extension of $\alpha$ to a diagonal matrix that accounts for scaling in individual leads is found to improve the cancellation performance further.

**Rotation**: Rotational changes of the heart in relation to the electrode locations are accounted for by the orthonormal, 3-by-3 matrix $\mathbf{Q}$; orthonormality implies that $\mathbf{Q}^T\mathbf{Q} = \mathbf{I}$ where $\mathbf{I}$ is the identity matrix. Rotational changes may be caused by respiration or body position changes, for example.

**Time synchronization**: Although $\mathbf{Y}$ is assumed to be reasonably well synchronized in time to $\mathbf{Y}_\mathrm{R}$ due to the preceding QRS detection, means for refining the time synchronization is introduced in the model by the shift matrix $\mathbf{J}_\tau$. Because of the larger size of $\mathbf{Y}_\mathrm{R}$, the observed loop $\mathbf{Y}$ can result from any of the $(2\Delta + 1)$ possible positions in $\mathbf{Y}_\mathrm{R}$. The shift matrix $\mathbf{J}_\tau$ is defined by the integer time shift $\tau$,

$$\mathbf{J}_\tau = \left[\mathbf{0}_{\Delta+\tau}\ \mathbf{I}\ \mathbf{0}_{\Delta-\tau}\right] \tag{1}$$

where $\tau = -\Delta, \ldots, \Delta$. The dimensions of the left and right zero matrices in Eq. 1 are equal to $N$-by-$(\Delta + \tau)$ and $N$-by-$(\Delta - \tau)$, respectively. One of the zero matrices vanishes when $\tau$ is $\pm\Delta$. The identity matrix $\mathbf{I}$ is $N$-by-$N$. By estimating the parameters that characterize these transformations, it will be possible to reduce the influence of extracardiac activities and thus to improve the alignment of $\mathbf{Y}$ to $\mathbf{Y}_\mathrm{R}$.

The preceding scaling, rotation, and time synchronization parameters are embraced by the following observation model:

$$\mathbf{Y} = \alpha \mathbf{J}_\tau \mathbf{Y}_\mathrm{R} \mathbf{Q} + \mathbf{W} \tag{2}$$

The transformed reference loop is assumed to be additively disturbed by white, Gaussian noise (represented by the $N$-by-3 matrix $\mathbf{W} = [\mathbf{w}_1\ \mathbf{w}_2\ \mathbf{w}_3]$). Furthermore, the noise is assumed to be uncorrelated from lead to lead and with identical variance, $\sigma_w^2$, in all leads. The noise probability density function is then given by

$$p_w(\mathbf{W}) = \prod_{l=1}^{3} p_w(\mathbf{w}_l) = \frac{1}{(2\pi)^{3N/2}\sigma_w^{3N}} e^{-\frac{1}{2\sigma_w^2}\sum_{i=1}^{3} \mathbf{w}_l \mathbf{w}_l^T} \tag{3}$$

$$= \frac{1}{(2\pi)^{3N/2}\sigma_w^{3N}} e^{-\frac{1}{2\sigma_w^2}\mathrm{tr}(\mathbf{w}\mathbf{w}^T)} \tag{4}$$

where tr denotes the matrix trace.

## 2.2. Maximum Likelihood Estimation

The joint ML estimator of the parameters $\alpha$, $\mathbf{Q}$, and $\tau$ is derived by maximizing the log-likelihood function [6], that is,

$$\frac{\partial}{\partial \alpha} \frac{\partial}{\partial \mathbf{Q}} \frac{\partial}{\partial \tau} \ln p_w(\mathbf{Y}|\alpha, \mathbf{Q}, \tau) = 0 \tag{5}$$

It can be shown that the calculation of Eq. 5 is equivalent to the minimization of the Frobenius norm $\epsilon^2$ between $\mathbf{Y}$ and $\mathbf{Y}_\mathrm{R}$ [3],

$$\epsilon^2_{\min} = \min_{\alpha, \mathbf{Q}, \tau} \left\| \mathbf{Y} - \alpha \mathbf{J}_\tau \mathbf{Y}_\mathrm{R} \mathbf{Q} \right\|_\mathrm{F}^2 \tag{6}$$

The Frobenius norm for an $m$-by-$n$ matrix $\mathbf{X}$ is defined by

$$\|\mathbf{X}\|_\mathrm{F}^2 = \mathrm{tr}(\mathbf{X}\mathbf{X}^T) = \sum_{i=1}^{m} \sum_{j=1}^{n} |x_{ij}|^2 \tag{7}$$

The minimization in Eq. 6 is performed by first finding closed-form expressions for the estimates $\alpha$ and $\mathbf{Q}$ under the assumption that $\tau$ is fixed. The optimal estimates of $\alpha$, $\mathbf{Q}$, and $\tau$ are then determined by evaluating the error $\epsilon^2$ for all values of $\tau$ in the interval $[-\Delta, \Delta]$.

The estimate of $\mathbf{Q}$ is obtained by first rewriting the error in (6) such that

$$\epsilon^2 = \mathrm{tr}(\mathbf{Y}\mathbf{Y}^T) + \alpha^2\,\mathrm{tr}\big(\mathbf{J}_\tau\mathbf{Y}_\mathrm{R}\mathbf{Y}_\mathrm{R}^T\mathbf{J}_\tau^T\big) - 2\alpha\,\mathrm{tr}(\mathbf{Y}_\mathrm{R}^T\mathbf{J}_\tau^T\mathbf{Y}\mathbf{Q}^T) \tag{8}$$

and then noting that (8) is minimized by choosing $\mathbf{Q}$ such that the last term $\mathrm{tr}(\mathbf{J}_\tau^T\mathbf{Y}_\mathrm{R}^T\mathbf{Q}^T\mathbf{Y})$ is maximized. The key step in finding the optimal $\mathbf{Q}$ is to use the singular value decomposition (SVD) [7]. In general, the SVD provides a decomposition of the $M$-by-$N$ matrix $\mathbf{Z}$ into the orthonormal matrices $\mathbf{U}$ ($M$-by-$M$) and $\mathbf{V}$ ($N$-by-$N$) and the diagonal matrix $\mathbf{\Sigma}$ with the singular values ($\mathbf{\Sigma} = \mathrm{diag}(\sigma_1, \ldots, \sigma_l, 0, \ldots 0)$, $l = \min(M, N)$),

$$\mathbf{Z} = \mathbf{U}\mathbf{\Sigma}\mathbf{V}^T \tag{9}$$

In particular, by defining the 3-by-3 matrix $\mathbf{Z}$ such that

$$\mathbf{Z}_\tau = \mathbf{Y}_\mathrm{R}^T\mathbf{J}_\tau^T\mathbf{Y} \tag{10}$$

the last term on the right-hand side of (8) can be rearranged and expressed as

$$\mathrm{tr}\big(\mathbf{Y}_\mathrm{R}^T\mathbf{J}_\tau^T\mathbf{Y}\mathbf{Q}^T\big) = \mathrm{tr}(\mathbf{Z}_\tau\mathbf{Q}^T) \tag{11}$$

By choosing $\mathbf{Q}$ such that $\mathbf{U}\mathbf{V}^T\mathbf{Q}^T = \mathbf{I}$, the expression in (11) is maximized and the resulting ML estimate is given by

$$\hat{\mathbf{Q}}_\tau = \mathbf{U}\mathbf{V}^T \tag{12}$$

The index $\tau$ has been attached in (12) because this estimate is optimal for only one particular value of $\tau$.

The estimate of $\alpha$ can be calculated when $\hat{\mathbf{Q}}_\tau$ is available,

$$\hat{\alpha}_\tau = \frac{\mathrm{tr}(\mathbf{Y}_\mathrm{R}^T\mathbf{J}_\tau^T\mathbf{Y}\hat{\mathbf{Q}}_\tau^T)}{\mathrm{tr}(\mathbf{J}_\tau\mathbf{Y}_\mathrm{R}\mathbf{Y}_\mathrm{R}^T\mathbf{J}_\tau^T)} \tag{13}$$

The discrete-valued time synchronization parameter $\tau$ is estimated by means of a grid search for the allowed set of values,

$$\hat{\tau} = \arg\min_\tau \|\mathbf{Y} - \hat{\alpha}_\tau\mathbf{J}_\tau\mathbf{Y}_\mathrm{R}\hat{\mathbf{Q}}_\tau\|_\mathrm{F}^2 \tag{14}$$

Finally, the estimate $\hat{\tau}$ determines which of the estimates in the set of estimates $\hat{\alpha}_\tau$ and $\hat{\mathbf{Q}}_\tau$ that should be selected. It is noted that the preceding estimation procedure always yields the optimal estimates because $\tau$ belongs to a finite set of values; the continuous-valued estimates $\hat{\alpha}$ and $\hat{\mathbf{Q}}$ are obtained conditioned on $\tau$. Furthermore, it should be noted that the resulting ML estimator, as defined by (12), (13), and (14), exhibits a nonlinear structure although the observation model in (2) has a linear characteristic.

Application of the preceding ML estimation procedure requires that a reference loop $\mathbf{Y}_\mathrm{R}$ has been first defined. In the simplest case of aligning two loops, $\mathbf{Y}_\mathrm{R}$ can be taken as any of the two available loops. When several loops $\{\mathbf{Y}_i\}_{i=1}^M$ are to be aligned, implying that the observation model is extended to $\mathbf{Y}_i = \alpha_i \mathbf{J}_{\tau,i} \mathbf{Y}_\mathrm{R} \mathbf{Q}_i + \mathbf{W}_i$ for $i = 1, \ldots, M$, a variety of definitions of $\mathbf{Y}_\mathrm{R}$ are possible.

The samples of the matrices $\mathbf{Y}$ and $\mathbf{Y}_\mathrm{R}$ are assumed to be appropriately centered around the QRS complex. The $2\Delta$ samples are equally divided into samples being prepended and appended to the QRS-centered interval, respectively. It should be noted that the noise variance $\sigma_w^2$ does not enter the ML estimation procedure and thus no estimate is required of this parameter.

Before demonstrating the effect of loop alignment, it should be pointed out that $\hat{\mathbf{Q}}$ can be used for retrieving information related to respiration. Such respiratory-related patterns are made more obvious by decomposing $\hat{\mathbf{Q}}$ into a product of three planar rotation matrices, where the angles $\varphi_X$, $\varphi_Y$, and $\varphi_Z$ define the rotation around each lead axis. These angles can be estimated from [3]:

$$\begin{aligned} \hat{\varphi}_Y &= \arcsin(\hat{q}_{13}) \\ \hat{\varphi}_X &= \arcsin\left(\frac{\hat{q}_{12}}{\cos \hat{\varphi}_Y}\right) \\ \hat{\varphi}_Z &= \arcsin\left(\frac{\hat{q}_{23}}{\cos \hat{\varphi}_Y}\right) \end{aligned} \tag{15}$$

where the estimate $\hat{q}_{kl}$ denotes the $(k, l)$th element of $\hat{\mathbf{Q}}$. The idea of studying angular changes in axis orientation as a basis for estimating the respiratory rate was suggested in the mid-1980s (e.g., see Refs. 8 and 9); in those studies, however, the angles were estimated with techniques different from that presented here.

## 3. LOOP ALIGNMENT AND MORPHOLOGIC VARIABILITY

The ML loop alignment method can, as pointed out in Section 1, compensate for certain limitations associated with the analysis of subtle beat-to-beat variations in QRS morphology. This type of analysis has received clinical attention due to its potential value for diagnosing myocardial ischemia and acute infarction [10–12]. It has been hypothesized that subtle morphologic variations may reflect, for example, islets of ischemic tissue or variations in myocardial contraction patterns. Straightforward computation of the standard deviation for a time-aligned ensemble of beats has been suggested as a means for describing such morphologic beat-to-beat variability [13]. Unfortunately, few techniques have been presented in the literature that are aimed at reducing the undesirable influence of respiration. This is somewhat surprising because it is well known that the electrical axis can vary as much as 10° in the transversal plane during inspiration [14].

The following examples illustrate the effect of ML loop alignment in terms of morphologic beat-to-beat variability. In each example, 50 consecutive sinus beats were selected from a high-resolution ECG recording using an orthogonal lead configuration ($X$, $Y$, and $Z$). The sampling rate was equal to 1000 Hz and the amplitude resolution was 0.6 $\mu$V. The recordings were selected from a database of subjects with previous myocardial infarction and/or episodes of sustained ventricular tachycardia. Noisy and aberrant beats were excluded from further analysis. The reference beat was simply selected as the first one out of the 50 beats. Finally, the ensemble standard deviation was employed as a measure of morphologic variability and was computed both before and after loop alignment.

The effect of loop alignment is demonstrated by the example in Figure 1a–c; the corresponding parameter estimates of $\alpha$, $\varphi_X$, $\varphi_Y$, $\varphi_Z$, $\epsilon$ and $\epsilon_{\min}$ are shown in Figure 1d–f as functions of time. It is obvious from Figure 1e and f that reduction in variability is related to scaling as well as to rotation of the loops. For ease of interpretation, the results in Figure 1 a–c are presented for individual leads although the alignment is an inherently spatiotemporal process.

Oscillatory patterns found in the error norm are likely to be related to respiratory activity. Such oscillations can be discerned in Figure 1d both before and after loop alignment, although the oscillations are less pronounced after alignment. However, in certain cases the model parameters are able to account very well for the oscillatory component in $\epsilon_{\min}$; see Figure 2d. It is noted that the variability in lead $X$ is essentially removed after loop alignment while the reduction in the other two leads is less dramatic (see the variation in the angle $\varphi_X$).

## 4. SENSITIVITY OF LOOP ALIGNMENT TO NOISE

Loop alignment may be of interest to use in noisy situations for the purpose of QRST complex cancellation (see Section 5) or for the analysis of data acquired during exercise. In this section, the noise properties of the rotation matrix estimate are studied in terms of accuracy of estimated rotation angles. Based on these results, the concept of a *breakdown noise level* is introduced to describe an essential characteristic of the angle estimates. This concept is then used to investigate the effect of different loop morphologies in the alignment process [4].

### 4.1. Parameter Estimation

A "signal-plus-noise" simulation model similar to that in [2] is adopted here in order to get an appreciation of how noise influences alignment performance. A VCG loop was selected from a healthy individual as the signal part of the model after reduction of the inherent noise level by conventional signal averaging (the loop was included in the databased considered in Section 4.2). The loop is then subjected to rotation on a sample-to-sample basis in order to account for respiratory-induced noise; that is, the matrix $\mathbf{Q}$ is a function of time $k$. The resulting simulation model is defined by

$$\begin{bmatrix} z_1(k) \\ z_2(k) \\ z_3(k) \end{bmatrix} = \mathbf{Q}(k) \begin{bmatrix} y_1(k) \\ y_2(k) \\ y_3(k) \end{bmatrix} + \begin{bmatrix} v_1(k) \\ v_2(k) \\ v_3(k) \end{bmatrix} \tag{16}$$

where $(y_1(k), y_2(k), y_3(k))$ constitute the original VCG loop at sample $k$. The additive noise $v_i(k)$ is assumed to be white Gaussian with variance equal to $\sigma_v^2$ and with no interlead correlation.

The matrix $\mathbf{Q}(k)$ in (16) is assigned a specific structure by the angles that characterize the three planar rotation matrices of $\mathbf{Q}(k)$. It is assumed that the angular

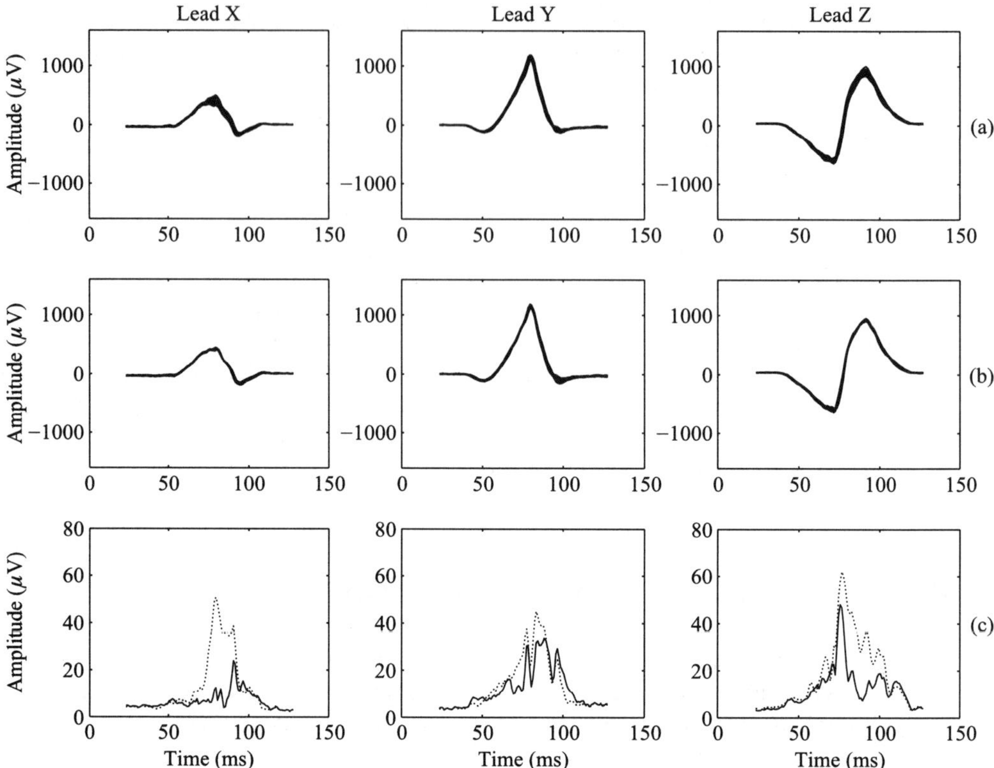


**Figure 1** An example of 50 superimposed beats (a) before loop alignment, (b) after alignment, and (c) the corresponding ensemble standard deviation [dotted and solid lines correspond to (a) and (b), respectively], (d) spatial variability before (dotted line) and after loop alignment [$\epsilon_{\min}$ in (6); solid line], (e) the scaling estimate $\hat{\alpha}$, and (f) the angle estimates $\hat{\varphi}_X$, $\hat{\varphi}_Y$, and $\hat{\varphi}_Z$ (solid, dotted, and dashed/dotted lines, respectively).

variation in each lead is proportional to the amount of air in the lungs during a respiratory cycle. A simplistic way to model this property is to use the product of two sigmoidal functions to describe the inspiratory and expiratory phases, respectively. The angular variation in lead X is defined by

$$\varphi_X(k) = \eta_X\left(\frac{1}{1+e^{\lambda_{in}(k-\kappa_{in})}}\right)\left(\frac{1}{1+e^{\lambda_{ex}(k-\kappa_{ex})}}\right) \tag{17}$$

where the durations of inspiration and expiration are determined by $\lambda_{\text{in}}$ and $\lambda_{\text{ex}}$, respectively, and the time delays of the sigmoidal functions by $\kappa_{\text{in}}$ and $\kappa_{\text{ex}}$. The parameter $\eta_X$ is an amplitude factor. The angular variations in leads $Y$ and $Z$ are defined in an analogous manner.

Signals at various stages in the simulation model in (16) are shown in Figure 3 and the corresponding rotation pattern is shown in Figure 4a. In this example, rotation is introduced only around the $X$ axis and therefore the morphology of the beats in Figure

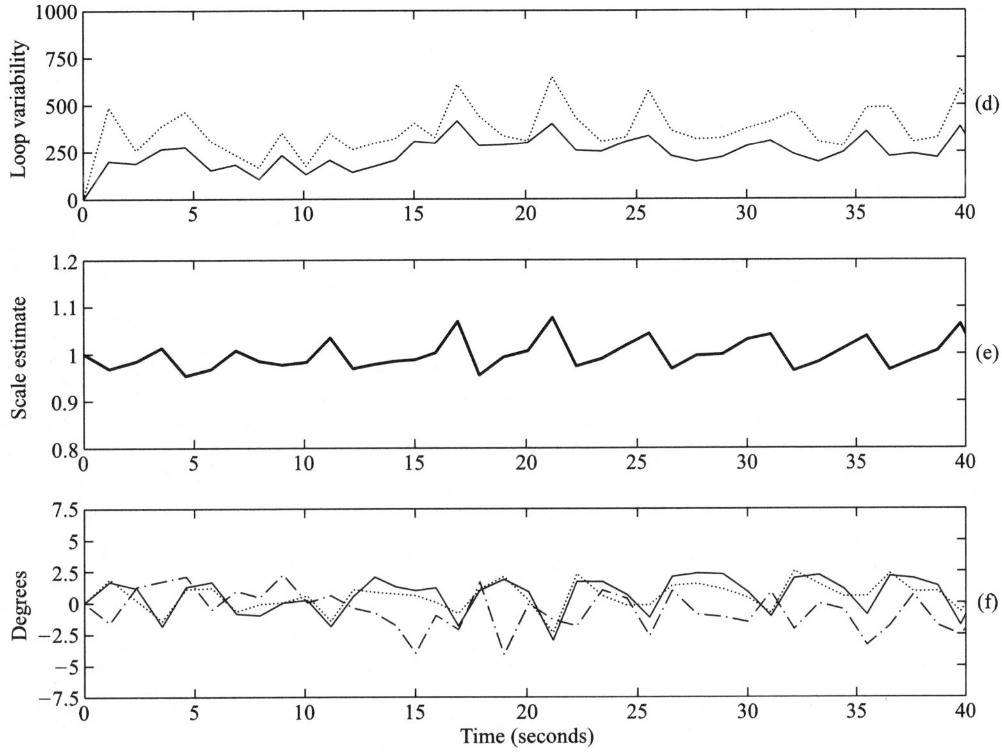


**Figure 1** *(continued)*

3b changes only in leads $Y$ and $Z$. The rotation angles were estimated from the noisy signal using (15) and are presented in Figure 4b.

A comparison of the original angles and the estimated ones must account for the fact that the original angles are time varying while only one angle estimate is obtained for the entire QRS complex. Therefore, the average of $\varphi(k)$ during the QRS complex was used as the reference value. The root-mean-square error measure for the angle estimates is then given by

$$\delta = \sqrt{\frac{1}{B}\sum_{i=1}^{B}(\hat{\varphi}_i - \bar{\varphi}_i)^2} \tag{18}$$

where $B$ denotes the number of beats and $\hat{\varphi}_i$ and $\bar{\varphi}_i$ denote the angle estimate and the corresponding average reference value, respectively, of the $i$th beat. It should be noted that the error present in $\delta$ due to the once-per-beat estimate of the loop alignment method is negligible for realistic choices of respiratory rates.

Figure 4c presents the error measure $\delta$ for each of the $X$, $Y$, and $Z$ leads as a function of the noise level $\sigma_v$. An interesting behavior can be observed in lead $Z$, where a distinct noise level exists (approximately 18 $\mu$V) above which the performance rapidly deteriorates and large estimation error results. Since this threshold behavior was

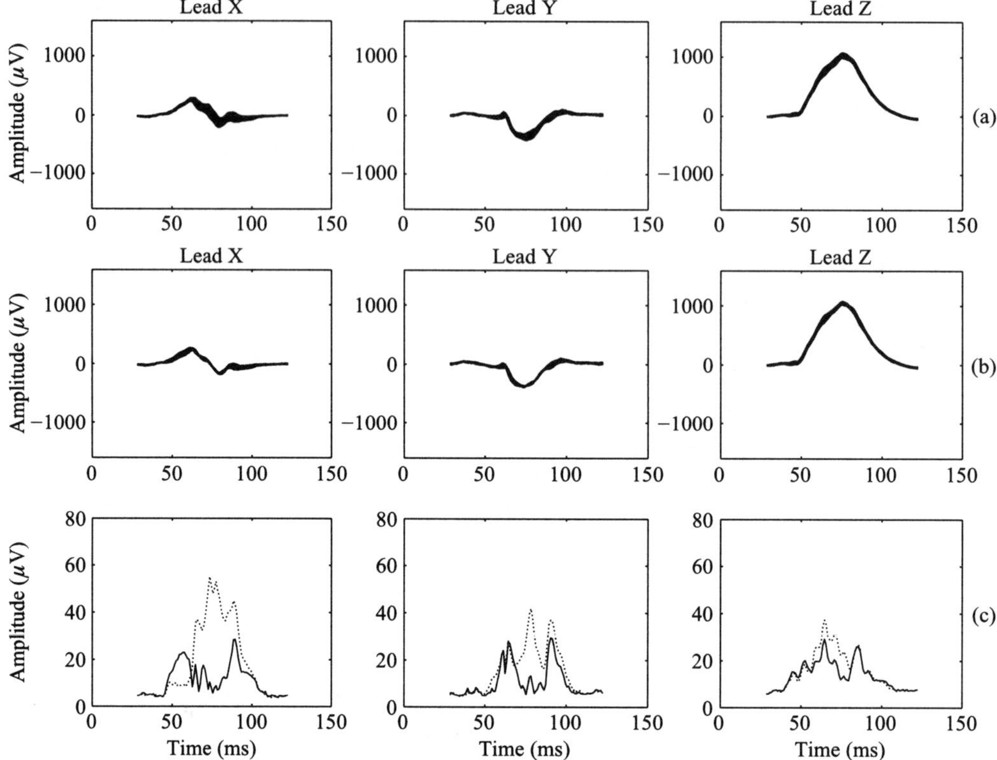


**Figure 2** An example with large morphologic variability in which the ML loop alignment dramatically reduces the oscillatory component in $\epsilon$. Again, 50 superimposed beats are shown (a) before loop alignment, (b) after alignment, and (c) the corresponding ensemble standard deviation [dotted and solid lines correspond to (a) and (b), respectively], (d) spatial variability before (dotted line) and after loop alignment [$\epsilon_{\min}$ in (6); solid line], (e) the scaling estimate $\hat{\alpha}$, and (f) the angle estimates $\hat{\varphi}_X$, $\hat{\varphi}_Y$, and $\hat{\varphi}_Z$ (solid, dotted, and dashed/dotted lines, respectively).

observed in all other VCG recordings analyzed in this study (see the results in Section 4.2), it seems well motivated to use the concept of breakdown noise level. In the present example, the behavior can also be observed in the other leads, although the decrease in performance is not as drastic as in lead $Z$. The original angle pattern and the corresponding estimated pattern in Figure 4a and b exemplify the outlier angle estimates that occur at three points in time.

## 4.2. Noise and Loop Morphology

The results in the previous section suggest that it may be of interest to investigate the relation between breakdown noise level and loop morphology. In order to investigate this aspect, a database was used with 34 nonselected individuals referred for myocardial scintigraphy [15]. These individuals had no signs of ischemia or infarction. The ECG signals were recorded during rest for 5 minutes using a standard 12-lead

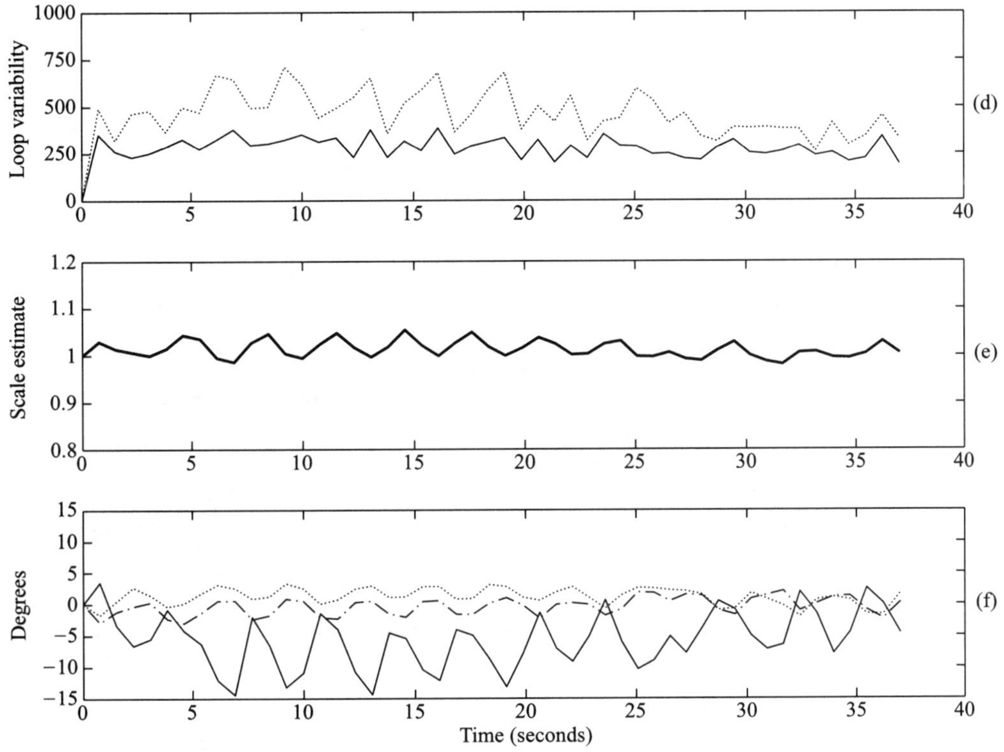


**Figure 2** *(continued)*

configuration. The VCG signals were then synthesized by means of linear combination of the 12 leads using the inverse Dower weighting matrix [16].

In order to characterize loop morphology, an overall measure was considered that reflects the planarity of a VCG loop. The measure is defined as the ratio between the minimum and the maximum singular values of the loop matrix **Y**, that is,

$$\rho = \frac{\sigma_{\min}}{\sigma_{\max}} \tag{19}$$

The limiting values of $\rho$ are 0 (a loop that is perfectly planar) and 1 (a loop that extends equally in all three dimensions).

The breakdown noise level, denoted $\tilde{\sigma}_v$, is defined as the noise level $\sigma_v$ that causes angle estimation errors in any lead to exceed a certain threshold $\chi$. The choice of $\chi$ was based on the observation that the estimation error is small below a certain noise level and then rapidly increases to a considerably larger error value. By setting $\chi$ equal to $\pi/10$, proper identification of the noise level at which angle estimates became anomalous was achieved.

Figure 5 shows that the accuracy of loop alignment with regard to noise level is strongly dependent on loop planarity; the breakdown noise level $\tilde{\sigma}_v$ actually ranges from 5 to 70 $\mu$V. This result suggests that an essentially linear relationship exists

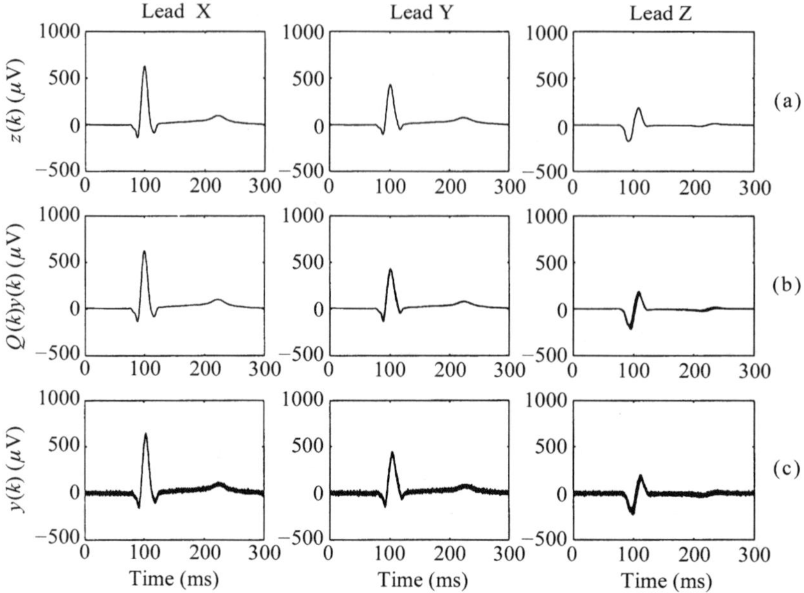


**Figure 3** An example of 50 superimposed beats at various stages in the simulation model. (a) Without added noise, (b) after transformation with **Q**, and (c) with noise added ($\sigma_v = 10\,\mu$V).

between $\rho$ and $\tilde{\sigma}_v$. It can be concluded that aligning planar loops is much more vulnerable to noise than is the alignment of a loop extending into all three dimensions.

It is well known that normal individuals in general have VCG loops that are more planar than those from patients with, for example, myocardial infarction. Myocardial damage is often associated with loops that include abnormal transitions ("bites") or sharp edges and therefore decreases planarity [17]. Such differences in loop characteristics may thus imply that alignment, in general, is more robust in infarct patients than in normal individuals.

## 5. SPATIOTEMPORAL ALIGNMENT AND QRST CANCELLATION

The characterization of atrial fibrillation (AF) using the surface ECG is facilitated by the derivation of a signal in which the ventricular activity has been first canceled. Since the atrial and the ventricular activities overlap spectrally, techniques based on linear filtering are less suitable. Instead, an average beat subtraction (ABS) method has been suggested which makes use of the fact that AF is uncoupled to ventricular activity. Subtraction of the average beat, which thus reflects the ventricular activity, produces a residual ECG containing essentially the fibrillation waveforms, that is, the f waves [18], [19], [20].

The performance of the ABS method relies on the assumption that the average beat can represent each individual beat accurately. However, as pointed out earlier in this chapter, the QRS complexes are often subject to beat-to-beat changes in morphology that, due to the single-lead nature of the ABS method, can cause large QRS-related

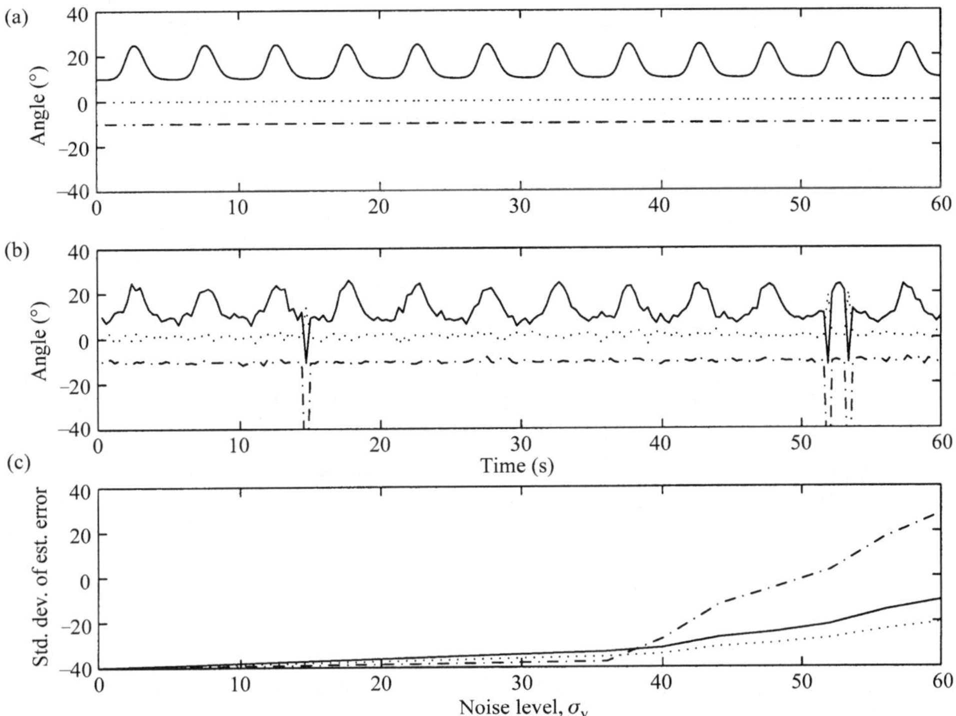


**Figure 4** Example of (a) angular variation pattern and (b) the corresponding estimates obtained from a signal disturbed by noise with variance $\sigma_v^2 = 10\,\mu$V. The angle patterns are plotted with a 5° displacement for each lead in order to improve legibility. (c) The angle estimation error $\delta$ as a function of noise variance.

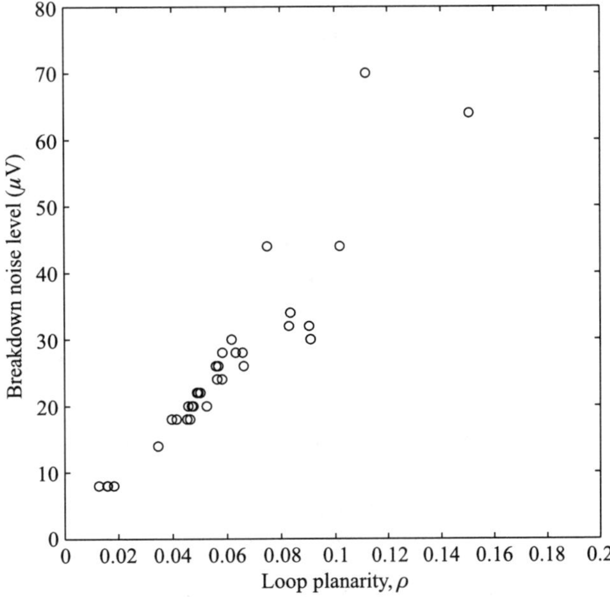


**Figure 5** The relationship between loop planarity and breakdown noise level.

residuals. The present loop alignment technique is further developed in the following in order to improve the cancellation of QRST complexes [5].

### 5.1. Signal Model with Lead-Dependent Amplitude Scaling

The QRST cancellation problem is based on a model similar to that in (2) but with the amplitude scaling factor $\alpha$ being replaced by a diagonal matrix $\mathbf{D}$ that accounts for scaling in individual leads,

$$\mathbf{Y} = \mathbf{J}_\tau \mathbf{Y}_\mathrm{R} \mathbf{D} \mathbf{Q} + \mathbf{F} \tag{20}$$

The average beat (the "reference beat") is denoted by $\mathbf{Y}_\mathrm{R}$ and the fibrillation activity by $\mathbf{F}$.

Again, the objective is to find the model parameter values that provide the best fit of the model to the observed signal. An estimate of the fibrillation signal $\mathbf{F}$ could be obtained by subtracting a scaled, rotated, and time-shifted version of $\mathbf{Y}_\mathrm{R}$ from $\mathbf{Y}$,

$$\mathbf{F} = \mathbf{Y} - \mathbf{J}_\mathrm{T} \mathbf{Y}_\mathrm{R} \hat{\mathbf{D}} \hat{\mathbf{Q}} \tag{21}$$

The Frobenius norm to be minimized is, in an expanded format, equal to

$$\epsilon^2 = \mathrm{tr}(\mathbf{Y}\mathbf{Y}^T) + \mathrm{tr}(\mathbf{J}_\tau \mathbf{Y}_\mathrm{R} \mathbf{D}\mathbf{D}^T \mathbf{Y}_\mathrm{R}^T \mathbf{J}_\tau^T) - 2\mathrm{tr}(\mathbf{D}^T \mathbf{Y}_\mathrm{R}^T \mathbf{J}_\tau^T \mathbf{Y} \mathbf{Q}^T) \tag{22}$$

Unfortunately, the minimization with respect to rotation and amplitude scaling can no longer be performed independently as was the case in Section 2.2. Because the exact solution is difficult to find, an alternating iterative approach is used in which the error in (22) for a fixed $\mathbf{D}$ is minimized with respect to $\mathbf{Q}$ by maximizing the last term. In order to find $\hat{\mathbf{Q}}$, the SVD is again employed but now operating on a different matrix,

$$\mathbf{Z} = \mathbf{D}^T \mathbf{Y}_\mathrm{R}^T \mathbf{J}_\tau^T \mathbf{Z} = \mathbf{U} \boldsymbol{\Sigma} \mathbf{V}^T \tag{23}$$

As in Section 2.2, the estimate of $\mathbf{Q}$ is given by the product of the matrices $\mathbf{U}$ and $\mathbf{V}^T$ containing the left and right singular vectors, respectively; see the expression in (12).

When an estimate of $\mathbf{Q}$ is available, (22) can be written as

$$\epsilon^2 = \mathrm{tr}\big(\mathbf{Y}\mathbf{Q}^{-1} - \mathbf{J}_\tau \mathbf{Y}_\mathrm{R} \mathbf{D}\big)\mathbf{Q}\mathbf{Q}^T \big(\mathbf{Y}\mathbf{Q}^{-1} - \mathbf{J}_\tau \mathbf{Y}_\mathrm{R} \mathbf{D}\big)^T \tag{24}$$

where the introduction of $\mathbf{Y}_2 = \mathbf{Y}\mathbf{Q}^{-1}$ yields

$$\epsilon^2 = \|\mathbf{Y}_2 - \mathbf{J}_\tau \mathbf{Y}_\mathrm{R} \mathbf{D}\|_\mathrm{F}^2 \tag{25}$$

Equation 25 can now be rearranged as

$$\epsilon^2 = \mathrm{tr}(\mathbf{Y}_2\mathbf{Y}_2^T) + \mathrm{tr}(\mathbf{D}\mathbf{D}^T\mathbf{Y}_\mathrm{R}^T\mathbf{J}_\tau^T\mathbf{J}_\tau\mathbf{Y}_\mathrm{R}) - 2\mathrm{tr}(\mathbf{D}^T\mathbf{Y}_\mathrm{R}^T\mathbf{J}_\tau^T\mathbf{Y}_2) \tag{26}$$

which is minimized by setting the derivative with regard to $\mathbf{D}$ to zero,

$$\frac{d\epsilon^2}{d\mathbf{D}} = 2\mathbf{D}\mathbf{Y}_\mathrm{R}^T\mathbf{J}_\tau^T\mathbf{J}_\tau\mathbf{Y}_\mathrm{R} - 2\mathbf{Y}_\mathrm{R}^T\mathbf{J}_\tau^T\mathbf{Y}_2 = 0 \tag{27}$$

The constraint of $\mathbf{D}$ as a diagonal matrix implies that (27) should be evaluated for individual leads. The diagonal entries in $\mathbf{D}$ can therefore be estimated by

$$\hat{d}_l = \left([\mathbf{J}_\tau\mathbf{Y}_\mathrm{R}]_l^T[\mathbf{J}_\tau\mathbf{Y}_\mathrm{R}]_l\right)^{-1}\left([\mathbf{J}_\tau\mathbf{Y}_\mathrm{R}]_l^T[\mathbf{Y}_2]_l\right) \tag{28}$$

For a given $\mathbf{Q}$, this expression estimates the scale factors of the average beat before rotation. Based on the new scaling factors, an improved rotation matrix can then be estimated.

Typically, a solution is desired that implies small rotation/scaling; that is, $\mathbf{Q}$ and $\mathbf{D}$ are close to $\mathbf{I}$. The alternating, iterative procedure for finding the parameter estimates is therefore initialized by $\mathbf{D}_0 = \mathbf{I}$. The rotation at step $k$, $\mathbf{Q}_k$, is then calculated based on $\mathbf{D}_{k-1}$. Since

$$\|\mathbf{Y} - \mathbf{J}_\tau\mathbf{Y}_\mathrm{R}\mathbf{D}_{k-1}\mathbf{Q}_k\|_\mathrm{F}^2 \le \|\mathbf{Y} - \mathbf{J}_\tau\mathbf{Y}_\mathrm{R}\mathbf{D}_{k-1}\mathbf{Q}_{k-1}\|_\mathrm{F}^2 \tag{29}$$

the error will be less than or equal to that in the previous step. When $\mathbf{Q}_k$ is known, $\mathbf{D}_k$ can be calculated. Accordingly,

$$\|\mathbf{Y} - \mathbf{J}_\tau\mathbf{Y}_\mathrm{R}\mathbf{D}_k\mathbf{Q}_k\|_\mathrm{F}^2 \le \|\mathbf{Y} - \mathbf{J}_\tau\mathbf{Y}_\mathrm{R}\mathbf{D}_{k-1}\mathbf{Q}_k\|_\mathrm{F}^2 \tag{30}$$

This procedure is then repeated until convergence is achieved. The algorithm will converge because the minimization with regard to both $\mathbf{Q}$ and $\mathbf{D}$ for each step according to (29) and (30) will improve the fit in terms of $\epsilon^2$ [21].

One difficulty in performing the above alignment is that the presence of AF influences the signal amplitude during the QRS interval. It is therefore desirable to remove the fibrillatory waveforms before the estimation of $\mathbf{Q}$, $\mathbf{D}$, and $\tau$ is done. This approach is obviously a contradiction: to get an estimate of $\mathbf{F}$, it must already be known. Our solution to this dilemma is to use a "quick-and-dirty" method to produce an AF estimate to be subtracted from $\mathbf{Y}$ prior to QRST cancellation. A method is used that "fills in" the AF waveforms in the QRS interval by interpolation based on the AF activity contained in the adjacent T-Q intervals; for further details see Ref. 5.

### 5.2. Example of QRST Cancellation

The performance of the spatiotemporal QRST cancellation method ("QD alignment") is illustrated with an ECG recording from leads $V_1$–$V_3$; see Figure 6. The residual ECGs were computed using the ABS and the QD alignment method; see Figure 7. It is obvious from Figure 7 that the QRS-related residuals are much better

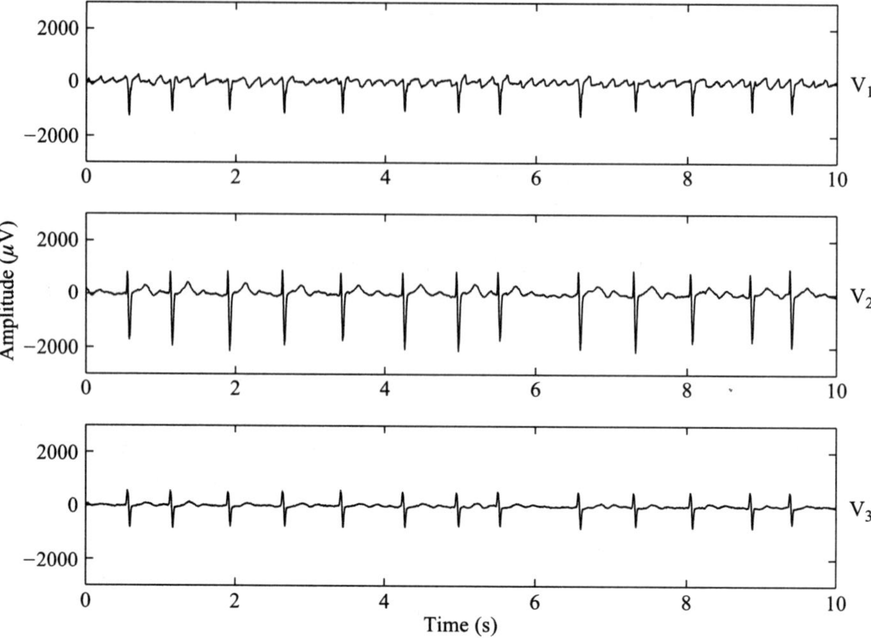


**Figure 6** Example of an ECG with atrial fibrillation.

canceled with the QD alignment technique. In this case the improvement is most striking in the leads with weaker AF, that is, $V_2$ and $V_3$. In Figure 7a, the periodically alternating polarity of the QRS-related residuals in lead $V_2$ of the ABS method suggests that these errors are caused by respiratory-induced variations in QRS complex morphology; these changes are efficiently handled by the QD method (see Figure 7b).

## 6. CONCLUSIONS

The problem of VCG loop alignment has been revisited by means of developing a statistical signal model to which ML estimation was applied. The resulting nonlinear estimation method is feasible from an implementational point of view while it still ensures that the optimal alignment parameter values (scaling, rotation, and synchronization) are always found.

The loop alignment was applied to the analysis of subtle beat-to-beat variability in QRS morphology where the cancellation of respiratory-induced variations is important for accurate morphologic measurements. Two examples illustrated that the effects of respiration on morphologic variability can be dramatically reduced by the new technique. Another application is found in the analysis of atrial fibrillation in the surface ECG where QRST cancellation is required The residual ECG produced by the present alignment method is better suited for, for example, time–frequency AF analysis than that of the ABS method because of the smaller QRS-related residuals.

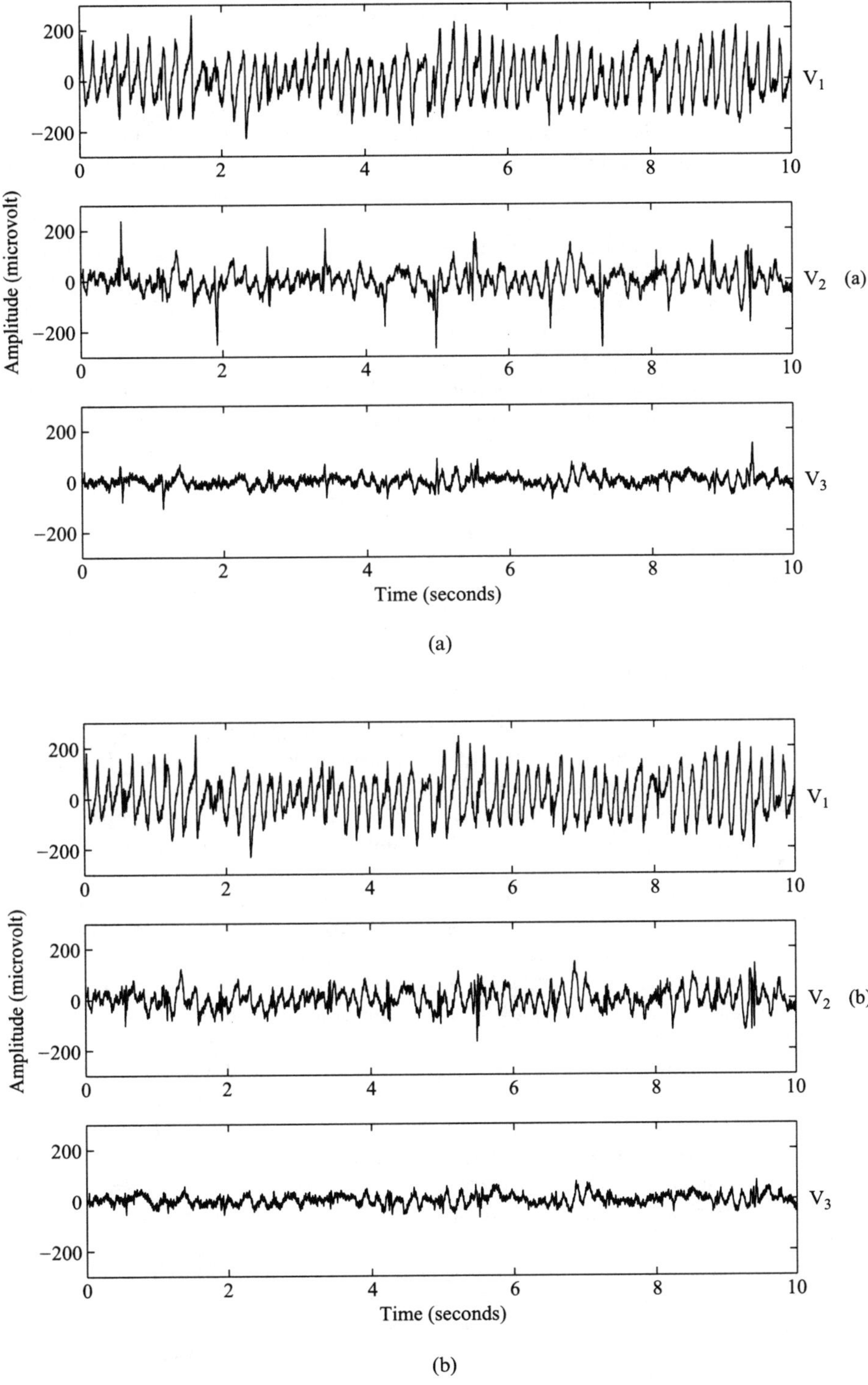


**Figure 7** QRST cancellation of the ECG signal shown in Figure 6 using (a) average beat subtraction and (b) spatiotemporal alignment.

## REFERENCES


[1] J. Fayn, P. Rubel, P. Arnaud, C. Flaemens, M. Frachon, and S. Hacker, New serial VCG comparison techniques. Application to the detection of myocardial infarction after bypass surgery. *Proceedings of Computers in Cardiology*, pp. 533–536, IEEE Computer Society, 1986.

[2] J. Fayn, P. Rubel, and N. Mohsen, An improved method for the precise measurement of serial ECG changes in QRS duration and QT interval. *J. Electrocardiol.* 24(Suppl):123–127, 1989.

[3] L. Sörnmo, Vectorcardiographic loop alignment and morphologic beat-to-beat variability. *IEEE Trans. Biomed. Eng.* 45:1401–1413, 1998.

[4] M. Åström, E. Carro, L. Sörnmo, B. Wohlfart, and P. Laguna, Vectorcardiographic loop alignment, morphologic beat-to-beat variability and the presence of noise. *IEEE Trans. Biomed. Eng.* 47:497–506, 2000.

[5] M. Stridh and L. Sörnmo, Spatiotemporal QRST cancellation techniques in the surface ECG for improved characterization of atrial fibrillation. *Proc. IEEE Eng. Med. Biol. Soc.* (CD-ROM), 1997.

[6] C. W. Therrien, *Discrete Random Signals and Statistical Signal Processing*. Englewood Cliffs, NJ: Prentice-Hall, 1992.

[7] G. Golub and C. van Loan, *Matrix Computations*, 2nd ed. Baltimore: Johns Hopkins University Press, 1989.

[8] F. Pinciroli, R. Rossi, and L. Vergani, Detection of electrical axis variation for the extraction of respiratory information. *Proceedings of Computers in Cardiology*, pp. 499–502. IEEE Computer Society Press, 1985.

[9] R. Pallas-Areny and F. C. Riera, Recovering the respiratory rhythm out of the ECG. *Med. Biol. Eng. Comput.* 23:338–339, 1985.

[10] S. Ben-Haim, A. Gil, Y. Edoute, M. Kochanovski, O. Azaira, E. Kaplinsky, and Y. Palti, Beat to beat variation in healed myocardial infarction. *Am. J. Cardiol.* 68:725–728, 1991.

[11] J. Nowak, I. Hagerman, M. Ylén, O. Nyquist, and C. Sylvén, Electrocardiogram signal variance analysis in the diagnosis of coronary artery disease—A comparison with exercise stress test in an angiographically documented high prevalence population. *Clin. Cardiol.* 16:671–682, 1993.

[12] H. Kestler, M. Hoeher, G. Palm, M. Kochs, and V. Hombach, Time domain variability of high resolution beat-to-beat recordings classified by neural networks. *Proceedings of Computers in Cardiology*, pp. 317–320. IEEE Computer Society, 1996.

[13] K. Prasad and M. Gupta, Phase-invariant signature algorithm. A noninvasive technique for early detection and quantification of ouabain-induced cardiac disorders. *Angiology* 30:721–732, 1979.

[14] J. Malmivuo and R. Plonsey, *Bioelectromagnetism*. Oxford: Oxford University Press, 1995.

[15] L. Sörnmo, B. Wohlfart, J. Berg, and O. Pahlm, Beat-to-beat qrs variability in the 12-lead ecg and the detection of coronary artery disease. *J. Electrocardiol.* 31:336–344, 1998.

[16] L. Edenbrandt and O. Pahlm, Vectorcardiogram synthesized from a 12-lead ECG: superiority of the inverse Dower matrix. *J. Electrocardiol.* 21:361–367, 1988.

[17] P. MacFarlane, L. Edenbrandt, and O. Pahlm, *12-Lead Vectorcardiography*. Oxford: Butterworth-Heinemann, 1994.

[18] J. Slocum, E. Byrom, L. McCarthy, A. Sahakian, and S. Swiryn, Computer detection of atrioventricular dissociation from surface electrocardiograms during wide QRS complex tachycardia. *Circulation* 72:1028–1036, 1985.

[19] M. Holm, S. Pehrsson, M. Ingemansson, L. Sörnmo, R. Johansson, L. Sandhall, M. Sunemark, B. Smideberg, C. Olsson, and B. Olsson, Non-invasive assessment of atrial refractoriness during atrial fibrillation in man—Introducing, validating and illustrating a new ECG method. *Cardiovasc. Res.* 38:69–81, 1998.

[20] S. Shkurovich, A. Sahakian, and S. Swiryn, Detection of atrial activity from high-voltage leads of implantable ventricular defibrillators using a cancellation technique. *IEEE Trans. Biomed. Eng*. 45:229–234, 1998.

[21] M. Koschat and D. Swayne, A weighted Procrustes criterion. *Psychometrica* 2:229–239, 1991.

Chapter 9

# DETECTING NONLINEAR DYNAMICS IN SYMPATHETIC ACTIVITY DIRECTED TO THE HEART

Alberto Porta, Giuseppe Baselli, Nicola Montano, Alberto Malliani, and Sergio Cerutti

## 1. INTRODUCTION

The analysis of complex nonlinear dynamics is often based on the calculation of indexes such as correlation dimension [1], Lyapunov exponents [2], Kolmogorov entropy [3], and the exponent $\alpha$ of the $1/f^{\alpha}$ process [4]. These parameters measure the effective number of degrees of freedom, the exponential divergence or convergence of nearby trajectories, and the amount of information carried by the process and its fractal properties, respectively, and quantify important features of the underlying dynamics. However, they need to be calculated over long data sequences (several thousand points) to be reliably estimated. Unfortunately, most of the data series obtained from biological experiments are short (a few hundred points) or, at best, medium-sized sets (around 1000 points). The size of the biological data sets is limited mainly by the difficulty in maintaining the stable experimental conditions over a large time window.

The present chapter reviews several methods that are found useful for detecting and classifying over short and medium-sized data segments the nonlinear interactions among spontaneous low- and high-frequency (LF, around 0.1 Hz, and HF, around ventilatory rate) rhythms in the beat-to-beat variability series of the sympathetic discharge and ventilation in decerebrate artificially ventilated cats [5–7]. The qualitative tools provide a graphical description of the underlying dynamics, and the quantitative ones produce numerical indexes. Whereas the former ones need an attentive interpretation, the latter ones are independent of the researcher's experience. The qualitative tools (superposition plot, recurrence map, space–time separation plot, frequency tracking locus) were helpful in describing the effect of ventilation as a perturbing periodic signal on spontaneous sympathetic activity [5, 6]. The quantitative tools (spectrum and bispectrum) were used to quantify, respectively, harmonic relationships and nonlinear coupling between spectral components in the sympathetic discharge [5, 6]. The corrected conditional entropy is a quantitative tool utilized to measure the sympathetic discharge regularity [7]. In order to indicate the capabilities of these tools and stress differences among the tools, the proposed techniques are applied to four examples of dynamics arising from nonlinear interferences between the sympathetic discharge and ventilation. The examples are (a) quasiperiodic dynamics (the HF oscillation in the sympathetic discharge is independent of ventilation), (b) 1:2 dynamics (the LF rhythm in the sympathetic discharge occurs every two ventilatory cycles), (c) 1:4 dynamics with HF rhythms (the LF oscillation grows every four ventilatory cycles while the HF oscillation is unlocked to ventilation), and (d) aperiodic dynamics characterized by LF and HF spectral peaks emerging from a broadband spectrum.

## 2. METHODS

### 2.1. Superposition Plot

When the forcing signal (i.e., ventilation) $v = \{v)i), i = 1, \ldots, N\}$ and the perturbed activity (i.e., sympathetic activity) $s = \{s(i), i = 1, \ldots, N\}$, where $N$ is the series length and $i$ the progressive cardiac beat number, are plotted as a function of the heartbeat on the same graph, the nonlinear interactions between the sympathetic discharge and ventilation are enhanced [5, 6].

### 2.2. Recurrence Map

Given a pair of points $P_i = (v(i), s(i))$ and $P_j = (v(j), s(j))$ in the plane $(v, s)$, their normalized Euclidean distance is defined as

$$d_{ij}^2 = \frac{(v(i) - v(j))^2}{\text{var}[v]} + \frac{(s(i) - s(j))^2}{\text{var}[s]} \qquad i \neq j \tag{1}$$

where var[$v$] and var[$s$] represent the variance of $v$ and $s$ signals. The recurrence map [8] is a plot in a time versus time domain where a dot is drawn at the point $(i, j)$ if the Euclidean distance $d_{i,j}$ between the pair of points $P_i$ and $P_j$ is below a threshold $d$. Usually this reference value is set to be a user-defined percentage of the maximum distance found in the plane $(v, s)$. The use of a plane $(v, s)$ allows the coupling between the two signals to be enhanced [5].

The lines parallel to the main diagonal (their equation is $j = i \pm \tau$) indicate that the trajectories in the plane $(v, s)$ return close after a fixed delay $\tau$. If the line is complete, close returns after a time $\tau$ are found in the whole set of data. Otherwise, if the line is broken and a sequence of segments appears, close returns occur only in a specific subsequence of the series. These lines are periodic if close returns are found at multiples of time $\tau$. When points in the plane $(v, s)$ remain close, a square block appears in the recurrence map with a side equal to the duration of the time interval during which the points stay close.

The recurrence plot is directly related to the correlation integral [1]

$$C(d) = \frac{\text{number of pairs } (i, j) \text{ whose distance } |P_i - P_j| < d}{N(N-1)} \qquad i \neq j \tag{2}$$

representing the percentage of points closer than $d$ independent of their separation in time. Indeed, given a recurrence map with a threshold $d$, the correlation integral can be obtained by counting the dots in the plot and dividing the results by $N(N-1)$.

### 2.3. Space–Time Separation Plot

The space–time separation plot [9] is designed to establish whether the closeness between two points $P_i$ and $P_j$ is maintained at a specific value of the time separation

$\tau = |i - j|$. If this occurs for $\tau \neq 0$, close returns after a time $\tau$ are detected in the plane $(v, s)$. Therefore, for each pair of points $P_i$ and $P_j$ their Euclidean distance $d_{i,j}$ (space separation) is plotted versus their time separation $\tau$. This plot is usually given as a contour map [9] representing the fraction of pairs of points closer than $d$ at a given $\tau$ as a function of the time separation $\tau$. Usually a limited number of contour curves is sufficient to detect close returns in the plane $(v, s)$ (e.g., three curves representing 5%, 50%, and 95% of points). Close returns after $\tau$ are enhanced by the decrease to zero of all three curves at $\tau$, thus evidencing nonlinear interferences between $v$ and $s$ [6].

The recurrence plot and the space–time separation plot are closely related. Indeed, in the recurrence map, as the diagonal lines parallel to the main diagonal represent time indexes separated by $|j - i| = \tau$, the percentage of points along these lines represents the probability that two points $P_i$ and $P_j$, closer than the threshold $d$, are separated in time by $\tau$.

## 2.4. Frequency Tracking Locus

The frequency tracking locus [10] represents the peak response of the sympathetic discharge to each ventilatory cycle as a vector $\rho = |\rho|e^{j\varphi}$. The vector magnitude $|\rho|$ is obtained by normalizing the maximum amplitude of the forced sympathetic response by the maximum amplitude of ventilation as

$$|\rho| = \frac{s(i_{\max})}{v(nT_v)} \qquad (n-1)T_v \leq i_{\max} \leq nT_v \tag{3}$$

where $T_v$ represents the period of the perturbing signal expressed in cardiac beats, $(n-1)T_v$ and $nT_v$ are the time occurrences of the peaks defining the $n$th forcing cycle, and $i_{\max}$ is the time occurrence of the maximum of the sympathetic signal during the $n$th ventilatory cycle. The vector phase is calculated as

$$\varphi = \frac{2\pi(nT_v - i_{\max})}{T_v} \tag{4}$$

and varies from 0 to $2\pi$. These two values are found when the forced sympathetic response is synchronous with the second and the first peak defining the $n$th ventilatory cycle, respectively.

A forced sympathetic response is detected in the $n$th forcing cycle if the maximum value of the sympathetic discharge exceeds a user-defined threshold. All the forced responses found in the considered ventilatory cycle are represented. Each arrow starts from the tip of the preceding one, thus maintaining the total phase shift. If no forced sympathetic response is detected, an open circle (instead of an arrow) is drawn and the number of cycles during which no forced response is found is written near the symbol. Therefore, the frequency tracking locus appears as a sequence more or less regular, according to the interactions between ventilation and sympathetic activity, of arrows and open circles [6].

## 2.5. Autoregressive Power Spectral Analysis

The autoregressive (AR) power spectral analysis [11] describes the beat-to-beat series of the sympathetic discharge $s$ as an AR process

$$s(i) = \sum_{k=1}^{p} a_k s(i-k) + w(i) \tag{5}$$

where $a_k$ $(k = 1, \ldots, p)$ are the coefficients of the model and $w$ is a zero-mean white noise with variance $\lambda^2$. The coefficients $a_k$ and the variance $\lambda^2$ are estimated via Levinson–Durbin recursion and the model order $p$ is chosen according to the Akaike figure of merit. After writing the transfer function generating the AR process as

$$H(z) = \frac{1}{1 - \sum_{k=1}^{p} a_k z^{-k}} \tag{6}$$

where $z^{-1}$ represents the one delay operator in the $z$-domain, the AR power spectral density can be calculated as

$$S(f) = T \cdot H(z)H(z^{-1})\Big|_{z=\exp(j2\pi fT)} \tag{7}$$

where $T$ is the mean heart period. Calculation of the frequencies of the $p$ poles of the AR transfer function $H(z)$ is useful to check whether harmonic relationships between oscillations in the sympathetic discharge are present [6].

## 2.6. Nonparametric Bispectrum and Bicoherence

The bispectrum or third-order spectrum of the sympathetic discharge is estimated directly from the Fourier transform of the signal $s$ instead of being derived from the Fourier transform of the estimate of the third-order cumulant [5, 12]. The signal $s = \{s(i),\ i = 1, \ldots, N\}$ is segmented in $K$ frames of length $M$. The frames are 50% overlapped and windowed [5, 12]. The Fourier coefficients over frame $k$ are obtained as

$$S_k(f) = \sum_{i=0}^{M-1} s_k(i) e^{-\frac{j2\pi fi}{M}} \tag{8}$$

The spectrum

$$P(f) = \frac{T}{KM} \sum_{k=1}^{K} S_k(f) S_k^*(f) \tag{9}$$

where $*$ stands for the complex conjugation operator and $T$ is the mean heart period over the frame $k$ and the bispectrum

$$B(f_1, f_2) = \frac{T^2}{KM^2} \sum_{k=1}^{K} S_k(f_1) S_k(f_2) S_k^*(f_1 + f_2) \tag{10}$$

are directly derived by averaging the spectral and bispectral estimates over all $K$ frames. The averaging procedure allows to reduce the variance of the estimate and to obtain more readable and smoother functions. Eleven frames of 128 samples are used in our studies [5]. The square bicoherence function is defined as the normalization of the bispectrum modulus $|B(f_1, f_2)|$

$$\text{bic}^2(f_1, f_2) = \frac{M}{T} \frac{|B(f_1, f_2)|^2}{P(f_1)P(f_2)P(f_1 + f_2)} \tag{11}$$

A peak close to 1 in the bicoherence function at the pair $(f_1, f_2)$ means that the frequency $f = f_1 + f_2$ is the result of a quadratic coupling between $f_1$ and $f_2$; therefore, a nonlinear interaction between $f_1$ and $f_2$ is detected. No peak in the bicoherence function or peaks less than 0.5 at the pair $(f_1, f_2)$ mean that the two frequencies are linearly superposed or a coupling different from a quadratic one has to be tested [13].

## 2.7. Corrected Conditional Entropy

Given the sympathetic discharge signal $s$, $N - L + 1$ patterns of $L$ samples can be extracted as $s_L = (s(i), s(i-1), \ldots, s(i-L+1))$. The sequence $s_L$ can be written as $s_L = (s(i), s_{L-1}(i-1))$. The conditional entropy (CE) [14]

$$\text{CE}(L) = \sum_{L-1} p(s_{L-1}) \sum_{i/L-1} p(s(i)/s_{L-1}) \log p(s(i)/s_{L-1}) \tag{12}$$

where $p(s_{L-1})$ and $p(s(i)/s_{L-1})$ represent the joint probability of the pattern $s_{L-1}$ and the probability of the value $s(i)$ conditioned by the sequence $s_{L-1}$, respectively. The CE measures the information carried by the most recent sample $s(i)$ of the pattern $s_L$ when the previous $L-1$ samples $s_{L-1}$ are known. Therefore, if the sympathetic discharge signal is strictly periodic, the CE decreases and reaches the zero value when the knowledge of the $L-1$ past samples allows complete prediction of the current value $s(i)$. On the contrary, if $s$ is a white noise, the CE is constant as a function of $L$ because each new sample is unpredictable independent of the number of samples on which the prediction is based.

Unfortunately, when calculated over a limited amount of samples, the CE of any process decreases to zero very suddenly while increasing $L$ [7]. Therefore, a high degree of regularity can be always found provided that the CE is evaluated at large $L$. This erroneous detection of regularity comes from the approximation of the conditional probability $p(s(i)/s_{L-1})$ with the conditional sample frequency $\hat{p}(s(i)/s_{L-1})$. Indeed, $\hat{p}(s(i)/s_{L-1}) = 1$ when the conditioning pattern $s_{L-1}$ is found only once in the data set. The corrected conditional entropy (CCE) [7]

$$\text{CCE}(L) = \text{CE}(L) + \text{perc}(L)E(s(i)) \tag{13}$$

overcomes this bias by introducing a term proportional to the percentage of patterns $s_L$ found only once in the data (perc($L$)). The proportionality constant

$$E(s(i)) = \sum_i p(s(i)) \log p(s(i)) \tag{14}$$

is the Shannon entropy of the signal $s$ (i.e., the information carried by each sample when no relationship with previous samples is hypothesized). The corrective term replaces the false certainty determined by the pattern $s_L$ found only once in the data set by the maximum uncertainty estimated by $E(s(i))$. As a function of $L$, the CCE can: (1) decrease to zero if the sympathetic discharge signal is strictly period; (2) initially diminish if some degree of regularity is found and, afterward, increase when no robust statistic can be performed; and (3) remain constant if no regularity is found in the data set as it occurs when a white noise realization is analyzed. The minimum of the CCE is taken as a measure of regularity over short segments of data [7]: the lower the minimum, the larger the regularity.

## 3. EXPERIMENTAL PROTOCOL

The sympathetic discharge was recorded from the third (T3) left thoracic white ramus communicans in decerebrate artificially ventilated cats. While the ventilatory frequency (VF) was kept fixed (18 breaths per minute), the ventilatory volume was adjusted to maintain arterial $PO_2$, $PCO_2$, and PH at normal levels. The raw sympathetic activity was counted on a temporal basis of 20 ms. The counted neural discharge was low-pass filtered at 1 Hz and sampled with ventilation once per cardiac beat (about 3 Hz in the cat) at the QRS complex detected on the ECG, thus obtaining the beat-to-beat series of the sympathetic discharge and sampled ventilatory signal. The sympathetic discharge was expressed in spikes/sec and ventilation in arbitrary units (a.u.). More details about the experimental protocol can be found in Ref. 15. Ventilation perturbed the sympathetic discharge, thus resulting in various types of dynamics according to the experimental condition; a complete classification of the interaction patterns can be found in Ref. 6. The four examples of dynamics analyzed in this study were detected at control (quasi-periodic dynamics), during the sympathetic activation induced by inferior vena cava occlusion (1:2 and 1:4 dynamics), and after separation of the supraspinal centers form the spinal ones by spinalization at C1 level (aperiodic dynamics).

## 4. RESULTS

Figure 1 represents four examples of dynamics found in the beat-to-beat variability series of the sympathetic activity: quasi-periodic dynamics (Figure 1a), 1:2 (Figure 1b) and 1:4 (Figure 1c) periodic dynamics, and aperiodic dynamics (Figure 1d). The superposition plot between the sympathetic discharge and ventilation enhanced the coupling between the signals. In the quasi-periodic dynamics (Figure 1a), a progressive shift in the phase of the sympathetic discharge (solid line) with respect to ventilation (dotted line) could be observed, thus evidencing a weak link between the signals. Clear examples of periodic dynamics can be observed in Figure 1b and c. The sympathetic activity grew with fixed phase every two and four ventilatory cycles, respectively: the LF rhythm was locked to a subharmonic of ventilation. In 1:4 dynamics (Figure 1c) an HF sympathetic activity was still present but unlocked to ventilation with a frequency slightly lower than VF. On the contrary, HF oscillations interacted with the LF oscillation because, as observed in Figure 1c, the HF rhythm appeared to be synchronized at the rising edge of the LF rhythm. When the sympathetic dynamics became aperiodic (Figure 1d), bursts stimulated by ventilation were still present but their occurrence was irregular. Moreover, the sympathetic bursts were not evident every ventilatory cycle and their appearance was erratic.

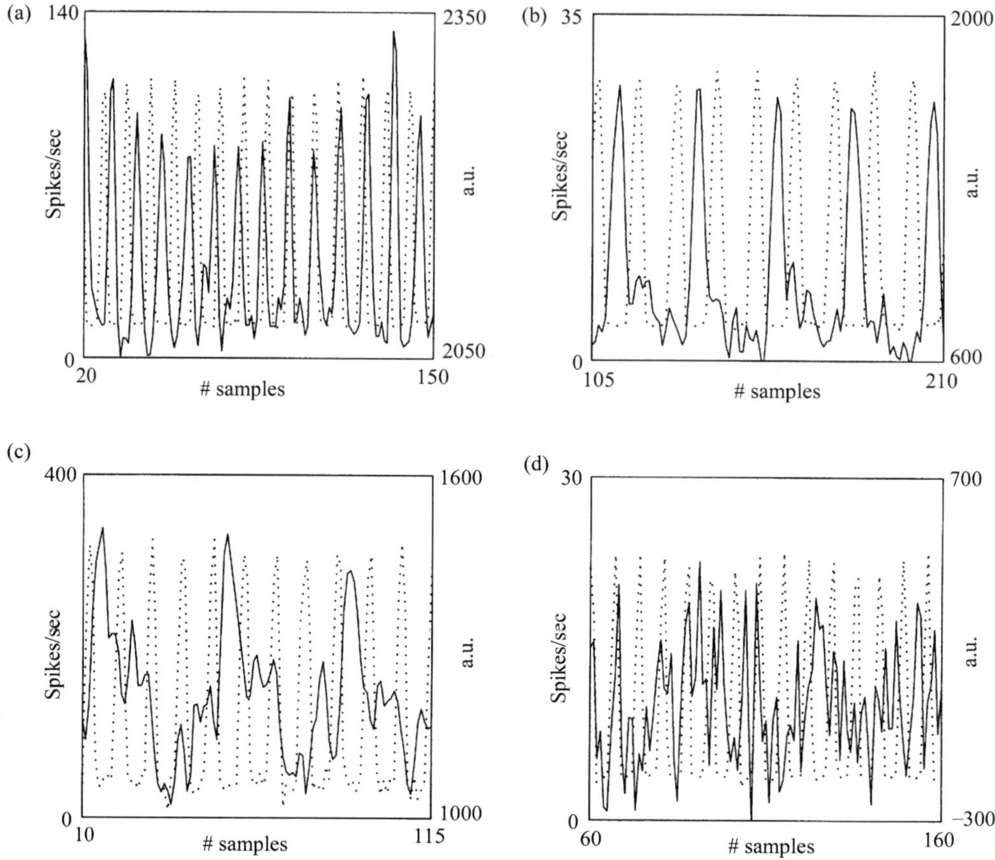


**Figure 1** Superposition plots of the beat-to-beat series of the sympathetic discharge (solid line) and ventilation (dotted line) are drawn in (a) quasi-periodic dynamics, (b) 1:2 dynamics, (c) 1:4 dynamics, and (d) aperiodic dynamics.

Figure 2 shows the recurrence map performed on the four types of dynamics shown in Figure 1. In the quasi-periodic dynamics the gradual phase shift between the sympathetic discharge and ventilation was captured by the recurrence map as lines parallel to the main diagonal that faded and, then, reappeared after several periods [the trajectories in the plane $(v, s)$ periodically moved apart and then became close again, Figure 2a]. The 1:2 dynamics appeared clearly as lines parallel to the main diagonal separated by two ventilatory cycles (Figure 2b). Square blocks lay on these lines as a result of pauses in ventilation lasting half of a ventilatory cycle (Figure 1b) during which small changes in the sympathetic discharge occurred. The recurrence map of the 1:4 dynamics appeared as an unclear ensemble of blocks of several shapes (not only square) that were due to the interference between the VF and the unlocked spontaneous HF rhythm. These blocks were aligned on lines parallel to the main diagonal separated by four ventilatory cycles (Figure 2c). The aperiodic dynamics produced a recurrence map in which diagonal lines and square blocks were not observed (Figure 2d). However, small segments parallel to the main diagonal could be detected, indicat-

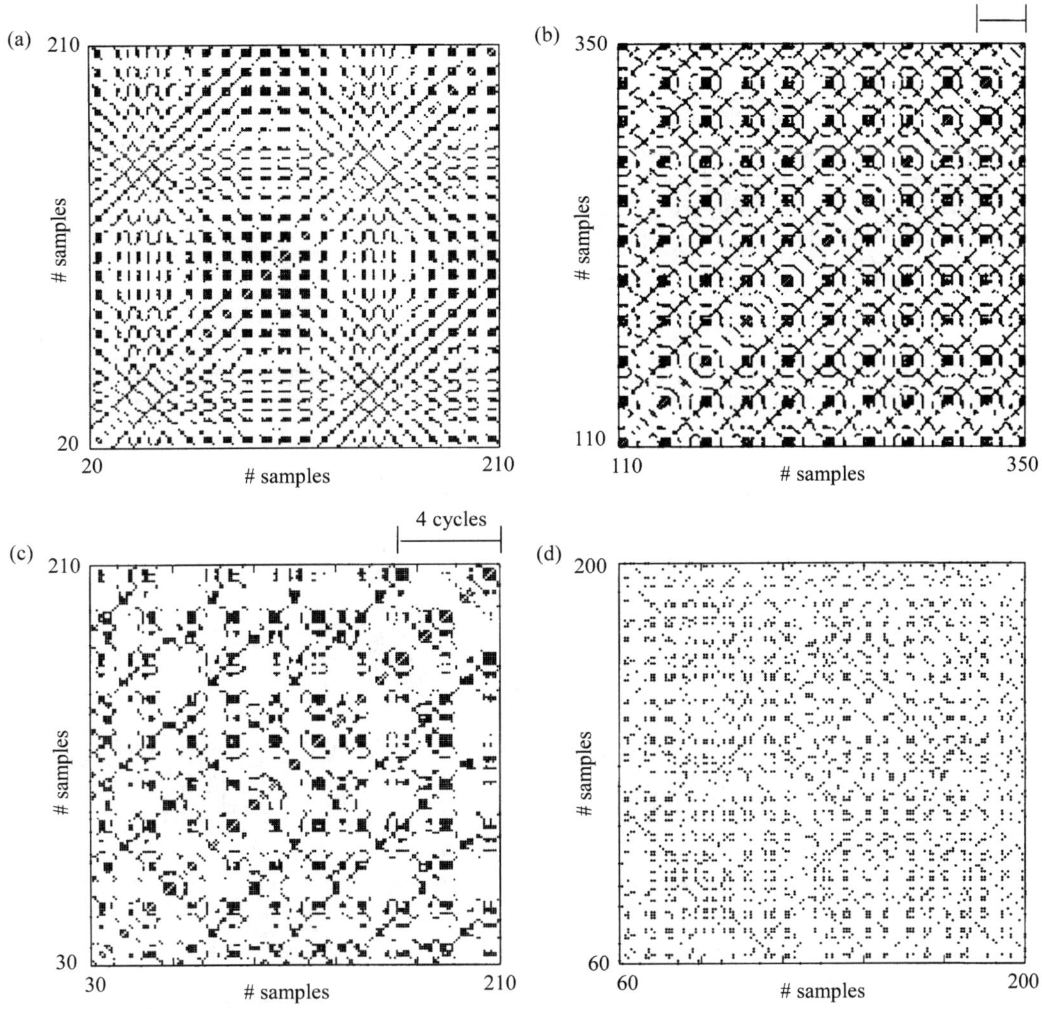


**Figure 2** Recurrence maps carried out over the quasi-periodic dynamics, over the 1:2 and 1:4 periodic dynamics, and over the aperiodic dynamics depicted in Figure 1 are drawn in (a)–(d), respectively.

ing that a certain degree of determinism was still present and produced close returns over small time intervals.

Figure 3 shows the space–time separation plots carried out on the dynamics depicted in Figure 1. In the quasi-periodic dynamics the regular phase shift between the signals produced the regular modulation of the space separation while increasing the time separation (Figure 3a). The space–time separation plots performed on 1:2 and 1:4 dynamics (Figure 3b and c) were characterized by a periodic decrease of the space separation of all three contour curves toward small values every two and four ventilatory cycles, respectively. No clear decrease in the space separation was observed in aperiodic dynamics (i.e., no close return after a time $\tau$ was detected). Therefore, the three curves were clearly distinct (Figure 3d).

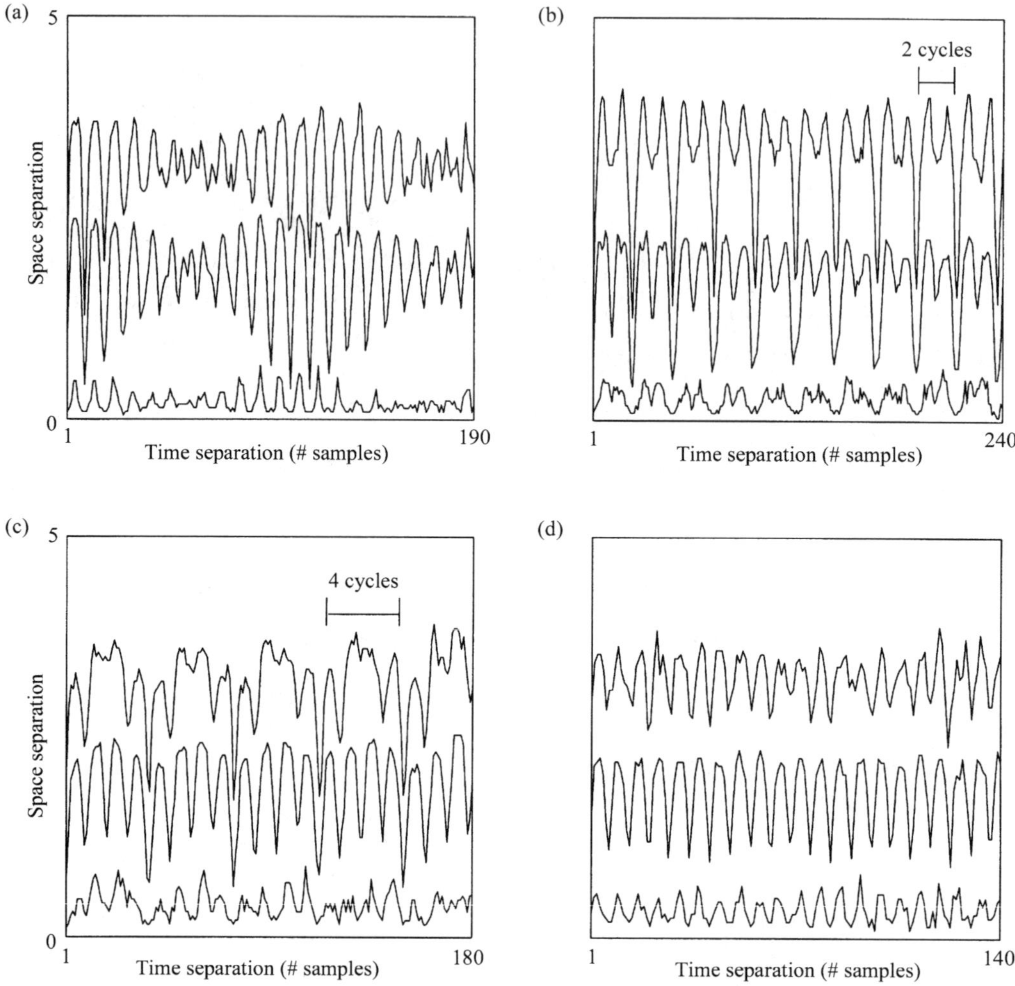


**Figure 3** Space–time separation plots performed over the quasi-periodic dynamics, over the 1:2 and 1:4 periodic dynamics, and over the aperiodic dynamics depicted in Figure 1 are drawn in (a)–(d), respectively.

The frequency tracking loci are shown in Figure 4. The quasi-periodic dynamics was clearly detected as a sequence of vectors progressively changing their phase (Figure 4a). The 1:2 dynamics was represented as a sequence of one vector followed by an open circle labeled with the number one (Figure 4b). The vector phases were characterized by small variability. In the 1:4 dynamics the missed responses were three (therefore the number three was written near the open circle) and the vector phases were fixed (Figure 4c). No clear and repetitive pattern was present in the frequency tracking locus relevant to aperiodic dynamics (Figure 4d). Sympathetic responses to the ventilatory cycle still occurred but their amplitude and phase (the vector magnitude and phase) exhibited large variabilities. Missed responses were detected, suggesting the presence of slow oscillations in the LF band in the sympathetic discharge. As missed responses did

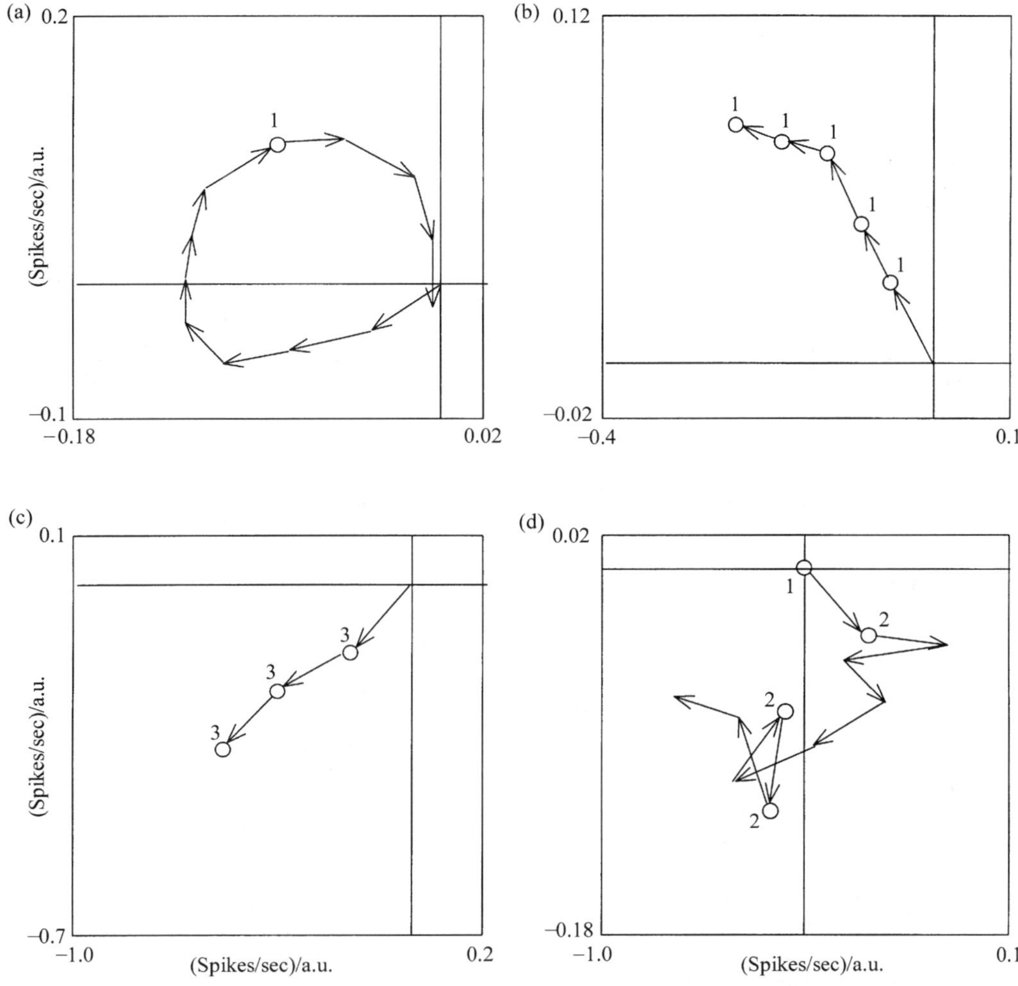


**Figure 4** Frequency tracking loci carried out over the quasi-periodic dynamics, over the 1:2 and 1:4 periodic dynamics, and over the aperiodic dynamics depicted in Figure 1 are drawn in (a)–(d), respectively.

not occur periodically, LF rhythms were irregular and unlocked to ventilation. The presence of LF and HF irregular oscillations was responsible for the complexity of this dynamics.

The AR power spectra performed on the sympathetic discharge (solid line) and ventilation (dotted line) of Figure 1 are depicted in Figure 5. In the quasi-periodic dynamics the main spectral peak of the sympathetic activity had a frequency close to but not coincident with that of ventilation (Figure 5a). Moreover, no rational ratio between these frequencies was found. The power spectral analysis performed on the 1:2 dynamics pointed out as the dominant peak of the sympathetic discharge (Figure 5b) was exactly at half of the ventilatory rate and some harmonics were even present. The power spectral density calculated on the 1:4 dynamics evidenced a clear important peak with a period four times slower than that of ventilation (Figure 5c). In addition, a clear

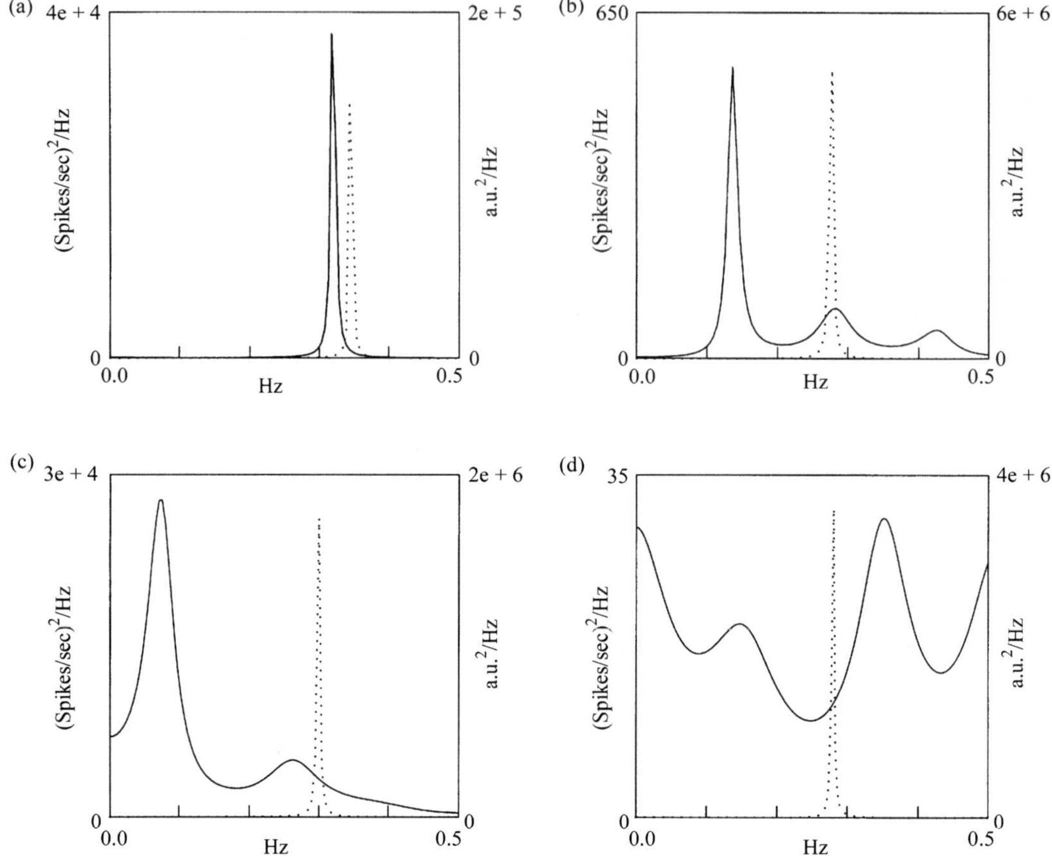


**Figure 5** The AR power spectra of the beat-to-beat series of the sympathetic discharge (solid line) and of ventilation (dotted line) for the data shown in Figure 1 are overlapped. In (a) (quasi-periodic dynamics) the HF peak of the sympathetic discharge is not coincident with that of ventilation. In (b) (1:2 dynamics) the LF oscillation in the sympathetic discharge is located at half of the ventilatory frequency. In (c) (1:4 dynamics) the ventilatory frequency is four time faster than the LF sympathetic oscillation and the HF peak is not synchronous with that of ventilation. In (d) (aperiodic dynamics) a broadband spectrum with emerging rhythmicities in the LF and HF bands can be observed.

component at a frequency different from the ventilatory rate suggested that the HF rhythm was still present in the sympathetic activity but did not interfere with ventilation. The power spectral density of the aperiodic dynamics was broadband although some rhythms emerged (Figure 5d). These rhythms were in the LF and HF bands and they were not synchronous with ventilation or its subharmonics.

In the quasi-periodic dynamics the contour plot (Figure 6a) suggested the presence of a clear peak in the bicoherence function around the pair of frequencies (HF, HF). This peak proved the presence of a harmonic of the HF rhythm at 2HF. This oscillation was nonlinearly related to the HF one. However, because of quasi-periodicity bearing regular phase shift and amplitude modulations, a cross-shaped contour area around the

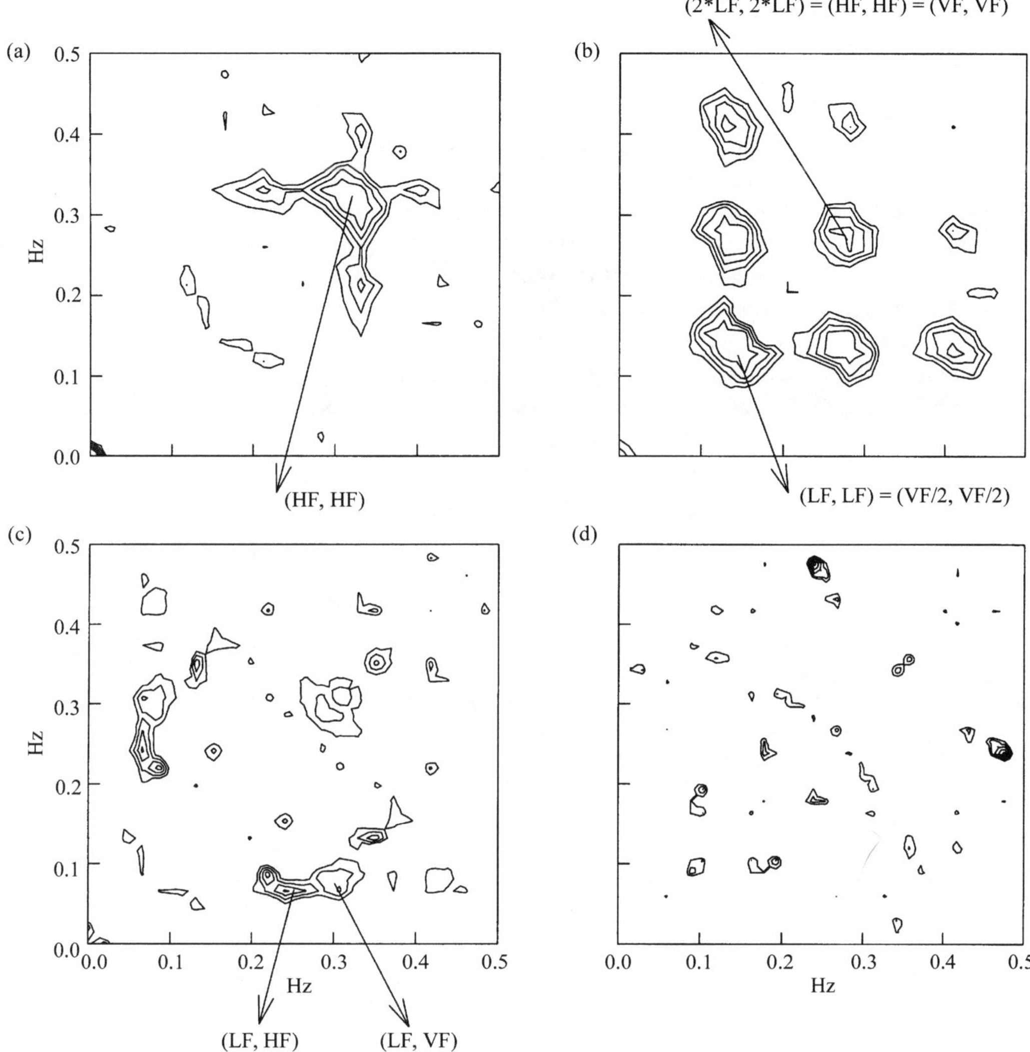


**Figure 6** Contour plots of the bicoherence function performed over the quasi-periodic dynamics, over the 1:2 and 1:4 periodic dynamics, and over the aperiodic dynamics depicted in Figure 1 are drawn in (a)–(d), respectively. Contour lines from 0.5 to 1 every 0.1 are given. VF, ventilatory frequency.

pair of frequencies (HF, HF) could be noted, indicating that all these pairs of frequencies interfered. The contour map in 1:2 dynamics (Figure 6b) confirmed that the LF oscillation was locked to a frequency half of the ventilation rate by evidencing that the bicoherence had a peak at the pair (LF, LF), where LF + LF = VF. The high degree of coupling between the LF rhythm and its harmonics was revealed by the bicoherence peaks at the pairs (LF, 2LF), (LF, 3LF), (LF, 4LF). The coupling was even maintained between the harmonics. Indeed, bicoherence peaks were detected at (2LF, 2LF) and (2LF, 3LF). In the 1:4 dynamics the coupling between the LF oscillation at a frequency four time slower than the ventilatory rate and ventilation was evidenced by the presence of the peak at the pair (LF, VF) (Figure 6c). As the HF rhythm in sympathetic discharge was synchronized at the rising edge of the LF oscillation, the bicoherence at the pair (LF, HF) is high. No important peak was detected at (HF, VF) because of the uncoupling between these oscillations. Also, the peaks at the pairs (LF, LF) and (HF, HF) were missed: no important harmonic of the LF and HF oscillations was present in the sympathetic discharge. When an example of aperiodic dynamics was considered, no significant bicoherence peak was clearly detected (Figure 6d); the pairs of frequency exhibiting coupling were scattered on the frequency plane. However, their distribution did not appear completely random.

The analysis of the sympathetic discharge regularity pointed out that both quasi-periodic and periodic 1:2 and 1:4 dynamics were regular (Figure 7a–c). Indeed, the CCE function exhibited a well-defined minimum. In the 1:2 dynamics the minimum (0.87)

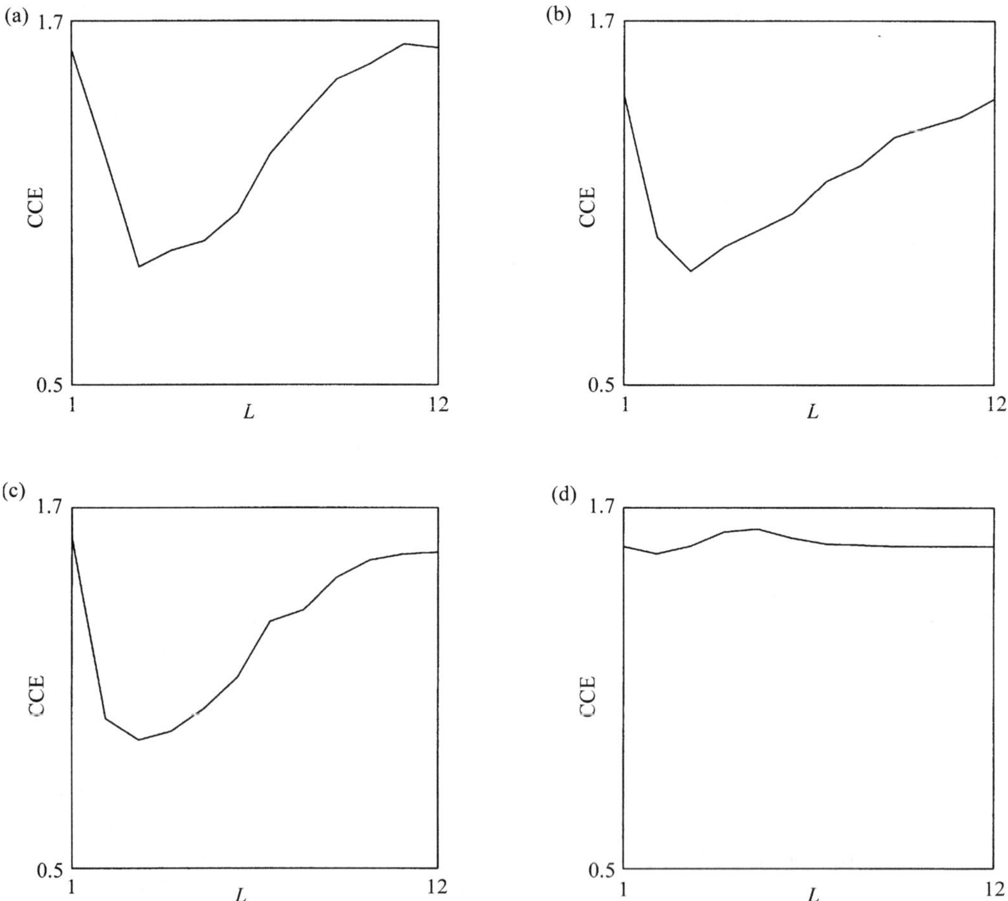


**Figure 7** Plots of the corrected conditional entropy as a function of the pattern length $L$ carried out over the quasi-periodic dynamics, over the 1:2 and 1:4 periodic dynamics, and over the aperiodic dynamics depicted in Figure 1 are drawn in (a)–(d), respectively.

was slightly smaller than in the quasi-periodic dynamics (0.89) as a result of the phase locking of the LF oscillation with ventilation. The slight increase of the CCE minimum (0.94) in 1:4 dynamics was due to the presence of an irregular rhythm HF unlocked to ventilation. When aperiodic dynamics was considered, the CCE function was flat (Figure 7d) and no regularity was detected.

## 5. DISCUSSION

The proposed qualitative tools (superposition plot, recurrence map, space–time separation plot, and frequency tracking locus), by exploiting directly both ventilation (i.e., the perturbing signal) and sympathetic activity (i.e., the perturbed signal), provide graphical evidence of the nonlinear interference between signals.

The quasi-periodic dynamics, which cannot be easily recognized by looking only at the sympathetic discharge, is enhanced by the contemporaneous analysis of both signals and by the use of nonlinear, even though graphical, methods. Indeed, the regular shifting between the two signals is captured by the superposition plot as a progressive drifting of the sympathetic discharge with respect to ventilation, by the recurrence map as a gradual loss of dots along the diagonal lines followed by a progressive recovery, by the space–time separation plot as a progressive increase of the space separation followed by a gradual decrease, and by the frequency tracking locus as the gradual change of vector phases. With the graphical tools one can easily recognize periodic dynamics. In the superposition plot an $N:M$ coupling pattern appears as $N$ different sympathetic responses every $M$ ventilatory cycles. In the recurrence map the $N:M$ periodic dynamics produces lines parallel to the main diagonal spaced by $M \cdot T_v$ samples. In the space–time separation plot the three percentile lines periodically decrease toward zero with a period $M \cdot T_v$. A repetitive sequence of $M$ symbols (a mix of arrows and circles) is detected in the frequency tracking locus. In the presence of aperiodic dynamics, none of these structures can be found. In the recurrence map the dots scatter uniformly, in the space–time separation plot the three percentile curves are clearly separated, and in the frequency tracking locus the magnitude and phase of the vectors change irregularly.

These nonlinear tools work better than the linear ones (e.g., the coherence function) if the task of detecting nonlinear coupling between the sympathetic discharge and ventilation is addressed. In all the considered examples of dynamics the squared coherence between the sympathetic discharge and ventilation is low (Figure 8a–d). Indeed, the dominant oscillation in the sympathetic discharge has a frequency different from that of ventilation and is not linearly related to it. On the contrary, as pointed out by all the nonlinear tools in 1:2 and 1:4 dynamics, the link between the sympathetic discharge and ventilation is strong.

The major drawback of the graphical tools is that they can be used only when the interaction scheme between the sympathetic discharge and ventilation is clearly detectable. Indeed, it becomes difficult to interpret them when the coordination between the signals is not rigidly fixed as an effect of a dynamics producing change in the coupling ratio (i.e., sliding dynamics) or of irregular unlocked rhythmicities. For example, the presence of irregular HF rhythms, in addition to the LF oscillation locked to ventilation, almost prevents the detection of the presence of diagonal lines spaced by four ventilatory cycles in the recurrence map (Figure 2c). As 1:4 dynamics is detectable by

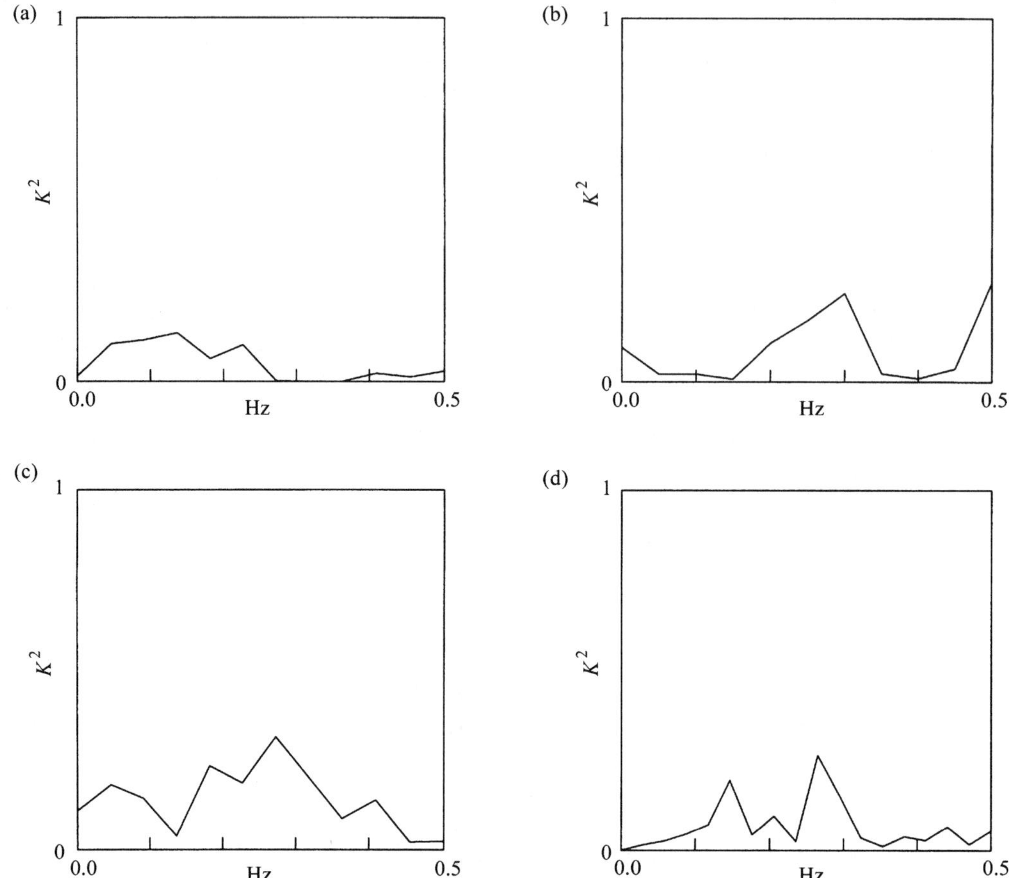


**Figure 8** Coherence functions between the sympathetic discharge and ventilation estimated via the fast Fourier transform (FFT) technique over the quasi-periodic dynamics, over the 1:2 and 1:4 periodic dynamics, and over the aperiodic dynamics depicted in Figure 1 are drawn in (a)–(d), respectively. The square coherence function is defined as the cross-spectrum square modulus normalized by the product of the two spectra. The frequency resolution and windowing procedure are the same as in Figure 6. No significant coherence value (> 0.5) is found.

using the space–time separation plot (Figure 3c) and the frequency tracking locus (Figure 4c), these two tools seem more robust for detecting periodic dynamics [6].

To characterize the sympathetic discharge dynamics and its relationship with ventilation, a linear tool such as the AR power spectrum estimate is useful. Indeed, it makes it possible to distinguish between periodic and complex dynamics. The power spectrum of periodic dynamics appears as a series of sharp spectral peaks characterized by harmonic or subharmonic relationships among them (Figure 5b and c). On the contrary, the spectrum of complex dynamics is broadband (Figure 5d). As power spectral analysis treats a process as a sum of independent oscillations at different frequencies and the phase is disregarded, the presence of harmonic relationships among frequencies is not a sufficient condition to conclude that nonlinear interactions between frequencies are present. In addition, based on the power spectrum only, chaotic dynamics cannot be distinguished from white noise with embedded important rhythmicities. Although separating chaos from randomness is a difficult task [16, 17], the detection of the nonlinear coupling between frequencies can be performed by the bicoherence function [13]. For example, in 1:2 dynamics, the degree of locking of the LF rhythm to a subharmonic of ventilation is measured by the value of the bicoherence function at the frequency pair (LF, LF) (Figure 6d). Similarly, the bicoherence function at the pairs (LF, 2LF) and (LF, 3LF) quantifies the degree of coupling between the LF rhythm and the harmonics of ventilation (Figure 6b). When (Figure 6c) the LF oscillation is locked to ventilation (1:4 dynamics) and the HF rhythm is coupled to the LF oscillation at its rising edge, peaks in the bicoherence function at the pairs (LF, VF) and (LF, HF) can be detected.

Tools based on entropy rates are useful for measuring signal complexity [18]. Among these tools, a procedure based on the minimum of a function referred to as CCE has quantified the regularity of the sympathetic discharge over a short segment of data [7]. This minimization procedure allows one to calculate a regularity index without imposing a priori the length of the pattern whose regularity is to be measured. In other words, this procedure does not fix the embedding dimension [7] and searches for that, allowing detection of the maximum degree of regularity in the data. Regular dynamics (Figure 7) cause a smaller minimum than irregular dynamics. Quasi-periodic dynamics appears less regular than 1:2 periodic dynamics. The 1:4 periodic dynamics with HF oscillations unlocked to ventilation is more complex than the 1:2 dynamics. Therefore, this tool is able to distinguish among different types of dynamics (periodic and aperiodic dynamics give different regularity indexes) but also among complexities associated with the same type of dynamics (1:2 and 1:4 dynamics are periodic but characterized by different degrees of recurrence of the dominant pattern).

## 6. CONCLUSIONS

Graphical tools are useful for detecting and classifying different types of nonlinear interferences between a forcing signal and perturbed activity. The bicoherence function is a quantitative tool that can measure the degree of quadratic coupling between rhythms. The corrected conditional entropy is able to measure regularity (i.e., the degree of recurrence of a specific pattern) of variability signals. All these tools give promising results even when utilized on short sequences of data (from a few hundred to a thousand samples). Therefore, these methods are proposed when shortness of the data sequence prevents the use of other more complex tools to detect and quantify complexity in biological series.

## REFERENCES

[1] P. Grassberger and I. Procaccia, Measuring the strangeness of strange attractors. *Physica D* 9:189–208, 1983.

[2] M. Sano and Y. Sawada, Measurement of the Lyapunov spectrum from a chaotic time series. *Phys. Rev. Lett.* 55:1082–1085, 1985.

[3] A. Cohen and I. Procaccia, Computing the Kolmogorov entropy from time signals of dissipative and conservative dynamical systems. *Phys. Rev. Lett* 31:1872–1882, 1985.

[4] M. S. Keshner, $1/f$ noise. *Proc. IEEE* 70:212–218, 1982.

[5] A. Porta, G. Baselli, N. Montano, T. Gnecchi Ruscone, F. Lombardi, A. Malliani, and S. Cerutti, Non linear dynamics in the beat-to-beat variability of sympathetic activity in decerebrate cats. *Methods Inform. Med.* 33:89–93, 1994.

[6] A Porta, G. Baselli, N. Montano, T. Gnecchi-Ruscone, F. Lombardi, A. Malliani, and S. Cerutti, Classification of coupling patterns among spontaneous rhythms and ventilation in the sympathetic discharge of decerebrate cats. *Biol. Cyber.* 75:163–172, 1996.

[7] A. Porta, G. Baselli, D. Liberati, N. Montano, C. Cogliati, T. Gnecchi-Ruscone, A. Malliani, and S. Cerutti, Measuring regularity by means of a corrected conditional entropy in sympathetic outflow. *Biol. Cybern.* 78:71–78, 1998.

[8] J. P. Eckman, S. O. Kamphorst, and D. Ruelle, Recurrence plots of dynamical systems. *Europhys. Lett.* 4:973–977, 1987.

[9] A. Provenzale, L. A. Smith, R. Vio, and G. Morante, Distinguishing between low-dimensional dynamics and randomness in measured time series. *Physica D* 58:31–49, 1992.

[10] R. I. Kitney and S. Bignall, Techniques for studying short-term changes in cardio-respiratory data. In *Blood Pressure and Heart Rate Variability*, Di Rienzo et al., eds, pp. 1–23. Amsterdam: IOS Press, 1992.

[11] S. M. Kay and S. L. Marple, Spectrum analysis: A modern perspective. *Proc. IEEE* 69:1380–1418, 1981.

[12] C. L. Nikias and M. R. Raghuveer, Bispectrum estimation: A digital signal processing framework. *Proc. IEEE* 75:869–891, 1987.

[13] M. R. Raghuveer, Time-domain approaches to quadratic phase coupling estimation. *IEEE Trans. Autom. Control* 35:48–56, 1990.

[14] A. Papoulis, *Probability, Random Variables and Stochastic Processes.* New York: McGraw-Hill, 1984.

[15] N. Montano, F. Lombardi, T. Gnecchi-Ruscone, M. Contini, M. L. Finocchiaro, G. Baselli, A. Porta, S. Cerutti, and A. Malliani, Spectral analysis of sympathetic discharge, RR interval and systolic arterial pressure in decerebrate cats. *J. Auton. Nerv. Syst.* 40:21–32, 1992.

[16] G. Sugihara and R. May, Nonlinear forecasting as a way of distinguishing chaos from measurement error in time series. *Nature* 344:734–741, 1990.

[17] A. A. Tsonis and J. B. Elsner, Nonlinear prediction as a way of distinguishing chaos from random fractal sequences. *Nature* 358:217–220, 1992.

[18] S. M. Pincus, Approximated entropy as a measure of system complexity. *Proc. Natl. Acad. Sci. USA* 88:2297–2311, 1991.

# Chapter 10 ASSESSMENT OF NONLINEAR DYNAMICS IN HEART RATE VARIABILITY SIGNAL

Maria G. Signorini, Roberto Sassi, and Sergio Cerutti

Nonlinear analysis of time series provides a parameter set that quantifies the characteristics of the system attractor even when we do not know the model structure. This is particularly useful in the study of heart dynamics, as it has to start from external observation of the system. The chapter deals with the nonlinear analysis of cardiovascular system dynamics considering the heart rate variability (HRV) time series. The reconstruction of a single time series by the time delay method in an embedding space of dimension $m$ allows measurement of the fractal dimension and Lyapunov exponents of the whole system. To do this, correctly, it is necessary to follow a set of procedures that ensure the correspondence between the estimated parameter values and the system structure. We introduce methods for performing the reconstruction of a time series in an embedding space and quantify the amount of irregularity and the complexity properties of HRV time series. Examples are shown for normal adults and newborns, heart-transplanted and myocardial infarction patients, as well as patients in the intensive care unit. A nonlinear model of heartbeat generation is presented as its results establish a link between complexity and pathophysiological heart conditions. The goal is to contribute to verification of the usefulness of nonlinear analysis for the diagnosis and prediction of cardiovascular disease episodes in human.

## 1. INTRODUCTION

Proper processing of biological signals is a fundamental step to understanding dynamic transitions from physiological to pathological conditions of various living systems. Parameters that can be obtained by this approach may increase the knowledge about pathophysiological mechanisms and contribute to identifying the causes of disease. Results presented in literature reinforced the opinion that it is worth investigating any tool that can better show the real structure of biological systems. In fact, the nonlinear deterministic dynamic approach provides a description of many complex phenomena of the real world, showing that apparently erratic behavior can be generated even by a simple deterministic system with nonlinear structure [1, 2]. Moreover, the intrinsic unpredictability of these systems can be explained by their strong dependence on the initial conditions [3]. *Deterministic chaos* refers to the theoretical description of this behavior. Dynamical system states manifesting these characteristics are called strange attractors. They are subspaces of the state space that trajectories tend to, with finite but noninteger (fractal) dimension. The evolution in time of the system state variables generates a flux of trajectories that converge to the attractor and remain bounded on it, such as happens for a classical attractor set like a fixed point, a torus, or a limit cycle. This phenomenon generates folds within folds ad infinitum, providing the fractal structure of the attractor. The geometric property of the system in state space is due to its temporal evolution, that is, a dynamical effect [3].

We know that some systems possess nonlinear dynamics with deterministic chaotic characteristics. They have been recognized by the observation of physical phenomena such as air motion, fluid turbulence, or chemical reactions [4]. For some of these systems we may generally have the entire set of differential equations because all variables generating the process are clearly identified. This is not the case for many experimental systems whose model is unknown and we may only have access to the measurement of some variables in time, as happens for human life systems. Many papers have adopted the approach derived from chaos theory to try to model and describe the cardiovascular system, finding evidence of chaotic signs [5].

The control mechanism of the cardiovascular system can be studied by analyzing one of the system's observed variables, mainly the electrocardiographic (ECG) signal and in particular the series of RR duration, called the heart rate variability (HRV) signal, which is directly derived from it [6]. The analysis of the cardiovascular variability signal through classical linear methods, in either the time or the frequency domain, provided two main outcomes: (1) it has been possible to quantify some important properties of the regulating action of the autonomic nervous system (ANS) by measuring the dynamic balance between sympathetic and vagal components regulating the HRV signal [7] and (2) the kind of information carried by this signal may not be totally explained by a linear approach [8]. The second statement confirms the importance of the analysis of cardiovascular variability signals through nonlinear methods.

In the following we briefly introduce an overview of methods for extracting from a time series and after a projection in an *embedding* space, the invariant characteristics of the system attractor: the *fractal dimension* ($D$) and the *Lyapunov exponents* (LEs), which quantify the rate of divergence of the system trajectories [3]. As the model structure is unknown in experimental signals, we introduce the *false nearest neighbor* method for estimating the system dimension [9]. *Nonlinear filtering* is performed to remove noise disturbances in the state space [10]. The *Hurst self-similarity coefficient* quantifies fractal properties of the time series. In particular, we discuss the clinical and prognostic value of the signal $1/f$ power distribution with the evaluation of the $\alpha$ *spectrum slope* [11]. Finally, *approximate entropy* (ApEn) provides a regularity index for the time series [12].

We will show how the measured beat-to-beat RR series, as detected in the ECG signal, can be considered as the output of the cardiovascular system, in which nonlinear properties can be recognized by starting from the signal structure itself. The aim of the chapter is to show whether a nonlinear dynamic model can explain the regulating mechanisms of the cardiovascular system. Results of HRV signal analysis will illustrate how changes in some nonlinear parameters can lead to a classification of different pathological states. Examples of experimental data refer to normal subjects and patients after heart transplantation, subjects who had an anterior myocardial infarction (MI) within 6 months, and newborn infants within 3 days after birth [13, 14]. Some data for patients in the intensive care unit (ICU) and with cardiovascular diseases confirm the diagnostic power of self-similarity parameters [15]. The chapter also contains a short introduction to the nonlinear modeling of heartbeat dynamics. We want to show how an approach based on bifurcation analysis can introduce new interpretative results explaining heart pathophysiology through a simple nonlinear model structure generating complex behaviors.

## 2. ASSESSMENT OF NONLINEAR PROPERTIES IN BIOLOGICAL SYSTEMS: THE TIME SERIES ANALYSIS

### 2.1. Review of the Principal Methods for the Estimation of Nonlinear System Characteristics

Under experimental conditions, the measurement of a physical or biological system results in the evaluation of an observed time series $x(n)$ $(n = 1, N)$ of length $N$ as the complete system structure is often unknown [9].

If the signal is a superimposition of sine waves at different frequencies and amplitudes, then its characteristics will be completely described by Fourier coefficients and the linear approach is adequate and powerful. However, the signal can be generated by a nonlinear system. It looks like noise but its irregular waveforms hide deterministic components and may be chaotic. In this case, other methods are required to analyze the signal.

Figure 1A–F illustrates step by step the estimation of invariant characteristics of the system attractor from a time series obtained from the first and most famous chaotic attractor, discovered by Lorenz in the 1960s by simplifying (from 11 to 3 equations) a nonlinear model of atmospheric motion [1]. The procedure can be extended to any time series. The time series in this case (Figure 1A) is the system variable $z$ sampled at $f_s = 100\,\text{Hz}$.

The reconstruction of the system trajectory in the space state uses time-delayed versions of the observed scalar quantities $x(t_0 + n\Delta t) = x(n)$, where $t_0$ is the initial time and $\Delta t$ is the sampling interval. These new points become the coordinates to reconstruct the system dynamics in an *embedding* space. The multivariate vectors we obtain in an $m$-dimensional space are defined as

$$\mathbf{y}(n) = \{x(n), x(n+\tau), \ldots, x(n+(m-1)\tau)\} \qquad n = 1, N-(m-1)\tau \tag{1}$$

where $\tau = n\Delta t$ is an appropriate time lag and $m$ is the dimension of the embedding space [9].

The embedding theorem theoretically applies for almost any value of $\tau$, but this is not true experimentally. $\tau$ can be fixed ($\tau = 1$), based on the first zero of the autocorrelation function or corresponding to the first local minimum of the average mutual information content $I(\tau)$ [16].

Figure 1B shows the two-dimensional delay map of the Lorenz attractor reconstructed from the time series. Its morphology reminds one of the original Lorenz system even it is quite modified. Nevertheless, the reconstructed attractor maintains the invariant properties of the original one: fractal dimension and Lyapunov exponents [17].

A correct time series reconstruction implies the definition of two parameters: embedding dimension ($m$) and state space dimension ($d$). The Mãné–Takens theorem [17–19] guarantees that a value $m > 2d_A$ ($d_A$ is the attractor dimension) resolves all false

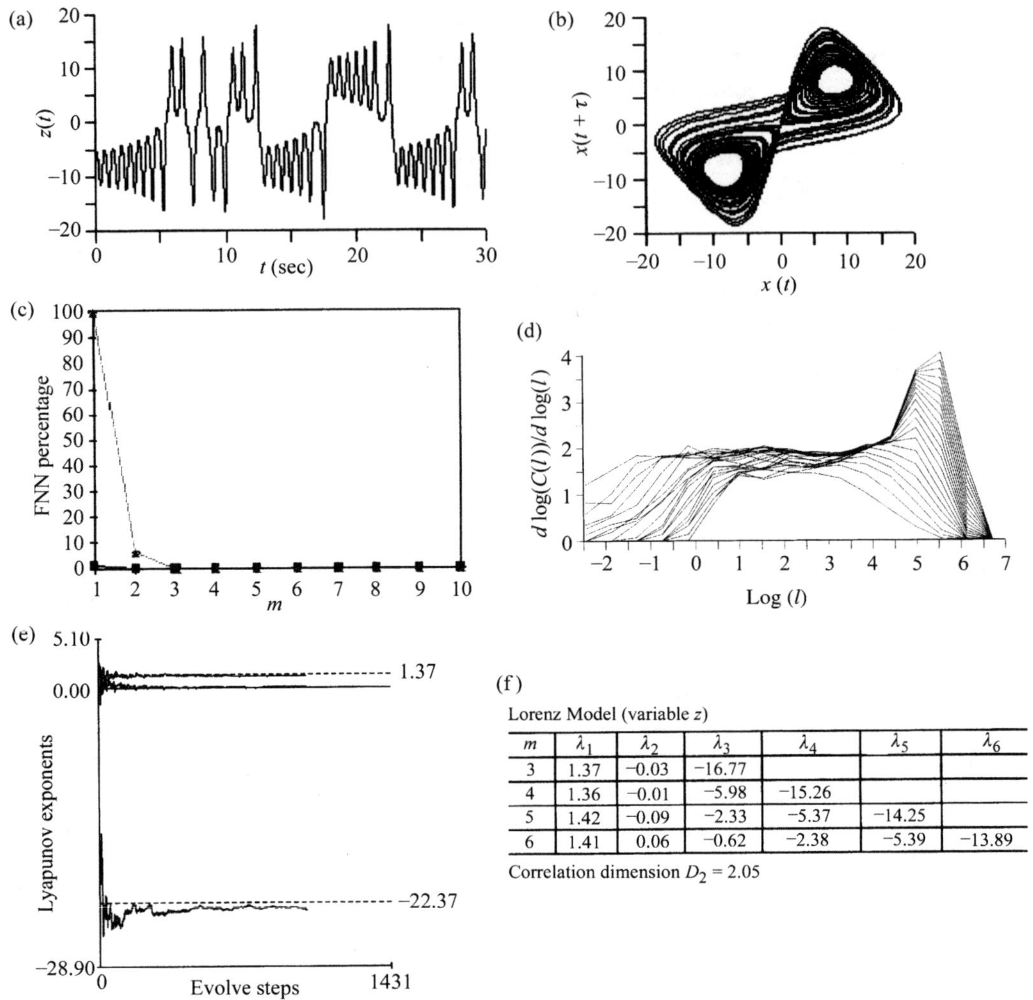


Lorenz Model (variable $z$)

| $m$ | $\lambda_1$ | $\lambda_2$ | $\lambda_3$ | $\lambda_4$ | $\lambda_5$ | $\lambda_6$ |
|---|---|---|---|---|---|---|
| 3 | 1.37 | −0.03 | −16.77 | | | |
| 4 | 1.36 | −0.01 | −5.98 | −15.26 | | |
| 5 | 1.42 | −0.09 | −2.33 | −5.37 | −14.25 | |
| 6 | 1.41 | 0.06 | −0.62 | −2.38 | −5.39 | −13.89 |

Correlation dimension $D_2 = 2.05$

**Figure 1** Estimation of nonlinear time series characteristics of the Lorenz attractor (variable $z$) through quantitative parameters. (a) Example of the time series from $z$ variable (fc = 100 Hz); (b) reconstructed attractor; (c) evaluation of the system space dimension $m$ through false nearest neighbors estimation; (d) graph of the correlation dimension; (e) Lyapunov exponents spectrum; (f) dependence of the numerical estimation of Lyapunov exponents on $m$ values.

self-crossings of the trajectory introduced by an improper projection. Unfortunately, the theorem gives only a *sufficient* condition and the problem is how to identify the appropriate value for the *minimum* embedding dimension. Moreover, if data are noisy, this procedure adds a certain number of spurious dimensions in the attractor projection as noise uniformly populates the state space, miming the presence of a high-dimensional, very complex system.

The *false nearest neighbors (FNN)* method determines whether two trajectory points are close to each other as they belong to the same trajectory converging a dimension for the system state space and the minimum $m$ value for its reconstruction

[9, 20]. In Figure 1C the number of FNN goes to zero for $m = 3$, which is the topologic state space dimension for the Lorenz system. The application of the FNN method primarily helps to distinguish among low- and high-dimensional systems, thus making a useful separation between nonlinear deterministic processes and stochastic ones, the latter indicated simply as "noise" [9, 20].

### 2.2. Estimation of Invariant Properties of the System Attractor

The correlation dimension parameter ($D_2$) performs the estimation of the fractal dimension for experimental data sets. $D_2$ provides an inferior bound for the fractal dimension $D$ of the system attractor ($D_2 \leq D$) [21]. The flat zone in the graph of Figure 1D supplies the fractal dimension of the Lorenz system: $D_2 = 2.04$.

A large class of stochastic processes can show finite values for $D_2$. This does not indicate the presence of deterministic dynamics but rather particular characteristics of the signal. For this reason $D_2$ calculation requires validation through surrogate data analysis [22].

Lyapunov exponents are quantitative measures of the attractor *stretching and folding* mechanism [3]. A sufficient condition to recognize a chaotic system is the presence of at least one positive exponent. Traditional algorithms compute a number of exponents as large as the embedding dimension $m$ [23, 24]. Figure 1E illustrates the LE spectrum. An original criterion to evaluate the dimension of the LE spectrum proposes a comparison between the value of the correlation dimension $D_2$ and the Lyapunov or Kaplan–Yorke dimension ($D_L$). $D_L$ directly derives from the knowledge of the LE spectrum by means of the Kaplan–Yorke conjecture [25, 26]. When $D_2 \leq D_{L(m)}$ this can be identified as the correct choice for the LE spectrum dimension $m$ with a consistent saving of computing time. Figure 1F summarizes numerical values of LE as a function of $m$: if $m$ is too large (bigger than the topological system dimension), it also affects the true exponent values.

### 2.3. Nonlinear Noise Filtering in the Space State

Signals generated by nonlinear systems show broadband spectra. In that case, noise reduction through linear filters introduces heavy signal distortion. We propose to analyze HRV time series by working directly in the reconstructed system state space [10]. Examples of HRV time series will show how it can help the extraction of complex signal characteristics.

The algorithm starts from the reconstruction of the system dynamics from experimental time series in an embedding space on the basis of the Takens theorem [17]. As a second step, we cover the system attractor with neighborhoods (each one containing 200 points) by making a random selection of the first reference point $\mathbf{x}_0^1$ from the vector series. Then we find its $v-1$ nearest points that form the neighborhoods $\mathrm{U}_1 = \{\mathbf{x}_0, \mathbf{x}_1, \ldots, \mathbf{x}_{v-1}\}$. This operation is repeated until the entire attractor is covered.

Correction of noisy trajectories takes place by locally projecting neighborhoods on subspaces that constitute a good approximation of the original surface of the local attractor underlying the system. To ensure that the neighborhoods have nonempty intersections, an already projected point will not be projected again if it is found to belong to other neighborhoods. The neighborhood points are projected in a linear subspace $H_k$, which is defined by the $k$ eigenvectors ($\mathbf{w}_1, \mathbf{w}_2, \ldots, \mathbf{w}_k$) corresponding to the largest $k$ eigenvalues ($\lambda_1 \geq \lambda_2 \geq \ldots \geq \lambda_k$ ($k < d$)) of the local covariance system matrix $C$ by starting from those that have more points. After that we come back to a time series and the method is iterated. The algorithm is applied in a recursive way to a time series $V(t)$ and produce at each step a $\tilde{V}(t)$ noise-corrected signal [27]. In this procedure noise is what does not contribute to the deterministic structure of the time series.

Figure 2 shows an HRV signal and relevant delay maps before (a and c) and after (b and d) the illustrated denoise procedure. Data refer to a normal subject. The reduc-

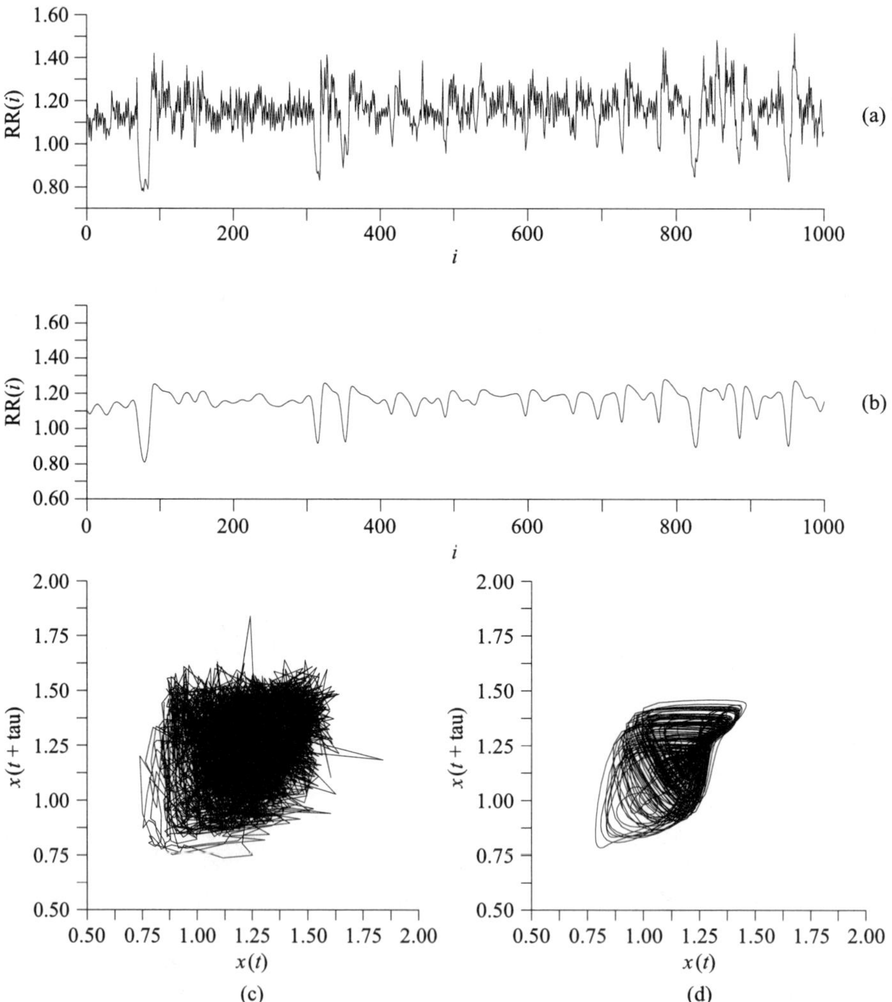


**Figure 2** Nonlinear noise filtering procedure applied to the HRV signal of a normal subject during the night. The upper panel refers to the relevant tachogram time series (1000 points) before and after the denoise procedure (a and b). The lower panel shows the delay maps before and after noise reduction (c and d)

tion of disturbances enhances the signal characteristics as maps of (a) and (b) clearly show.

## 2.4. Self-Similarity Parameters

The HRV signal does not only contain linear harmonic contributions (traditionally identified through spectral analysis techniques). It possesses a fractal-like geometry characterized by many rhythmic components interacting over different scales. A fractal object is defined as one that shows self-similarity or self-affinity independent scaling.

HRV time series can show fractal characteristics in their patterns as well as in the temporal scales. The time series, under different degrees of magnification of the temporal step, show patterns that possess self-similar characteristics to a greater or lesser extent. The *Hurst exponent* $H$ [28] characterizes the level of *self-similarity*, providing information on the recurrence rate of similar patterns in time at different scales. Several methods are available to estimate the $H$ parameter: the more traditional one is based on the *periodogram*, given as $P(f) = f^{-\alpha}$. Signals can show a broadband spectrum with power values that scale with the frequency, according to the law, where $\alpha$ is a constant.

The $\alpha$ parameter was obtained as the slope of the regression line on the power spectrum. For the HRV signal the range is taken between $10^{-4}$ and $10^{-2}$ Hz corresponding to the very low frequency components of the signal. Figure 3 shows two log–log spectra calculated for 24-hour HRV signals of a normal subject (a) and a patient after heart transplantation (b). For this method $H = 2 - \alpha$ where $\alpha$ is the spectrum slope.

Together with the periodogram, some other estimators of the Hurst exponent in experimental time series exist: *absolute value; aggregated variance; Higuchi, Fano, and Allan factor*; and *detrended fluctuation analysis (DFA)* methods [29–31]. They can be regarded as strong tools for characterizing biological time series, which generally show long-range positive correlation.

## 2.5. Approximate Entropy

ApEn is a statistical index that quantifies regularity and complexity and appears to have potential application to a wide variety of relative short (> 100 values) and noisy time series data [12]. It depends on two parameters, $m$ and $r$: $m$ is the length of compared runs and $r$ is a percentage of the signal STD that fixes a filtering level. In practice, we evaluate within a tolerance $r$ the regularity, or frequency, of patterns similar to a given pattern of window length $m$.

For a numerical series $u(1), u(2), \ldots, u(N)$ and fixed $m \in N$, ApEn measures the likelihood that runs of patterns that are close for $m$ observations remain close on the next incremental comparison. It classifies both deterministic and stochastic signals requiring a reduced amount of points and is robust against noise contamination. The presence of regularity, that is, the greater likelihood of the signal remaining close, produces smaller ApEn values and vice versa. The number $N$ of input data points ranges from 75 to 5000. Both theoretical analysis and clinical applications showed that $m = 1, 2$ and $r$ between 0.1 and 0.25 SD of the input data produce good statistical validity of ApEn $(m, r, N)$.

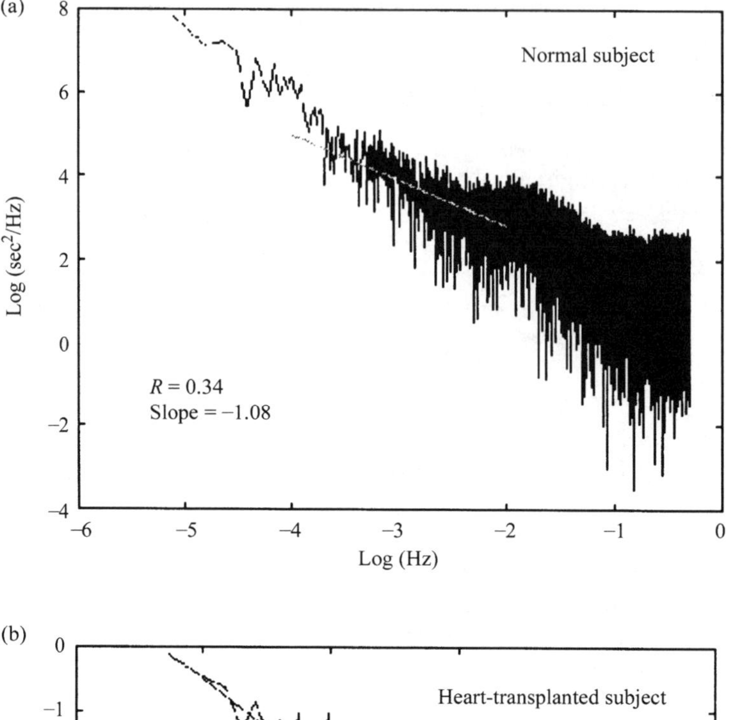


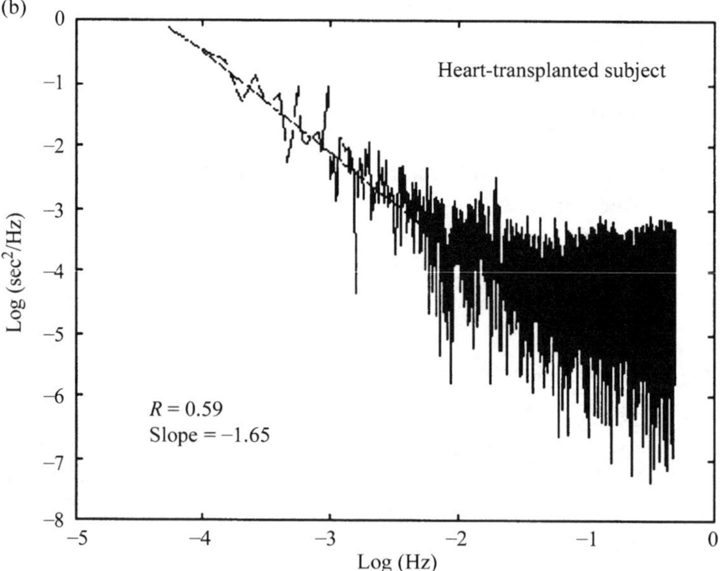


**Figure 3** Log–log spectra calculated from 24-hours HRV signals in a normal subject (a) and in a patient after heart transplantation (b). The $\alpha$ slope is higher in the transplanted patient.

# 3. NONLINEAR PHYSIOLOGICAL MODELS FOR THE GENERATION OF HEARTBEAT DYNAMICS

## 3.1. Realization and Study of a Nonlinear Model of the Heart

The interest in the description of different manifestations of pathophysiological cardiac rhythms by the dynamical systems approach has been constantly growing. Nonlinear models presented in the literature are of both microscopic and macroscopic nature [32].

The microscopic ones try to replicate mechanisms of ionic exchange at the cellular membrane level in the cardiac fibers. Macroscopic models consider the average behavior of cardiac muscles and of principal pacemaker sites. As an example of the powerful tool represented by nonlinear modeling, we illustrate here a recent development of a historical approach to dynamical system behavior replication and study with application to heartbeat generation.

Most cardiac cells spontaneously contract. The contraction must take place in an ordered way ensuring the strength of the cardiac action. This property is guaranteed by the conduction system. A network of cells whose behavior is similar to that of neural cells composes the entire system that generates the electrical pulse that gives rise to the cardiac cycle. The main pacemaker is the sinoatrial (SA) node imposing its rhythm on the whole heart (about 70 beats/min). Figure 4 (left) shows the mechanism of electrical pulse propagation. From the SA node, the pulse travels to the atrioventricular (AV) node and, through the His–Purkinje fibers, it reaches the whole ventricle. This activity

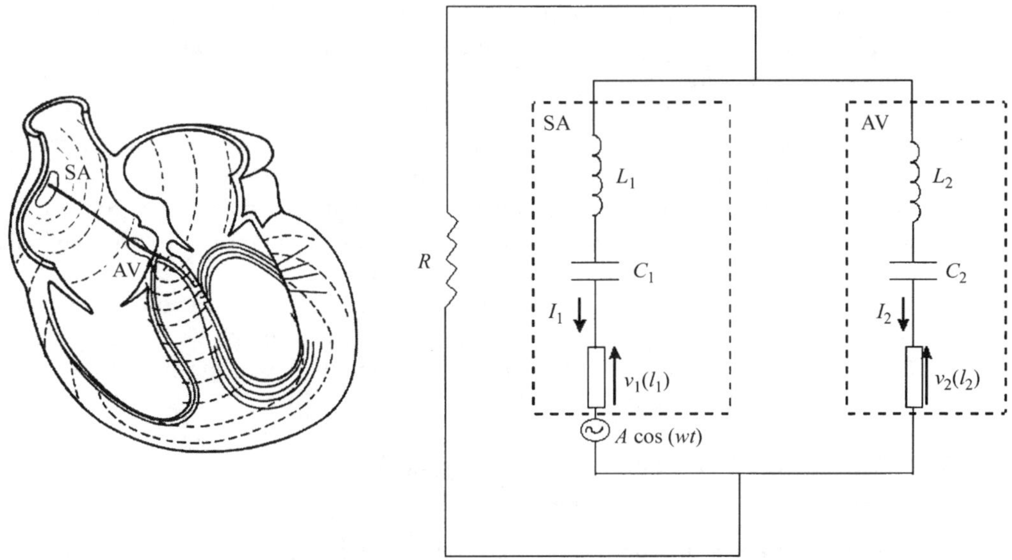


**Figure 4** Anatomy of the electrical conduction system in the heart (left). Structure of the model coupling two van der Pol modified oscillators through a resistance $R$ (right). The oscillators represent sinoatrial (SA) and atrioventricular (AV) nodes.

generates an electric field, which yields the ECG trace as a voltage signal between two surface electrodes on the body. This description uses the classical physiologic interpretation in which the AV node is just a passive conduit instead of an active pacemaker region. It may happen that the AV node can substitute for the SA node in its dominant pacemaker role under particular pathophysiological conditions. For this reason, according to recent physiological results, the models discussed in Refs. 5 and 33 and the present one, which is proposed in Figure 4 (right), attribute an active role to the AV node. This assumption improves the understanding of heart physiology and of a certain number of pathological conditions.

Since the model of van der Pol and van der Mark [34] was introduced, many papers have tried to describe the dynamics of the heartbeat as generated by coupled nonlinear oscillators. This model class includes those based on limit cycle oscillators. It is difficult, however, to find a correspondence between their parameters and physiological quantities. Moreover, some of these models were created to simulate a specific type of arrhythmia and not to explain the general behavior of the cardiac conduction system.

One of the general models of this group, which also attributes an active role to the AV node, is the model of West et al. [33]. Nevertheless, the limited number of variables describing the cardiac system (SA and AV node) was preventing aperiodic and chaotic solutions against the hypothesis that cardiac activity can stem from complex deterministic dynamics even of a chaotic nature [5, 8]. The model did not consider the study of complex solutions that we know could also be generated by low-dimensional nonlinear circuits. It could not even simulate some arrhythmias such as atrial bigeminy.

The model shown at the right in Figure 4, although using some important results of the previous ones, can overcome their limitations. In this model we hypothesize that a bidirectional coupling exists between the SA and AV nodes. In physiological conditions, the first oscillator (situated in the SA node) is the dominant pacemaker, and it is coupled to a second one (situated in the AV node) so that they show 1:1 phase entrainment [35]. Figure 4 shows the two modified van der Pol–like oscillators representing the SA and AV nodes that are supposed to be the two pacemakers of the heart [35].

The bifurcation analysis explains the asymptotic behavior of a dynamical system through the identification of its various steady-state solutions. In fact, each solution is associated with a different attractor depending upon system parameter variation [36] that can induce changes of the trajectory map. System subspaces in which the system behavior remains the same are structurally stable and characterized by the same attractor for any parameter perturbation. For a parameter variation, a *bifurcation* represents a transition of the system to a new equilibrium state. The bifurcation curves are lines marking the separation between zones characterized by different attractors. Bifurcation analysis has been performed for the dynamical system representing the model in Figure 4 by the LOCBIF numerical tool [37].

### 3.2. An Example of the Model Analysis: The Stability of the Equilibrium Point

The system has a unique equilibrium point in which the two variables representing the action potentials are zero. In this case, every heart oscillation is off and the only possible state is cardiac arrest.

We investigate which zones of the parameter space can lead to this condition. Figure 5 shows the competition of different attractors near the equilibrium point.

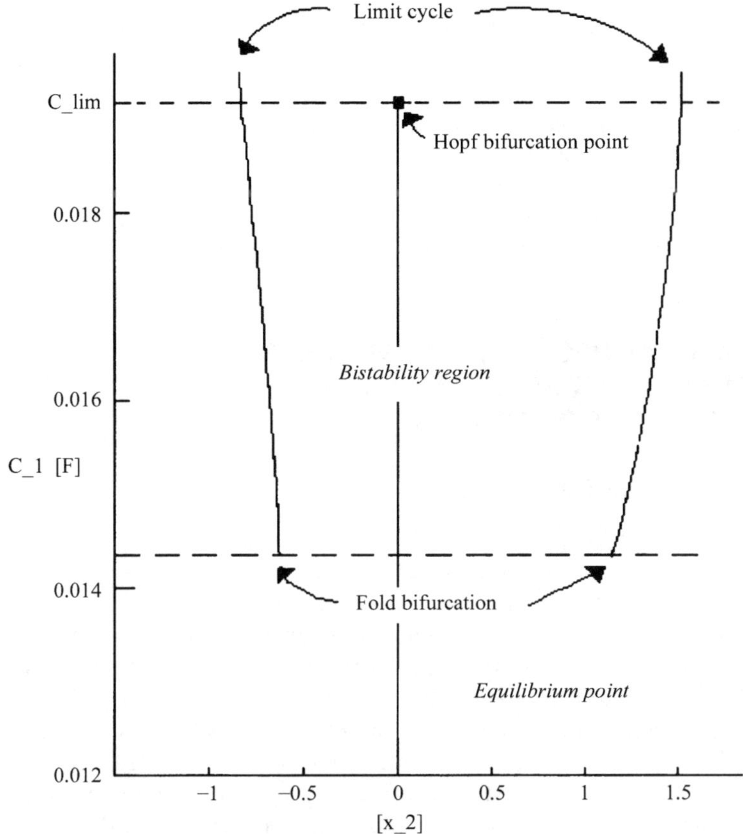


**Figure 5** Subcritical Hopf bifurcation of the nonlinear model in Figure 4. The bistability region is shown as $C_1$ varies for $R = 1.1\,\Omega$. The system passes through a bistability region and can fall in the equilibrium point in the model that represents the cardiac arrest condition (see text).

The $R$ and $C_1$ coupling parameters are varied, where $R$ is the coupling "strength" between SA and the AV nodes. When $R$ increases the coupling between the two oscillators increases, as less current flows through the resistance. In the model, a change of $C_1$ increases or decreases the depolarization frequency of the SA node.

To study the bifurcations of the equilibrium point of our system, we calculate the eigenvalues of the Jacobian matrix. When $R = 1$, a value of $C_1$ exists for which all four eigenvalues are pure complex numbers. This suggests that a double Hopf bifurcation takes place. When $R > 1$, two eigenvalues become pure complex numbers for a specific value of $C_1$ [36]. Figure 5 shows the sequence of bifurcation leading the system to fall in the equilibrium points with no more oscillatory activity. We notice that the Hopf bifurcation is subcritical, so the unstable limit cycle folds back and becomes stable.

We can make a physiologic interpretation of the bifurcation scenario shown in Figure 5. When the intrinsic frequency of the SA node increases (that is, $C_1$ decreases in the model) beyond a critical value, the AV node may no longer be able to fire at the high frequency dictated by the SA node (in Figure 5 this corresponds to the bistability region).

The more the SA node frequency increases, the more likely it is that the heart will stop: the bistability region ends and the only remaining stable attractor is the critical point.

The numerical analysis detected different kinds of bifurcations corresponding to different heart rhythms [35]. The nonlinear model, maintaining relative simplicity in its structure, is able to classify pathologies, such as several classes of arrhythmic events, as well as to suggest hypotheses on the mechanisms that induced them. Moreover, the results obtained showed that the mechanisms generating the heartbeat obeyed a complex law.

## 4. EXPERIMENTAL RESULTS: EXAMPLES IN CARDIOVASCULAR PATHOLOGIES AND PHYSIOLOGICAL CONDITIONS

### 4.1. Myocardial Infarction

As previously discussed, nonlinear parameters can help in the classification of different levels of cardiovascular pathologies. In this case we show the nonlinear analysis of HRV signals for 24 hours in 18 patients who recently had an MI episode. The study was performed by means of a nonlinear approach based on the multiparametric analysis of some invariant properties of the dynamical system generating the time series [14]. The analysis follows the guidelines that have been described by reconstructing the system in an embedding space from the HRV time series. We propose to estimate the correct dimension of the state space in which the system attractor is embedded through the false nearest neighbors (FNN) criterion. We then calculate the correlation dimension ($D_2$) and Lyapunov exponent parameters together with the alpha slope of the $1/f$ power spectrum. Results show that nonlinear dynamics are certainly involved in the long-period HRV signal. Furthermore, $D_2$ and alpha perform a significant classification ($p < 0.05$) among MI subjects, distinguishing between the group of subjects who maintain good performance of the cardiac pump (normal ventricular ejection function, NEF) after MI and the group who show an alteration of this function (reduced ventricular EF) after MI. Figure 6 summarizes the results in terms of a bar graph. Alpha values are $1.19 \pm 0.25$ (LEF) versus $0.98 \pm 0.16$ (NEF) and $D_2$ values in the day epoch are $5.2 \pm 1.0$ (LEF) versus $6.2 \pm 0.7$ (NEF). The variance of the HRV series was not able to separate the NEF and LEF group significantly. Therefore it appears necessary to investigate further the nonlinear characteristics involved in the HRV control mechanism as they can assume clinical and predictive relevance in management of cardiovascular pathology.

#### *4.1.1. Normal Subjects versus Patients with a Transplanted Heart*

For this population we estimate the invariant attractor parameters with and without the noise reduction procedure [38, 39]. The estimation of the correlation dimension confirms that normal patients have higher $D_2$ values than transplanted patients ($D_2 = 7.12 \pm 1.33$ vs. $4.5 \pm 1.47$ average $\pm$SD). In this way, they seem to show greater HRV complexity. Figure 7 shows that the noise reduction induces a decrease of $D_2$ values ($D_2 = 5.23 \pm 1.80$ vs. $3.06 \pm 0.89$ avg. $\pm$ SD) and makes more evident the lower

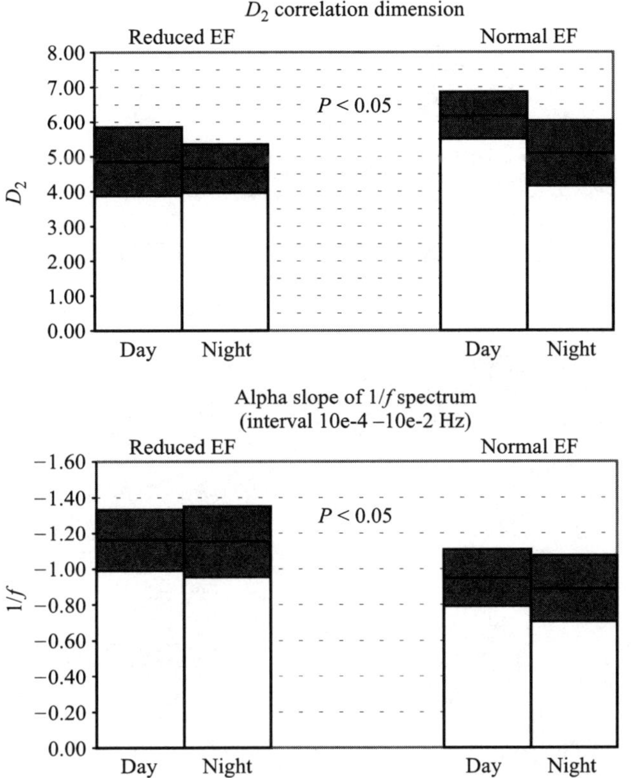


**Figure 6** Nonlinear analysis of HRV series of myocardial infarction patients with normal (NEF) and reduced (REF) ejection fraction (9 + 9 subjects). The upper panel shows the bar graph of the $D_2$ correlation dimension estimation during the day and night periods (20,000–30,000 points corresponding to 6–8 hours). The lower panel shows values of the $\alpha$ slope estimated in the same intervals.

complexity of transplanted patients, increasing the difference between the average values for the two populations. In some cases, only after noise reduction is it possible to observe the $D_2$ parameter saturation.

Even the estimation of attractor dimension with the self-similarity parameter $H$ ($D_H = 1/H$) gives values greater for normal patients than for transplanted ones ($D = 9.11 \pm 3.51$ vs. $3.46 \pm 2.09$). After nonlinear noise reduction, a marked decrease is observed in the two populations ($D = 6.54 \pm 2.65$ vs. $2.29 \pm 0.75$). Moreover, after noise reduction, the $\alpha$ coefficient of the $1/f^{\alpha}$ spectrum has very small variations. This coefficient depends on the signal components over a long period, which is not influenced by the filtering algorithm (before noise filtering $0.96 \pm 0.15$ vs. $1.67 \pm 0.46$; after filtering $1.16 \pm 0.15$ vs. $1.8 \pm 0.42$).

Transplanted patients show $\alpha$ values that are always greater than those of normal patients, confirming that $\alpha$ values $> 1$ are correlated with pathological conditions. Starting from this observation, we have also calculated the Hurst parameter by different methods. Signals were 24-hour HRV series of 11 normal and 7 heart-transplanted

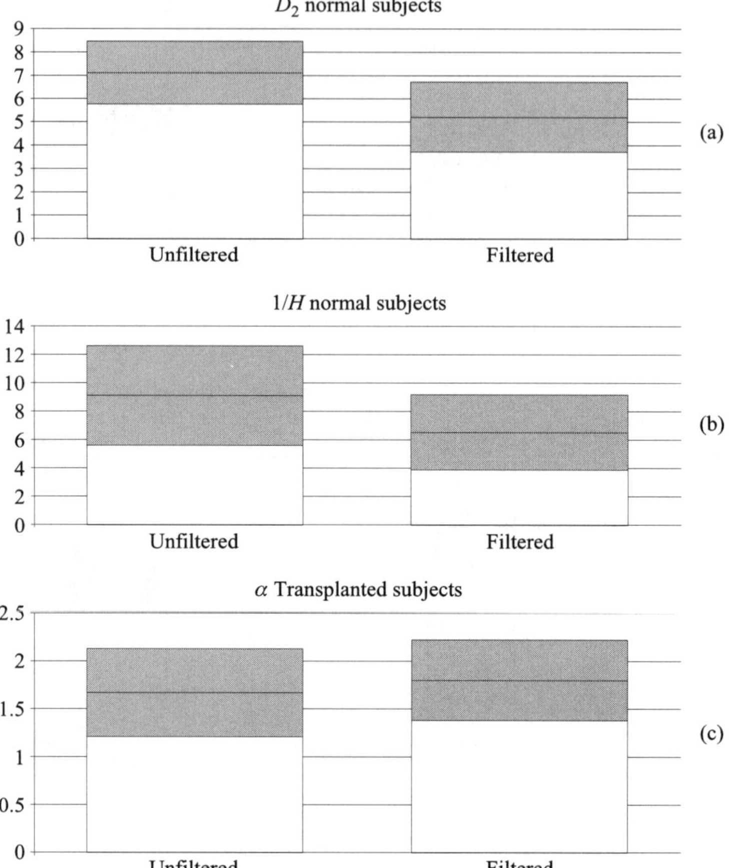


**Figure 7** Effects of noise reduction in nonlinear parameters in normal subjects and transplanted patients. (a) Noise reduction induces a decrease of $D_2$ values ($D_2 = 5.23 \pm 1.80$ vs. $3.06 \pm 0.89$ avg. $\pm$ STD). (b) $D_H = 1/H$ is greater for normal patients before nonlinear noise reduction ($9.1 \pm 3.5$ vs. $6.5 \pm 2.6$). (c) The $\alpha$ coefficient of the $1/f^{\alpha}$ spectrum shows very small variations ($1.67 \pm 0.46$ vs. $1.8 \pm 0.42$ in transplanted).

subjects. The length $N$ of the RR interval series varied from 70,000 to 130,000 points. Even in this case, healthy subjects showed a higher value of the slope, which means a smaller value in terms of $H$. For the $H$ parameter, we obtained the range 0.12–0.22 with the *periodogram* ($p < 0.02$).Results are in agreement with other measurements obtained with different nonlinear parameters (correlation dimension, entropy, and Lyapunov exponents). They always confirm the role of HRV nonlinear dynamics as a global marker of cardiovascular pathology [13].

## 4.2. ICU Patients: Classification of Death Risk through Hurst Coefficient

Self-similarity $H$ parameters obtained by different approaches have been used for the assessment of the status of patients in ICU. Sixteen patients were studied. Data were extracted from the IMPROVE Data Library (DL) [15]. Under normal conditions, the index $\alpha$ in 24-hour HRV signals shows values near 1, confirming the broadband nature of the spectrum, and it increases in the presence of pathological cardiovascular events. As previously show, the power law regression parameter, $\alpha$ slope, may predict the risk of death in patients after MI. In this case long-term parameters evidenced a significant difference in the $\alpha$ slope between survivors and nonsurvivors after an ICU period.

The $\alpha$-slope parameter computed over the 24-hour spectrum of RR interval series differed in the patient groups who died (D) and survived (S) ($D \rightarrow 1.44 \pm 0.35$ vs. $S \rightarrow 1.13 \pm 0.10$, respectively; $p < 0.05$). On the contrary, the slopes of the systolic arterial pressure ($D \rightarrow 1.21 \pm 0.17$ vs. $S \rightarrow 1.27 \pm 0.21$; ns) and pulmonary arterial pressure times series ($D \rightarrow 1.05 \pm 0.25$ vs. $S \rightarrow 1.18 \pm 0.35$; ns) did not show differences between the two groups.

Comparison of survivors and nonsurvivors shows for the $H$ parameters 0.15 versus 0.29 with the periodogram ($p < 0.05$), 0.2 versus 0.3 with the absolute value ($p < 0.005$), and 0.15–0.21 with Higuchi's method ($p < 0.05$). Results agree with other measurements obtained with different nonlinear parameters (correlation dimension, entropy). They confirm the role of nonlinear dynamics in the HRV signal as a global marker of cardiovascular pathologies. These preliminary results in ICU patients indicate that long-period indexes seem to be strongly related to the prediction of both positive and negative patient outcomes [15].

## 4.3. Newborn State Classification

The ApEn parameter performs a significant separation in the groups composed of newborns after spontaneous delivery and after cesarean section. Figure 8 shows results from HRV signals obtained from Holter recordings (2 hours) in healthy newborns

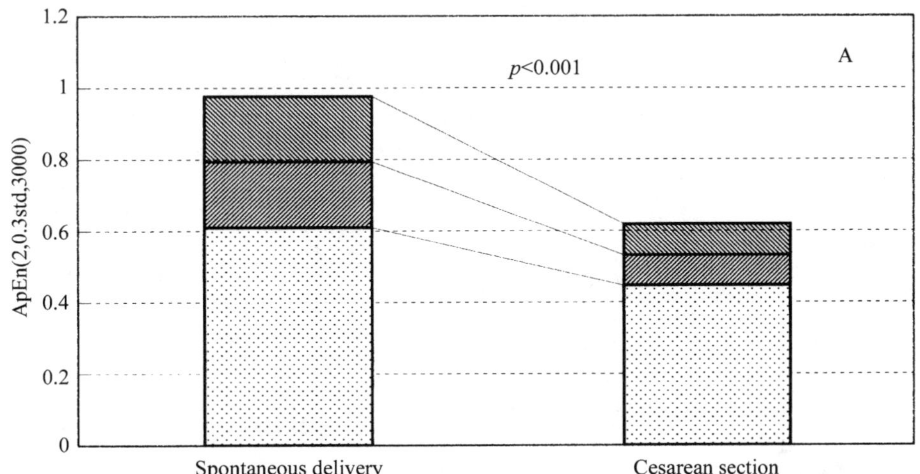


**Figure 8** Results from HRV signals in healthy newborns within 12 hours after birth by spontaneous delivery and cesarean section. ApEn with a reduced point number makes it possible to distinguish the two populations with a high level of confidence ($p < 0.001$).

within 12 hours after birth. Bar graphs showing the ApEn parameter with a reduced point number allow distinguishing the two populations with high confidence level ($p < 0.001$, $t$-test for two populations assuming unequal variance). The ApEn parameter was calculated with $N = 3000$, $m = 2$, $r = 0.2$ std. This is the parameter set that produces the best performance in the separation of the two groups ($p < 0.001$). The analysis was repeated for $N = 300$ and $N = 1000$. ApEn values are always different for the spontaneous and cesarean babies. ApEn is in the range $1.07 \pm 0.3$–$0.95 \pm 0.01$ (normal) and $0.74 \pm 0.23$–$0.9 \pm 0.07$ (cesarean). Data were evaluated from 2-hour recordings during sleep after the meal in the first and third days after birth. For the same experimental epochs, the correlation dimension ($D_2$) and Hurst parameter have been estimated. Figure 8 shows a comparison of the two parameters in the spontaneous and cesarean newborn populations. As with the newborn HRV signal and even the MI population, ApEn has been able to separate the two groups significantly with less computational effort with respect to $D_2$ parameter estimation.

## 5. CONCLUSION

The methods presented in this chapter perform a classification of the nonlinear characteristics of a system for which the model is unknown by starting from a single time series extracted from it. The analysis procedure is powerful and allows capturing system characteristics that cannot be obtained by other methodological approaches. The main disadvantages are the very complicated structure of several approaches and the large amount of data needed to obtain reliable estimates. Nevertheless, as has been shown in several pathophysiological examples, this approach seems particularly adequate for biological signal and system analysis.

The results confirm the ability of the nonlinear parameters to separate different patterns in the HRV time series, at the same time giving a strong indication of possible clinical applications of the calculated parameters. The use of various proposed parameters provided differentiation and classification of pathologic populations. In particular, we show how to separate MI patients who had different performances of the cardiac pump with a higher connected risk of sudden cardiac death. Estimation of correlation dimension and ApEn confirms that normal patients show $D_2$ values greater than pathological ones, showing, in these terms, a greater amount of complexity in the HRV signal.

The $\alpha$ coefficient of the $1/f^{\alpha}$ spectrum seems to classify efficiently the population with a higher cardiovascular death risk as that with reduced ejection fraction values. Values of $\alpha > 1$ are often correlated with pathological conditions and with cardiovascular risk.

A healthy cardiovascular system is strongly characterized by fractal properties in the time behavior of its variables. On the contrary, for a cardiovascular system with disease these complex patterns are reduced.

The results of the ApEn analysis suggest that this index could be helpful in integrating the diagnostic information on heart pathology. Its employment, together with other more classical indicators of the cardiac neural control function, could improve the understanding of physiological heart dynamics. In many biological systems, greater regularity can correspond to a greater autonomy and greater isolation of the system components.

Some authors hypothesize that healthy systems have good lines of communication and systems in disease states can show reduced speed of crucial biological messages transfer and reception until they become unable to connect with the rest of the system components. In this case, the greater regularity corresponds to decreased complexity of the experimentally measured signals in time. If this behavior does not indicate per se the presence of chaos, the observed system characteristics do comply with a nonlinear model controlling many biological system and heart dynamics. Advances in nonlinear modeling of heartbeat dynamics clearly show that a replication and a physiopathological explanation of heart dynamics can be correctly obtained by adopting a nonlinear modeling approach.

Complexity in cardiovascular control can be related to a nonlinear model driving the system dynamics. Knowledge of these system properties introduces new insight into the heart pathophysiology together with more sensitive predictive parameters. The proposed multiparametric approach to HRV signal classification may contribute diagnostic and prognostic tools in which nonlinear parameters could complement more traditional HRV measures.

## REFERENCES

[1] E. N. Lorenz, Deterministic non-periodic flow. *J. Atmos. Sci.* 20:130–141, 1963.

[2] A. V. Holden and M. A. Muhamad, *Chaos*. Manchester: Manchester University Press, 1986.

[3] J. P. Eckmann and D. Ruelle, Ergodic theory of chaos and strange attractors. *Rev. Mod. Phys.* 57:617–656, 1985.

[4] T. S. Parker and L. O. Chua, Chaos: A tutorial for engineers. *Proc. IEEE* 75:982–1008, 1987.

[5] B. J. West, *Fractal Physiology and Chaos in Medicine*. Singapore: World Scientific, 1990.

[6] Task Force of ESC and NASPE, Heart rate variability, standards of measurements, physiological interpretation and clinical use. *Circulation* 93:1043–1065, 1996.

[7] M. Pagani, F. Lombardi, S. Guzzetti, O. Rimoldi, R. Furlan, P. Pizzinelli, G. Sandrone, G. Malfatto, S. Dell'Orto, E. Piccaluga, M. Turiel, G. Baselli, S. Cerutti, and A. Malliani, Power spectral analysis of heart rate and arterial pressure variabilities as a marker of sympatho-vagal interaction in man and conscious dog. *Circ. Res.* 59:178–193, 1986.

[8] G. Schmidt and G. Morfill, Complexity diagnostic in cardiology: Methods. *PACE* 17:2336–2341, 1994.

[9] H. D. I. Abarbanel, R. Brown, J. J. Sidorowich, and L. S. Tsimiring, The analysis of observed chaotic data in physical systems. *Rev. Mod. Phys.* 65:1331–1392, 1993.

[10] E. J. Kostelich and T. Schreiber, Noise reduction in chaotic time series: A survey of common methods. *Phys. Rev. E* 48:1752–1753, 1990.

[11] A. R. Osborne and A. Provenzale, Finite correlation dimension for stochastic systems with power-law spectra. *Physica D* 35:357–381, 1989.

[12] S. M. Pincus, Approximate entropy (ApEn) as a complexity measure. *Chaos* 5:110–117, 1995.

[13] S. Guzzetti, M. G. Signorini, C. Cogliati, et al. Deterministic chaos indices in heart rate variability of normal subjects and heart transplanted patients. *Cardiovasc. Res.* 31:441–446, 1996.

[14] F. Lombardi, G. Sandrone, A. Mortara, D. Torzillo, M. T. La Rovere, M. G. Signorini, S. Cerutti, and A. Malliani, Linear and nonlinear dynamics of heart rate variability after acute myocardial infarction with normal and reduced left ventricular ejection fraction. *Am. J. Cardiol.* 77:1283–1288, 1996.

[15] L. T. Mainardi, A. Yli-Hankala, I. Korhonen, M. G. Signorini, et al. Monitoring the autonomic nervous system in the ICU through cardiovascular variability signals. *IEEE-EMB Mag.* 16:64–75, June 1997.

[16] A. M. Fraser and H. L. Swinney, Independent coordinates for strange attractors from mutual information. *Phys. Rev. A* 33:1134–1140, 1986.

[17] F. Takens, Detecting strange attractors in turbulence. In *Dynamical Systems and Turbulence*, D. A. Rand and L. S. Young, eds., Warwick 1980, Lecture Notes in Mathematics, Vol. 898, pp. 366–381. Berlin: Springer-Verlag, 1981.

[18] R. Mãné, On the dimension of the compact invariant set of certain nonlinear maps. In *Dynamical Systems and Turbulence*, D. A. Rand and L. S. Young, eds., Warwick 1980, Lecture Notes in Mathematics, Vol. 898, pp. 230–242. Berlin: Springer-Verlag, 1981.

[19] T. Sauer, J. A. Yorke, and M. Casdagli, Embedology. *J. Stat. Phys.* 65:579–616, 1991.

[20] M. B. Kennel, R. Brown, and H. D. I. Abarbanel, Determining embedding dimension for phase-space reconstruction using a geometrical construction. *Phys. Rev. A* 45:3403–3411, 1992.

[21] P. Grassberger and I. Procaccia, Measuring the strangeness of strange attractors. *Physica D* 9:189–208, 1983.

[22] J. Theiler, Some comments on the correlation dimension of $1/f^{\alpha}$ noise. *Phys. Lett. A* 155:480–493, 1991.

[23] M. Sano and Y. Sawada, Measurement of the Lyapunov spectrum from a chaotic time series. *Phys. Rev. Lett.* 55:1082–1085, 1985.

[24] World Wide Web at http://www.zweb.com.apnonlin/ and at http://www.cquest.com.-chaos.html

[25] J. L. Kaplan and J. A. Yorke, Chaotic behavior of multidimensional difference equations. In *Lecture Notes in Mathematics*, Vol. 730. Berlin: Springer, 1978.

[26] S. Cerutti, G. Carrault, P. J. M. Cluitmans, A. Kinie, T. Lipping, N. Nikolaidis, I. Pitas, and M. G. Signorini, Nonlinear algorithms for processing biological systems. *Comp. Meth Progr. Biomed.* 51:51–73, 1996.

[27] R. Cawley and G. H. Hsu, Local-geometric-projection method for noise reduction in chaotic maps and flows. *Phys. Rev. A* 46:3057–3082, 1992.

[28] J. Beran, *Statistics for Long-Memory Processes*. New York: Chapman & Hall, 1994.

[29] R. G. Turcott and M. C. Teich, Fractal character of the electrocardiogram: Distinguishing heart failure and normal patients. *Ann. Biomed. Eng.* 24:269–293, 1996.

[30] M. S. Taqqu, V. Teverovsky, and W. Willinger, Estimators for long-range dependence: an empirical study, *Fractals,* vol. 4:785–798, March 1996.

[31] C. K. Peng, S. Havlin, J. M. Hausdorff, J. E. Mietus, H. E. Stanley, and A. L. Goldberger, Fractal mechanisms and heart rate dynamics, long-range correlations and their breakdown with disease, *Journal of Electrocardiology* 28(Suppl):59–64, 1995.

[32] L. Glass, P. Hunter, and A. McCulloch, eds, *Theory of heart. Biomechanics, biophysics and nonlinear dynamics of cardiac function*, New York: Springer-Verlag, 1991.

[33] B. J. West, A. L. Goldberger, G. Rovner, and V. Bhargava, Non linear dynamics of the heartbeat. The AV junction: passive conduit or active oscillator?, *Physica D*, vol. 17:198–206, 1985.

[34] B. van der Pol and J. van der Mark, The Heartbeat considered as a Relaxation Oscillation and an Electrical Model of the Heart, *Phil. Mag.,* vol. 6:763–775, 1928.

[35] M. G. Signorini, D. di Bernardo, and S. Cerutti, Simulation of heartbeat dynamics: a nonlinear model, *Int. J. of Bifurcation & Chaos*, vol. 8:1725–1731, 1998.

[36] Y. Kuznetsov, *Elements of Applied Bifurcation Theory*. Applied Mathematical Sciences Series. New York: Springer-Verlag, 1995.

[37] Y. Khibnik, Y. Kuznetsov, V. Levitin, and E. Nicolaev, *Interactive LOCal BIFurcation Analyzer*. User Manual, version 1.1, Research Computing Centre, Academy of Sciences, Pushchino, 1991.

[38] M. G. Signorini, F. Marchetti, A. Cirigioni, and S. Cerutti, Nonlinear noise reduction for the analysis of heart rate variability signals in normal and heart transplanted subjects. *Proceedings 19th International Conference IEEE Engineering in Medicine & Biology Society*. CD-ROM, pp. 1402–1405, 1997.

[39] M. G. Signorini and S. Cerutti, Nonlinear properties of cardiovascular time series. In *Methodology and Clinical Applications of Blood Pressure and Heart Rate Analysis*, M. Di Rienzo, G. Parati, A. Zanchetti, and A. Pedotti, eds. Amsterdam: IOS Press, 1998.

Chapter 11

# NONLINEAR DETERMINISTIC BEHAVIOR IN BLOOD PRESSURE CONTROL

Nigel Lovell, Bruce Henry, Branko Celler, Fernando Camacho, Drew Carlson, and Martha Connolly

## 1. INTRODUCTION

Much of what is known about physiological systems has been learned using linear system theory. However, this cannot explain the known nonlinear behavior in arterial blood pressure. Arterial blood pressure exhibits all the hallmarks of a deterministic chaotic system. It is repetitive but unpredictable, there is underlying order in the apparently random behavior, and there is sensitive dependence on initial conditions.

The complex interplay of control systems that gives rise to blood pressure variability has been known for many centuries. Initial reports date back to the 1730s, with Hales demonstrating the occurrence of blood pressure variations related to respiratory activity [1]. More than a quarter of a century ago (1972), Guyton et al. [2] presented a complex system diagram of circulatory function with over 350 components. Although research to date has delved into the structural and functional aspects of many if not all of these control elements and more elements have been added and modified, it is with some irony that the focus of this chapter and indeed nonlinear determinism deals with the representation of systems of low dimensionality demonstrating complex nonlinear behavior.

In a previous chapter, a description of the main techniques used in nonlinear time series analysis was presented. In this chapter, these techniques will be used in the analysis of arterial blood pressure recordings, specifically to investigate further the incidence of nonlinear deterministic behavior in the control mechanisms associated with arterial blood pressure. Included is a brief review of the literature pertaining to blood pressure variability and nonlinear determinism in the control of blood pressure.

## 2.. CHAOS IN THE CARDIOVASCULAR SYSTEM

Nonlinear dynamical behavior (and deterministic chaos) has been observed in a number of physiological parameters and pathophysiological states (see Refs. 3–7 for general reviews). Excitable cells were perhaps the first area to receive attention with excellent experimental and mathematical analyses done by Guevara and colleagues [8, 9] in embryonic chick heart cell aggregates. Subsequently, Michaels et al. [10] described a mathematical model of a vagally driven sinoatrial node that clearly demonstrated chaotic dynamics. Nonlinear deterministic behavior has also been reported in the whole heart in terms of fluctuations in heart interval [11, 12] and in terms of electrical activity in the intact dysrhythmic heart [13]. Interestingly, there has been some conjecture as to whether ventricular fibrillation should be regarded as chaotic activity. Kaplan and Cohen [14] argued against nonlinear determinism due to the inability to detect a low-dimension attractor. They also noted the difficulty in obtaining sufficient points in a time series because of the inherent nonstationarity of the biological processes caused by tissue ischemia. This point belies one of the main difficulties in nonlinear deterministic analysis, namely obtaining sufficient data points to estimate a correlation dimension reliably. An estimate by Ruelle [15] that suggests a minimum number of points in a data set is given by $N_{min} = 10^{(D/2)}$, where $D$ is the fractal dimension, is discussed in detail in a previous chapter.

Other investigators have demonstrated chaotic activity during sympathetic nerve discharge [16], renal nerve activity [17], and renal tubular pressure and flow [18]. Zweiner et al. [19] examined the incidence of nonlinear deterministic behavior in both heart rate fluctuations and blood pressure fluctuations. This study highlights the role of parasympathetic pathways in their mediation, underlying the broad role of the autonomic nervous system as a mediator of the nonlinear deterministic behavior [20]. Figure 1 illustrates a phase-space reconstruction of a 20-minute cardiac period recording in a conscious dog. The apparent structure in the reconstruction is one of the indicators of nonlinear deterministic behavior. As anticipated, ganglionic blockade with hexamethonium reduces the variability in the cardiac period and abolishes the chaotic activity [21].

Pressor control mechanisms are central for the maintenance of homeostasis in an organism. Spontaneous fluctuations in hemodynamic parameters reflect the operation of these control mechanisms. The variability of heart interval and blood pressure has been thoroughly documented from the viewpoint of linear processes and from examination of data in the frequency domain [22]. The reader is referred to Ref. 23 for an excellent review of blood pressure variability. Notwithstanding, literature on the applicability and use of nonlinear dynamical techniques for analysis of blood pressure fluctuations has been sparse [24]. Possibly the most thorough analysis has been presented by Wagner and colleagues [25], who demonstrated significant changes in the nonlinear deterministic characteristics of blood pressure after sinoaortic denervation. A significant study examining pathophysiology was conducted by Almog et al. [26]. They reported on differences in the blood pressure correlation dimension in two strains of rats: spontaneously hypertensive rats (SHR) and their age-matched normotensive progenitors, Wistar-Kyoto rats (WKY).

## 3. CAROTID BAROREFLEX AND CHAOTIC BEHAVIOR

It is now well known that low-dimensional nonlinear dynamical systems can generate seemingly random time series data. However, random fluctuations in the time series data for an observable dynamical variable may also be due to other sources such as measurement errors, unpredictable environmental changes, and coupling to large numbers of other dynamical variables.

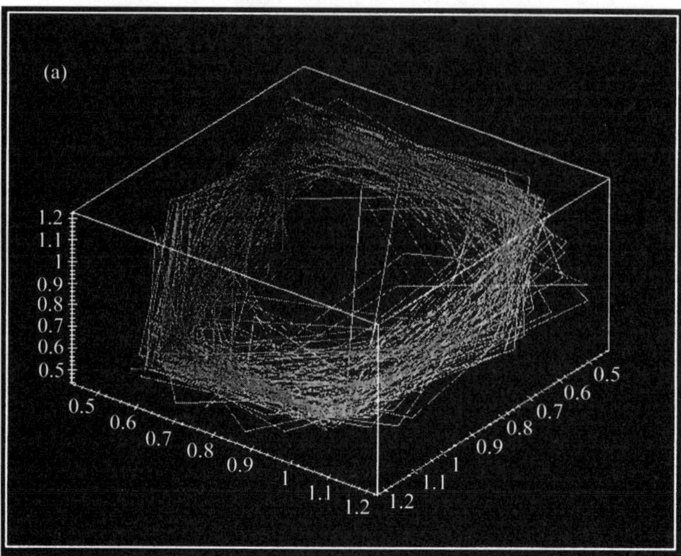


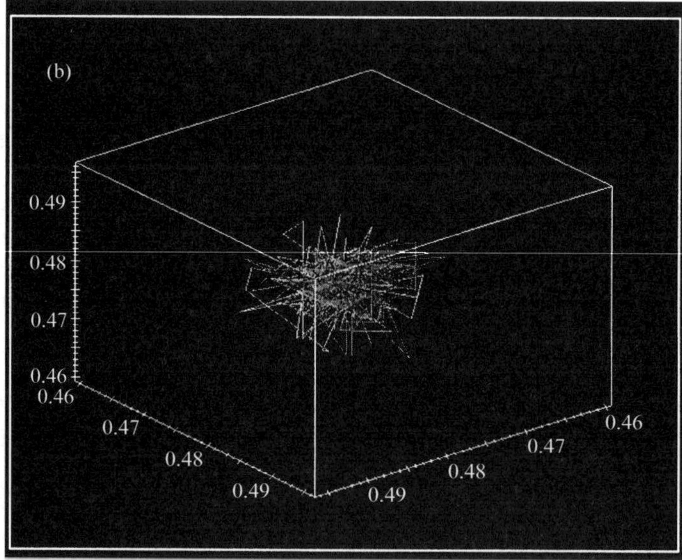


**Figure 1** Reconstructed phase-space trajectories of cardiac period (in units of seconds) in a conscious dog (a) in a control animal and (b) after complete ganglionic blockade with hexamethonium. In both cases, the time delay was one cardiac period.

The purpose of the work described here was to determine whether time series data from blood pressure measurements could be described within the framework of low-dimensional nonlinear dynamics. The role of the carotid baroreflex in the complexity and predictability of the dynamics was assessed quantitatively through measurement of the correlation dimension and the largest Lyapunov exponents.

The two principal disease states in which derangement of arterial pressure and its control are observed are hemorrhagic hypovolemia and hypertension. The carotid baroreflex is one of the predominant mechanisms by which regulation of blood pressure is achieved. It is known that the carotid baroreflex is limited in its ability to buffer completely against long-term changes in arterial pressure in hemorrhagic hypovolemia and hypertension. It may be the loss of regulatory ability at the high and low limits of blood pressure that contributes to the genesis of these pathophysiological states.

The principal data set employed in this study involves a carotid sinus blind sac preparation. The carotid sinus region was isolated using a preparation used extensively by the authors [27]. Briefly, catheters were placed in both lingual arteries for the measurement and control of carotid sinus pressure. Hydraulic vascular occluder cuffs (In Vivo Metrics, Inc) were placed around the distal external carotid artery and the proximal common carotid artery. All other arteries in the region were ligated, and the preparation was tested to ensure that no leaks were present in the blind sac preparation. The lingual arteries were connected via a feedback amplifier to a syringe pump mechanism. Autonomic tone was thus manipulated via modulation of the baroreflex response. Using this procedure, the baroreflex was investigated in an open-loop situation by altering the carotid sinus pressure. From the physiological viewpoint, this procedure is much more realistic than electrical nerve stimulation of either the stellate ganglion or vagus nerve or pharmacological interventions, as it involves direct manipulation of the baroreflex arc.

In a separate surgery an aortic flow probe (Transonics) and femoral arterial catheter were implanted. The experiments were performed approximately 2–5 days after the animal had recovered from the surgical interventions.

During the experiment, the vascular occluders were inflated and the lingual arteries were connected via a feedback amplifier to a syringe pump mechanism to maintain a constant carotid sinus pressure (CSP). Arterial pressure, peak aortic flow, CSP, and heart interval (RR) were measured for 20 minutes. Data were sampled at 250 Hz per channel and stored in digitized form for later analysis using a program developed in ASYST (ASYST Technologies). In random order we measured a control record (baroreceptors closed loop) and a record with CSP fixed flow, midrange, and high (approximately 75, 125, and 150 mm Hg, respectively) (intervention group).

In software, mean arterial pressure (MAP) was extracted on a beat-by-beat basis using the maximal rate of upstroke of the pressure wave as a fiducial reference point. The data were then resampled at 4 Hz using a zero-order hold interpolation.

As a first step in the analysis, we used the method of time delays [28] to construct phase-space trajectories from the time series data. Reconstructions were carried out over a range of the two parameters: delay time and embedding dimension. Delay time was estimated as the first zero crossing of the autocorrelation of the time series. The correlation dimension of the reconstructed trajectories was measured using the Grassberger–Procaccia algorithm [29]. In a deterministic chaotic system, the correlation dimension of the reconstructed trajectory remains constant as a function of the embedding dimension, after a threshold embedding dimension is exceeded.

In the cases where the correlation dimension converged, we used linear forecasting techniques [30] to predict the future path on the strange attractor. By comparing forecasts using the early part of the data with actual data in later segments of the same record, we were able to confirm the deterministic nature of the signal. A mean correlation coefficient (for forecast times extending from 1 to 10 samples) was calculated by averaging 100 forecasts over the time series.

Additional evidence of chaotic behavior was provided by performing for each data record a surrogate data analysis [31]. Eight surrogate data sets (consisting of phase-randomized versions of the original time series data) were qualitatively compared for correlation dimension convergence and forecasting ability. Statistical measures and power spectra of the original and surrogate data were compared to ensure minimal change in the linear properties of the data sets.

A typical recording of the arterial pressure and the beat-by-beat calculated MAP is shown in Figure 2. In six of seven dogs in the control and intervention groups, the correlation dimension converged. In the seventh dog, the dimension of the attractor did not clearly converge over the range of embedding dimensions possible with our approximate data record size of 5000 points. No further analysis was performed on this dog. A typical phase-space reconstruction of the MAP data for a delay time of three samples and an embedding dimension of 3 is shown in Figure 3A for a control dog. The characteristic doughnut shape is indicative of deterministic dynamics. In Figure 3B, a surrogate data set was constructed from the data of Figure 3A. There is an obvious loss of structure in the trajectory.

Surrogate data analysis was applied to both the correlation integral and nonlinear forecasts of the data. Qualitative comparisons of the effects of phase randomization were performed. In all cases, the surrogate data set demonstrated an increase in correlation dimension for increasing embedding dimensions (i.e., loss of convergence), as is

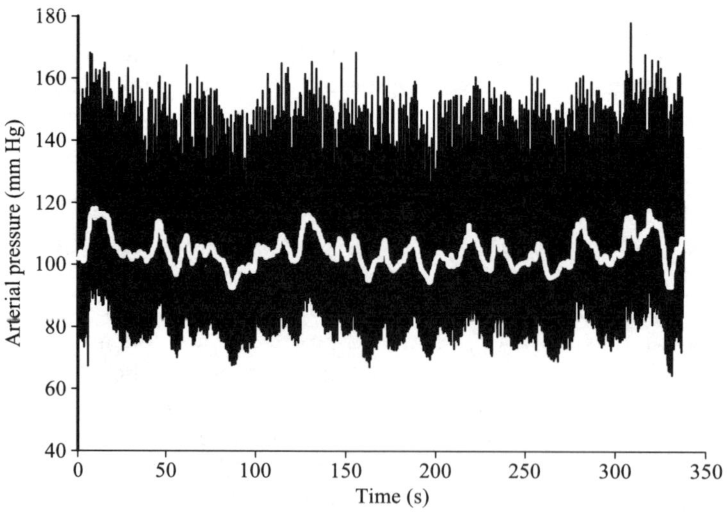


**Figure 2** Arterial pressure recorded in a conscious dog with intact carotid sinus. Waveform in black digitized at 250 Hz. Waveform in white represents the mean arterial pressure calculated on a beat-by-beat basis and interpolated at 4 Hz with a zero-order hold interpolation.

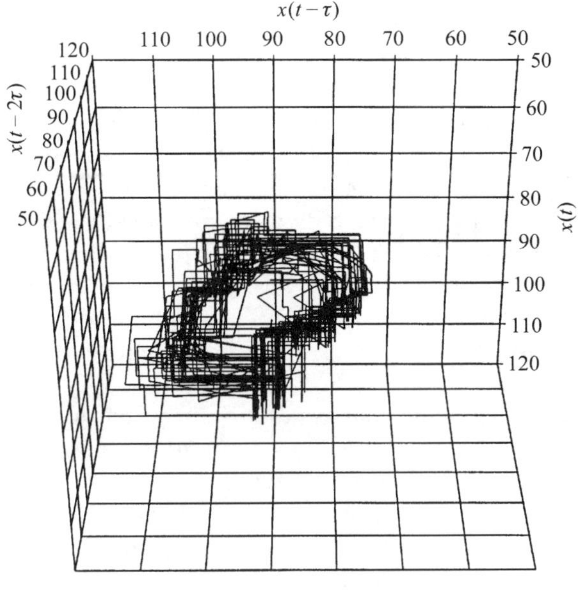


(a) Reconstructed MAP data

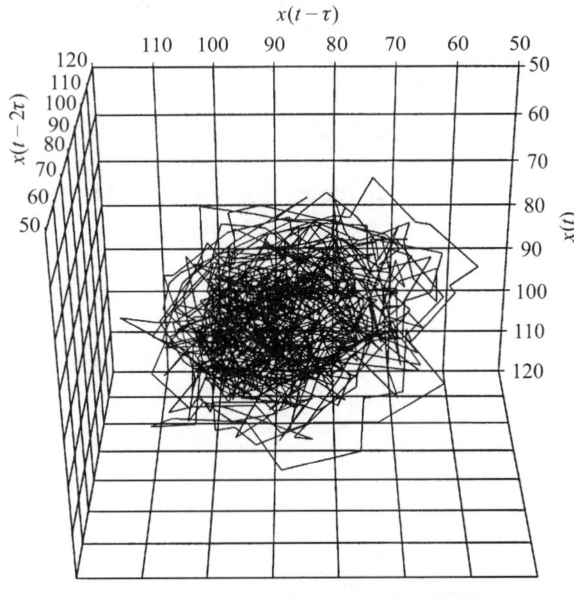


(b) Reconstructed surrogate data (FFT)

**Figure 3** Reconstructed phase-space trajectories using a time delay of three samples and an embedding dimension of 3 for MAP data in (a) control animal with the carotid sinus intact. (b) Phase randomization (surrogate) of the MAP data in (a)—note the obvious loss of structure in the trajectories.

the case for random processes. A typical plot of correlation dimension versus embedding dimension from an original data record from the control group and for eight sets of surrogate data derived form this original record is shown in Figure 4. The surrogate data points represent mean ± SE of the eight data sets.

Data forecasting using a delay time of three samples and an embedding dimension of 4 also confirmed the existence of deterministic chaos. Correlation coefficients averaged for 100 forecasts within each data set confirmed the reliability of the forecasts at differing forecast times (Figure 5). Furthermore, the correlation coefficients decreased with forecasting time, in agreement with the description of the data as deterministic chaos rather than uncorrelated noise [30]. As an additional check, surrogate data were also used in the forecasting process. In this case, the correlation coefficients show no dependence on forecast time.

Table 1 summarizes the changes that occur when the carotid sinus is intact versus isolated (open loop). Manipulating the CSP results in a change in MAP [32] in a fashion similar to that shown in Figure 6. A significant ($p < 0.01$) increase in the correlation dimension was measurable between the control group ($3.2 \pm 0.4$) ($n = 6$) and the midrange CSP intervention group ($4.2 \pm 0.2$) ($n = 6$). In all cases the Lyapunov exponents were positive, suggestive of chaotic activity. The largest Lyapunov exponents revealed a significant ($p < 0.01$) decrease in the midrange intervention group ($0.26 \pm 0.03$) compared with the control group ($0.40 \pm 0.04$). Similar statistically significant results were apparent for the low-range CSP intervention compared with the control. For the high-range CSP intervention, there were no significant changes compared with the control. However, it was not possible to lower the MAP pressure significantly below control for the 20 minutes necessary to perform the recording. This was because the dog frequently became agitated and aroused as the MAP dropped, thus overriding the carotid baroreflex drive. Table 1 also reveals the expected result that the statistical variability (standard deviation) of the MAP increases significantly in all three intervention groups when compared with the control.

From the data, it could be argued that the change in correlation dimension is directly related to the observed change in MAP. However, the evidence of Almog et al. [26] would support a more basic difference in the blood pressure control system. They showed that there was a significant increase in the correlation dimension between

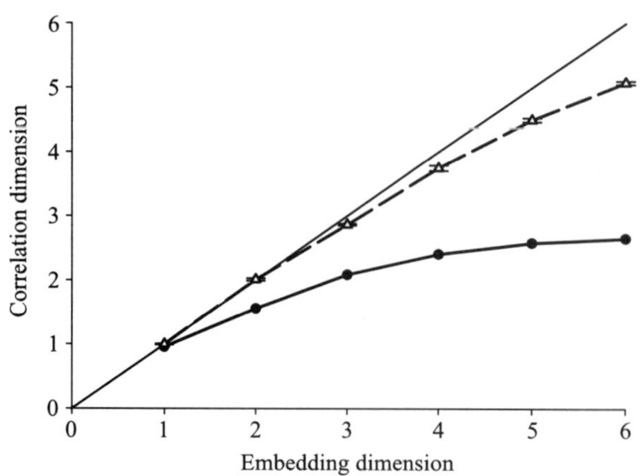


**Figure 4** Correlation dimension versus embedding dimension for control data (circle and solid line) and for eight surrogate data sets (triangle and dashed line). The mean and SE values for the surrogate data are shown, as is a line of identity. The correlation dimension of a random signal would be expected to increase with increasing embedding dimension.

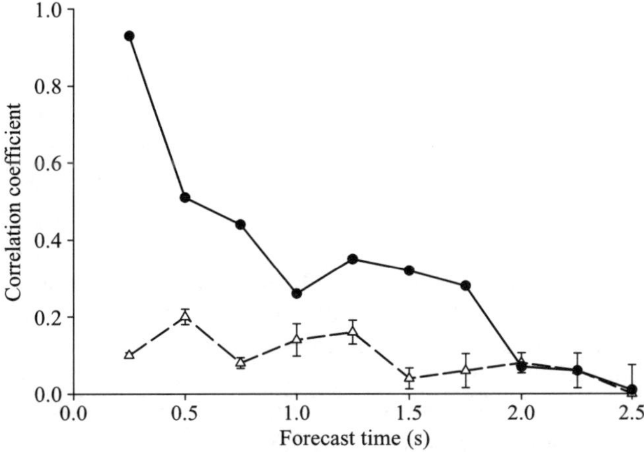


**Figure 5** Correlation coefficient versus forecast time for control data (circle and solid line) and for eight surrogate data sets (triangle and dashed line). The mean and SE values for the surrogate data are shown. The correlation coefficient was averaged for 100 forecasts within each data set.

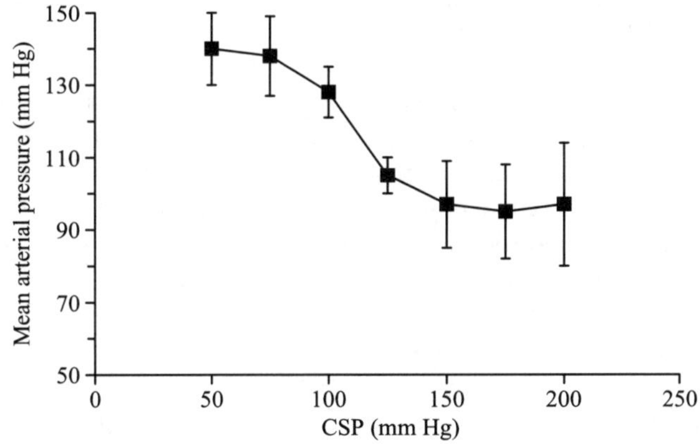


**Figure 6** Relationship between carotid sinus pressure and mean arterial pressure in the conscious dog ($n = 7$). Measurements were taken 30 seconds after reaching a steady-state CSP.

**TABLE 1** Effect of Open-Loop Carotid Sinus Interventions at Low-, Mid-, and High-Range Carotid Sinus Pressures Compared with the Intact Control.[a]

| | Control | Intervention (low CSP) | Intervention (mid-range CSP) | Intervention (high CSP) |
|---|---|---|---|---|
| MAP (mm Hg) | $94.9 \pm 4.3$ | $134.4 \pm 11.2^{**}$ | $118.2 \pm 10.0^{*}$ | $97.9 \pm 12.1$ |
| SD (mm Hg) | $7.1 \pm 1.1$ | $10.8 \pm 1.0^{*}$ | $12.8 \pm 2.0^{*}$ | $15.3 \pm 4.4^{*}$ |
| Correlation dimension | $3.2 \pm 0.4$ | $4.6 \pm 0.2^{**}$ | $4.2 \pm 0.2^{**}$ | $3.7 \pm 0.6$ |
| Lyapunov exponent | $0.40 \pm 0.04$ | $0.26 \pm 0.03^{**}$ | $0.29 \pm 0.03^{**}$ | $0.34 \pm 0.04$ |

[a] Student's $t$-test at the $^{*}p > 0.05$ and $^{**}p > 0.01$ significance levels.

WKY and SHR rats even at the age of 1 month before the SHR had demonstrated measurable hypertension.

Wagner et al. [25], in their analysis of chaotic activity and the baroreflex, reported for arterial pressure a correlation dimension (CD) of $3.05 \pm 0.23$ in an intact conscious dog preparation. This value for correlation dimension is in reasonable agreement with our value of $3.2 \pm 0.4$ considering differences in experimental protocol. The major difference was that Wagner and co-workers analyzed a downsampled version of the actual pressure waveform, whereas we extracted beat-by-beat measures of MAP.

Wagner also reported that complete surgical denervation caused a reduction in the CD to a value of $1.74 \pm 0.2$, implying a loss of complexity in the signal. Allied with this decrease in complexity was an increase in predictability evidenced by a decrease in the largest Lyapunov exponent from $1.85 \pm 0.18$ in the controls to $0.74 \pm 0.08$ in the denervated group. In agreement with our results, the variability as measured by the standard deviation of the signal increased from $8.3 \pm 1.4$ to $22.2 \pm 3.1$ ($n = 7$).

Although our data support an increase in predictability (as the Lyapunov exponent decreased significantly), they show opposite trends in terms of complexity. By opening the baroreflex loop and supplying a fixed pressure to the carotid sinus, one may argue that complexity should decrease and predictability increase in a manner similar to denervation. As this is not evidenced in our data [33], it is obvious that from the viewpoint of nonlinear dynamical behavior, the cardiovascular control system responds significantly differently to an open-loop carotid baroreflex as opposed to complete sinoaortic and cardiopulmonary denervation. Further comparison is made difficult by the fact that the denervation protocol used by Wagner et al. [25] was staged over several weeks, possibly resulting in longer term compensatory mechanisms influencing the nonlinear response. However, it may be argued that the differences lie in the adaptation process. Freeman and colleagues [34, 35] examine the adaptive brain processes under new unlearned processes. Elbert et al. [4] argue that to manipulate the environment, homeostatic and deterministic systems need to be "destabilized" to produce new activity patterns. Baroreceptor feedback serves in part as a monitor for the central nervous pathways that govern the gains of the various systems that control blood pressure. When the arterial pressure control is normal in the intact animal, the pathways that adjust gain in the system are likely to be relatively quiet. Opening the baroreflex loop may well cause the system to become active in an attempt to "see" an "expected" pressure signal. This activity could be evidenced as an observable increase in variability and increase in the correlation dimension. In the case of Wagner's experiments [25], the chronic removal of feedback could lead these systems to shut down as they would "sense" that no adjustments were effective. Such adaptation would be likely to reduce the correlation dimension.

## 4. CONCLUSIONS

The presence of deterministic behavior is a landmark of a healthy physiologic system and is altered or absent in diseases, indicating that the processes that are responsible for the generation of the chaotic behavior may be affected by disease.

In this study, a number of indicators, including observance of structure in the phase-space trajectories, the convergence of the correlation dimension, positive Lyapunov exponents, the ability to perform nonlinear forecasting, and the loss of convergence and forecasting ability with surrogate data sets, all support the hypothesis that the mean arterial blood pressure of the conscious dog exhibits chaotic behavior. This study demonstrated that there was a significant increase in the "complexity" of the chaotic activity between the intact and isolated carotid sinus preparations. Future studies could examine the characterization of responses at a number of different carotid sinus pressures and the effect of supplying a "chaotic" MAP to the carotid sinus.

Conversely, the largest Lyapunov exponents demonstrate increased "predictability" in the open-loop, chronically isolated carotid sinus group of animals when compared with a control group of animals with an intact baroreflex. The decreased Lyapunov exponent evidenced in our data reflects the slower rate at which the system explores the possible states on the attractor.

It may be possible in the future to improve diagnosis of cardiovascular disease and identification of risk factors by the quantification of various aspects of nonlinear dynamical control of the cardiovascular system. Skinner et al. [36] have demonstrated this fact for sudden cardiac death, but no predictive studies have been reported in the human for predisposition to chronic hypertension. The work described herein represents one of the first necessary stages of this process, whereby the nonlinear aspects of the short-term arterial pressure control are analyzed after experimentally induced changes in the carotid baroreflex.

The presence of deterministic behavior is a landmark of a healthy physiologic system and is often altered or absent in diseases, indicating that the systems and control processes that are responsible for the generation of the chaotic behavior may be affected by disease. Study of these fundamental nonlinear characteristics may lead to a greater understanding of the causes of these pathophysiologic states. At the very least, the proposed analyses represent an exciting and promising direction in integrative physiology, with the strong possibility that nonlinear dynamical theory will be to the next 30 years what linear control has been for the last 30.

## REFERENCES


[1] S. Hales, *Statistical Assays: Containing Haemastaticks.* London: Innys, Manby and Woodward, 1733.

[2] A. C. Guyton, T. G. Coleman, and H. J. Granger, Circulation: Overall regulation. *Ann. Rev. Physiol.* 34:13–46, 1972.

[3] P. B. Persson and C. D. Wagner, General principles of chaotic dynamics. *Cardiovasc. Res.* 31:332–341, 1996.

[4] T. Elbert, W. J. Ray, Z. J. Kowalik, J. E. Skinner, K. E. Graf, and N. Birbaumer, Chaos and physiology: Deterministic chaos in excitable cell assemblies. *Physiol. Rev.* 74(1):1–47, 1994.

[5] T. A. Denton, G. E. Diamond, R. H. Halfant, S. Khan, and H. Karagueuzian, Fascinating rhythms: A primer on chaos theory and its application to cardiology. *Am. Heart J.* 120:1419–1440, 1990.

[6] J. E. Skinner, A. L. Goldberger, G. Mayer-Kress, and R. E. Ideker, Chaos in the heart: Implications for clinical cardiology. *Biotechnology* 8:1018–1033, 1990.

[7] A. L. Goldberger and J. W. Bruce, Application of nonlinear dynamics to clinical cardiology. *Ann. N. Y. Acad. Sci.* 504:195–212, 1987.

[8] M. R. Guevara, L. Glass, and A. Shrier, Phase locking, period doubling bifurcations, and irregular dynamics in periodically stimulated cardiac cells. *Science* 214:1350–1352, 1981.

[9] M. R. Guevara, A. Shrier, and L. Glass, Phase-locked rhythms in periodically stimulated heart cell aggregates. *Am. J. Physiol.* 254:H1–H10, 1988.

[10] D. C. Michaels, D. R. Chialvo, E. P. Matyas, and J. Jalife, Chaotic activity in a mathematical model of the vagally driven sinoatrial node. *Circ. Res.* 65:1350–1360, 1989.

[11] K. M. Stein, N. Lippman, and P. Kligfield, Fractal rhythms of the heart. *J. Electrocardiol.* 24(Suppl):72–76, 1992.

[12] J. E. Skinner, C. Carpeggiani, C. E. Landisman, and K. W. Fulton, The correlation-dimension of the heartbeat is reduced by myocardial ischemia in conscious pigs. *Circ. Res.* 68:966–976, 1991.

[13] D. R. Chialvo, R. F. Gilmore Jr., and J. Jalife, Low dimensional chaos in cardiac tissue. *Nature* 343:653–656, 1990.

[14] D. Kaplan and R. J. Cohen, Is fibrillation chaos? *Circ. Res.* 67:886–900, 1990.

[15] D. Ruelle, Where can one hope to profitably apply the ideas of chaos? *Phys. Today* July:24–30, 1994.

[16] Z. S. Huang, G. L. Gebber, S. Zhong, and S. Barman, Forced oscillations in sympathetic nerve discharge. *Am. J. Physiol.* 263:R564–R571, 1991.

[17] T. Zhang and E. J. Johns, Chaotic characteristics of renal nerve peak interval sequence in normotensive and hypertensive rats. *Clin. Exp. Pharm. Physiol.* 25:896–903, 1998.

[18] K. P. Yip, N. H. Holstein-Rathlou, and D. J. Marsh, Chaos in blood flow control in genetic and renovascular hypertensive rats. *Am. J. Physiol.* 261:F400–F408, 1991.

[19] U. Zwiener, D. Hoyer, R. Bauer, B. Luthke, B. Walter, K. Schmidt, S. Hallmeyer, B. Kratzsch, and M. Eiselt, Deterministic-chaotic and periodic properties of heart rate and arterial pressure fluctuations and their mediation in piglets. *Cardiovasc. Res.* 3:455–465, 1996.

[20] D. Hoyer, B. Pompe, H. Herzel, and U. Zwiener, Nonlinear coordination of cardiovascular autonomic control. *IEEE Eng. Med. Biol. Mag.* 17(6):17–21, 1998.

[21] N. H. Lovell, B. Henry, B. G. Celler, and M. J. Brunner, Chaotic behavior of blood pressure and heart rate in the conscious dog. *18th Annual International Conference of the IEEE Engineering in Medicine and Biology Society*, Amsterdam, October 31–November 3, 1996.

[22] G. S. Baselli, S. Cerutti, D. Civardi, F. Lombardi, A. Malliani, and M. Pagani, Spectral and cross-spectral analysis of heart rate and blood pressure variability signals. *Comp. Biomed. Res.* 19:520–534, 1986.

[23] G. Mancia, G. Parati, M. Di Rienzo, and A. Zanchetti, Blood pressure variability. In *Handbook of Hypertension*, Vol. 17: *Pathophysiology of Hypertension*, A. Zanchetti and G. Mancia, eds. 1997.

[24] C. D. Wagner, B. Nafz, and P. B. Persson, Chaos in blood pressure control. *Cardiovasc. Res.* 31:380–387, 1996.

[25] C. D. Wagner, R. Mrowka, B. Nafz, and P. B. Persson, Complexity and "chaos" in blood pressure after baroreceptor denervation of conscious dogs. *Am. J. Physiol.* 269:H1760–H1766, 1995.

[26] Y. Almog, S. Eliash, O. Oz, and S. Akselrod, Nonlinear analysis of BP signal: Can it detect malfunctions in BP control? *Am. J. Physiol.* 27:H396–H403, 1996.

[27] R. S. Wolfer, N. H. Lovell, and M. J. Brunner, Exogenous vasopressin does not enhance carotid baroreflex control in the conscious dog. *Am. J. Physiol.* 266:R1510–H1516, 1994.

[28] F. Takens, Detecting strange attractors in turbulence. In *Dynamical Systems and Turbulence*, D. A. Rand and L-S Young, eds., pp. 366–381. Berlin: Springer, 1980.

[29] P. Grassberger and I. Procaccia, Measuring the strangeness of a strange attractor. *Physica D* 9:189–208, 1983.

[30] G. Sugihara and R. M. May, Nonlinear forecasting as a way of distinguishing chaos from measurement error in time series. *Nature* 344:734, 1990.

[31] J. Theiler, S. Eubank, A. Longtin, B. Galdrikian, and J. D. Fraser, Testing for nonlinearity in time series: The method of surrogate data. *Physica D* 58:77–94, 1992.

[32] N. H. Lovell, D. Carlson, A. Park, and M. J. Brunner, Baroreflex control of left ventricular contractility in the conscious dog. *17th Annual International Conference of the IEEE Engineering in Medicine and Biology Society*, Montreal, September 20–23, 1995.

[33] N. H. Lovell, B. Henry, A. Avolio, B. G. Celler, D. E. Carlson, and M. J. Brunner, Nonlinear chaotic dynamics of mean arterial pressure after carotid baroreceptor isolation. *19th International Conference of the IEEE Engineering in Medicine and Biology Society*, Chicago, October 30–November 2, 1997.

[34] W. J. Freeman, The physiology of perception. *Sci. Am.* 264:78–85, 1991.

[35] C. A. Skarda and W. J. Freeman, How brains make chaos in order to make sense. *Behav. Brain Sci.* 10:161–195, 1987.

[36] J. E. Skinner, C. M. Pratt, and T. Vybiral, A reduction in the correlation dimension of heartbeat intervals precedes imminent ventricular fibrillation in human subjects. *Am. Heart J.* 125:731–743, 1993.

Chapter 12

# MEASUREMENT AND QUANTIFICATION OF SPATIOTEMPORAL DYNAMICS OF HUMAN EPILEPTIC SEIZURES

L. D. Iasemidis, J. C. Principe, and J. C. Sackellares

## 1. INTRODUCTION

Since its discovery by Hans Berger in 1929, the electroencephalogram (EEG) has been the signal most utilized to assess brain function clinically. The enormous complexity of the EEG signal, in both time and space, should not surprise us because the EEG is a direct correlate of brain function. If the system to be probed is complex and our signal is a reliable descriptor of its function, then we can also expect complexity in the signal. Unfortunately, traditional signal processing (SP) theory is based on very simple assumptions about the system that produced the signal (e.g., linearity assumption). Hence the application of SP methodologies to quantify the EEG automatically has met the challenge with varying degrees of success. We can say that today the trained electroencephalographer or neurologist is still the "gold standard" in characterizing phasic events in the EEG such as spikes, the quantification of background activity (as in sleep staging), and identification and localization of epileptic seizures.

The status quo is not going to change overnight, but this chapter will report on an exciting and promising new methodology to quantify the EEG that is rooted in the theory of nonlinear dynamics. EEG characteristics such as alpha activity and seizures (limit cycles [1]), instances of bursting behavior during light sleep, amplitude-dependent frequency behavior (the smaller the amplitude, the higher the EEG frequency), and the existence of frequency harmonics (e.g., under photic driving conditions) are extensively reported in the clinical literature. All these characteristics also belong to the long catalog of typical properties of nonlinear systems. By applying techniques from nonlinear dynamics, several researchers have provided evidence that the EEG is a nonlinear signal with deterministic and perhaps chaotic properties (for reviews see Refs. 1–9).

This chapter will address the quantification of EEG in epilepsy. Epilepsy is a group of disorders characterized by recurrent paroxysmal electrical discharges of the cerebral cortex that result in intermittent disturbances of brain function [10]. The bulk of research into human epilepsy has emphasized the description and categorization of the clinical and electroencephalographic features of seizures, defining clinical features of various epileptic syndromes, and correlating clinical and electroencephalographic features with anatomical lesions of the brain or with genetic disorders. However, all this work has not addressed the essential feature of epilepsy, which is the fact that seizures occur intermittently; that is, seizure generation (and eventually epilepsy) is a dysfunction of brain dynamics.

Our group, employing analytical techniques developed for the analysis of complex nonlinear systems, was the first to demonstrate and quantify specific spatiotemporal dynamical changes in the EEG that begin *several minutes* before and end several minutes after a seizure [11–13]. These changes appear to evolve in a characteristic pattern, culminating in a seizure. Our current work indicates that, in the near future, we will obtain for the first time a signal processing methodology capable of quantifying the EEG more accurately and in a finer detail than it is possible to do through clinical evaluation. This should not come as a surprise because we are employing an analysis of dynamical patterns on a spatiotemporal trajectory, information that is not directly accessible from observing the EEG with the naked eye. As the theory of nonlinear dynamics clearly shows [14], state space is the natural domain in which to quantify properties of nonlinear dynamical systems. These properties may be undetectable in the time domain of the system output, the EEG tracing.

The measures we have employed do not assume any particular nonlinear model, nor do they depend upon a priori detailed understanding of the underlying mechanisms responsible for the generation of the data to be analyzed. They belong therefore to the class of nonparametric approaches. We have to state further that the general theory of nonlinear dynamical systems [14] does not rigorously describe brain dynamics. The brain is not an autonomous system. Its interaction with the environment creates a never-ending sequence of dynamical states in which the brain internal dynamics are the overwhelming defining factor. Therefore, we have to realize that there is *a physiological time scale for brain dynamics*. Dynamical measures should therefore be computed within this time scale. Failing to realize this fact leads to the averaging of temporal dynamical information and the subsequent loss of information. An analogy helps our point. Speech researchers understood many years ago that speech is a locally stationary signal. Hence, speech quantification is limited to windows in time where the signal is relatively stationary (10-ms windows). If we used long-term statistics to quantify speech, all the local information would be washed away. Likewise, in brain dynamics, the dynamical measures should not be computed over long time intervals assuming stationarity as required by the dynamical theory of autonomous systems. We believe that this point is one of the distinguishing characteristics of our methodology and one of the pillars of its success.

The chapter is organized as follows. We start by a brief description of the short-term largest Lyapunov exponent ($STL_{\max}$), the nonlinear dynamical measure utilized in our research. We then apply the STL algorithm to EEG tracings, recorded from several brain sites, to create a set of $STL_{\max}$ time series containing local (in time and in space) information about dynamical mixing in the brain. We further create and quantify spatial maps of $STL_{\max}$ time series. At this level of analysis we are quantifying the brain spatiotemporal dynamics. We will show that it is at this level of analysis that reliable predictive measures of epileptic seizures are being derived.

# 2. METHODS FOR NONLINEAR DYNAMICAL ANALYSIS

## 2.1. Application to the EEG

The standard methods for time series analysis (e.g., power analysis, linear orthogonal transforms, parametric linear modeling) not only fail to detect the critical features of a time series generated by a nonlinear system but also may falsely suggest that most of the series is random noise [15]. In recent years, the methods developed for the dynamical analysis of complex series have been applied to the investigation of signals produced by real biological systems, such as the EEG. The EEG can be conceptualized as a series of numerical values (voltages) over time. Such a series is called a "time series," and an example of an EEG from different stages of an epileptic seizure is shown in Figure 1. The ictal stage starts at the seizure's onset and ends at the seizure's end. The preictal stage is the period preceding the seizure onset. The postictal stage is the period following the seizure end. The period between the postictal stage of one seizure and the preictal stage of the next seizure is called the interictal stage.

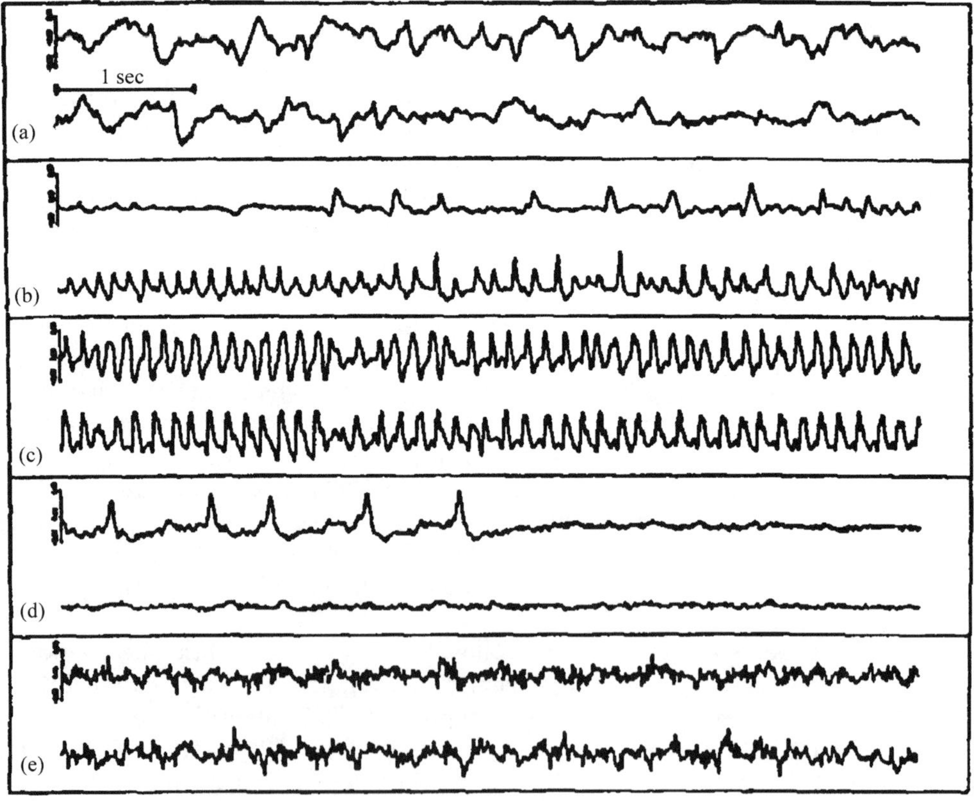


**Figure 1** Five EEG segments from different stages of a human epileptic seizure (patient 1) in the time domain (each segment is of 12 seconds duration and scaled separately). (a) Preictal; (b) transition from preictal to ictal; (c) ictal; (d) transition from ictal to postictal; (e) postictal.

The EEG, being the output of a multidimensional system, has statistical properties that depend on both time and space [16]. Components of the brain (neurons) are densely interconnected and the EEG recorded from one site is inherently related to the activity at other sites. This makes the EEG a multivariable time series. A well-established technique for visualizing the dynamical behavior of a multidimensional (mutivariable) system is to generate a phase space portrait of the system. A phase space portrait is created by treating each time-dependent variable of the system as a component of a vector in a multidimensional space, usually called the state or phase space of the system. Each vector in the phase space represents an instantaneous state of the system. These time-dependent vectors are plotted sequentially in the phase space to represent the evolution of the state of the system over time. For many systems, this graphic display creates an object confined over time to a subregion of the phase space. Such subregions of the phase space are called "attractors." The geometrical properties of these attractors provide information about the global state of the system.

One of the problems in analyzing multidimensional systems is knowing which observable (variable of the system that can be measured) to analyze. In addition, experimental constraints may limit the number of observables that can be obtained. It turns out that when the variables of the system are related over time, which must be the case for a dynamical system to exist, proper analysis of a single observable can provide information about all variables of the system that are related to this observable. In principle, through the method of delays described by Packard et al. [17] and Takens [18], sampling of a single variable of a system over time can reproduce the related attractors of a system in the phase space. This technique for the reconstruction of the phase space from one observable can be used for the EEG. In Figure 2b a phase space portrait has been generated from an ictal EEG signal recorded by a single electrode located on the temporal cortex (Figure 2a). The characteristics of the formed epileptic attractor are typical for all of the seizures we have analyzed to date. This picture indicates the existence of limit cycles in the phase space [1] (corresponding to periodic behavior in time domain) with trajectories moving in and out of the main body of the attractor (the large excursions correspond to spikes in the time domain).

The geometrical properties of the phase portrait of a system can be expressed quantitatively using measures that ultimately reflect the dynamics of the system. For example, the complexity of an attractor is reflected in its dimension [19]. The larger the dimension of an attractor, the more complicated it appears in the phase space. It is important to distinguish between the dimension of the phase space (embedding dimension) and the dimension of the attractor. The embedding dimension ($p$) is the dimension of the phase space that contains the attractor and it is always a positive integer. On the other hand, the attractor dimension ($D$) may be a noninteger. $D$ is directly related to the number of variables of the system and inversely related to the existing coupling among them.

According to Takens [18], the embedding dimension $p$ should be at least equal to $(2 * D + 1)$ in order to correctly embed an attractor in the phase space. Of the many different methods used to estimate $D$ of an object in the phase space, each has its own practical problems [7, 19, 20]. The measure most often used to estimate $D$ is the phase space correlation dimension ($\nu$). Methods for calculating the correlation dimension from experimental data have been described [21, 22] and were employed in our work to approximate $D$ of the epileptic attractor.

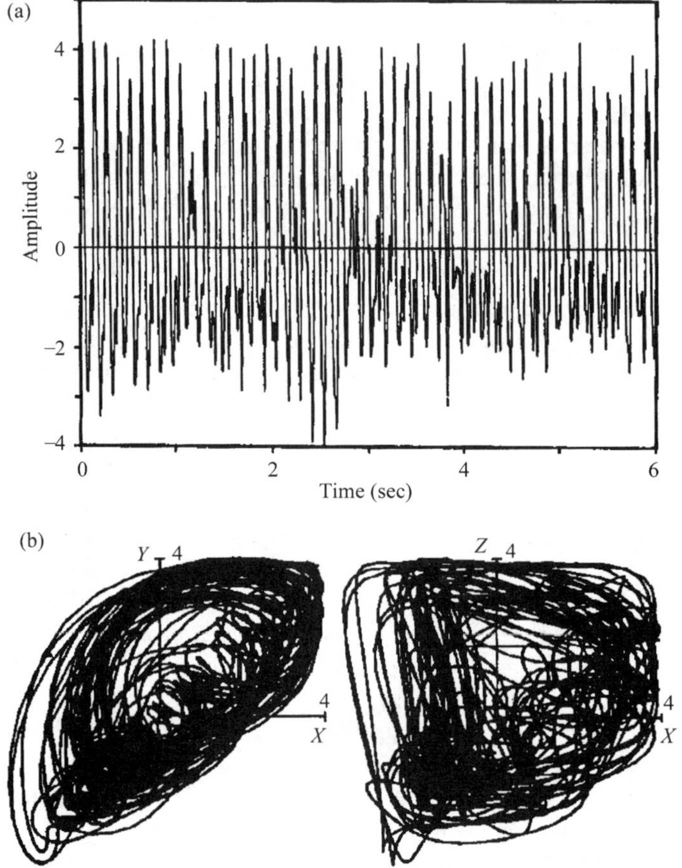


**Figure 2** (a) Ictal segment of raw EEG data from an epileptic seizure (patient 1). (b) The EEG data in the phase space.

A chaotic attractor is an attractor where, on the average, orbits originating from similar initial conditions (nearby points in the phase space) diverge exponentially fast (expansion process); they stay close together only for a short time. If these orbits belong to an attractor of finite size, they will fold back into it as time evolves (folding process). The result of these two processes is a layered structure [14]. The measures that quantify the chaoticity of an attractor are the Kolmogorov entropy ($K$) and the Lyapunov exponents [23–25]. For an attractor to be chaotic, the Kolmogorov entropy or at least the maximum Lyapunov exponent ($L_{\max}$) must be positive. The Kolmogorov (Sinai or metric) entropy ($K$) measures the uncertainty about the future state of the system in the phase space given information about its previous states (positions or vectors in the phase space). The Lyapunov exponents measure the average rate of expansion and folding that occurs along the local eigendirections within an attractor in phase space. If the phase space is of $p$ dimensions, we can estimate theoretically up to $p$ Lyapunov exponents. However, only ($[D] + 1$) of them will be real. The rest will be spurious [26]. Methods for calculating these dynamical measures from experimental data have been published [26–30]. The estimation of the largest Lyapunov exponent

($L_{max}$) in a chaotic system has been shown to be more reliable and reproducible than the estimation of the remaining exponents [20, 31], especially when $D$ changes over time as it does with nonstationary data. We will now summarize our algorithm for estimating $L_{max}$ from nonstationary data.

## 2.2. Estimation of Short-Term Largest Lyapunov Exponents ($STL_{max}$)

As we discussed in the introduction, a relevant time scale should always be used in order to quantify the physiological changes occurring in the brain. Furthermore, the brain being a nonstationary system, algorithms used to estimate measures of the brain dynamics should be capable of automatically identifying and appropriately weighing existing transients in the data. The method we developed for estimation of $STL_{max}$, an estimate of $L_{max}$ for nonstationary data, is explained in detail elsewhere [12, 27]. Herein we will present only a short description of our method.

Construction of the embedding phase space from a data segment $x(t)$ of duration $T$ is made with the method of delays. The vectors $X_i$ in the phase space are constructed as

$$X_i = (x(t_i), x(t_i + \tau) \cdots x(t_i + (p-1) * \tau))^T \tag{1}$$

where $\tau$ is the selected time lag between the components of each vector in the phase space, $p$ is the selected dimension of the embedding phase space, and $t_i \in [1, T - (p-1)\tau]$. If we denote by $L$ the estimate of the short-term largest Lyapunov exponent $STL_{max}$, then

$$L = \frac{1}{N_a \Delta t} \sum_{i=1}^{N_a} \log_2 \frac{|\delta X_{i,j}(\Delta t)|}{|\delta X_{i,j}(0)|} \tag{2}$$

with

$$\delta X_{i,j}(0) = X(t_i) - X(t_j) \tag{3}$$

$$\delta X_{i,j}(\Delta t) = X(t_i + \Delta t) - X(t_j + \Delta t) \tag{4}$$

where $X(t_i)$ is the point of the fiducial trajectory $\phi_t(X(t_0))$ with $t = t_i$, $X(t_0) = (x(t_0) \cdots x(t_0 + (p-1)\tau))^T$, $^T$ denotes the transverse, and $X(t_j)$ is a properly chosen vector adjacent to $X(t_i)$ in the phase space (see below).

$\delta X_{i,j}(0) = X(t_i) - X(t_j)$ is the displacement vector at $t_i$, that is, a perturbation of the fiducial orbit at $t_i$, and $\delta X_{i,j}(\Delta t) = X(t_i + \Delta t) - X(t_j + \Delta t)$ is the evolution of this perturbation after time $\Delta t$.

$t_i = t_0 + (i-1) * \Delta t$ and $t_j = t_0 + (j-1) * \Delta t$, where $i \in [1, N_a]$, and $j \in [1, N]$ with $j \neq I$.

$\Delta t$ is the evolution time for $\delta X_{i,j}$, that is, the time one allows $\delta X_{i,j}$ to evolve in the phase space. If the evolution time $\Delta t$ is given in seconds, then $L$ is in bits per second.

$t_0$ is the initial time point of the fiducial trajectory and coincides with the time point of the first data in the data segment of analysis. In the estimation of $L$, for a complete scan of the attractor, $t_0$ should move within $[0, \Delta t]$.

$N_a$ is the number of local $L_mathop$max's that will be estimated within a duration $T$ data segment. Therefore, if $Dt$ is the sampling period of the time domain data, $T = (N - 1)Dt = N_a \Delta t + (p - 1)\tau$.

The short-term largest Lyapunov exponent $STL_{\max}$ is computed by a modified version of the program proposed by Wolf et al. [27]. We call the measure *short term* to differentiate it from the ones in autonomous dynamical systems studies. We found that for small data segments with transients, as in epileptic data, a modification of this algorithm is absolutely necessary for a better estimation of $STL_{\max}$. This modification has to be made mainly in the searching procedure for a replacement vector at each point of a fiducial trajectory. For example, in our analysis of the EEG, we found that the crucial parameter of the $L_{\max}$ estimation procedure, in order to distinguish between the preictal, the ictal, and the postictal stages, was not the evolution time $\Delta t$ or the angular separation $V_{i,j}$ between the evolved displacement vector $\delta X_{i-1,j}(\Delta t)$ and the candidate displacement vector $\delta X_{i,j}(0)$ (as claimed in Frank et al. [32]). The crucial parameter is the *adaptive estimation in time and phase space of the magnitude bounds* of the candidate displacement vector (see Eq. 8) to avoid catastrophic replacements. Results from simulation data of known attractors have shown the improvement in the estimates of $L$ achieved by using the proposed modifications [11].

Our rules can be stated as follows:

- For $L$ to be a reliable estimate of $STL_{\max}$, the candidate vector $X(t_j)$ should be chosen such that the previously evolved displacement vector $\delta X_{(i-1),j}(\Delta t)$ is almost parallel to the candidate displacement vector $\delta X_{i,j}(0)$, that is,

$$|V_{i,j}| = | < \delta X_{i,j}(0), \delta X_{(i-1),j}(\Delta t) > | \leq V_{\max} \tag{5}$$

where $V_{\max}$ should be small and $|\langle \epsilon, \phi \rangle|$ denotes the absolute value of the angular separation between two vectors $\epsilon$ and $\phi$ in the phase space.

- For $L$ to be a reliable estimate of $STL_{\max}$, $\delta X_{i,j}(0)$ should also be small in magnitude in order to avoid computer overflow in the future evolution within very chaotic regions and to reduce the probability of starting up with points on separatrices [33]. This means

$$|\delta X_{i,j}(0)| = |X(t_i) - X(t_j)| < \Delta_{\max} \tag{6}$$

with $\Delta_{\max}$ assuming small values.

Therefore, the parameters to be selected for the estimation procedure of $L$ are:

- The embedding dimension $p$ and the time lag $\tau$ for the reconstruction of the phase space
- The evolution time $\Delta t$ (number of iterations $N_a$)
- The parameters for the selection of $X(t_j)$, that is, $V_{i,j}$ and $\Delta_{\max}$
- The duration of the data segment $T$

It is worth noting here that because only vector differences are involved in the estimation of $L$ (Eq. 2), any DC present in the data segment of interest does not influence the value of $L$. In addition, only vector difference ratios participate in the estimation of $L$. This means that $L$ is also not influenced by the scaling of the data (as long as the parameters involved in the estimation procedure, i.e., $\Delta_{max}$, assume not absolute values but relative ones to the scale of every analyzed data segment). Both points make sense when one recalls that $L$ relates to the entropy *rate* of the data [30].

### 2.2.1. Selection of p and $\tau$

We select $p$ such that the dimension $\nu$ of the attractor in phase space is clearly defined. In the case of the epileptic attractor [11, 28, 34], $\nu = 2, 3$, and according to Takens' theorem a value of $p \geq (2 \times 3 + 1) = 7$ is adequate for the embedding of the epileptic attractor in the phase space. This value of $p$ may be too small for the construction of a phase space that can embed all states of the brain interictally, but it should be adequate for detection of the transition of the brain toward the ictal stage if the epileptic attractor is active in its space prior to the occurrence of the epileptic seizure. The parameter $\tau$ should be as small as possible to capture the shortest change (i.e., highest frequency component) present in the data. Also, $\tau$ should be large enough to generate (with the method of delays) the maximum possible independence between the components of the vectors in the phase space. In the literature, these two conditions are usually addressed by selecting $\tau$ as the first minimum of the mutual information between the components of the vectors in the phase space or as the first zero of the time domain autocorrelation function of the data [26]. Theoretically, since the time span $(p-1)\tau$ of each vector in the phase space represents the duration of a state of the system, $(p-1)\tau$ should be at most equal to the period of the maximum (or dominant) frequency component in the data. For example, a sine wave (or a limit cycle) has $\nu = 1$; then a $p = 2 \times 1 + 1 = 3$ is needed for the embedding and $(p-1)\tau = 2\tau$ should be equal to the period of the sine wave. Such a value of $\tau$ would then correspond to the Nyquist sampling of the sine wave in the time domain. In the case of the epileptic attractor, the highest frequency present is 70 Hz (the EEG data are low-pass filtered at 70 Hz, which means that if $p = 3$, the maximum $\tau$ to be selected is about 7 ms. However, since the dominant frequency of the epileptic attractor (i.e., during the ictal period) was never more than 12 Hz, according to the preceding reasoning, the adequate value of $\tau$ for the reconstruction of the phase space of the epileptic attractor is $(7-1)\tau = 83$ ms, that is, $\tau$ should be about 14 ms (for more details see Ref. 12).

### 2.2.2. Selection of $\Delta t$

The evolution time $\Delta t$ should not be too large, otherwise the folding process within the attractor adversely influences $L$. On the other hand, it should not be too small, in order for $\delta X_{i,j}(\Delta t)$ to follow the direction of the maximum rate of information change. If there is a dominant frequency component $f_0$ in the data, $\Delta t$ is usually chosen as $\Delta t = \frac{1}{2} f_0$. Then, according to the previous arguments for the selection of $p$ and $\tau$, the $\Delta t \approx ((p-1)\tau)/2$, which for EEG results in $\Delta t \approx 42$ msec. In Ref. 12, it is shown that such a value is within the range of values for $\Delta t$ that can very well distinguish the ictal from the preictal state.

### 2.2.3. Selection of $V_{\max}$

It sets the limits for the variation of $V_{i,j}$. We start with an initial $V_{\max}$ (initial) $= 0.1$ rad. In the case that a replacement vector $X(t_j)$ is not found with $0 \leq |V_{i,j}| < V_{\max}$ (initial) and $|\delta X_{ij}(0)| < 0.1\Delta_{\max}$, we relax the bound for $|\delta X_{ij}(0)|$ and repeat the process with bounds up to $0.5\Delta_{\max}$. If not successful, we relax the bounds for $|V_{i,j}|$ by doubling the value of $V_{\max}$ and repeat the process with bounds for $V_{\max}$ up to 1 rad. Values of $V_{\max}$ larger than 0.8 rad never occurred in the procedure. If they do, the replacement procedure stops, a local $L(t_i)$ is not estimated at $t_i$, and we start the whole procedure at the next point in the fiducial trajectory.

### 2.2.4. Selection of $\Delta_{\max}$

In Wolf's algorithm, $\Delta_{\max}$ is selected as

$$\Delta_{\max} = \max_{j,i} |\delta X_{i,j}(0)| \tag{7}$$

where $j = 1, \ldots, N$ and $i = 1, \ldots N_a$.

Thus, $\Delta_{\max}$ is the global maximum distance between any two vectors in the phase space of a segment of data. This works fine as long as the data are stationary and relatively uniformly distributed in the phase space. With real data this is hardly the case, especially with the brain electrical activity, which is strongly nonstationary and nonuniform [35–37]. Therefore, a modification in the searching procedure for the appropriate $X(t_j)$ is essential. First, an adaptive estimation of $\Delta_{\max}$ is made at each point $X(t_i)$, and the estimated variable is

$$\Delta_{i,\max} = \max_{j} |\delta X_{i,j}(0)| \tag{8}$$

where $j = 1, \ldots, N$. By estimating $\Delta_{\max}$ as before, we take care of the nonuniformity of the phase space [$\Delta_{\max}$ is now a *spatially local quantity* of the phase space at a point $X(t_i)$] but not of the effect of existing nonstationarities in the data. We have attempted to solve the problem of nonstationarities by estimating $\Delta_{\max}$ also as a *temporarily local quantity*. Then a more appropriate definition for $\Delta_{\max}$ is

$$\Delta_{i,\max} = \max_{\mathrm{IDIST}_1 < |t_i - t_j| < \mathrm{IDIST}_2} |\delta X_{i,j}(0)| \quad \text{with } j \neq i \tag{9}$$

and

$$\mathrm{IDIST}_1 = \tau \tag{10}$$

$$\mathrm{IDIST}_2 = (p-1)\tau \tag{11}$$

where $\mathrm{IDIST}_1$ and $\mathrm{IDIST}_2$ are the upper and lower bounds for $|t_i - t_j|$, that is, for the temporal window of local search.

Thus, the search for $\Delta_{i,\max}$ is always made temporally about the state $X(t_i)$ and its changes within a period of the time span $(p-1)\tau$ of a state. According to the previous formulas, the values for the parameters involved in the adaptive estimation of $\Delta_{i,\max}$ in our EEG data are

$$\text{IDIST}_1 = \tau = 14\,\text{ms and IDIST}_2 = (p-1)\tau = 84\,\text{ms}$$

#### 2.2.5. Selection of $X(t_j)$

The replacement vector $X(t_j)$ should be spatially close to $X(t_i)$ in phase space (with respect to magnitude and angle deviation), as well as temporally not very close to $X(t_i)$ to allow selecting $X(t_j)$ from a nearby (but not the same) trajectory (otherwise, replacing one state with one that shares common components would lead to a false underestimation of $L$). These two arguments are implemented in the following relations:

$$0 \leq |V_{i,j}| < V_{i,j}\ (\text{initial}) = 0.1\,\text{rad} \tag{12}$$

$$b\Delta_{i,\max} \leq \delta X_{i,j}(0) \leq c\Delta_{i,\max} \tag{13}$$

$$|t_i - t_j| > \text{IDIST}_3 \approx (p-1)\tau \tag{14}$$

The parameter $c$ starts with a value of 0.1 and increases, with a step of 0.1, up to 0.5 in order to find a replacement vector $X(t_j)$ satisfying Eq. 12 through Eq. 14. The parameter $b$ must be smaller than $c$ and is used to account for the possible noise contamination of the data, denoting the distance below which the estimation of $L$ may be inaccurate (we have used $b = 0.05$ for our data [12, 27]). The temporal bound $\text{IDIST}_2$ should not be confused with the temporal bound $\text{IDIST}_3$, since $\text{IDIST}_2$ is used to find the appropriate $\Delta_{i,\max}$ at each point $X(t_i)$ (searching over a limited time interval), whereas $\text{IDIST}_3$ is used to find the appropriate $X(t_j)$ within a $\Delta_{i,\max}$ distance from $X(t_i)$ (searching over all possible times $t_j$).

#### 2.2.6. Selection of $T$

For data obtained from a stationary state of the system, the time duration $T$ of the analyzed segment of data may be large for the estimate of $L$ to converge to a final value. For nonstationary data we have two competing requirements: on the one hand we want $T$ to be as small as possible to provide local dynamic information, but on the other hand the algorithm requires a minimum length of the data segment to stabilize the estimate of $STL_{\max}$. Figure 3 shows a typical plot for the change of $STL_{\max}$ with the size of the window for segments in the preictal and ictal stages of a seizure. From this figure, it is clear that values of 10 to 12 seconds for $T$ are adequate to distinguish between the two extreme cases (preictal and ictal) in our data and for the algorithm to converge.

After extensive sensitivity studies with EEG data in epilepsy [11,12] we have concluded that the critical parameter of the algorithm just presented is the $\text{IDIST}_2$, that is, the parameter that establishes a neighborhood in time at each point in the fiducial trajectory for the estimation of the parameter $\Delta_{i,\max}$, which then establishes a spatial neighborhood for this point in the phase space (see Eq. 9). This is very clearly illustrated in Figure 4. It is obvious that, with values of $\text{IDIST}_2$ greater than 160 ms, one is not even able to distinguish between the preictal and the ictal state of a seizure based on the thus generated values of $L$!

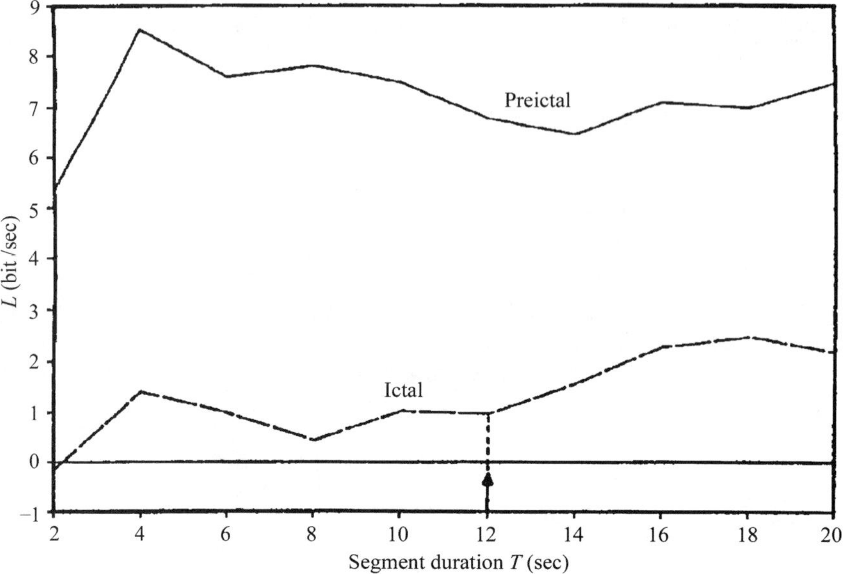


**Figure 3** The variation of $STL_{\max}$ with the length $T$ of the data segment for data in the preictal and ictal state of an epileptic seizure (patient 1). The rest of the parameters for the $STL_{\max}$ algorithm were $p = 7$, $\tau = 14\,\text{ms}$, $\Delta t = 42\,\text{ms}$, $\text{IDIST}_1 = 14\,\text{msec}$, $\text{IDIST}_2 = 84\,\text{ms}$, $\text{IDIST}_3 = 84\,\text{msec}$, $b = 0.05$, $c = 0.1$, $V_{i,j}(\text{initial}) = 0.1\,\text{rad}$.

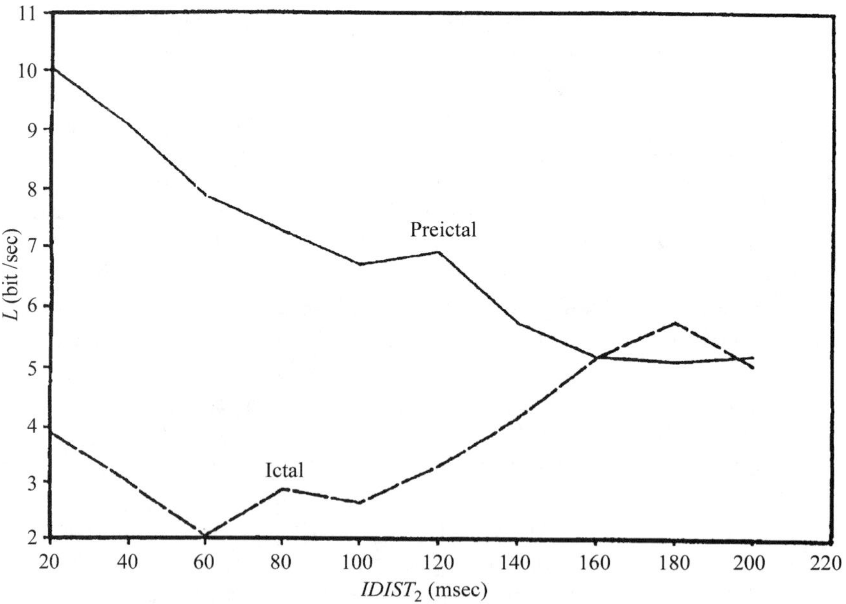


**Figure 4** The variation of $STL_{\max}$ with the $\text{IDIST}_2$ parameter for data in the preictal and ictal state of an epileptic seizure (patient 1). The rest of the parameters for the $STL_{\max}$ algorithm were $p = 7$, $\tau = 14\,\text{msec}$, $\Delta t = 42\,\text{ms}$, $\text{IDIST}_1 = 14\,\text{ms}$, $\text{IDIST}_3 = 84\,\text{ms}$, $b = 0.05$, $c = 0.01$, $V_{i,j}(\text{initial}) = 0.1\,\text{rad}$.

## 3. $STL_{MAX}$ TIME SERIES OF EEG DATA

We will now show the results of the application of our $STL_{max}$ algorithm over time from long electrocorticographic (ECoG) and scalp EEG data. Our studies have employed recordings obtained from implanted electrodes placed in the hippocampus and over the inferior temporal and orbitofrontal cortex (ECoG), as well as coverage of the temporal lobes by a limited number of scalp electrodes (scalp EEG). Figure 5 shows our typical 28-electrode montage used for subdural and depth recordings. Continuous EEG signals were typically analyzed for 1 hour before to 1 hour after a seizure, sampled with a sampling frequency of 256 Hz, and low-pass filtered at 70 Hz. Figure 6 depicts a typical ictal EEG recording, as manifested in scalp and depth electrodes, centered at the time of seizure onset.

By dividing the recorded EEG data from an electrode site into nonoverlapping segments, each 12 seconds in duration, and estimating $L$ from each of the thus created data segments, a profile of $L$ over time per electrode site was generated. A typical plot of $L$ over time, from the preictal to the postictal state of a seizure, is shown in Figure 7. The exponent $L$ is positive during the whole period of 19 minutes (10 minutes preictally, 2 minutes ictally, and 7 minutes postictally). We can then see that, even during the seizure where $L$ assumes its lowest value (the seizure can easily be detected from the lowest values of $L$), $L$ is still positive. This is consistent with an interpretation of the

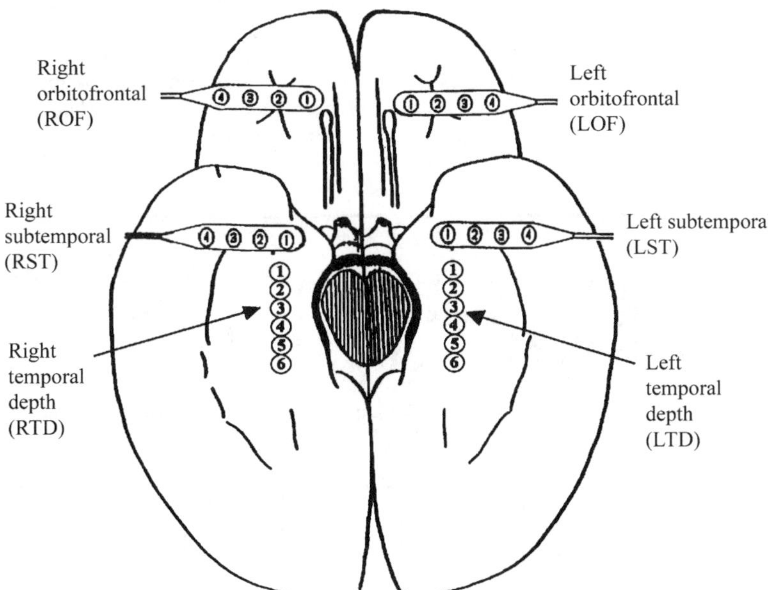


**Figure 5** Schematic diagram of the depth and subdural electrode placement. This view from the inferior aspect of the brain shows the approximate location of depth electrodes, oriented along the anterior–posterior plane in the hippocampi (RTD, right temporal depth; LTD, left temporal depth), and subdural electrodes located beneath the orbitofrontal and subtemporal cortical surfaces (ROF, right orbitofrontal; LOF, left orbitofrontal; RST, right subtemporal; LST, left subtemporal).

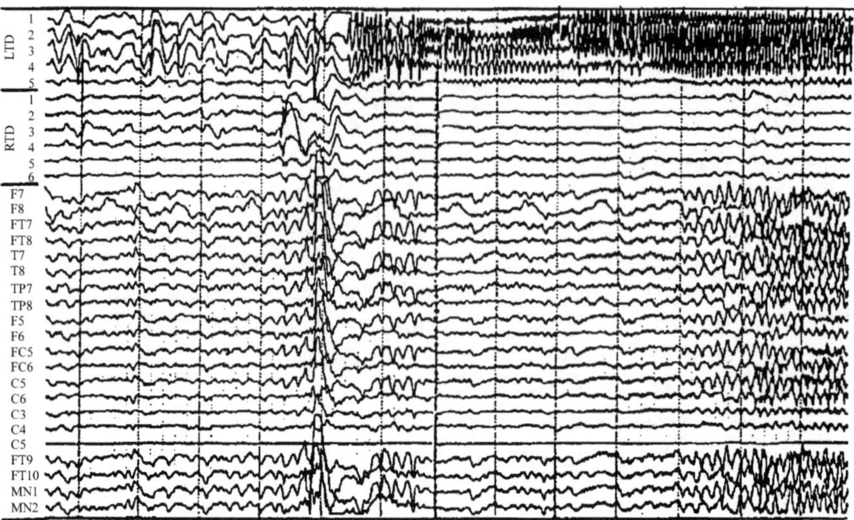


**Figure 6** A 13.6-second EEG segment at the onset of a left temporal lobe seizure (patient 2), recorded referentially to linked ears from bilaterally placed depth (hippocampal) electrodes (first 11 channels; nomenclature according to Figure 5), sphenoidal electrodes (last 2 channels), and scalp electrodes (remaining 19 channels; nomenclature according to AEEGS guidelines for scalp recordings, extended International 10-20 System). The ictal discharge begins in the left depth electrodes (LTD 1–4) as a series of high-frequency repetitive spikes, approximately 5 seconds into the record. The initial manifestation of the seizure in scalp/sphenoidal electrodes is a diffuse drop in voltage, with a definitive ictal discharge approximately 6 seconds after the initial ictal discharge in the left hippocampus. The ictal discharge spreads to the right hippocampus about 13 seconds later (not shown in this figure).

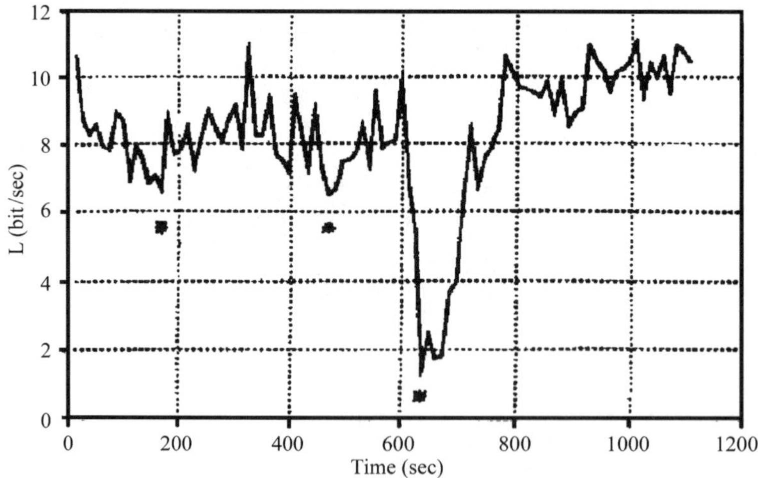


**Figure 7** Plot of the $STL_{max}$ over time derived from an EEG signal recorded at an electrode site not overlying the seizure focus (patient 1). The estimation of the $L$ values was made by dividing the signal into nonoverlapping segments of 12 seconds each.

ictal stage as a stage of fewer degrees of freedom than before, usually enough for the system to find a way out and the patient to recover [38]. Comparing the mean value of $L$ in the preictal state with that in the postictal state, we can deduce that the preictal state is less chaotic than the immediate postictal one. The seizure's onset corresponds to the maximum drop in the values of $L$ that were obtained at electrodes participating in the manifestation of the seizure. This was typical for all analyzed seizures across our subjects. Also, in the preictal state of Figure 7, one can notice two prominent drops in the values of $L$, one about 150 seconds and the other 450 seconds into the record. These preictal drops in $L$ can be explained as attempts of the system toward a phase transition much before (minutes) the actual seizure outburst. Such an explanation, as well as the answer to why the seizure did not occur at these preictal points but much later, required the consideration of the activity at other electrode sites as well.

In Figure 8a, we present the Lyapunov profile of the EEG recorded from an electrode site overlying the seizure focus for the same seizure as in Figure 6. We notice here that a prominent preictal drop in $L$ occurs 350 seconds into the record. We also notice that the preictal profile thereafter (segment A: 350 to 600 seconds in the record) is very similar in morphology to the immediate postictal profile (segment B: 700 to 1000 seconds in the record), the latter being more dilated. This becomes clear in Figure 8b, where the cross-correlation of segment B with the entire Lyapunov profile is shown. The global maximum of this cross-correlation occurs at about 720 seconds (approximately the starting point of segment B), as expected. In addition, we notice a prominent local maximum of the cross-correlation in the preictal period at about 350 seconds, a time lag that corresponds to the beginning of the preictal segment A. This suggests that A and B are strongly correlated, and it indicates that the brain exhibits cyclic behavior going into and out of the phase transition. This behavior was invisible from the raw EEG data or from the clinical symptoms of the patient before and after the seizure.

In all cases studied to date, the largest drop in the value of the $STL_{max}$ exponent occurred first at electrode sites located where the seizure discharged originated. Also, the chaoticity of the EEG signal (reflected by the value of $STL_{max}$) was highest during the postictal state, lowest during the seizure discharge, and intermediate in the preictal state. Thus, from a dynamical perspective, the onset of a seizure represents a temporal transition of our system from a chaotic state to a less chaotic one (more ordered, corresponding to the synchronized rhythmic firing pattern of neurons participating in the seizure discharge) and back to a more chaotic state. In the next section we will show that this transition is mainly a spatiotemporal chaotic transition.

## 4. SPATIOTEMPORAL $STL_{MAX}$ PROFILES

We will now present $STL_{max}$ data preceding, during, and following an epileptic seizure at various brain sites to illustrate the spatiotemporal dynamics involved in the manifestation of a seizure. The raw EEG data from each electrode site were divided into nonoverlapping segments of 12 seconds duration each. From each segment, the phase space was reconstructed and the maximum Lyapunov exponent was estimated with the algorithm described in Section 2. Thus, for each electrode site, profiles consisting of the values of $STL_{max}$ over time were generated.

Preictal fluctuations in the value of $L_{max}$ over time were observed at each recording site. Initially, the fluctuations among sites are out of phase and differ in their mean

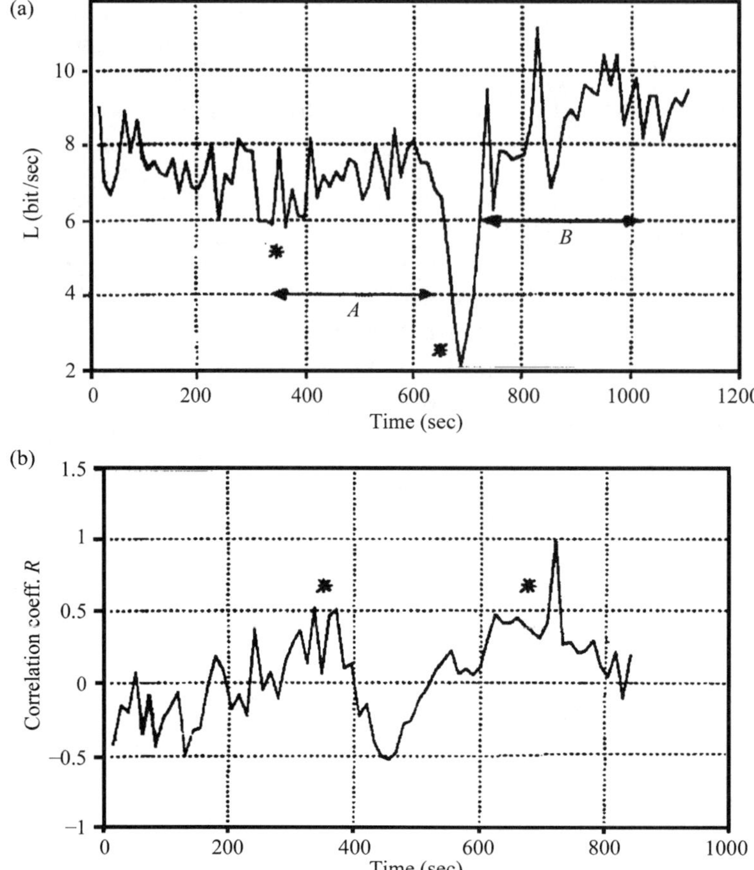


**Figure 8** (a) Plot of the Lyapunov exponent over time derived from an EEG signal recorded at an electrode site overlying the seizure focus (patient 1). (b) Cross-correlation of the postictal portion B of the $STL_{max}$ in (a) with the entire record of $STL_{max}$ in (a), after subtraction of the running mean of $L$.

values. Beginning tens of minutes prior to a seizure, the mean values and phases of $L_{max}$ at these sites start to converge. Several minutes prior to the seizure, there is a widespread locking among the $STL_{max}$ estimates (in both value and phase) from different cortical regions in both sides of the brain. Seconds prior to seizure onset, $STL_{max}$ values from most of the sampled regions of the brain are locked in an abrupt transition to a more ordered state. An example of such a situation is illustrated in Figures 9 and 10.

In Figure 9, the epileptic seizure occurred 46 minutes into the recording (the EEG profile of this seizure's onset is depicted in Figure 6). Locking between areas that overlie the epileptogenic focus (left temporal lobe in this patient) may exist for a long time prior to a seizure. For example, the locking of the left sphenoidal electrode (MN1) with the left frontotemporal scale electrode (FT9) was maintained for at least 46 minutes prior to the seizure (see Figure 9a). The locking of MN1 with the left hippocampus (LTD3) comes 21 minutes after its locking with the left frontotemporal lobe (FT9) (see

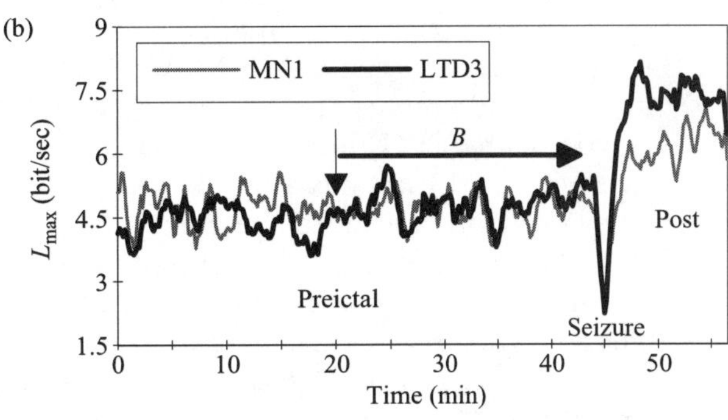


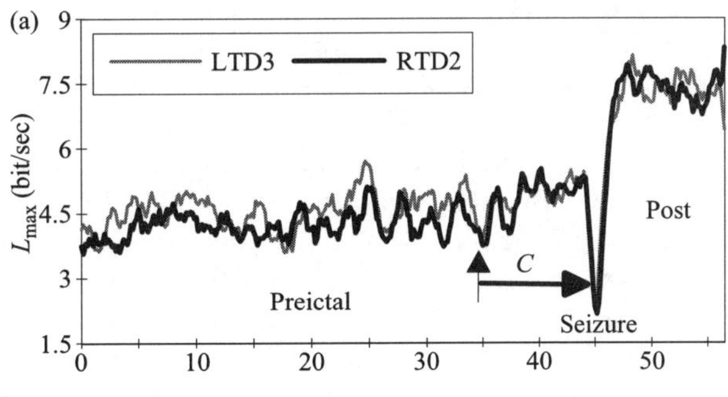


**Figure 9** Locking within the epileptogenic left temporal lobe. The seizure's onset is at 46 minutes into the recording and it is depicted in Figure 6. The patient was awake and this seizure lasted for 1.8 minutes (patient 2). (a) Sphenoidal (MN1) vs. frontotemporal scalp (FT9): the mean values of $L_{max}$ remain similar throughout the entire preictal trace (period A). The seizure is characterized by a simultaneous drop in the value of $L_{max}$ in both electrodes. (b) Sphenoidal (MN1) vs. left hippocampus (LTD3): the $L_{max}$ profiles become value and phase entrained 25 minutes prior to seizure, remain locked until the seizure (period B), and are unlocked postictally.

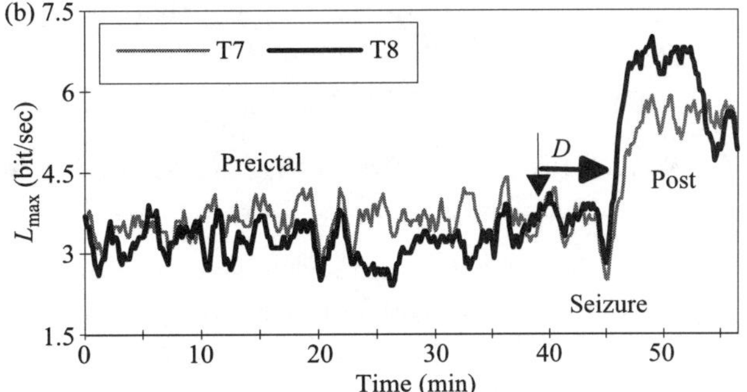


**Figure 10** Locking of the epileptogenic left temporal lobe with homologous contralateral sites. The seizure's onset is at 46 minutes into the recording and it is depicted in Figure 6. The patient was awake and this seizure lasted for 1.8 minutes (patient 2). (a) Long-term locking between the left and right hippocampus for approximately 10 minutes prior to seizure (period C). (b) Long-term locking between the left (T7) and right temporal (T8) scalp electrode sites for approximately 8 minutes prior to seizure (period D).

Figure 9b). Such long-term dynamical interactions, revealed from analysis of the EEG, have never been reported in the EEG literature. Figure 10 shows the development of long-term preictal locking of sites even across hemispheres: the epileptogenic left hippocampus with normal contralateral hippocampus (locked 10 minutes prior to seizure's onset—see Figure 10a) and the scalp electrode sites ipsilateral to the focus with homologous contralateral scalp electrode sites (locked 8 minutes prior to seizure's onset—see Figure 10b).

Thus, it is clear that the time of onset of locking more different sites may vary widely. For example, whereas the intrahemispheric locking ipsilateral to the focus could be detected as long as 45 minutes prior to this seizure (see Figure 9a) the interhemispheric locking (e.g., between left and right hippocampus) does not start until 10 minutes prior to the seizure (see Figure 10a). Another relevant point is of note here: although a 10-minute preictal locking between the hippocampi is evident from our analysis of the EEG, interaction of the two hippocampi is visually detectable in the raw EEG only 13 seconds after the seizure's onset (see Figure 6). We will return to this important point later in this section.

Four important characteristics of the locking process prior to an epileptic seizure can now be inferred. All these characteristics have regularly been observed in our analyses with depth and subdural electrodes and appear to be also detected with scalp recordings. The *first characteristic* is that a pair of sites may be locked for some period of time and subsequently may become unlocked without occurrence of a *clinical* seizure. For example, in Figure 10a, the left hippocampus locked with the right hippocampus about 18 and 26 minutes into the recording and then unlocked until 35 minutes into the recording, thereafter remaining locked up to the occurrence of the seizure (period C). The *second characteristic* is that a number of critical sites have to be locked with the epileptogenic focus over a common period of time in order for a seizure to take place. This forms the basis of our critical mass hypothesis. From Figures 9 and 10, it is clear that the order of preictal locking intervals for the analyzed seizure in this patient was $A \rightarrow B \rightarrow C \rightarrow D$, that is, left temporal cortex with left frontotemporal cortex (MN1–FT9) $\rightarrow$ left temporal cortex with left hippocampus (MN1–LTD3) $\rightarrow$ left hippocampus with right hippocampus (LTD3–RTD2) $\rightarrow$ left temporal cortex with right temporal cortex (T7–T8). The *third characteristic* is the progressiveness and the large time constants involved in the process of locking of brain sites. The *fourth characteristic* is the postictal resetting of the brain. It appears that epileptic seizures occur only under abnormal conditions of increased spatiotemporal order in the brain. It also appears that the seizure itself represents a resetting of the brain from the preictal, abnormal, state to a reduced spatiotemporal order, more normal, state. This theory of resetting is supported by our observations from the spatiotemporal preictal locking and the spatiotemporal postictal unlocking with the epileptogenic focus of $STL_{\max}$ profiles of various brain sites.

### 4.1. Quantification of the Spatiotemporal Dynamics

We will now examine the spatiotemporal dynamical changes in the EEG prior to, during, and after epileptic seizures by quantifying the observed progressive locking and unlocking of $STL_{\max}$ profiles over time and space. We employ the $T$ index (from the well-known $t$-test for comparisons of means of paired-dependent observations) as a measure of distance between the mean values of pairs of $STL_{max}$ profiles over time. The $T$ index at time $t$ between electrodes sites $i$ and $j$ is then defined as

$$T_{i,j}(t) = E\{|STL_{max,i}(t) - STL_{max,j}(t)|\} \div \frac{\sigma_{i,j}(t)}{\sqrt{N}} \tag{15}$$

where $E\{\}$ denotes the average of all absolute differences $|STL_{max,i}(t) - STL_{max,j}(t)|$ within a moving window $w_t(\lambda)$ defined as

$$w_t(\lambda) = 1 \text{ for } \lambda \in [t - N - 1, t] \text{ and } w(\lambda) = 0 \text{ for } \lambda \notin [t - N - 1, t]$$

where $N$ is the length of the moving window. Then $\sigma_{i,j}(t)$ is the sample standard deviation of the $STL_{max}$ differences between electrode sites $i$ and $j$ within the moving window $w_t(\lambda)$. The thus defined $T$ index follows a $t$-distribution with $N - 1$ degrees of freedom.

In the estimation of the $T_{i,j}(t)$ indices in our data we used $N = 30$ (i.e., averages of 30 differences of $STL_{max}$ exponents per moving window and pair of electrode sites). Since each exponent in the $STL_{max}$ profiles is derived from a 12-second EEG data segment, the length of the window used corresponds to approximately 6 minutes in real time units. Therefore, a two-tailed $t$-test with $N - 1 = 29$ degrees of freedom, at a statistical significance level $\alpha$, should be used to test the null hypothesis $H_0$ = "brain sites $i$ and $j$ acquire identical $STL_{max}$ values at time $t$." If we set $\alpha = 0.1$, the probability of a type I error, or better the probability of falsely rejecting $H_0$ if $H_0$ is true, is 10%. For the $T$ index to pass this test $T_{i,j}(t)$ should be less than 1.699.

We will show the results of such an analysis through an example. In Figure 11, the EEG of a seizure onset is depicted from a patient with an epileptogenic focus in the left hippocampus. Electrode site LTD2 in the left hippocampus appears to show the ictal onset prior to all other sites. Within seconds the discharge spreads to electrode sites in both hemispheres. No visual signs of the impending seizure from the EEG are present before the appearance of the initial discharge at LTD2. We will now show that this EEG picture is not fully representative of what is really occurring.

We first visually examine the $STL_{max}$ profiles in Figure 12 generated from our analysis of the EEG at seven electrode sites representing all six recording areas (LTD1 and LTD3 form the epileptogenic left hippocampus, RTD2 from the more normal right hippocampus, LST4 from the surface of the left temporal cortex, RST4 from the surface of the right temporal cortex, LOF3 from the surface of the left orbitofrontal cortex, ROF3 from the surface of the right orbitofrontal cortex). We make the following observations: (1) The LTD1 and LTD3 from the epileptogenic left hippocampus exhibit similar preictal entropy rates $STL_{max}$, being more or less locked for 60 minutes prior to this seizure onset, and start converging more at about 20 minutes prior to seizure onset. Postictally, this is the last pair to start unlocking, about 22 minutes after seizure ends. (2) LTD1 and RTD2 exhibit similar entropy rates and start converging more about 8 minutes prior to seizure onset. Postictally, they unlock by seizure end. (3) LST4 starts progressively locking to LTD1 about 20 minutes prior to seizure onset and finally locks to it about 5 minutes prior to seizure onset. Postictally, they start unlocking about 10 minutes after seizure end and are clearly unlocked 30 minutes after the seizure. (4) RST4 starts progressively locking to LTD1 about 20 minutes prior to seizure onset and does not lock to it until the seizure onset.

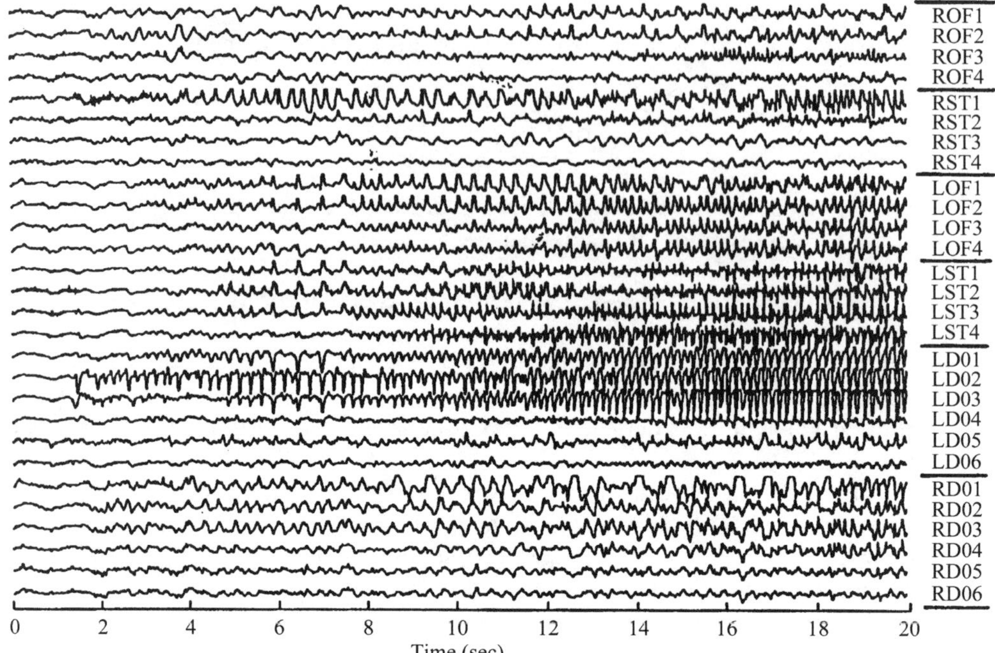


**Figure 11** A 20-second EEG segment at the onset of a left temporal lobe seizure (patient 3), recorded referentially to linked ears from bilaterally placed 12 depth (hippocampal) electrodes (last 12 channels in the figure), 8 subdural temporal electrodes, and 8 subdural orbitofrontal electrodes (nomenclature according to Figure 5). The ictal discharge begins as a series of high-amplitude sharp and slow wave complexes in the left depth electrodes (LTD 1–3, more prominently LTD2) approximately 1.5 seconds into the record. Within seconds, it spreads to RST1, the right hippocampus (RTD1–3), the left temporal and frontal lobe. It lasts for 80 seconds (not shown in this figure).

Postictally, they unlock by seizure end. (5) LOF3 starts progressively locking to LTD1 about 20 minutes prior to seizure onset and finally locks to it about 5 minutes prior to seizure onset. Postictally, they start unlocking 3 minutes after seizure end. (6) ROF3 starts progressively locking to LTD1 about 20 minutes prior to seizure onset and does not lock to it until the seizure onset. Postictally, they unlock by seizure end.

The observations from the preictal and postictal $STL_{\max}$ profiles of the six pairs of the electrode sites we considered are quantified with the $T(t)$ indices and presented in Figure 13. We can now order the pairs according to their preictal and postictal $T$ indices. From Figure 13 it is evident that, in this seizure, all recording areas exhibit preictal locking to the epileptogenic focus but differ in the degree and the timing of the initiation of the locking process. Preictally, the left and right hippocampi are the most strongly interacting areas (lowest value of $T$ index over time), left subtemporal cortex is the next strongly interacting area with the focus, right subtemporal cortex and left orbitofrontal are less strongly interacting with the focus, and, finally, the right orbitofrontal area interacts most weakly with the focus.

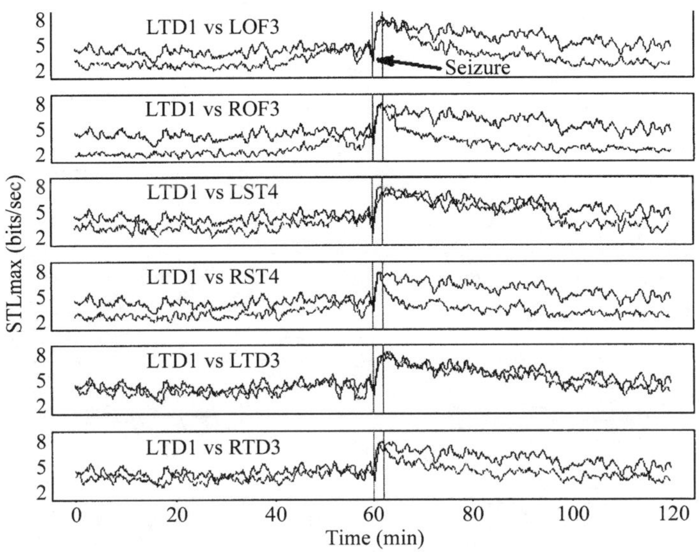


**Figure 12** Preictal locking and postictal unlocking of the epileptogenic left hippocampal site LTD1 to sites from the rest of the recording areas. Pairs of 5-point smoothed $STL_{max}$ profiles from LTD1 and one electrode site per recording area are shown. The seizure onset was depicted in Figure 11 (patient 3) and it is located 60 minutes into the record shown here, corresponding to the maximum drop of the unsmoothed $STL_{max}$ values at LTD1 and LTD2. Although the seizure's duration was only 80 seconds, in this figure it is denoted as 1 minute and 80 seconds (time between the vertical green lines) since this is the time the seizure itself influences the $STL_{max}$ values due to its duration and the 1 minute moving average smoothing of the $STL_{max}$ profiles. The most prominent preictal locking is observed within the epileptogenic hippocampus as well as between the epileptogenic and contralateral hippocampi. The rest of the recording areas lock to the epileptogenic hippocampus in the last 5 minutes prior to seizure's onset. Postictally, the areas in the contralateral hemisphere (RTD, RST, ROF) unlock from the epileptogenic focus first. The areas in the ipsilateral hemisphere unlock from the epileptogenic focus last (LST, LOF).

Postictally, the most impressive observation is that all areas in the right, more normal, hemisphere (RTD, RST, ROF) were the first to unlock (within seconds after the seizure) from the left epileptogenic hippocampus. The LOF area unlocks next (within 3 minutes from seizure's end). The two remaining areas, LST and LTD, are closer to the focus and are the last to unlock preictally from the focus (10 and 22 minutes, respectively).

Finally, in Figure 14, we show that the spatiotemporal interaction during the preictal-to-ictal transition is even more striking when we consider larger groups of electrode sites. From Figures 12 and 13 we saw that (LTD1, RTD2) and (LTD1, LTD3) were the most closely locked electrode pairs preictally. It is interesting to see how the remaining pairs of electrode sites from the two hippocampi interact prior to this seizure. There are 36 pairs of interaction between the six LTD and the six RTD recording sites of each hippocampus. In Figure 14 we present the average $T$ index profile estimated by averaging all 36 $T$ index profiles (one $T$ index profile per electrode

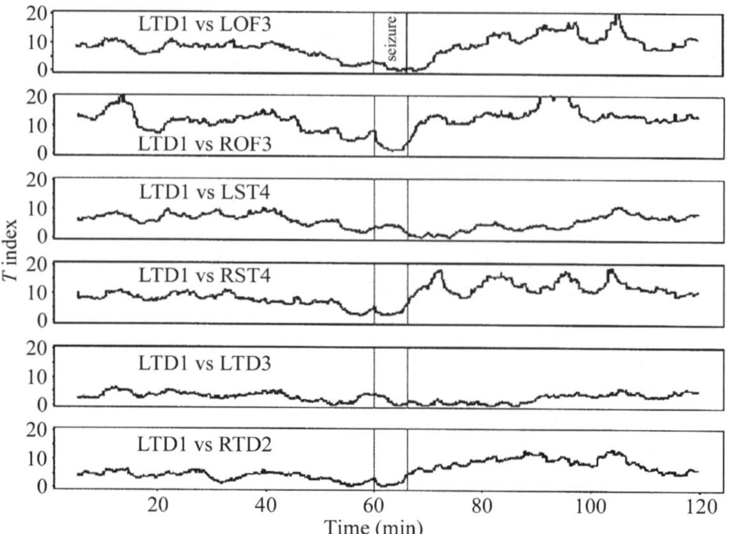


**Figure 13** Quantification of the observed long-term locking in Figures 12 between the epileptogenic left hippocampus and ipsilateral as well as contralateral sites from all recording areas. Although the seizure duration was only 80 seconds, in this figure it is denoted as 7.4 minutes (time between the vertical green lines) since this is the time the seizure itself influences the $STL_{max}$ values due to its duration and the 6 minute moving window for the estimation of the T-index profiles. The longest preictal and postictal locking (lowest values of T-index) is observed within the epileptogenic hippocampus (LTD1 vs. LTD3). The shortest preictal and postictal locking (highest values of T-index) is observed between the epileptogenic hippocampus (LTD1) and the contralateral orbitofrontal region of the cortex (ROF3).

pair between the left and the right hippocampus). The important observation here is that, using the average $T$ index, we can detect a preictal transition (negative slope in the average $T$ index profile over an 80-minute preictal period) between the two hippocampi. That was not possible when we were considering pair interactions. For example, the preictal $T$ index for the pair (LTD1, RTD2) (depicted in Figure 13) hardly exhibits any negative slope over the 60-minute preictal period (exhibiting similar entropy rates and developing a warning for an impending seizure only 8 minutes prior to seizure onset). Group interactions may reveal a hidden global interaction or they may obscure a critical, local (i.e., pair) interaction. Identification of the critical electrode pair interactions is a challenging problem. Although pair interactions can detect a critical local interaction, they may miss the global picture (e.g., compare Figures 13 and 14).

These findings indicate that the occurrence of an epileptic seizure represents a spatiotemporal phase transition involving widespread areas of the hippocampus and neocortex. This transition appears to occur over a much longer time scale than can be explained by current theories of epileptogenesis. However, these findings are quite consistent with the theories of synergetics and chaos. $STL_{max}$ may well serve as one of the order parameters of the epileptic transition [39]. Further study is needed to identify and characterize control parameters that govern this transition.

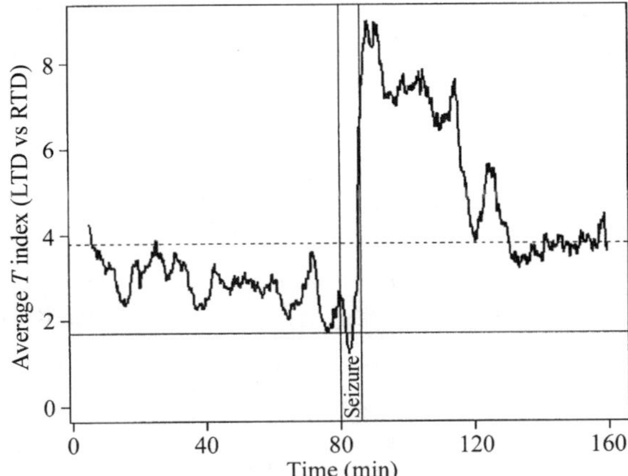


**Figure 14** Average locking of all 6 electrode sites of the epileptogenic left hippocampus to the homologous 6 electrode sites of the more normal hippocampus. The seizure is the one depicted in Figure 11 (patient 3) and it is located 80 minutes into the record shown here. The average T-index of this interaction shows that onset of preictal locking precedes seizure's onset for at least 80 minutes. The red horizontal line denotes the value of the T-index that corresponds to the t-test's $\alpha = 0.1$ significance level. This line is approached by the preictal T-index values only a few minutes prior to seizure's onset and could be used as a threshold line for a short-term preictal prediction of this seizure. The dotted black horizontal line shows the postictal level of the T-index long after the end of the seizure (60 to 80 minutes). This line, in conjunction with the preictal average T-index values, could be used as a threshold line for a long-term preictal prediction of this seizure.

## 5. CONCLUSIONS

The methodology for the estimation of $STL_{max}$ profiles from the EEG (Section 2), quantification of the observed locking and unlocking of brain sites, and examples of the preictal locking and postictal unlocking (Sections 3 and 4) have been presented. By analyzing over time the $STL_{max}$ values of EEG signals recorded from subdural electrodes placed over the epileptogenic temporal lobe and regions of ipsilateral frontal and parietal cortex, our group discovered that, beginning tens of minutes before seizure onset, regions of the anterior temporal cortex and regions more distant from the epileptogenic focus become locked with respect to their content of chaos. This observation indicated that, several minutes before a seizure, large regions of the cortex become dynamically locked as measured by the convergence of their $STL_{max}$ profiles. The same locking process was subsequently demonstrated in recordings from hippocampal depth electrodes and scalp electrodes. After a seizure, there is a resetting of the preictal locking process.

Perhaps the most exciting discovery to emerge from dynamical analysis of the EEG in temporal lobe epilepsy is that human epileptic seizures are preceded by dynamical changes in the EEG signal. This phenomenon could not be detected by visual inspection of the original EEG signal or by other more traditional methods of signal processing.

The nature of the phenomenon indicates that it may be possible to localize the epileptogenic focus more accurately as well as to predict the onset of a seizure in time to intervene with abortive therapy.

## ACKNOWLEDGMENT

This research is supported by a research grant from the National Institute of Health (RO1 NS31451) and a Merit Review grant from the Department of Veterans Affairs, USA.

## REFERENCES

[1] L. D. Iasemidis, H. P. Zaveri, J. C. Sackellares, and W. J. Williams, Linear and nonlinear modeling of ECoG in temporal lobe epilepsy. *Proceedings of the 25th Annual Rocky Mountains Bioengineering Symposium*, Vol. 24, pp. 187–193, 1988.

[2] L. D. Iasemidis, J. C. Principe, and J. C. Sackellares, Spatiotemporal dynamics of human epileptic seizures. In *3rd Experimental Chaos Conference*, R. G. Harrison, W. Lu, W. Ditto, L. Pecora, M. Spano, and S. Vohra, eds., pp. 26–30. Singapore: World Scientific, 1996.

[3] M. Casdagli, L. D. Iasemidis, J. C. Sackellares, S. N. Roper, R. L. Gilmore, and R. S. Savit, Characterizing nonlinearity in invasive EEG recordings from temporal lobe epilepsy. *Physica D* 99:381–399, 1996.

[4] M. Casdagli, L. D. Iasemidis, R. L. Gilmore, S. N. Roper, R. S. Savit, and J. C. Sackellares, Nonlinearity in invasive EEG recordings from patients with temporal lobe epilepsy. *Electroencephalogr. Clin. Neurophysiol.* 102:98–105, 1997.

[5] A. Babloyantz and A. Destexhe, Low dimensional chaos in an instance of epilepsy. *Proc. Natl. Acad. Sci. USA* 83: 3513–3517, 1986.

[6] B. H. Jansen, Is it and so what? A critical review of EEG-chaos. In *Measuring Chaos in the Human Brain*, D. W. Duke and W. S. Pritchard, eds., pp. 83–96. Singapore: World Scientific, 1991.

[7] T. Elbert, W. J. Ray, J. Kowalik, J. E. Skinner, K. E. Graf, and N. Birbaumer, Chaos and physiology: Deterministic chaos in excitable cell assemblies. *Physiol. Rev.* 74:1–47, 1994.

[8] W. J. Freeman, Strange attractors that govern mammalian brain dynamics shown by trajectories of electroencephalography (EEG) potentials. *IEEE Trans. Cas.* 35:781–784, 1988.

[9] C. Kock, R. Palovcik, and J. C. Principe, Chaotic activity during iron induced epileptiform discharges in rat hippocampal slices. *IEEE Trans. Biomed. Eng.* 39:1152–1160, 1992.

[10] A. S. Gevins and A. R. Remond, *Handbook of Electroencephalography and Clinical Neurophysiology*. Amsterdam: Elsevier, 1987.

[11] L. D. Iasemidis, *On the dynamics of the human brain in temporal lobe epilepsy*. Ph.D. dissertation, University of Michigan, Ann Arbor, 1991.

[12] L. D. Iasemidis and J. C. Sackellares, The temporal evolution of the largest Lyapunov exponent on the human epileptic cortex. In *Measuring Chaos in the Human Brain*, D. W. Duke and W. S. Pritchard, eds., pp. 49–82. Singapore: World Scientific, 1991.

[13] L. D. Iasemidis and J. C. Sackellares, Chaos theory in epilepsy. *Neuroscientist* 2:118–126, 1996.

[14] A. V. Holden, *Chaos—Nonlinear Science: Theory and Applications*. Manchester: Manchester University Press, 1986.

[15] A. V. Oppenheim, Signal processing in the context of chaotic signals. *IEEE Int. Conf. ASSP* 4:117–120, 1992.

[16] F. Lopes da Silva, EEG analysis: theory and practice; Computer-assisted EEG diagnosis: Pattern recognition techniques. In *Electroencephalography; Basic Principles, Clinical Applications and Related Fields*, E. Niedermeyer and F. Lopes da Silva, eds., pp. 871–919. Baltimore: Urban & Schwarzenburg, 1987.

[17] N. H. Packard, J. P. Crutchfield, J. D. Farmer, and R. S. Shaw, Geometry from time series. *Phys. Rev. Lett.* 45:712–716, 1980.

[18] F. Takens, Detecting strange attractors in turbulence. In *Dynamical Systems and Turbulence*. Lecture Notes in Mathematics. D. A. Rand and L. S. Young, eds., pp. 366–381. Heidelburg: Springer-Verlag, 1981.

[19] P. Grassberger and I. Procaccia, Measuring the strangeness of strange attractors. *Physica D* 9:189–208, 1983.

[20] P. Grassberger, T. Schreiber, and C. Schaffrath, Nonlinear time sequence analysis. *Int. J. Bifurc. Chaos* 1:521–547, 1991.

[21] G. Mayer-Kress, ed., *Dimension and Entropies in Chaotic Systems*. Berlin: Springer-Verlag, 1986.

[22] E. J. Kostelich, Problems in estimating dynamics from data. *Physica D* 58:138–152, 1992.

[23] A. Oseledec, A multiplicative ergodic theorem—Lyapunov characteristic numbers for dynamical systems (English translation), *IEEE Int. Conf. ASSP* 19:179–210, 1968.

[24] J. Pesin, Characteristic Lyapunov exponents and smooth ergodic theory. *Russ. Math. Surv.* 4:55–114, 1977.

[25] P. Walters, *An Introduction to Ergodic Theory*. Springer Graduate Text in Mathematics, Vol. 79. Berlin: Springer-Verlag, 1982.

[26] H. D. I. Abarbanel, *Analysis of Observed Chaotic Data*. New York: Springer-Verlag, 1996.

[27] A. Wolf, J. B. Swift, H. L. Swinney, and J. A. Vastano, Determining Lyapunov exponents from a time series. *Physica D* 16:285–317, 1985.

[28] L. D. Iasemidis, J. C. Sackellares, H. P. Zaveri, and W. J. Williams, Phase space topography of the electrocorticogram and the Lyapunov exponent in partial seizures. *Brain Topogr.* 2:187–201, 1990.

[29] J. P. Eckmann, S. O. Kamphorst, D. Ruelle, and S. Ciliberto, Lyapunov exponents from time series. *Phys. Rev. A* 34:4971–4972, 1986.

[30] M. Palus, V. Albrecht, and I. Dvorak, Information theoretic test for nonlinearity in time series. *Phys. Lett A.* 175:203–209, 1993.

[31] J. A. Vastano and E. J. Kostelich, Comparison of algorithms for determining Lyapunov exponents from experimental data. In *Dimensions and Entropies in Chaotic Systems: Quantification of Complex Behavior*, G. Mayer-Kress, ed., pp. 100–107. Berlin: Springer-Verlag, 1986.

[32] W. G. Frank, T. Lookman, M. A. Nerenberg, C. Essex, J. Lemieux, and W. Blume, Chaotic time series analyses of epileptic seizures. *Physica D* 46:427–438, 1990.

[33] A. Wolf and J. A. Vastano. Intermediate length scale effects in Lyapunov exponent estimation, In *Dimensions and Entropies in Chaotic Systems: Quantification of Complex Behavior*, G. Mayer-Kress, ed., pp. 94–99. Berlin: Springer-Verlag, 1986.

[34] L. D. Iasemidis, H. P. Zaveri, J. C. Sackellares, and W. J. Williams, Phase space analysis of EEG. *IEEE EMBS, 10th Annual Int. Conf.*, pp. 1201–1203, 1988.

[35] J. S. Barlow, Methods of analysis of nonstationary EEGs with emphasis on segmentation techniques: A comparative review. *J. Clin. Neurophysiol.* 2:267–304, 1985.

[36] G. Ferber, Treatment of some nonstationarities in the EEG. *Neuropsychobiology* 17:100–104, 1987.

[37] B. H. Jansen and W. K. Cheng, Structural EEG analysis: An explorative study. *Int. J. Biomed. Comput.* 23:221–237, 1988.

[38] L. D. Iasemidis, J. C. Principe, J. M. Czaplewski, R. L. Gilmore, S. N. Roper, and J. C. Sackellares, Spatiotemporal transition to epileptic seizures: A non-linear dynamical analysis of scalp and intracranial EEG recordings. In *Spatiotemporal Models in Biological and Artificial Systems*, F. Lopes da Silva, J. C. Principe, and L. B. Almeida, eds., pp. 81–88. Amsterdam: IOS Press, 1996.

[39] H. Haken, *Principles of Brain Functioning: A Synergetic Approach to Brain Activity, Behaviour and Cognition*. Springer series in Synergetics, Vol. 67. Berlin: Springer-Verlag, 1996.

# Chapter 13 RHYTHMS AND CHAOS IN THE STOMACH

Z. S. Wang, M. Abo, and J. D. Z. Chen

## 1. INTRODUCTION

Nonlinear feedback induces oscillation, whereas dynamic equilibrium between positive and negative nonlinear feedback generates rhythms [1]. Physiological rhythms are central to life [1, 2]. However, due to the extreme variability and complexity of the underlying system, rhythms in life usually are variable and unstable [1–3]. As a measure of variability and complexity, a new fast-developing interdisciplinary chaos theory has been applied to life science and its attractive potentials are being explored [1–15]. The extreme sensitivity to initial conditions and the superposition of a very large (infinite) number of periodic motions (orbits) are two essential characteristics of chaos [16–20].

The superposition of an infinite number of periodic motions means that a chaotic system may spend a short time in an almost periodic motion and may then move to another periodic motion with a frequency much higher than the previous one. Such evolution from one unstable periodic motion to another would give an overall impression of randomness, but in a short-term analysis it would show an order.

The extreme sensitivity to initial conditions can be portrayed by a famous metaphor of the butterfly effect. It goes as follows: a small breeze generated from a butterfly's wings in China may eventually mean a difference between a sunny day and a torrential rainstorm in California. Let us consider a chaotic system with a large number of unstable periodic orbits in a bounded space; the trajectories corresponding to each of the unstable periodic orbits are packed very closely together. In such close proximity, a small change in the input or in the initial condition of the system would push it from one orbit to another. This sensitivity of the system has presented both a problem and a hope for those who deal with chaotic systems. This is a problem because it makes the underlying system unstable and unpredictable. There is a hope because a small change in the system can cause the system to be controlled into a desired behavior. This has been theoretically proved by Ott, Grebogi, and Yorke [17]. Their chaos control scheme, called the OGY method, has been successfully applied to controlling cardiac chaos [10–12].

The two characteristics of chaos just described are observed in many life phenomena and biological experiments. Many biological electrical activities exhibit such a pattern, with slow oscillations omnipresent and intermittently superimposed with spiky activities (fast oscillations). The spikes are of much higher frequencies and their occurrence is discontinuous and random. That is, the spikes and slow oscillations have different periodic orbits and the periodic orbits of the spikes are unstable [3, 4]. The occurrence of the spikes is usually attributed to some transitions in the biological process and is controlled by complex neural and chemical mechanisms, such as neurotransmitters and hormones. These mechanisms are often sensitive to the initial condition or small outside perturbations. For example, the normal cardiac motion has a range of rhythms and the electrocardiogram (ECG) reflecting its electrical activity looks like a periodic signal. The periodicity of the ECG is, however, unstable. It has been extensively reported that the heart rate variability (HRV) is actually chaotic [6–12]. Some studies even suggested that loss of complexity in the heart rate would increase the mortality rate of cardiac patients [7–9]. Experiments on rat brain slices showed that preserving chaos might be more desirable in the process of terminating epileptic seizures [13]. Similarly, experiments on the brain electrical activity in humans indicated that the human memory is chaotic [14, 15]. The memory capability increases with increasing dimension of chaos [14, 15]. In other words, some life systems require chaos and/or complexity in order to function properly.

Motility is one of the most important physiological functions in the human gut. Without coordinated motility, digestion and absorption could not take place [21–23]. To accomplish its functions effectively, the gut needs to generate not only simple contractions but also coordinated contractions (peristalsis) [22]. For the stomach, coordinated gastric contractions are necessary for emptying of the stomach. Impairment of gastric motility results in delayed emptying of the stomach and leads to symptoms of nausea, vomiting, abdominal pain, and discomfort [22]. Gastric motility disorders are common in clinical gastroenterology and associated diseases including gastroparesis (diabetic, idiopathic or postsurgical), idiopathic cyclic nausea and vomiting syndrome, and functional dyspepsia [22]. Whereas the symptoms of functional dyspepsia may not be entirely caused by a motility disorder, gastroparesis is definitively attributed to motility disorders. With the accomplishment of the digestive process or motor function of the stomach, from mixing, stirring, and agitating to propelling and emptying, an ordered spatiotemporally coordinated and organized electric pattern is formed. This pattern is called gastric myoelectrical activity (GMA) [22–26]. It includes temporal evolution from endogenous rhythmic oscillating (electrical control activity, or ECA) originating in the pacemaker with a rhythm of 3 cycles per minute (cpm) in humans and spikes (electrical response activity, or ERA) directly associated with contractions and modulated by neurochemical mechanisms [22–24, 27–36]. GMA controls gastric motility activity and can be measured by using one of the three methods: serosal electrodes implanted intraoperatively, intraluminal electrodes mounted on a flexible intraluminal tube, and abdominal cutaneous electrodes [22–25]. The serosal recording provides the most reliable information on GMA and the data presented in this chapter were obtained using serosal electrodes.

Studies in our laboratory have shown that GMA is chaotic and its chaotic behavior may have potential applications in the diagnosis and treatment of gastrointestinal (GI) motility [27–29]. As discussed previously, GMA consists of two components: omnipresent slow waves (or ECA) and intermittent spikes (ERA). The normal ECA is rhythmic, whereas the ERA is unpredictable, generated via a bifurcation, controlled by neurochemical mechanisms, and sensitive to a microperturbation in these mechanisms [22, 25–29]. It is obvious that GMA satisfies all the characteristics of chaos [29]. Under normal conditions, the ECA is rhythmic in temporal evolution and ordered in spatial propagation, but in some diseased states, the frequency and spatial coordination of ECA are disrupted. The disruption in frequency regularity of the ECA is called dysrhythmia or temporal disorder. Three categories of dysrhythmia are usually observed: bradygastria (lower than normal rhythm), tachygastria (higher than normal rhythm), and arrhythmia (no dominant rhythm) [22, 37–48]. The disruption in the spatial coordination of ECA is called discoordination or spatial disorder, implying that the contractions of the stomach would no longer begin in the corpus and propagate to the antrum. The spatial disorder of contractions may mean largely mixing and little propulsion. Both the temporal disorder and spatial disorder lead to gastric motility disorders such as temporal dysrhythmia and spatial propagation failure (retrograde propagation and uncoupling) [22, 41–45, 47]. Spatiotemporal disorders of the ECA may be spatiotemporal chaotic [28].

In this chapter, we will prove that GMA is indeed chaotic and that its chaotic characteristics can be used to differentiate the physiological states of the stomach and the response of the stomach to an exogenous intervention.

## 2. RHYTHM AND CHAOS IN THE STOMACH

### 2.1. Gastric Migrating Myoelectrical Complex (MMC)

It is known that ECA is omnipresent and controls the frequency and propagation of gastric contractions and that the spikes are directly associated gastric contractions. In the fed state, the stomach contracts in a regular fashion with maximum frequency. That is, every wave of the ECA is superimposed with a burst of spikes. The fed pattern of gastric contractions lasts for 2–4 hours.

The contractile pattern of the stomach in the fasting state is, however, different and characterized by three distinctive phases: phase I, no contractions; phase II, intermittent and irregular contractions; phase III, regular contractions with the maximum frequency, similar to the fed pattern. The cycle of these three phases is called the migrating myoelectrical complex (MMC) and repeats every 90–120 minutes [32]. The MMC starts from the stomach and migrates along the small intestinal to the ileum. It functions as a housekeeper by transporting indigestible solids from the stomach to the colon.

Accordingly, GMA during phase I consists of only regular slow waves with a frequency of 3 cpm in humans and about 5 cpm in dogs. Irregular bursts of spikes are superimposed on the regular slow waves during phase II of the MMC. During phase III, every slow wave is accompanied by a burst of spikes. The aim of the following experiment was to investigate whether GMA was chaotic and, if it was, whether the chaotic characteristics were able to differentiate the three phases of the MMC.

### 2.2. Subjects and Methods

#### *2.2.1. Subjects*

The study was performed in eight healthy female hound dogs (15–22 kg). Each dog was implanted with four pairs of bipolar electrodes via laparotomy. The bipolar electrodes were sutured on the serosal surface of the stomach along the greater curvature. The most distal pair was 2 cm above the pylorus and the distance between every two adjacent pairs of electrodes was 4 cm. The electrodes in each pair were 1 cm apart. The electrodes were 28-gauge wires (A&E Medical, Farmingdale, NJ) and brought out through the anterior abdominal wall, channeled subcutaneously along the right side of the trunk, and placed outside the skin for recording. The four-channel data were derived by connecting the wires to a data recording system. The system consisted of an analog-to-digital (AD) converter, an MP100 amplifier module (BIOPAC Systems) and an IBM-compatible computer. The channel corresponding to the most distal pair (close to the pylorus) was used for this study. The study was initiated about 15 days after the surgery to guarantee that the dog was completely recovered from the operation. The study protocol was approved by the animal committee of the Veterans Affairs Hospital, Oklahoma City, Oklahoma.

### 2.2.2. Data Acquisition

After the dog was fasted overnight, one complete cycle (90–120 minutes) of the gastric MMC was recorded. The data were sampled at 20 Hz. This sampling frequency was high enough to acquire spike activities exactly, because gastric slow waves in dogs have a frequency range of 0-12 cpm, whereas the frequency range of spike activities is 2–4 Hz. The original data were acquired and stored using a software package from the manufacturer (acqKnowledge, BIOPAC Systems). Further data analysis was performed using MATLAB for Windows.

### 2.2.3. Data Analysis Methods

***Spike Detection and Phase Identification of MMC.*** Traditionally, the spikes superimposed on the slow waves in the GMA recording are extracted using a high-pass or band-pass filter [22, 27]. Because of the sharp rise (high-frequency components) of the depolarization edge of the slow wave, the conventional method is not able to provide accurate results. In this study, the blind source separation (BSS) method developed in our laboratory [27] was used to extract spikes and identify the phases of the MMC.

Figures 1 and 2 show an example. Figure 1a presents a 1-minute tracing of the GMA recording, which was composed of both slow waves and spikes. Applying the BSS method to this signal, we obtained the totally separated slow waves and spikes, as shown in Figure 1b and c, respectively. Applying this method to a 180-minute GMA recording that consisted of two complete cycles of the MMC, we obtained the phase identification results as shown in Figure 2. In Figure 2, (a) illustrates the detected number of spike bursts per minute (NBPM); (b) illustrates the number of spikes per minute (NSPM); and (c) illustrates the energy of spikes per minute (ESPM) within 180 minutes. These figures are indicative of the different phases of the MMC. For instance, in Figure 2a, the period from 0 to 40 minutes reflects phase I, the period from 41 to 80 minutes reflects phase II, the period from 81 to 90 minutes reflects phase III (most spikes, i.e., strongest contractions of stomach), and the period from 91 to 120 minutes reflects phase I of the second cycle of the MMC.

***Reconstruction of Attractors.*** Using Takens' embedding theorem [16], two parameters reflecting chaotic behavior, the attractor and the Lyapunov exponent of the myoelectrical recording, were reconstructed and computed, respectively. Statistical analysis was performed to investigate the difference in the Lyapunov exponents among different phases of the MMC.

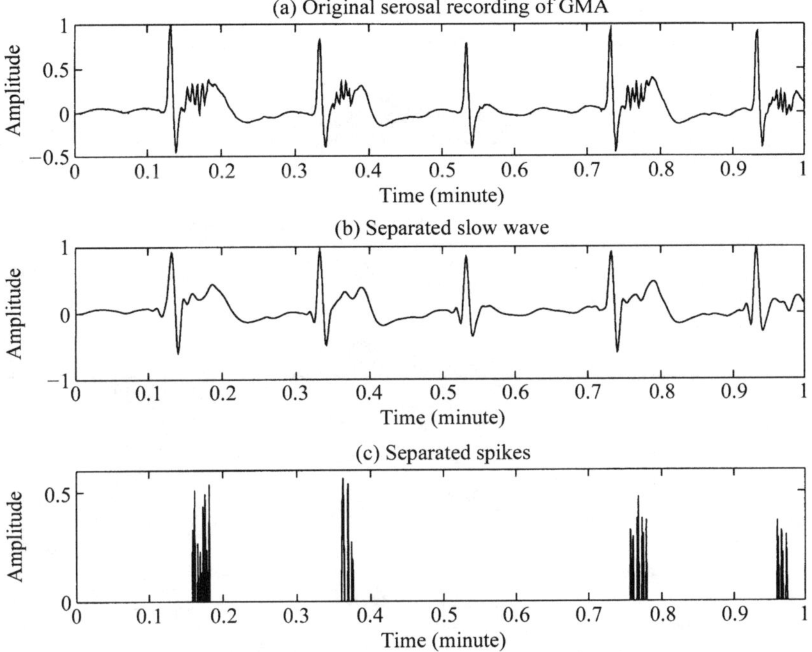


**Figure 1** Separation of slow wave and spikes from GMA using the BSS method. (a) Original serosal recording of GMA. (b) Separated slow wave. (c) Separated spikes.

We denote a given MMC time sequence with a length of $N$ as $x(n)$. Based on the nonlinear theory, this sequence can be described by the following nonlinear differential equation:

$$x(n+1) = F(x(n)) \tag{1}$$

where $F$ is a nonlinear mapping. Another form is

$$x(n) = F(x(n - d_{\mathrm{E}}\tau), x(n - (d_{\mathrm{E}} - 1)\tau), \ldots, x(n - \tau)) \tag{2}$$

where $d_{\mathrm{E}}$ is the embedding dimension and $\tau$ is the time delay. Based on Takens' embedding theorem, to reconstruct a two-dimensional attractor of $x(n)$, we need a pair of sequence $\Phi(n) = [x(n), x(n - \tau)]$, which constructs a two-dimensional state space. Because of the limitation of the length of the observed data, the time delay $\tau$ should be chosen properly. If $\tau$ is too small, $x(n)$ and $x(n - \tau)$ are basically the same measurement. If $\tau$ is too large, $x(n)$ and $x(n - \tau)$ are random with respect to each other. We use the average mutual information method to choose $\tau$.

Let **A** be the ensemble of values $x(n)$ that are the first components of the evolution vector $\Phi(n)$. Thus, $\mathbf{A} = \{x(n)|n = 1, 2, \ldots, N\}$. Also, let $a$ denote an arbitrary element of **A**. Likewise, let **B** be the ensemble of values that are the second components of the evolution vectors $\Phi(n)$, that is, $\mathbf{B} = \{x(n - \tau)|n = 1, 2, \ldots N\}$. And $b$ denotes an arbitrary element of **B**. Let $P_{\mathrm{A}}(a)$ denote the probability of choosing $a$ when making a

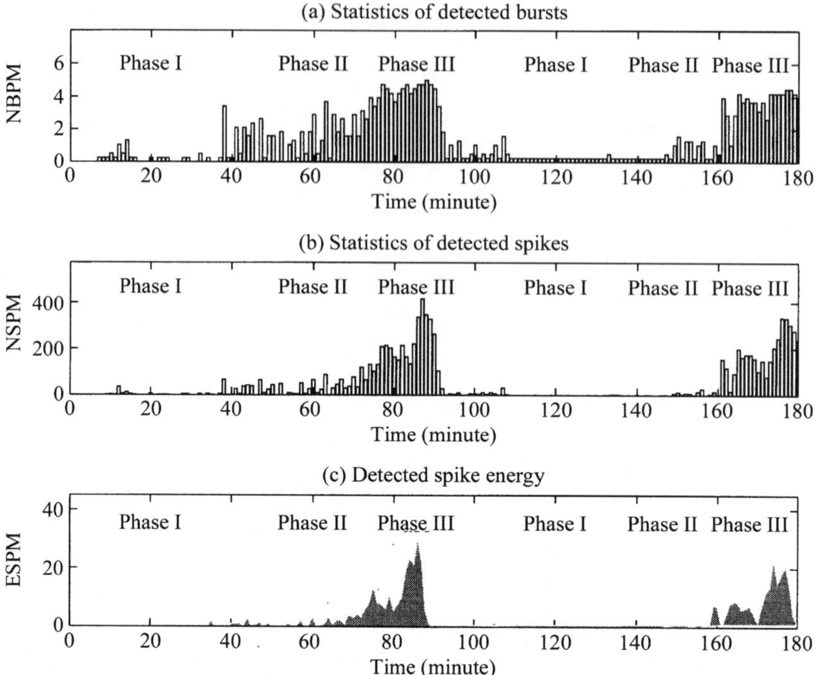


**Figure 2** Accurate detection of two cycles of the MMC using the BSS method. (a) Detected MMC pattern based on number of bursts per minute (NBPM). (b) Detected MMC pattern based on number of spikes per minute (NSPM). (c) Detected MMC pattern based on energy of spikes per minute (ESPM).

selection from set **A** and $P_B(b)$ denote the probability of choosing $b$ when making a selection from set **B**. $P_{A,B}(a, b)$ denotes the joint probability distribution, which is the probability of getting $a$ as the first component and $b$ as the second component of $\Phi(n)$. As a function of the time delay $\tau$, the average mutual information $I(\tau)$ is defined by

$$I(\tau) = \sum_{\alpha \in A, b \in B} P_{A,B}(a, b) \log_2 \left[ \frac{P_{A,B}(a, b)}{P_A(a) P_B(b)} \right] \tag{3}$$

There are many kernel density estimators available to estimate these probability distributions, such as the Gaussian kernel, Epanechnikov kernel, and Dirac delta function. In this chapter, we use the delta function as the approximations of the three probability distributions. For instance, the delta function of the joint probability distribution $P_{A,B}$ $(a, b)$ is

$$P_{A,B}(a, b) = \frac{1}{N} \sum_{n=1}^{N} \delta(a - x(n)) \delta(b - x(n - \tau)) \tag{4}$$

Figure 3 presents the average mutual information of an MMC recording as a function of the time delay $\tau$ computed using Eq. 3. The first minimum mutual information occurs at $\tau = 5$. Having obtained $\tau$, we are ready to reconstruct the attractors of the GMA recording during different phases of the MMC. Figure 4 illustrates four attractors of the one-cycle MMC, corresponding to phase I (Figure 4a), phase II (Figure 4b), phase III (Figure 4c), and phase I of the next cycle (Figure 4d). It is shown that the attractors of the MMC during three phases are strange attractors and presents identifiable shapes for different phases (e.g., phase II is fatter than phase I, and phase III is even fatter than phase II).

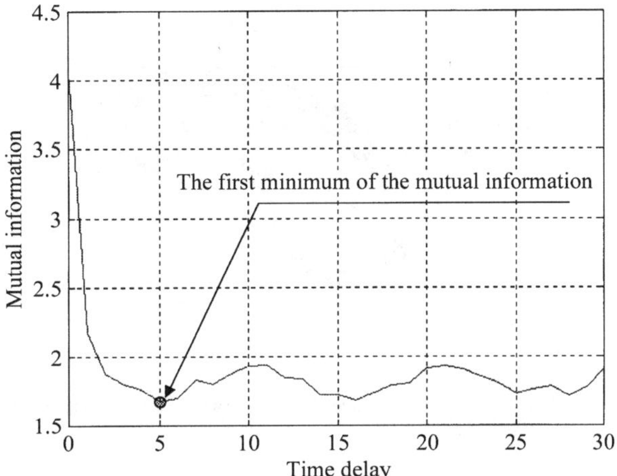


**Figure 3** Determination of the optimal time delay $\tau$. The first minimum of the mutual information is at a time delay of 5, thus $\tau = 5$.

***Computation of Lyapunov Exponents.*** Lyapunov exponents measure the rate at which nearby orbits converge or diverge [19, 20]. There are as many Lyapunov exponents as there are dimensions in the state space of the system, but the largest is usually the most important [20]. Roughly speaking, the distance between two nearby orbits evolves with time $t$ and can be expressed as an exponential function in the form $e^{\lambda t}$, where the constant $\lambda$ is the largest Lyapunov exponent. Obviously, if $\lambda$ is negative, the orbits converge in time, and the dynamical system is insensitive to initial conditions. If $\lambda$ is positive, the distance between nearby orbits grows exponentially in time, and the system exhibits a sensitive dependence on initial conditions. In other words, the system is chaotic.

The algorithm developed by Wolf et al. [19] is well suited for computing the largest Lyapunov exponent $\lambda$ from an observed time series. Before utilizing Wolf's algorithm, two important parameters have to be determined. One is the time delay, which has been discussed previously. The other is the embedding dimension. According to Takens' embedding theorem, a proper embedding dimension of a time series can be determined by computing its correlation dimension versus a number of possible values of the embedding dimension [16, 20]. The proper embedding dimension $D_E$ is the value at which the correlation dimension reaches a saturation point as shown in Figure 5a, where the correlation dimension reaches the saturation of 4.7734 when the embedding dimensions $d_E$ equals 6.

Having obtained $\tau$ and $d_E$, we use Wolf's algorithm to compute the largest Lyapunov exponent of the GMA recording. Figure 5b illustrates the iteration process

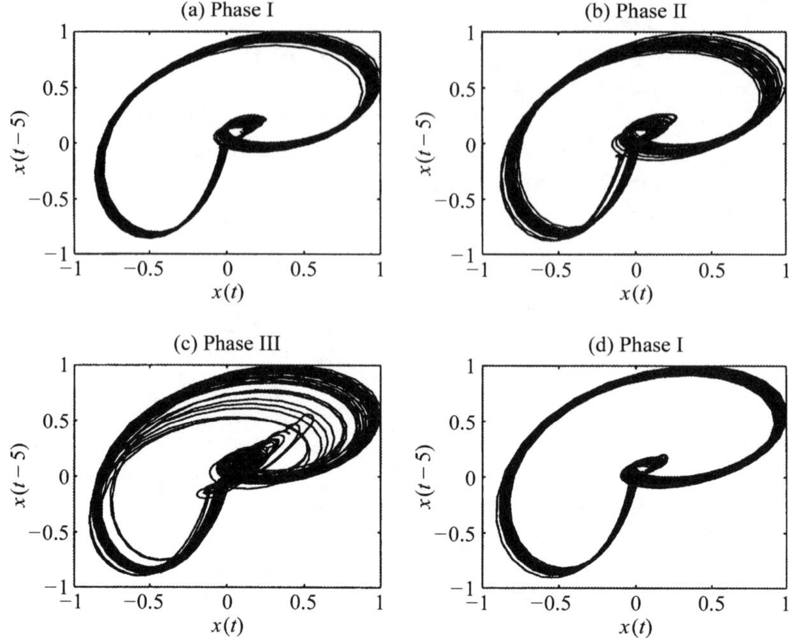


**Figure 4** Attractors of the gastric MMC during different phases.

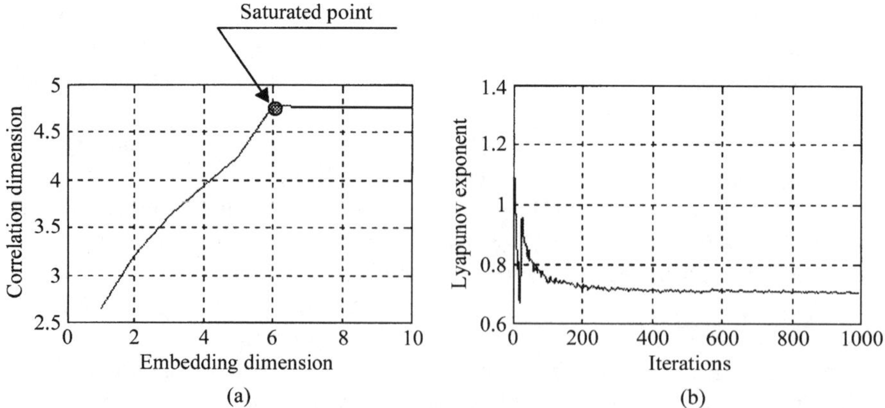


**Figure 5** Determination of the optimal embedding dimension $d_E$. The correlation dimension reaches the saturated value when the embedding dimension $= 6$, thus $d_E = 6$.

of Wolf's algorithm in computing the largest Lyapunov exponent. One segment of MMC data during phase II is used and the algorithm converges to a stable value of 0.7083 after 200 iterations.

## 2.3. Results

Having computed the Lyapunov exponents of all GMA signals recorded from the eight dogs, we perform statistic analysis on these numeral characteristics. All results are presented as mean ± SE. The Lyapunov exponents are 0.3418 ± 0.0594 during phase I, 0.7242 ± 0.0531 ($p < 0.0001$ vs. phase I) during phase II, 0.9779 ± 0.1674 ($p < 0.003$ vs. phase I, $p < 0.002$, vs. phase II) during phase III. The analysis of variance (ANOVA) reveals a significant difference in the Lyapunov exponents among three different phases ($p < 0.00002$). The statistic analysis results are also illustrated in Figure 6.

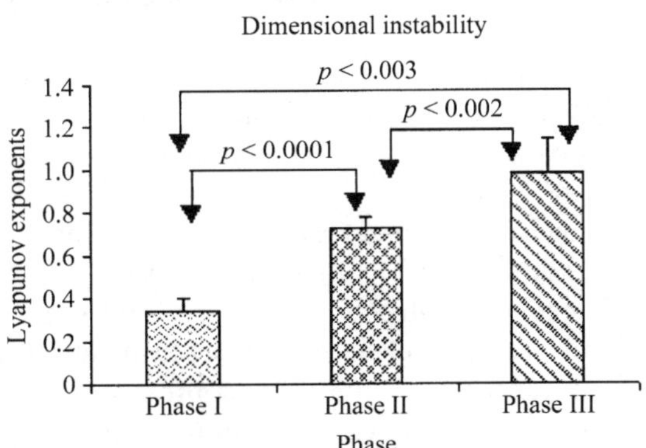


**Figure 6** Lyapunov exponents during different phases of the MMC data. Lyapunov exponents among three phases are significantly different (ANOVA, $p < 0.00002$). The Lyapunov exponent during phase III is significantly larger than that during II (0.9779 ± 0.1674 vs. 0.7242 ± 0.0531, $p < 0.02$). The Lyapunov exponent during phase II is significantly larger than that during phase I (0.7242 ± 0.0531 vs. 0.3418 ± 0.0594, $p < 0.001$).

# 3. INVESTIGATION OF CHAOS IN THE STOMACH VIA MODELING OF GMA

In the previous section, we investigated the chaotic behavior of long-term GMA in the fasting status, that is, gastric MMC. The fact that all Lyapunov exponents in all three phases are larger than zero shows that GMA is chaotic. In addition, the Lyapunov exponent during phase III is significantly larger than that during phase I and phase II. This indicates that motility is complex or chaotic and the motility needs complexity to maintain. In this section, we will further investigate this fact and attempt to find the underlying mechanisms by mathematical modeling of GMA. The typical route to chaos—period-doubling bifurcation of the GMA model—will be presented.

It is believed that a well-established mathematical model is able to make up for the limitations of medical observations and may be used to gain deep insight into the core of some biological phenomena or as a guide to design new medical experiments to reveal these phenomena [3, 4, 8]. Strenuous attempts have been made to simulate GMA using relaxation oscillators–based models [50, 51]. However, it becomes increasingly clear that these models are not very useful because they do not include the unique parameters that govern the behavior of specific ionic channels or the cable properties of smooth muscle cells. Accordingly, they could not be used to investigate the ionic mechanisms governing GMA, especially the spiky phenomena resulting from bifurcation due to interactions in cells. Incorporating the electrophysiological characteristics, Miftakhov et al. [52–55] presented a mathematical model of the small intestinal myoelectrical activity (SIMA) based on the modified Hodgkin–Huxley equations. Using this model, they successfully exhibited various kinds of patterns in SIMA and investigated ionic mechanisms inducing abnormal SIMA. In our previous work, we presented a model of GMA based on modified Hodgkin–Huxley equations [28]. In this chapter, we will use this model to present GMA patterns and investigate its chaotic bifurcation phenomena.

## 3.1. Modeling of GMA via Modified Hodgkin–Huxley Equations

### 3.1.1. Electrophysiological Background

The interior of smooth muscle cells is maintained at a negative potential with respect to the extracellular fluid by ionic transport mechanisms. This negative potential is called the resting membrane potential. This potential exhibits periodic depolarization that determines the excitability of the smooth muscle to contract by triggering the inward movement of calcium ions or calcium release from the internal stores. The smooth muscle contracts only when the membrane potential depolarization exceeds a threshold potential. This threshold is modulated by neurochemical mechanisms. The release of an excitatory neurotransmitter, such as acetylcholine, from the postsynaptic motor neurons stimulates contractions, whereas the release of an inhibitory neurotransmitter, such as vasoactive intestinal polypeptide (VIP), inhibits contractions (enlarges the threshold).

### 3.1.2. Mathematical Model

We propose the circuit model shown in Figure 7 to describe qualitatively and quantitatively the intracellular electrical activity of the human stomach. In Figure 7, $V$ denotes the membrane potential, $C_M$ the equivalent membrane capacity, $I_{Ca}^{slow}$ and $I_{Ca}^{fast}$ the slow and fast inward calcium currents via voltage-dependent $Ca^{2+}$ ion channels, respectively, $I_K$ and $I_{Ca-K}$ the outward calcium-activated potassium current and

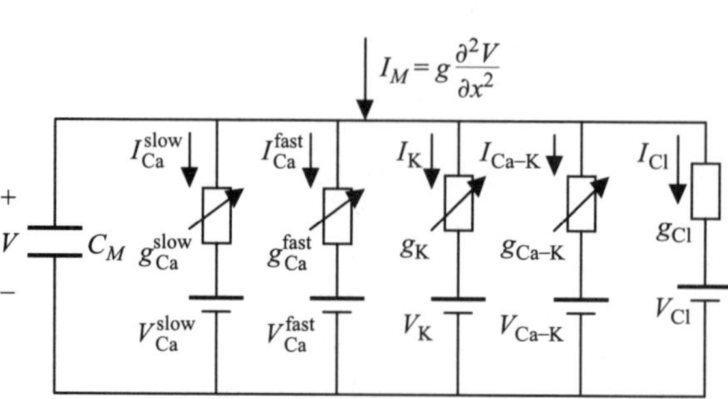


**Figure 7** Single-cell GMA model. $C_M$, equivalent membrane capacity; $V$, membrane potential; $I_M$, membrane current; $g_{Ca}^{slow}$, variable conductance of slow $Ca^{2+}$ channel; $V_{Ca}^{slow}$, rest potential of slow $Ca^{2+}$ channel; $g_{Ca}^{fast}$, variable conductance of fast $Ca^{2+}$ channel; $V_{Ca}^{fast}$, rest potential of fast $Ca^{2+}$ channel; $g_K$, variable conductance of $K^+$ channel; $V_K$, rest potential of $K^+$ channel; $g_{Ca-K}$, variable conductance of $Ca^{2+}$-activated $K^+$ channel; $V_{Ca-K}$, rest potential of $Ca^{2+}$-activated $K^+$ channel; $g_{Cl}$, conductance of $Cl^-$ channel; $V_{Cl}$, rest potential of $Cl^-$ channel.

voltage-activated potassium current, respectively, $I_{\rm Cl}$ the leakage current, and $I_{\rm M}$ the membrane current. Based on the circuit law, we have

$$I_{\rm M} = C_{\rm M}\frac{dV}{dt} + \sum I_{\rm ionic} = C_{\rm M}\frac{dV}{dt} + I_{\rm Ca}^{\rm slow} + I_{\rm Ca}^{\rm fast} + I_{\rm K} + I_{\rm Ca-K} + I_{\rm Cl} \tag{5}$$

where $I_{\rm Ca}^{\rm slow} = g_{\rm Ca}^{\rm slow}(V - V_{\rm Ca}^{\rm slow})$, $I_{\rm Ca}^{\rm fast} = g_{\rm Ca}^{\rm fast}(V - V_{\rm Ca}^{\rm fast})$, $I_{\rm K} = g_{\rm K}(V - V_{\rm K})$, $I_{\rm Ca-K} = g_{\rm Ca-K}(V - V_{\rm Ca-K})$, and $I_{\rm Cl} = g_{\rm Cl}(V - V_{\rm Cl})$. In these equations, $V_{\rm Ca}^{\rm slow}$ and $V_{\rm Ca}^{\rm fast}$ denote the slow and fast threshold potentials, respectively, and $V_{\rm K}$, $V_{\rm Ca-K}$, and $V_{\rm Cl}$ are the reversal potentials for the respective currents. Except that the leakage conductance $g_{\rm Cl}$ is a constant, all the other conductances, $g_{\rm Ca}^{\rm slow}$, $g_{\rm Ca}^{\rm fast}$, $g_{\rm K}$, and $g_{\rm Ca-K}$, are voltage and time dependent and modulated by some neurochemical mechanisms such as excitatory postsynaptic potential.

We propose the following conductance model: $g_{\rm Ca}^{\rm slow} = \bar{g}_{\rm Ca}^{\rm slow} P_{\rm e}^{\rm slow} P_{\rm i}^{\rm slow}$, $g_{\rm Ca}^{\rm fast} = \bar{g}_{\rm Ca}^{\rm fast} P_{\rm e}^{\rm fast} P_{\rm i}^{\rm fast}$, where $\bar{g}_{\rm Ca}^{\rm slow}$ and $\bar{g}_{\rm Ca}^{\rm fast}$ are the maximal conductance constants of the slow and fast channels, respectively, and $P_{\rm e}^{\rm slow}$ and $P_{\rm i}^{\rm slow}$, $P_{\rm e}^{\rm fast}$ and $P_{\rm i}^{\rm fast}$ are two probability pairs, which denote the probabilities that the slow or fast channel cell is in the excitatory or inhibitory status, respectively. In fact, the excitatory or inhibitory probability pair plays a role of positive feedback or negative feedback. It is the dynamical equilibration of the positive and negative feedbacks that induces the rhythmic oscillation of the gastrointestinal myoelectrical activity. The varying rates of the probability pairs have a linear relationship with their current values, which yields the following ordinary derivative equation (ODE):

$$\frac{dy}{dt} = \alpha_y(1 - y) - \beta_y y \tag{6}$$

where $y$ denotes $P_{\rm e}^{\rm slow}$, $P_{\rm i}^{\rm slow}$, $P_{\rm e}^{\rm fast}$, or $P_{\rm i}^{\rm fast}$, and $\alpha_y$ and $\beta_y$ are the two voltage-dependent nonlinear functions, similar to those defined in the Hodgkin–Huxley equations. When the two probability pairs reach stable status, that is, when $dy/dt = 0$, from Eq. 6 one can get the stable value $y_\infty = \alpha_y/(\alpha_y + \beta_y)$.

Assume the relaxation time $\tau_y = 1/(\alpha_y + \beta_y)$, and Eq. 6 can be written in another form as follows

$$\frac{dy}{dt} = (y_\infty - y)/\tau_y \tag{7}$$

where $y_\infty$ can be a piecewise linear function, cubic function, or sigmoid function of the membrane potential. We investigate the sigmoid function, $y_\infty = 1/(1 + e^{\sigma_y(V_y - V)})$, where $V_y$ and $\sigma_y$ are model parameters, which will be adjusted to generate different patterns of GMA. $\tau_y$ is defined as the hyperbolic secant function of the membrane potential $V$ and model parameters $V_y$ and $\sigma_y$, that is, $\tau_y = K_y \,\mathrm{sech}(0.5\sigma_y(V_y - V))$, where $K_y$ is also an adjustable model parameter with time dimension. The other two conductances in Eq. 5, $g_{\rm K}$ and $g_{\rm Ca-K}$, are defined as $g_{\rm K} = \bar{g}_{\rm K} P_{\rm K}$ and $g_{\rm Ca-K} = \bar{g}_{\rm Ca-K} P_{\rm Ca-K}$, where, as mentioned before, $\bar{g}_{\rm K}$ and $\bar{g}_{\rm Ca-K}$ are two maximal conductance constants, and $P_{\rm K}$, $P'_{\rm Ca-K}$ are two nonlinear functions. But unlike the two probability pairs of slow and fast channels, $P_{\rm K}$, $P_{\rm Ca-K}$ have different action mechanisms and $P_{Ca-K}$ is defined as $P_{Ca-K} = [Ca^{2+}]/(1 + [Ca^{2+}])$ where $[Ca^{2+}]$ is the intracellular concentration of calcium ions. $P_K$ and $[Ca^{2+}]$ are voltage and time dependent and their first-order differential equations take the form of Eq. 7. For a single cell, $I_M = 0$; thus, we have the following membrane potential differential equation to describe human gastric electrical activity of a single cell:

$$\begin{aligned} -C_M \frac{dV}{dt} &= \bar{g}_{slow} P_e^{slow} P_i^{slow}(V - V_{slow}) + \bar{g}_{fast} f\left(V_p - V_p^T\right) P_e^{fast} P_i^{fast}(V - V_{fast}) \\ &+ \bar{g}_K P_K (V - V_K) + \bar{g}_{Ca-k}\left[Ca^{2+}\right]/\left(1 + \left[Ca^{2+1}\right]\right)(V - V_{Ca-K}) \\ &+ g_{Cl}(V - V_{Cl}) \end{aligned} \tag{8}$$

where $f(V_p - V_p^T)$ is the neural and chemical weighted term to modulate the fast wave occurrence, which reflects the fact that the electrical response activity is controlled by some neurochemical mechanisms. $V_p$ denotes the membrane potential of the postsynaptic motor neuron cell to stimulate contractions; $V_p^T$ denotes the threshold potential to determine the contraction occurrence. Since $V_p$ reflects the neural cell potential, we can obtain it by using standard Hodgkin–Huxley equations [52]. $f(\cdot)$ is a threshold function, which is proposed as a sigmoid function or Heaviside step function [$H(x) = 1$ if $x \geq 0$, $H(x) = 0$ if $x < 0$]. We propose to use the Heaviside function as the threshold function. Under MATLAB, we solve Eqs. 7 and 8 and a group of Hodgkin–Huxley equations to define $V_p$ by using the stiff ODE solver ODE15S. The results are illustrated in Figure 8, where Figure 8a illustrates 3 cpm slow

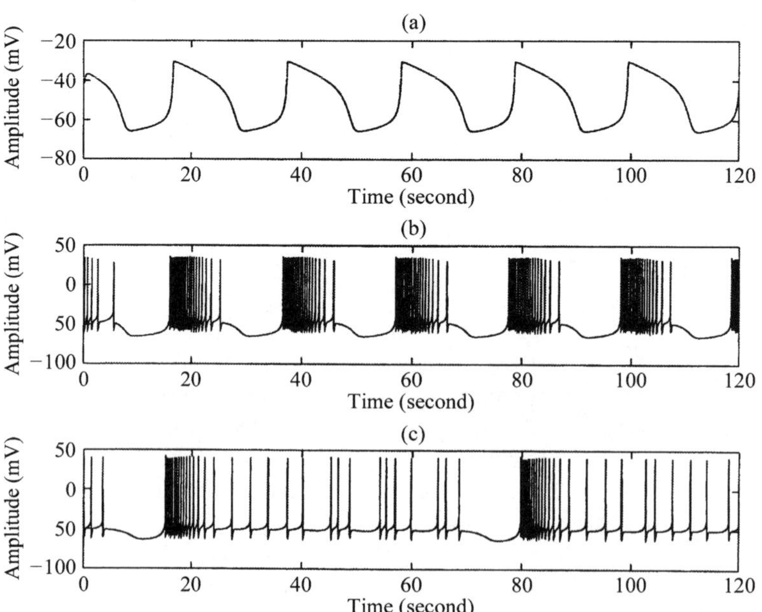


**Figure 8** Simulated GMA patterns. (a) Regular slow waves. (b) Regular slow waves with regular spikes. (c) Chaotic GMA.

waves without any spikes when the threshold potential of the postsynaptic motor neuron cell $V_{\rm p}^{\rm T} = 40\,{\rm mV}$ and the model parameter $K_{P_i}^{\rm fast} = 10\,{\rm ms}$, which suggests that when $V_{\rm p}^{\rm T}$ is too big and the inhibitory relaxation time is too small, no spikes occur. Figure 8b illustrates that when $V_{\rm p}^{\rm T} = 30\,{\rm mV}$ and $K_{P_i}^{\rm fast} = 200\,{\rm ms}$, regular bursting of spikes occurs and they are superimposed on slow waves. Figure 8c illustrates that when $V_{\rm p}^{\rm T} = 30\,{\rm mV}$ and $K_{P_i}^{\rm fast} = 520\,{\rm ms}$, chaotic bursting of spikes occurs because with the increment of $K_{P_i}^{\rm fast}$, the inhibitory mechanism of the fast channel becomes weak and the dynamic equilibrium between the nonlinear positive and negative feedbacks is broken. Such nonequilibrium status results in bifurcation and, finally, chaos occurs.

### 3.2. Bifurcation Analysis of GMA Model

By adjusting some model parameters, such as concentration of calcium ions and relaxation time, we have obtained different myoelectrical activity patterns from the proposed model. The following simulations will show that chaotic electrical patterns can be obtained by Hopf bifurcation analysis. In order to investigate the Hopf bifurcation of the proposed models of the GMA, we first consider the dynamic system taking the form $d\mathbf{x}/dt = \mathbf{F}(\mathbf{x}, \mu)$, where $\mathbf{x} = [x_1, x_2, \ldots, x_n]^T \in \mathbf{R}^n$ is the $n$-dimensional state variable, $\mathbf{F} = [F_1, F_2, \ldots, F_n]^T$ is the vector of $n$ nonlinear mappings, and $\mu \in R$ is the key parameter in $\mathbf{F}$, which may induce the system bifurcation. We will develop an efficient algorithm to investigate the system Hopf bifurcation phenomena based on the following theorem:

Assume that the system $d\mathbf{x}/dt = \mathbf{F}(\mathbf{x}, \mu)$ has an equilibrium point at the origin for all values of the real parameter $\mu$. Furthermore, assume that the eigenvalues, $\lambda_1(\mu)$, $\lambda_2(\mu), \ldots, \lambda_m(\mu)$ of the Jacobian matrix $\mathbf{J}_0 = \mathbf{DF}(\mathbf{0}, \mu)$ are purely imaginary for $\mu = \mu^*$. If the real part of the eigenvalues, $\mathcal{R}\{\lambda_1(\mu)\}$, satisfies

$$\frac{d}{d\mu}\mathcal{R}\{\lambda_1(\mu)\}\big|_{\mu=\mu^*} > 0 \tag{9}$$

and the origin is an asymptotically stable equilibrium point for $\mu = \mu^*$, then (1) $\mu = \mu^*$ is a bifurcation point of the system; (2) for $\mu \in (\mu_1, \mu^*)$, the origin is a stable focus; (3) for $\mu \in (\mu^*, \mu_2)$, the origin is an unstable focus surrounded by a stable limit cycle, whose size increase with $\mu$. For the model given in Eq. 8, let

$$\frac{dV}{dt} = F_V\left(V, P_{\rm e}^{\rm slow}, P_{\rm i}^{\rm slow}, P_{\rm e}^{\rm fast}, P_{\rm i}^{\rm fast}, P_{\rm K}, [{\rm Ca}^{2+}]\right) =$$

$$-\frac{1}{C_{\rm M}}\Bigg(\bar{g}_{\rm slow}P_{\rm e}^{\rm slow}P_{\rm i}^{\rm slow}(V - V_{\rm slow}) + \bar{g}_{\rm fast}f(V_{\rm p} - V_{\rm p}^{\rm T})P_{\rm e}^{\rm fast}P_{\rm i}^{\rm fast}(V - V_{\rm fast})$$

$$+\bar{g}_{\rm K}P_{\rm K}(V - V_{\rm K}) + \bar{g}_{\rm Ca\text{-}K}\frac{[{\rm Ca}^{2+}]}{1 + [{\rm Ca}^{2+}]}(V - V_{\rm Ca\text{-}K}) + g_{\rm Cl}(V - V_{\rm Cl})\Bigg)$$

and $dy/dt = f_y(V, y) = (y_\infty(V) - y)/\tau_y(V)$, where $y = P_{\rm e}^{\rm slow}$, $P_{\rm i}^{\rm slow}$, $P_{\rm e}^{\rm fast}$, $P_{\rm i}^{\rm fast}$, $P_{\rm K}$ or $[{\rm Ca}^{2+}]$. We have the following Jacobian matrix for the dynamic system in Eq. 8:

$$\mathbf{J} = \begin{bmatrix} \partial F_v/\partial V & \partial F_v/\partial P_e^{\text{slow}} & \partial F_v/\partial P_i^{\text{slow}} & \partial F_v/\partial P_e^{\text{fast}} & \partial F_v/\partial P_i^{\text{fast}} & \partial F_v/\partial P_K & \partial F_v/\partial[\mathrm{Ca}^{2+}] \\ \partial f_{p_e^{\text{slow}}}/\partial V & \partial f_{p_e^{\text{slow}}}/\partial P_e^{\text{slow}} & 0 & 0 & 0 & 0 & 0 \\ \partial f_{p_i^{\text{slow}}}/\partial V & 0 & \partial f_{p_i^{\text{slow}}}/\partial P_i^{\text{slow}} & 0 & 0 & 0 & 0 \\ \partial f_{p_e^{\text{fast}}}/\partial V & 0 & 0 & \partial f_{p_e^{\text{fast}}}/\partial P_e^{\text{fast}} & 0 & 0 & 0 \\ \partial f_{P_l^{fast}}/\partial V & 0 & 0 & 0 & \partial f_{P_l^{fast}}/\partial P_i^{fast} & 0 & 0 \\ \partial f_{P_K}/\partial V & 0 & 0 & 0 & 0 & \partial f_{P_K}/\partial P_k & 0 \\ \partial f_{\mathrm{Ca}^{2+}}/\partial V & 0 & 0 & 0 & 0 & 0 & \partial f_{\mathrm{Ca}^{2+}}/\partial[\mathrm{Ca}^{2+}] \end{bmatrix}$$

Let $\mathbf{x} = [V, P_e^{\text{slow}}, P_i^{\text{slow}}, P_e^{\text{fast}}, P_i^{\text{fast}}, P_K, [\mathrm{Ca}^{2+}]]^T$ and $\theta$ denote some model parameter, such as the maximal conductance $\bar{g}_K$ or $\bar{g}_{\mathrm{Ca-K}}$ or the maximal relaxation time $\tau_y$. The Jacobian matrix corresponding to some particular parameter $\theta$ can be denoted by $\mathbf{J}(\mathbf{x}, \theta)$. The bifurcation analysis for the system in Eq. 8 can be performed by investigating the dynamic properties of the eigenvalues of the matrix $\mathbf{J}_0 = \mathbf{J}(\mathbf{0}, \theta)$ based on the bifurcation theorem. By adjusting the model parameter $K_{\mathrm{Ca}^{2+}}$ in Eq. 8, we can obtain the double-period bifurcation process as shown in Figure 9 for (a) one period ($K_{\mathrm{Ca}^{2+}} = 1/34.5$), (b) a double period ($K_{\mathrm{Ca}^{2+}} = 1/36.2$), (c) a period of four ($K_{\mathrm{Ca}^{2+}} = 1/41$), and (d) chaos ($K_{\mathrm{Ca}^{2+}} = 1/500$) whose time domain waveform was shown in Figure 8c. It is shown that the chaotic attractor generated by the proposed model is similar to that of MMC recordings (both double scroll).

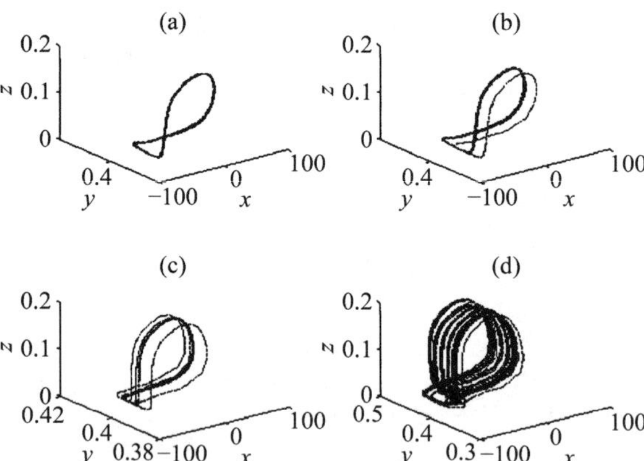


**Figure 9** Bifurcation process from periodic motions to chaos of the proposed GMA model. (a) Single period. (b) Double period. (c) Period of four. (d) Chaos (double scroll attractor).

## 4. EFFECTS OF WATER DISTENTION ON CHAOTIC BEHAVIOR OF GMA IN DOGS

In Sections 2 and 3, we investigated chaotic behavior of the GMA. In this section, we will present an application of the chaotic behavior. The GMA signals were recorded from 10 healthy dogs for 30 minutes in the fasting state and 30 minutes after the ingestion of 150 ml of water. The 150 ml of water can be viewed as a small distention or, mathematically, a perturbation on the stomach. If there exists chaos in the stomach, this small perturbation can induce a large variability in GMA. This variability may be assessed by computing the largest Lyapunov exponent of GMA. In the following, we will present the results obtained.

### 4.1. Measurements of GMA in Dogs before and after Water Distention

The study was performed in 10 healthy female hound dogs (15–22.5 kg). Subject preparation and data acquisition were similar to those described in Section 2 for the MMC recording. Each dog was implanted with four pairs of bipolar electrodes via laparotomy along the greater curvature from the pacemaker region to the distal antrum. The signal obtained from the fourth electrode corresponding to the distal antrum was used for this study.

### 4.2. Data Analysis

Takens' embedding theorem was used to reconstruct the attractors. Wolf's algorithm was used to compute the largest Lyapunov exponents of all the data from the 10 dogs before and after water distention. The time delay $\tau$ and embedding dimension $d_E$ were determined by computing the mutual information and correlation dimension of the GMA recording, respectively.

### 4.3. Results

Figure 10 illustrates two 4-minute time domain tracings in one dog before and after water distention. Compared with Figure 10a, Figure 10b presents more spikes due to water distention. The Lyapunov exponents are listed in Table 1. The Student test shows that 150 ml water distention induces a significant increase of the Lyapunov exponent ($1.193909 \pm 0.263229$ vs. $0.901158 \pm 0.322475$, $p < 0.03$).

## 5. DISCUSSION AND CONCLUSIONS

There may be two obstacles to accepting the fact that some biological rhythms are chaotic. One is that a rhythm is viewed as a single periodic oscillation, for instance, a sinusoidal-like wave. In fact, the rhythms we can obtain from most biological signals are only the dominant frequency components based on Fourier analysis. A signal with a dominant frequency may contain infinitely many other frequency components with relatively lower energies. That is, the signal may be low-dimensional chaotic. For example, Rossler's chaos has a dominant frequency component and its time domain waveform looks like a periodic signal. In many bioelectric signals, such a pattern is frequently observed: an omnipresent slow oscillating wave and an unstable bursting of spikes randomly superposed on the slow oscillating wave. If we simply apply Fourier spectral analysis to such a signal, the dominant frequency we obtain is only that of its slow wave, because the spike components occur discontinuously and thus have relatively lower energies in the frequency domain. The underlying system generating the spikes is usually nonlinear and involves complex physiological mechanisms, such as neural and chemical mechanisms. The rhythm of such a signal is unstable, and its frequency often jumps from one value to another (can be much higher). Mathematically, this is a bifurcation solution from one period to another. A bifurcation may induce chaos.

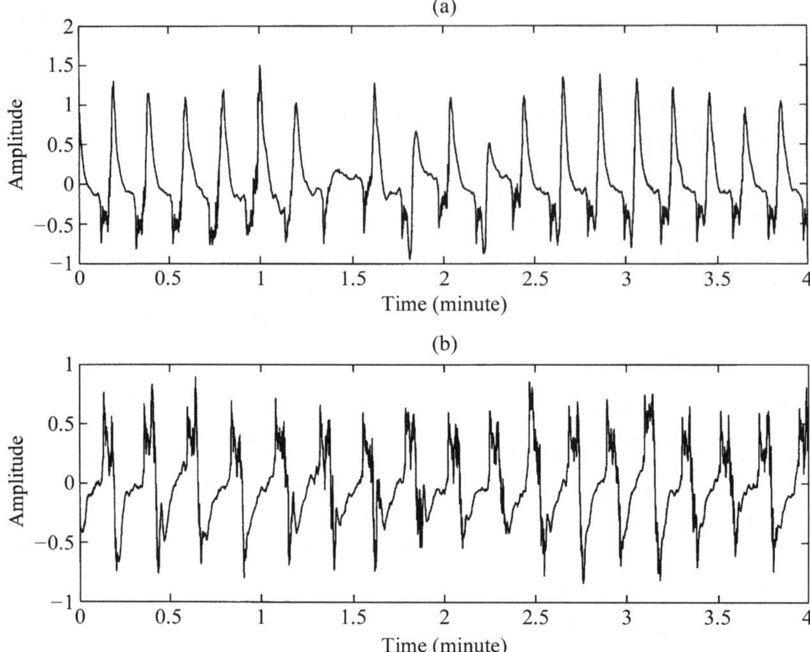


**Figure 10** Time domain waveforms of the GMA serosal recordings in the fasting state and after water ingestion in one dog. (a) Four-minute tracing of the fasting GMA recording with normal ECA and few spikes. (b) Four-minute tracing of the water distention GMA recording with normal ECA and more spikes.

**TABLE 1** Lyapunov Exponents of Water Distention versus Baseline in Dogs

| | Lyapunov Exponents | |
|---|---|---|
| Dog number | Baseline (fasting) | 150 ml Water distention |
| 1 | 0.920354 | 1.042321 |
| 2 | 0.300852 | 1.013037 |
| 3 | 0.912064 | 1.230978 |
| 4 | 0.616789 | 1.679327 |
| 5 | 0.633915 | 0.833895 |
| 6 | 1.300858 | 1.414062 |
| 7 | 1.19794 | 1.417311 |
| 8 | 0.862508 | 0.89366 |
| 9 | 1.314533 | 1.286048 |
| 10 | 0.951767 | 1.128455 |
| Mean | 0.901158 | 1.193909 |
| SD | 0.322475 | 0.263229 |
| $p$ (significance) | 0.022961 (water distention vs. baseline) | |

The other obstacle is the misunderstanding of chaos. Before chaos was recognized as a separate class of dynamics, it was thought of as randomness and indeterminacy. In fact, this notion was held steadfastly despite many observations of systems that actually evolved in an almost periodic manner for a short time before returning to a more random behavior. An easily accepted definition of chaos is the superposition of an infinite number of periodic motions. This means that a chaotic system may spend a short time in an almost periodic motion and then move to another periodic motion with a frequency much higher than the previous one. Such evolution from one unstable periodic motion to another would give an overall impression of randomness, but in a short-term analysis it would show an order.

It has been believed that bioelectrical activities originate from biological oscillators. When two or more oscillators are coupled together so that their feedbacks influence each other, chaotic behavior often occurs. Chaos theory has been used successfully to explain abnormal patterns in bioelectrical activities and has even motivated new therapeutic tools. The work conducted by Garfinkel et al. showed that based on chaos control methods, critically timed perturbations could have much better effects on cardiac pacing than periodic impulses.

In this chapter, chaotic characteristics of the gastric MMC have been investigated. It has been shown that different phases of the MMC are characterized by distinguishable shapes of the attractors and different values of the Lyapunov exponents. The chaotic behavior of the gastric MMC has great potential for the identification of different phases.

A mathematical model of GMA has been established and its bifurcation process has been presented in this chapter. Using this model, neural and chemical mechanisms associated with the occurrence of spikes are simulated. The occurrence of spikes in real GMA is irregular. This case takes place in the model when bifurcation induces chaos. The ionic mechanisms that induce chaos can be used as a guide to design experiments to maintain or control chaos in the stomach or reveal some gastric physiological and pathological mechanisms.

As an application of chaotic characteristic analysis, data from a water distention study were used to perform Lyapunov exponent analysis. It was shown that a small volume of water distention induces significant instability in Lyapunov exponents, which suggests that small outside perturbations may induce significant changes in neural and chemical mechanisms governing contractions of the stomach. This further validates that GMA is chaotic because its underlying system is chaotic.

In conclusion, gastric myoelectrical activity is chaotic. The chaotic characteristic of GMA is different in the different states of the stomach. In addition, the chaotic parameter (Lyapunov exponent) is sensitive to exogenous stimulation of the stomach and therefore may find applications in clinical and physiological studies of the stomach. Furthermore, chaos theory may be used to control impaired GMA or impaired gastric motility disorders. Further studies are, however, necessary.

## REFERENCES

[1] L. Glass and M. C. Mackey, *From Clocks to Chaos: The Rhythms of Life*. Princeton, NJ: Princeton University Press, 1988.

[2] B. J. West, *Patterns, Information and Chaos in Neural Systems*, SNPLS2, Singapore: World Scientific, 1993.

[3] Y. S. Lee, T. R. Chay, and T. Ree, On the mechanism of spiking and bursting in excitable cells. *J. Biophys. Chem.* 18:25–34, 1983.

[4] T. R. Chay and Y. S. Fan, Bursting, spiking, chaos, fractals and universality in biological rhythms. *Int. J. Birfurcation Chaos* 5:595–635, 1995.

[5] P. S. Chen, A. Garfinkel, J. N. Weiss, and H. S. Karagueuzian, Spirals, chaos, and new mechanisms of wave propagation. *Pace* 20(Part II):414–421, 1997.

[6] D. R. Chialyo and J. Jalife, Non-linear dynamics of cardiac excitation and impulse propagation. *Nature* 330:749–752, 1987.

[7] D. R. Chialvo, J. Gilmour, and J. Jalife, Low dimensional chaos in cardiac tissue. *Nature* 343:653–657, 1990.

[8] J. Jalife, Mathematical approaches to cardiac arrhythmias. *Ann. the N. Y. Acad. Sci.* 591:351–391, 1990.

[9] T. R. Chay, Bifurcations in heart rhythms. *Int. J. Bifurcation Chaos* 5:1439–1486, 1995.

[10] A. Garfinkel, M. L. Spano, W. L. Ditto, and J. N. Weiss, Controlling cardiac chaos. *Science* 257:1230–1235, 1992.

[11] A. Garfinkel, J. N. Weiss, W. L. Ditto, and M. L. Spano, Chaos control of cardiac arrhythmias. *Trends Cardiovasc. Med.* 5:76–80, 1995.

[12] J. N. Weiss, A. Garfinkel, M. L. Spano, and W. L. Ditto, Chaos and chaos control in biology. *J. Clin. Invest.* 93:1355–1360, 1994.

[13] A. Celletti and A. E. P. Villa, Low-dimensional chaotic attractors in the rat brain. *Biol. Cybern.* 74:387–393, 1996.

[14] D. W. Duke and W. S. Pritchard, eds., *Measuring Chaos in the Human Brain.* Singapore: World Scientific, 1991.

[15] A. Accardo, M. Affinito, M. Carrozzi, and F. Bouquet, Use of the fractal dimension for the analysis of electroencephalographic time series. *Biol. Cybern.* 77:339–350, 1997.

[16] F. Takens, On the numerical determination of the dimension of an attractor. In *Dynamical Systems and Bifurcation.* Lecture Notes in Mathematics, Vol. 20, pp. 167–192. New York: Springer-Verlag, 1971.

[17] E. Ott, C. Grebogi, and J. A. Yorke, Controlling chaos. *Phys. Rev. Lett.* 64:1196–1199, 1989.

[18] W. L. Ditto, M. L. Spano, and J. F. Lindner, Techniques for the control of chaos. *Physica* D86:198–211, 1995.

[19] A. Wolf, J. B. Swift, H. L. Swinney, and J. A. Vastano, Determining Lyapunov exponents from a time series. *Physica* 16D:285–317, 1985.

[20] S. Haykin and X. B. Li, Detection of signals in chaos. *Proc. IEEE* 83:95–122, 1995.

[21] A. Dubois, The stomach. In *A Guide to Gastrointestinal Motility*, J. Christensen and D. L. Wingate, eds., pp. 101–127. Bristol: Wright, 1983.

[22] M. M. Schuster, *Atlas of Gastrointestinal Motility in Health and Disease.* Baltimore: Willliams & Wilkins, 1993.

[23] S. K. Sarna, K. L. Bowes, and E. E. Daniel, Gastric pacemakers. *Gastroenterology* 70:226–231, 1976.

[24] S. K. Sarna, Gastrointestinal electrical activity: Terminology. *Gastroenterology* 68:1631–1635, 1975.

[25] S. K. Sarna and E. E. Daniel, Vagal control of gastric electrical control activity and motility. *Gastroenterology* 68:301–308, 1975.

[26] E. Bozler, The action potentials of the stomach. *Am. J. Physiol.* 144:693–700, 1945.

[27] Z. S. Wang and J. D. Z. Chen, Detection of spikes from chaos-like gastrointestinal myoelectrical activity. *Proc. IEEE Biomedical Eng.* VI:3234–3237, 1998.

[28] Z. S. Wang, J. Y. Cheung, S. K. Gao, and J. D. Z. Chen, Spatio-temporal nonlinear modeling of gastric myoelectrical activity. *Third International Workshop on Biosignal Interpretation (BSI99)*, Chicago, pp. 59–62, June 12–14, 1999.

[29] Z. S. Wang, J. Y. Cheung, S. K. Gao, and J. D. Z. Chen, Strange attractor and dimensional instability motor complex of the stomach. *Third International Workshop on Biosignal Interpretation (BSI99)*, Chicago, pp. 111–114, June 12–14, 1999.

[30] N. Sperelakis, Electrical field model: An alternative mechanism for cell-to-cell propagation in cardiac muscle and smooth muscle. *J. Gastrointest. Motil.* 3:63–84, 1991.

[31] J. H. Szurszewski, Mechanism of action of pentagastrin and acetylcholine on the longitudinal muscle of the canine antrum. *J. Physiol. (Lond.)* 252:335–361, 1975.

[32] J. H. Szurszewski, Electrical basis for gastrointestinal motility. In *Physiology of the Gastrointestinal Tract*, L. R. Johnson, ed., pp. 383–422. New York: Raven, 1987.

[33] E. E. Daniel and K. M. Chapman, Electrical activity of the gastrointestinal tract as an indication of mechanical activity. *Am. J. Dig. Dis.* 54–102, 1963.

[34] E. E. Daniel, Gap junctions in smooth muscle. In *Cell-to-Cell Communication*. New York: Plenum, 1987.

[35] T. Y. El-Sharkawy, K. G. Morgan, and J. H. Szurszewski, Intracellular electrical activity of canine and human gastric smooth muscle. *J. Physiol. (Lond.)* 279:291–307, 1978.

[36] G. Gabelia, Structure of smooth muscle. In *Smooth Muscle: An Assessment of Current Knowledge*, E. Bulbring, A. F. Brading, A. W. Jones, and T. Tomita, eds., pp. 1–46. Austin: University of Texas Press, 1981.

[37] J. D. Z. Chen, J. Pan, and R. W. McCallum, Clinical significance of gastric myoelectrical dysrhythmia. *Dig. Dis.* 13:275–290, 1995.

[38] T. L. Abell and J. R. Malagelada, Glucagon evoked gastric dysrhythmias in humans shown by an improved electrogastrographic technique. *Gastroenterology* 88:1932–1940, 1985.

[39] C. H. Kim, F. Azpiroz, and J. R. Malagelada, Characteristics of spontaneous and drug-induced gastric dysrhythmias in a chronic canine model. *Gastroenterology* 90:421–427, 1986.

[40] C. J. Stoddard, R. H. Smallwood, and H. L. Duthie, Electrical arrhythmias in the human stomach. *Gut* 22:705–712, 1981.

[41] C. H. You, K. Y. Lee, and W. Y. Menguy, Electrogastrographic study of patients with unexplained nausea, bloating and vomiting. *Gastroenterology* 79:311–314, 1980.

[42] C. H. You and W. Y. Chey, Study of electromechanical activity of the stomach in human and in dogs with particular attention to tachygastria. *Gastroenterology* 86:1460–1468, 1985.

[43] M. Bortolotti, P. Sarti, L. Barara, and F. Brunelli, Gastric myoelectric activity in patients with chronic idiopathic gastroparesis. *J. Gastrointest. Motil.* 2:104–108, 1990.

[44] H. Geldof, E. J. van der Schee, V. van Blankenstein, and J. L. Grasuis, Electrogastrographic study of gastric myoelectrical activity in patients with unexplained nausea and vomiting. *Gut* 27:799–808, 1986.

[45] J. D. Z. Chen and R. W. McCallum, *Electrogastrography—Principles and Applications*. New York: Raven Press, 1994.

[46] Z. S. Wang, Z. Y. He, and J. D. Z. Chen, Optimized overcomplete signal representation and its applications to adaptive time–frequency analysis of EGG. *Ann. Biomed. Eng.* 26:859–869, 1998.

[47] Z. S. Wang, Z. Y. He, and J. D. Z. Chen, Filter banks and neural network–based features extraction and automatic classification of electrogastrogram. *Ann. Biomed. Eng.* 27:88–95, 1999.

[48] Z. S. Wang, J. Y. Cheung, and J. D. Z. Chen, Blind separation of multichannel EGGs. *Med. Biol. Eng. Comput.* 37:80–86, 1999.

[49] S. K. Sarna, E. E. Daniel, and Y. J. Kingma, Simulation of the electric-control activity of the stomach by an array of relaxation oscillators. *Am. J. Dig. Dis.* 17:299–310, 1972.

[50] N. G. Publicover and K. M. Sanders, Are relaxation oscillators an appropriate model of gastrointestinal electrical activity? *Am. J. Physiol.* 256:265–274, 1989.

[51] A. L. Hodgkin and A. F. Huxley, A quantitative description of membrane current and its application to conduction and excitation in nerve. *J. Physiol.* 117:500–544, 1952.

[52] G. R. Miftakhov and G. R. Abdusheva, Effects of selective $K^+$-channel agonists and antagonists on myoelectrical activity of a locus of the small bowel. *Biol. Cybern.* 75:331–338, 1996.

[53] G. R. Miftakhov and G. R. Abdusheva, Numerical simulation of excitation–contraction coupling in a locus of the small bowel. *Biol. Cybern.* 74:455–467, 1996.

[54] G. R. Miftakhov, G. R. Abdusheva, and D. L. Wingate, Model predictions of myoelectrical activity of the small bowel. *Biol. Cybern.* 74:167–179, 1996.

[55] G. R. Miftakhov and D. L. Wingate, Electrical activity of the sensory afferent pathway in the enteric nervous system. *Biol. Cybern.* 75:471–483, 1996.

# INDEX

# ABOUT THE EDITOR

**Metin Akay** is assistant professor of engineering at Dartmouth College, Hanover, New Hampshire. He received the B.S. and M.S. in electrical engineering from Bogazici University, Istanbul, Turkey, in 1981 and 1984, respectively, and the Ph.D. from Rutgers University, NJ, in 1990.

He is author or coauthor of 10 other books, including *Theory and Design of Biomedical Instruments* (Academic Press, 1991), *Biomedical Signal Processing*, (Academic Press, 1994), *Detection Estimation of Biomedical Signals* (Academic Press, 1996), and *Time Frequency and Wavelets in Biomedical Signal Processing* (IEEE Press, 1997). In addition to 11 books, Dr. Akay has published more than 57 journal papers, 55 conference papers, 10 abstracts, and 12 book chapters. He holds two U.S. patents. He is the editor of the IEEE Press Series on Biomedical Engineering.

Dr. Akay is a recipient of the IEEE Third Millennium Medal in Bioengineering. In 1997 he received the EMBS Early Career Achievement Award "for outstanding contributions in the detection of coronary artery disease, in understanding of early human development, and leadership and contributions in biomedical engineering education." He also received the 1998 and 2000 Young Investigator Award of the Sigma Xi Society, Northeast Region, for his outstanding research activity and the ability to communicate the importance of his research to the general public.

His research areas of interest are fuzzy neural networks, neural signal processing, wavelet transform, detection and estimation theory, and application to biomedical signals. His biomedical research areas include maturation, breathing control and respiratory-related evoked response, noninvasive detection of coronary artery disease, noninvasive estimation of cardiac output, and understanding of the autonomic nervous system.

Dr. Akay is a senior member of the IEEE and a member of Eta Kappa, Sigma Xi, Tau Beta Pi, The American Heart Association, and The New York Academy of Science. He serves on the advisory board of several international journals and organizations and the National Institute of Health study session and several NSF review panels.